Coffee Science and Biotechnology

Genomic Biochemical and Nanotechnological Perspectives

Editor

Erdi Can Aytar
Usak University, Faculty of Agriculture
Department of Horticulture, Turkey

CRC Press
Taylor & Francis Group
Boca Raton London New York

CRC Press is an imprint of the
Taylor & Francis Group, an **informa** business
A SCIENCE PUBLISHERS BOOK

First edition published 2027
by CRC Press
2385 NW Executive Center Drive, Suite 320, Boca Raton FL 33431

and by CRC Press
4 Park Square, Milton Park, Abingdon, Oxon, OX14 4RN

CRC Press is an imprint of Taylor & Francis Group, LLC

Library of Congress Cataloging-in-Publication Data (applied for)

ISBN: 978-1-041-12350-7 (hbk)
ISBN: 978-1-041-17371-7 (pbk)
ISBN: 978-1-003-68939-3 (ebk)

DOI: 10.1201/9781003689393

Typeset in Times New Roman
by Prime Publishing Services

Preface

Coffee is much more than a globally consumed beverage; it represents a complex and multidisciplinary research domain encompassing agriculture, genetics, biochemistry, health sciences, nanotechnology, environmental sustainability, and industrial innovation. In recent years, advances in coffee science have expanded from the preservation of genetic diversity and the molecular understanding of bioactive compounds to the development of advanced technologies and sustainable strategies for waste management and resource utilization.

This edited volume aims to provide a comprehensive and integrated perspective on contemporary coffee science. The book begins with an introduction to coffee science, addressing the biological and agronomic foundations of Coffea species, with particular emphasis on the conservation, utilization, and improvement of *Coffea canephora* genetic diversity for variety development. Advances in coffee genomics, including gene mapping and sequencing approaches, are presented to highlight their relevance to crop improvement, biotechnology, and health-related research.

The biochemical composition of coffee and detailed content analyses are extensively discussed, offering insight into key bioactive constituents and their functional properties. These topics are further connected to human health, tracing the role of coffee from traditional use to evidence-based pharmacological applications. Special attention is given to coffee genetics in the context of health, emphasizing molecular mechanisms underlying protective and therapeutic effects.

A major focus of this book lies at the intersection of coffee science and nanotechnology. Chapters explore innovative applications of nanotechnology in coffee research, including the synthesis and functionalization of nanoparticles derived from *Coffea arabica* and other coffee species. Emerging strategies for antimicrobial and anticancer therapies, as well as environmental and industrial applications of coffee-derived nanomaterials, are critically examined. In parallel, issues related to toxicity, safety, and risk assessment of coffee-based nanomaterials and their derivatives are addressed to support the development of safe and responsible technologies.

Additionally, this volume places strong emphasis on coffee waste management and valorization. The increasing global production and consumption of coffee generate substantial amounts of by-products and residues, such as spent coffee grounds, husks, and pulp. This book highlights sustainable strategies for the management and conversion of coffee waste into value-added products, including biomaterials, nanomaterials, energy resources, and environmentally friendly industrial inputs.

These approaches are discussed within the framework of circular economy and zero-waste concepts, underscoring their importance for environmental protection and sustainable development in the coffee sector.

This book is intended for researchers, academicians, graduate students, and industry professionals seeking an interdisciplinary and up-to-date reference on coffee science, nanotechnology, health, and sustainability. By integrating fundamental knowledge with applied research and emerging technologies, this volume aims to inspire further innovation and contribute to the sustainable and responsible utilization of coffee resources.

This book is dedicated, with love and gratitude, to my mother Serpil Aytar and my father Birol Aytar, whose constant support and encouragement have guided me throughout my academic journey.

Erdi Can Aytar

Contents

CHAPTER

1

Introduction to Coffee Science

Zhivka Goranova

Department of Food Technology, Institute of Food Preservation and Quality – Plovdiv, Agricultural Academy, Bulgaria

Coffee science has evolved into a highly interdisciplinary field that integrates plant genomics, biochemistry, processing technology, sensory science, health research, and sustainability studies. As one of the most widely consumed beverages worldwide, coffee represents a biologically complex and economically vital crop whose quality is shaped by interactions across the entire value chain—beginning with plant genetics and environmental conditions and extending through post-harvest processing, roasting, brewing, and human perception. Recent advances in genomics and multi-omics technologies have expanded our understanding of *Coffea* species diversity, identified molecular loci associated with flavour precursors and stress tolerance, and highlighted the impact of climate change on coffee physiology and global production systems. At the biochemical level, comprehensive profiling of green and roasted beans has revealed the intricate roles of carbohydrates, lipids, proteins, chlorogenic acids, and alkaloids in determining aroma, flavour, acidity, body, and antioxidant capacity.

Processing innovations, including controlled fermentation and precision drying, have demonstrated strong potential to enhance sensory quality and chemical consistency. Meanwhile, modern roasting analytics have provided new insights into Maillard pathways, caramelisation kinetics, and the formation of key volatile compounds. Advances in brewing science reveal how grinding physics, water chemistry, and extraction dynamics influence the solubility and release of flavour-active molecules. Sensory evaluation increasingly combines trained cupping with instrumental profiling techniques such as NMR and GC–MS, enabling more robust correlations between chemical markers and perceived quality. Parallel to these

E-mail: jivka_goranova@abv.bg

DOI: 10.1201/9781003689393-1

technological developments, emerging health research links coffee's bioactive compounds with cognitive, metabolic, and anti-inflammatory benefits, positioning coffee not only as a cultural beverage but also as a functional food with demonstrable physiological effects.

Given the rapid shifts in climate, consumer expectations, and global market structures, sustainability and circular bioeconomy strategies have become essential components of contemporary coffee research, emphasising ecological resilience, resource efficiency, and value recovery from by-products. This chapter provides a comprehensive introduction to these scientific domains, establishing a conceptual foundation for the integrated and technologically advanced perspectives that define modern coffee science and biotechnology.

1. Introduction

Coffee science has developed into a comprehensive interdisciplinary field that integrates plant biology, chemistry, sensory science, processing technology, health research, and sustainability studies. As one of the most consumed beverages globally, coffee's scientific relevance continues to expand through new insights enabled by genomics, metabolomics, and data-driven sensory evaluation (Medrano et al., 2025; Genovese, 2025). This growing complexity is further supported by recent research demonstrating how fermentation microbes shape volatile compounds (Amorocho-Cruz et al., 2025) and how advanced extraction technologies improve chemical yield (Kowalski, 2025).

The global demand for high-quality, sustainably produced coffee has intensified scientific interest in understanding the factors that shape flavour, aroma, and functional properties. Studies from 2023 to 2025 show that processing choices, roasting chemistry, and brewing dynamics interact strongly with bean genetics and environmental conditions, ultimately influencing perceived sensory attributes (Fitri et al., 2025; Méndez Harper and Hendon, 2024). This interconnectedness underscores the need for integrative frameworks that bridge molecular processes with sensory and consumer outcomes.

Coffee also plays a crucial socio-economic role, supporting the livelihoods of over 125 million people. Recent sustainability analyses have highlighted the vulnerability of coffee systems to climate change, particularly for Arabica, which is expected to face significant habitat loss by 2050 (Bilen et al., 2023; Grandez-Alberca et al., 2025). Technological innovation ranging from precision fermentation to low-water processing and climate-resilient cultivars has therefore become a central focus of contemporary research agendas (Uslu, 2025).

At the same time, interest in coffee's physiological effects has grown due to emerging evidence linking bioactive compounds with cognitive performance, metabolic health, and anti-inflammatory properties (Shi, 2025; Li, 2025). These findings reinforce the concept of coffee as both a culturally significant beverage and a potential functional food.

This chapter provides a foundational synthesis of the major scientific domains that define contemporary coffee research. Through a multidisciplinary lens, it examines the biological, chemical, technological, sensory, health-related, and sustainability dimensions that interact to produce the beverage's complexity. The chapter lays the conceptual groundwork for subsequent sections, which examine each domain in greater detail.

Coffee science has emerged as a mature interdisciplinary field that integrates biological, chemical, technological, and socio-economic knowledge to understand one of the world's most influential agricultural commodities. Coffee is not merely a beverage; it is a complex system shaped by plant genetics, environmental factors, biochemical transformations, human practices, and global economic dynamics. Its study encompasses a breadth of scientific domains—from plant physiology and microbial ecology to analytical chemistry, sensory science, public health, and sustainability—each contributing essential insights into the quality and functionality of the final product.

The global significance of coffee reinforces the need for a scientific foundation that extends beyond traditional agricultural or culinary perspectives. Coffee is cultivated in more than 70 tropical and subtropical countries, supporting the livelihoods of over 125 million people. It has become a cornerstone of the global beverage industry and a key component of social and cultural practices. The diversity of coffee consumption habits, preparation methods, and flavour expectations across societies has stimulated a corresponding diversification in research approaches. As a result, contemporary coffee science is characterised by a movement towards precision agriculture, controlled fermentation, advanced roasting analytics, instrument-assisted sensory evaluation, and data-driven quality assessment.

At the centre of coffee science lies the dynamic interaction between the plant's biological properties and the technological steps that define the transformation from cherry to cup. Genetic and environmental factors determine the chemical composition of raw coffee beans, while post-harvest processing, roasting, and brewing modify this composition through biochemical and thermochemical pathways. Understanding these transformations requires a holistic perspective that links molecular changes to sensory attributes and consumer preference.

In recent years, the rapid expansion of the specialty coffee sector has catalysed scientific innovation. Producers, roasters, and researchers have begun to emphasise traceability, terroir expression, controlled microbial fermentation, and precise roast profiles. At the same time, climate change, shifts in global supply chains, and socio-economic pressures have elevated the importance of sustainability science, prompting new research on climate-resilient cultivars, agroforestry systems, and circular bioeconomy strategies for coffee by-products.

Scientific inquiry into coffee has also expanded into the domain of human health. Beyond caffeine, a wide array of bioactive compounds—such as chlorogenic acids, diterpenes, trigonelline, and melanoidins—have been associated with antioxidant, anti-inflammatory, and metabolic effects. Epidemiological research continues to examine the role of coffee consumption in the risk of chronic diseases, cognitive performance, and gut health, providing further motivation for comprehensive and mechanistic studies. This introductory chapter, therefore, seeks to provide a broad yet scientifically rigorous overview of the field. Rather than exhaustively detailing each sub-discipline, the aim is to establish a conceptual framework that contextualises the biological origins, chemical foundations, technological processes, sensory principles, and socio-economic dimensions of coffee. Each subsequent section highlights the core concepts that underpin the deeper, more specialised analyses presented in later chapters of the volume.

By situating coffee within an integrated scientific framework, this chapter invites the reader to approach coffee research not as a collection of isolated topics, but as a cohesive system with interdependent components. Such a systems perspective is essential for addressing the contemporary challenges and opportunities facing the coffee sector, from ensuring sustainable production to meeting evolving consumer expectations and leveraging technological innovation.

2. Historical and cultural foundations of coffee science

The scientific study of coffee is closely entwined with its historical, cultural, and economic trajectory. Understanding these foundations is essential for contextualising contemporary research, as many of the technological, agronomic, and sensory paradigms in modern coffee science originate from historical practices and the global dissemination of *Coffea* species.

2.1 Origins and early domestication

The origins of *Coffea arabica* are widely traced to the montane forests of Ethiopia, where wild populations continue to exhibit substantial genetic diversity, a finding reinforced by recent genomic analyses (Medrano et al., 2025; Bilen et al., 2023). Ethnobotanical accounts suggest that early consumption of coffee cherries was for their stimulating effects, and although often anecdotal, cultural narratives, such as the legend of the goatherd Kaldi, align with biochemical evidence that caffeine functions as a natural defence compound, deterring herbivores and influencing plant–plant interactions (Bealu Girma, 2022).

The movement of coffee from Ethiopia to Yemen marks the beginning of deliberate cultivation and domestication. Monastic communities in Yemen developed early roasting and fermentation practices to prepare a beverage that could sustain long periods of prayer and study. Written records from the fifteenth century describe the use of roasted and brewed coffee in religious settings, foreshadowing later scientific interest in flavour development (Medrano et al., 2025). These historical practices laid the foundation for modern investigations into microbial fermentation, chemical transformation, and sensory precursor formation, now studied through genomic and biochemical methods (Lu, 2025; Collazos-Chaves, 2025).

2.2 Global dissemination and socio-economic transformation

From Yemen, coffee spread across the Arabian Peninsula and into the Ottoman Empire, where coffeehouses (*qahveh khaneh*) became prominent centres for cultural exchange, intellectual discussion, and political debate. The introduction of coffee to Europe in the seventeenth century coincided with the rise of empirical scientific inquiry, prompting early chemical analyses of the beverage and its physiological effects (Tran et al., 2016). Modern economic analyses confirm that this expansion shaped global trade patterns, labour systems, and agricultural development (Grandez-Alberca et al., 2025).

European colonial expansion further accelerated the globalisation of coffee cultivation. Dutch introduction of *C. arabica* to Java and the subsequent establishment of plantations in the Caribbean and Latin America enabled large-scale production, with Brazil emerging as the dominant producer by the nineteenth century. These developments reshaped landscapes and economies, establishing coffee as a major global commodity. Recent phytogeographic and population genetics studies indicate that these early translocations generated distinct genetic lineages, now reflected in regional sensory and biochemical differentiation (Reyes-Herrera et al., 2023).

From a scientific perspective, this geographic diversification exposed *C. arabica* and *C. canephora* to varied environmental pressures. Modern genomic and phenotypic studies now attribute considerable terroir-driven variability to these historical patterns of dispersal and adaptation (Medrano et al., 2025).

2.3 Evolution of scientific interest in coffee

Scientific inquiry into coffee initially centred on its stimulant effects. The isolation and structural characterisation of caffeine in the early nineteenth century established a foundation for subsequent biochemical studies. As analytical chemistry advanced, research broadened to include chlorogenic acids, lipids, proteins, and volatile compounds, all now recognised as significant determinants of flavour, aroma, and health-relevant properties (Abdelwahab et al., 2025).

The twentieth century marked a shift toward integrating agronomy, plant physiology, and microbiology. Epidemics of coffee leaf rust (*Hemileia vastatrix*) stimulated breeding programmes and molecular investigations into disease resistance, while advances in genetics—including recognition of *C. arabica* as an allotetraploid—supported the application of genomic tools for cultivar improvement (Bealu Girma, 2022; Bilen et al., 2023).

Recent decades have seen a rapid expansion in scope, driven by high-resolution analytical techniques such as LC–MS/MS, GC–MS, and NMR, which enable the detailed mapping of chemical, microbial, and sensory processes (Fitri et al., 2025; Uslu, 2025). Sensory–omics integration has provided robust links between chemical fingerprints and sensory descriptors, enhancing understanding of quality perception (Martinez et al., 2022; Genovese, 2025). In parallel, research has increasingly addressed sustainability, climate resilience, and precision agriculture, reflecting the broader transformation of coffee science into an interdisciplinary field that connects molecular insights with global economic and ecological systems.

3. Biological and botanical principles underpinning coffee science

The biological foundations of *Coffea* species constitute a central pillar of coffee science. Plant morphology, reproductive biology, genetic structure, and environmental interactions collectively determine the agronomic performance and biochemical potential of coffee beans. These biological factors underpin the sensory and functional attributes that emerge during processing, roasting, and brewing. A comprehensive understanding of coffee biology is therefore essential for integrating plant-level processes with technological and industrial considerations.

3.1 Coffea species and genetic diversity

The genus *Coffea* contains more than 120 species, although global production is dominated by *C. arabica* and *C. canephora* (Bilen et al., 2023). Their contrasting evolutionary histories and genomic architectures largely explain the divergent sensory, chemical, and agronomic characteristics associated with each species.

Arabica is an allotetraploid (2n = 4x = 44), originating from hybridisation between *C. eugenioides* and *C. canephora.* This genomic configuration confers high allelic diversity, contributing to *Arabica*'s superior flavour complexity, elevated sucrose and organic acid concentrations, and comparatively lower caffeine content (Bilen et al., 2023; Medrano et al., 2025). Wild Ethiopian populations exhibit the most significant genomic diversity, indicating that modern cultivars capture only a fraction of the available genetic reservoir. Advances in sequencing technologies have facilitated the identification of loci associated with flavour biogenesis, disease resistance, and abiotic stress tolerance, supporting the development of marker-assisted breeding strategies (Reyes-Herrera et al., 2023).

Robusta (*C. canephora*), a diploid species (2n = 2x = 22), displays greater resilience to heat, pests, and disease, and contains higher baseline levels of caffeine and chlorogenic acids. These traits underpin its agronomic robustness and its widespread use in espresso blends for bitterness and crema formation (Bealu Girma, 2022). Contemporary breeding programmes increasingly explore interspecific hybridisation to combine *Arabica*'s sensory quality with *Robusta*'s environmental tolerance (Medrano et al., 2025).

Less widely cultivated species, such as Liberica and Excelsa, are attracting renewed scientific interest due to their unique phenolic profiles and potential adaptive traits relevant to climate resilience. Recent whole-genome assemblies of elite cultivars, including Geisha, are facilitating detailed investigations into metabolic, physiological, and sensory pathways (Tran et al., 2016; Medrano et al., 2025).

3.2 Plant physiology and development

Environmental conditions strongly influence the physiology of the coffee plant. Photosynthesis, flowering, and fruit development are regulated through tightly coordinated hormonal and metabolic pathways. Temperature stress alters chlorophyll content, stomatal conductance, and carbohydrate allocation, ultimately influencing bean density and chemical composition (Reyes-Herrera et al., 2023). Fruit maturation in *Arabica* typically spans 6–9 months, during which sucrose accumulation, organic acid metabolism, and lipid biosynthesis determine the chemical potential of the beans.

Genes associated with lipid biosynthesis, linked to body and mouthfeel in brewed coffee, are particularly sensitive to heat stress and nutrient availability (Medrano et al., 2025). This molecular evidence supports long-standing observations that high-quality coffees originate from high-altitude environments characterised by moderate temperatures and slow cherry development (Grandez-Alberca et al., 2025).

Environmental factors exert measurable and often interrelated effects on coffee bean chemistry. Altitude is widely recognised as a key determinant of bean quality, as higher elevations enhance acidity and aromatic intensity through increased concentrations of chlorogenic acids and volatile aroma precursors (Abdelwahab et al., 2025). In addition to altitude, light availability plays a regulatory role in secondary metabolism; moderate shading conditions promote the accumulation of chlorogenic acids and trigonelline, indicating the influence of light quality on metabolic pathways (Bealu Girma, 2022). In contrast, sustained heat stress accelerates cherry maturation, disrupts normal lipid biosynthesis, and ultimately reduces lipid content, resulting in diminished body and flavour roundness in the final beverage.

Physiological responses to pests and diseases also influence bean quality. Infection by coffee leaf rust (*Hemileia vastatrix*) reduces photosynthetic capacity and disrupts nutrient allocation, leading to smaller beans and altered chemical profiles. *Robusta*'s higher caffeine and phenolic content may contribute to its enhanced defence responses (Bealu Girma, 2022). Integrating plant physiology with disease ecology and metabolomics has therefore become an essential focus in contemporary coffee research.

3.3 Biotic interactions and ecological roles

Coffee plants interact with diverse microbial, fungal, and insect communities that influence both plant health and post-harvest potential. Modern microbiome research indicates that endophytic microorganisms enhance nutrient uptake, stress tolerance, and overall plant vigour (Uslu, 2025). Conversely, pests such as the coffee berry borer (*Hypothenemus hampei*) and pathogens such as *Hemileia vastatrix* remain significant threats. Climate-driven modelling predicts expansion of these pest and pathogen ranges under future warming scenarios (Bilen et al., 2023; Grandez-Alberca et al., 2025), underscoring the importance of genomics-informed resistance breeding.

Agronomic practices also significantly influence bean quality through their impact on plant physiology and pre-harvest metabolism.

Pruning and canopy management regulate light interception within the plant architecture, thereby influencing photosynthetic efficiency and the synthesis of secondary metabolites. Soil fertility management plays a critical role in determining the bioavailability of essential nutrients such as nitrogen, magnesium, and potassium, which are closely associated with caffeine biosynthesis, chlorogenic acid accumulation, and bean size development. Irrigation strategies further modulate bean quality, as moderate water stress may enhance perceived sweetness, whereas excessive irrigation tends to increase bitterness and reduce overall yield. In addition, shade-grown agroforestry systems contribute to improved biodiversity and reduced heat stress, and may promote greater aromatic complexity in coffee beans.

These biotic and agronomic interactions determine the biochemical composition of the beans before processing—a critical consideration, as roasting cannot fully compensate for poor raw material quality. Table 1 summarises the major *Coffea*

Table 1: Major *Coffea* species and their scientific relevance

Species	Genomic characteristics	Agronomic traits	Chemical profile	Scientific relevance
C. arabica	Allotetraploid (2n = 4x = 44) (Bilen et al., 2023)	Prefers high altitude; disease-susceptible	Higher sucrose, organic acids; lower caffeine	Key species for speciality coffee; focus of genomics and flavour research
C. canephora (Robusta)	Diploid (2n = 2x = 22)	Heat and pest-tolerant; high-yielding	High caffeine and CGA; stronger, more bitter taste	Essential for resilience studies and mass-market blends
C. liberica	Diploid	Adapted to humid, lowland environments	Distinct smoky, woody notes; larger beans	Niche flavour attributes: potential for breeding
C. excelsa	Diploid (often grouped with Liberica)	Grows at mid-altitudes; moderate disease resistance	Fruity, tart, complex phenolic profile	Increasing interest in sensory diversification and blending

species, highlighting their genomic characteristics, agronomic traits, chemical profiles, and scientific relevance.

4. Chemical and biochemical composition of coffee

The chemical composition of coffee beans provides the molecular foundation for the flavour, aroma, colour, and physiological properties of the beverage. More than 1,000 compounds have been identified in coffee, including alkaloids, phenolic acids, lipids, carbohydrates, proteins, and volatile aroma compounds (Abdelwahab et al., 2025). Their relative abundance varies across species, environmental conditions, processing techniques, and roasting profiles. As a result, coffee chemistry occupies a central place in contemporary research, linking plant biology with sensory perception, technology, and potential health effects.

4.1 Green coffee bean chemistry

Green coffee beans contain a complex matrix of carbohydrates, lipids, alkaloids, polyphenols, and proteins that serve as precursors to flavour and aroma. Advances in LC–MS/MS profiling have expanded the known diversity of chlorogenic acids (CGAs), identifying more than 60 structural isomers that contribute to antioxidant activity and later flavour formation during roasting (Fitri et al., 2025). Comparative studies show that *Arabica* typically contains higher concentrations of sucrose and lipids, associated with sweetness and body, whereas *Robusta* contains more caffeine and CGAs, contributing to bitterness and perceived strength (Bealu Girma, 2022; Abdelwahab et al., 2025).

Trigonelline, a minor alkaloid, has gained renewed research interest due to its degradation into pyridines during roasting and its emerging association with neuroprotective properties (Shi, 2025).

A more detailed breakdown of major chemical groups is provided below.

Alkaloids – Caffeine and trigonelline: Caffeine occurs at ~1.0–1.5% in *Arabica* and 2.0–4.0% in *Robusta* (Bealu Girma, 2022). It contributes to bitterness and physiological stimulation and functions in the plant as a defence compound. Trigonelline plays a limited sensory role in green beans but converts into pyridines and other aroma-active compounds during the roasting process.

Chlorogenic acids (CGAs): CGAs are the dominant phenolic class, constituting 5–12% of green bean dry mass (Abdelwahab et al., 2025). They influence acidity, astringency, antioxidant capacity, and the generation of degradation products during roasting. Higher CGA concentrations in *Robusta* underpin its sharper acidity and greater bitterness.

Carbohydrates: Carbohydrates make up 50–60% of green bean dry weight, primarily as cellulose, galactomannans, and arabinogalactans. Sucrose is the principal low-molecular-weight carbohydrate and a key precursor for Maillard reactions and caramelisation. The higher sucrose content of *Arabica* contributes to its sweeter and more complex aromatic profile.

Lipids: Lipids account for 10–18% of *the dry mass of Arabica and 8–12% of the dry mass of Robusta.* They include triacylglycerols, diterpenes (cafestol and kahweol), sterols, and phospholipids. Lipids play essential roles in mouthfeel, crema formation, and the retention of volatile aroma compounds. Diterpenes have also been studied for their physiological effects on cholesterol metabolism (Medrano et al., 2025).

Proteins and amino acids: Proteins constitute 10–13% of green bean mass. During roasting, amino acids participate in Maillard reactions that generate melanoidins, pyrazines, furans, and other essential aroma compounds. The proportion of free amino acids and protein matrix structure strongly influences reaction rates and sensory outcomes.

4.2 Formation of aroma precursors

Roasting transforms green beans into a chemically complex product through the development of volatile and non-volatile compounds. Multi-omics studies indicate that roasting intensity modulates the formation of key volatile classes, including pyrazines, furans, aldehydes, and ketones, which are shaped mainly by Maillard and Strecker reactions (Genovese, 2025).

Fermentation before roasting significantly influences precursor availability. Yeast-driven fermentations enhance ester precursors, promoting fruity and floral aromatics, whereas lactic acid bacteria increase organic acid content, intensifying brightness and acidity (Amorocho-Cruz et al., 2025; Lu, 2025).

The main classes of aroma precursors include:

Maillard precursors: Reaction between reducing sugars (primarily sucrose degradation products) and amino acids generates a wide spectrum of Maillard

Table 2: Major chemical compounds in green coffee beans, their concentrations, and roles

Chemical class	**Representative compounds**	**Typical concentration range**	**Role in coffee quality**	**Notes/Recent findings (2023–2025)**
Alkaloids	Caffeine, Trigonelline	Caffeine: 1.0–1.5% (Arabica), 2–4% (Robusta)	Bitterness, stimulation, aroma precursors	New research has linked the degradation of trigonelline to the production of neuroprotective compounds following roasting (Shi, 2025).
Chlorogenic Acids (CGAs)	5-CQA, 3-CQA, 4-CQA	5–12% of bean dry mass	Acidity, astringency, and antioxidant activity	Over 60 isomers have been identified (Fitri et al., 2025); this number is higher in Robusta.
Carbohydrates	Sucrose, cellulose, arabinogalactans, galactomannans	~50–60% of dry mass	Sweetness, Maillard precursors, body	Arabica contains more sucrose → richer aroma (Genovese, 2025).
Lipids	Triglycerides, diterpenes (cafestol, kahweol)	10–18% (Arabica); 8–12% (Robusta)	Body, crema, aroma retention	Lipid oxidation has been shown to affect staling more strongly than previously believed (Uslu, 2025).
Proteins and Amino Acids	Free amino acids, storage proteins	10–13%	Maillard precursors, aroma compounds	Amino acids are strongly affected by environmental stress (Reyes-Herrera et al., 2023).
Organic Acids	Citric, malic, quinic acids	Varies	Acidity, sourness, balance	Higher acid content linked to high-altitude beans (Fitri et al., 2025).
Volatile Compounds (post-roast)	Pyrazines, furans, aldehydes, ketones	Trace levels	Aroma complexity	Controlled fermentation increases ester-based volatiles (Amorocho-Cruz et al., 2025).
Melanoidins	High-molecular Maillard polymers	Formed during roasting	Colour, body, antioxidant activity	2024–25 studies highlight prebiotic effects on gut microbiota (Uslu, 2025).

intermediates, browning polymers, and characteristic nutty, caramel-like notes. These reactions dominate in medium to dark roasts.

Lipid-derived volatiles: Thermal degradation and oxidation of lipids generate aldehydes, ketones, and lactones responsible for buttery, creamy, or fruity notes. Lipids also act as reservoirs that modulate volatile release during grinding and brewing.

Phenolic and quinic derivatives: The thermal decomposition of CGAs produces caffeic and quinic acids and phenolic volatiles associated with bitterness and acidity. These compounds strongly influence differences between light and dark roast profiles.

The major chemical classes present in green coffee beans, together with their typical concentration ranges, functional roles in coffee quality, and recent findings, are summarised in Table 2.

4.3 Biochemistry of roasting

Roasting involves a cascade of thermally induced physical and chemical events. High-resolution NMR and GC–MS studies indicate that melanoidin formation is closely associated with antioxidant activity and the body, whereas pyrolysis contributes to the formation of phenolic compounds responsible for smoky or burnt notes at higher temperatures (Fitri et al., 2025; Uslu, 2025). Research in 2024–2025 further emphasises lipid oxidation as a significant determinant of aroma stability and staling.

Key biochemical stages include:

- **Dehydration**, reducing moisture content, and enabling subsequent browning reactions.
- **Maillard reaction**, generating melanoidins and hundreds of flavour-active molecules, typically between 150–200°C.
- **Caramelisation**, contributing to sweetness and brown colour.
- **Pyrolysis**, occurring above ~200°C, produces smoky, spicy or bitter notes through the breakdown of complex macromolecules.
- **Degassing**, involving CO_2 release after roasting, which affects freshness and extraction dynamics.

Roasting is therefore both an art and a science: precision control of time–temperature profiles allows roasters to accentuate or suppress specific reactions and tailor sensory outcomes.

5. Processing, technology, and post-harvest science

Post-harvest processing is a crucial phase in the coffee value chain, significantly shaping the sensory, chemical, and physical properties of the final product. After harvesting, cherries undergo a series of biological, physicochemical, and mechanical transformations that determine their quality potential. These processes form a critical bridge between agronomic conditions and roasting chemistry, placing post-harvest science at the centre of modern coffee research (Grandez-Alberca et al., 2025).

5.1 Overview of post-harvest pathways

Contemporary research highlights the biochemical and microbial complexity of post-harvest processing. Washed, natural, and honey methods create distinct microbial ecologies that generate different volatile and non-volatile compounds, influencing flavour and aroma (Collazos-Chaves, 2025). Thus, processing choice is both a technological operation and a biochemical intervention with direct sensory consequences. Three canonical pathways dominate global practice:

Wet (washed) processing: Washed processing involves rapid depulping followed by controlled fermentation to remove mucilage through microbial metabolism. This method typically produces a clean, high-acidity cup profile with pronounced aromatic clarity. It requires substantial water use and careful monitoring to avoid over-fermentation or acetic defects (Abdelwahab et al., 2025).

Dry (natural) processing: Natural processing dries whole cherries intact, allowing microbial activity on the fruit surface and within the pulp. This approach yields fruity, complex aromatics but carries risks of mould development or inconsistent drying. It is most successful in dry, sunny climates and requires meticulous turning regimes (Grandez-Alberca et al., 2025).

Semi-dry (honey) processing: Honey processing removes the pulp while retaining a proportion of mucilage, whose sugars influence microbial dynamics and the kinetics of drying. Colour designations (yellow, red, black honey) reflect mucilage levels and correlate with sweetness, body, and acidity.

Across all pathways, biochemical and microbial transformations modify precursors for aroma and flavour. Increasingly, speciality coffee research applies controlled fermentations with selected yeast or bacterial strains to achieve predictable and desirable sensory outcomes (Tran et al., 2016).

5.2 Fermentation as a biochemical driver

Fermentation—whether spontaneous or inoculated—is a central driver of flavour development. A landmark study demonstrated that inoculating cherries with lactic acid bacteria and specific yeasts significantly enhanced sensory uniformity and increased the production of esters, higher alcohols, and organic acids (Amorocho-Cruz et al., 2025). Multi-omics work further demonstrates that microbial strain selection can modulate the formation of pyridines, acetate esters, and lactones, thereby enriching fruity and floral notes (Lu, 2025).

Microbial consortia comprising **yeasts** (*Saccharomyces*, *Hanseniaspora*), lactic acid bacteria (*Lactobacillus*, *Leuconostoc*), and acetic acid bacteria metabolise polysaccharides, pectins, and simple sugars into acids, alcohols, and volatile precursors (Pyrzynska et al., 2024).

Key biochemical outcomes of fermentation include the modulation of acidity through the production of organic acids such as malic, lactic, and acetic acids, as well as the generation of esters, aldehydes, and higher alcohols that serve as important aroma precursors. Fermentative enzymatic activity also promotes the degradation of mucilage, thereby facilitating subsequent washing and drying processes. However, inadequate control of fermentation conditions may increase the risk of sensory defects, including over-fermented, phenolic, or "winey" characteristics.

Precision fermentation parallels technological innovations in wine and cocoa, reflecting the growing sophistication of biotechnological approaches in coffee processing (Tran et al., 2016).

5.3 Drying and physical transformations

Drying governs the reduction of moisture, microbial stability, and the retention of chemical precursors. Engineered drying technologies, such as infrared and ultrasonic systems, are being evaluated for improving uniformity, reducing contamination, and enhancing energy efficiency (Uslu, 2025).

Moisture reduction from ~60% to 10–12% stabilises beans for storage. Drying conditions influence colour, water activity, microbial viability, and the preservation of volatile precursors. Suboptimal drying can trigger fungal growth, enzymatic degradation, or uneven chemical development. Two dominant strategies exist:

- **Sun drying (patios or raised beds):** Allows slow, even dehydration and can promote complex aromatics under controlled conditions.
- **Mechanical drying:** Ensures rapid drying in humid regions, reducing the risk of microbial spoilage, but requires careful temperature control to avoid heat damage.

Drying dynamics influence lipid oxidation, chlorogenic acid retention, and the integrity of sugars and amino acids—all of which affect roast behaviour and final cup quality (Abdelwahab et al., 2025).

5.4 Storage, grading, and stability

Lipid oxidation during storage results in the loss of fruity and floral volatiles, a process accelerated by temperature fluctuations and elevated humidity (Genovese, 2025). Modern preservation strategies, including vacuum packaging and climate-controlled storage, help maintain the integrity of chlorogenic acid and its aromatic stability.

After drying, beans undergo hulling to remove the outer layer, or hull, which is similar to the parchment or dried fruit skin. Optional polishing may improve appearance, though its chemical impact is minimal. Beans are then graded by density, size, and defect count in accordance with international standards.

Storage conditions exert strong biochemical effects. Exposure to oxygen, humidity, or heat increases oxidative degradation and reduces sensory quality. Optimal storage environments are cool and dry, with stable moisture, enabling preservation of green bean quality over extended periods.

5.5 Technological innovations

Coffee processing technology has advanced rapidly, integrating engineering, microbiology, and digital analytics. IoT-enabled fermentation tanks, automated demucilagers, and AI-assisted drying systems are increasingly implemented in

Table 3: Comparison of wet, dry, and honey processing methods and their chemical, microbial, and sensory outcomes

Processing method	Key steps	Microbial activity	Chemical effects	Sensory outcomes	Recent findings (2023–2025)
Wet (washed)	Depulping → controlled fermentation → washing → drying	High microbial control; dominated by *Lactobacillus*, *Saccharomyces*	Cleaner acid profile; reduced sugars; controlled degradation of mucilage	High acidity, clarity, and floral notes	Inoculated fermentation improves uniformity and increases ester formation (Amorocho-Cruz et al., 2025).
Dry (natural)	Whole cherries dried intact → hulling	Complex, diverse microbial consortia on fruit skin and pulp	Increased sugar-to-acid conversion; enhanced ester formation; higher risk of defects	Fruity, winey, sweet; heavier body	Microbiome studies show unique yeast-driven aromatic signatures (Collazos-Chaves, 2025).
Honey (semi-dry)	Depulping → partial mucilage retention → drying	Moderate, mucilage-rich microbial activity	Balanced sugar retention; moderate acid degradation	Sweetness and body between washed and natural	Colour grades (yellow/red/black) correlate with drying speed and volatile formation (Genovese, 2025).
Anaerobic fermentation	Airtight tanks; extended fermentation	Restricted oxygen → specialised microbial profiles	Elevated organic acids, fruit-forward esters	Very high-intensity flavours, sometimes "boozy."	2024–25 data indicate an increased risk of phenolic taints when fermentation exceeds 72 hours.
Carbonic maceration	Whole cherries in CO_2-rich tanks	Yeast-driven internal fermentation	Pronounced ester synthesis; selective metabolite pathways	Tropical fruit notes, high sweetness	Borrowed from wine science; now used in speciality coffee competitions (Lu, 2025).
Mechanically demucilaged (Eco-Wet)	Depulping → mechanical mucilage removal → minimal water	Minimal microbial impact	Lower fermentation aromas; preserves intrinsic bean chemistry	Clean cup, balanced acidity	Reduced water use; stable quality even in humid climates (Uslu, 2025).

advanced mills (Uslu, 2025). Hard tank extraction systems have demonstrated improved recovery of chlorogenic acids and antioxidants, suggesting a potential synergy between post-harvest pre-treatments and brewing optimisation (Kowalski, 2025).

Recent innovations in coffee processing include the adoption of digitally controlled fermentation systems that enable continuous monitoring of temperature, pH, and microbial populations, thereby improving process consistency and flavour predictability. Mechanised demucilagers have been increasingly implemented in washed processing methods to significantly reduce water consumption. Advances in drying technologies, such as infrared and ultrasonic drying, have been developed to enhance drying uniformity and improve energy efficiency. In parallel, IoT-based environmental tracking systems allow real-time monitoring of drying curves and moisture dynamics, providing greater control over post-harvest processes. Collectively, these developments reflect broader trends toward sustainability, resource efficiency, and increased scientific precision within the speciality coffee sector. The major post-harvest processing pathways used in coffee production, together with their associated microbial activity, chemical effects, and sensory outcomes, are compared in Table 3.

6. Grinding, brewing, and extraction

Brewing represents the final scientific interface between roasted coffee and the consumer, where chemical constituents are selectively dissolved, transformed, or retained. The interplay of grinding, water chemistry, temperature, pressure, and time governs extraction dynamics and ultimately shapes the sensory profile of the beverage. Modern coffee science, therefore, treats brewing as a physicochemical process governed by principles of mass transfer, solubility, surface area, flow behaviour, and thermodynamics (Méndez Harper and Hendon, 2024).

6.1 Grinding physics and particle behaviour

Grinding transforms roasted coffee into particles with vastly increased surface area, enabling rapid diffusion of soluble compounds during brewing. Particle size is the dominant variable influencing extraction kinetics:

- **Coarse grind** → slow extraction, suited to immersion methods (French press, cold brew).
- **Medium grind** → balanced extraction, ideal for pour-over and batch brewing.
- **Fine grind** → rapid extraction under pressure, essential for espresso.
- **Ultrafine (Turkish) grind** → suspension-like dispersions that create intense, full-bodied brews.

Uniformity of particle-size distribution is critical. Excessive fines and boulders create bimodal extraction: fines over-extract (bitterness, astringency), while large particles under-extract (sourness, lack of complexity). Burr grinders generate narrower distributions and greater reproducibility than blade grinders.

Recent research highlights the role of triboelectric charging, whereby particles accumulate electrostatic charge during grinding, influencing clumping, flow behaviour, and tamping performance. Controlling grinder temperature, ambient humidity, and burr materials can mitigate these effects and improve espresso consistency (Méndez Harper and Hendon, 2024).

Laser diffraction studies further confirm that grind size distribution strongly affects the permeability of the coffee bed, flow uniformity, and extraction efficiency across different brewing methods.

6.2 Principles of extraction

Extraction involves the dissolution and diffusion of water-soluble compounds, such as organic acids, sugars, alkaloids, phenolics, lipids, melanoidins, and volatile aromatics into the brew. Updated modelling studies show that extraction follows a non-linear diffusion curve shaped by particle porosity, saturation gradients, and local flow dynamics (Genovese, 2025). While an extraction yield of 18–22% is often proposed as a reference range, sensory balance remains the ultimate determinant of quality.

Coffee extraction is governed by a combination of physicochemical processes, including solubility, diffusion, and hydrodynamics. Solubility varies as a function of temperature, solvent polarity, and molecular structure, determining the extent to which individual compounds are released into the brew. Diffusion is driven by concentration gradients within coffee particles, while hydrodynamic conditions—ranging from laminar to turbulent flow—regulate solvent movement through porous coffee beds. Temperature and pressure further influence extraction by modulating reaction kinetics, solubility, and the volatilisation of aroma compounds.

Insufficient extraction typically results in sour, thin, or unbalanced brews dominated by organic acids, whereas excessive extraction leads to bitterness, harsh phenolic notes, and a drying mouthfeel due to the over-dissolution of alkaloids, polyphenols, and melanoidins. Contemporary sensory research consistently links these chemical profiles to distinct extraction regimes, underscoring the importance of precise control over brewing parameters.

6.3 Brewing methods and their scientific basis

Brewing methods create distinct physicochemical environments that influence the selectivity of extraction and sensory outcomes.

Espresso: A high-pressure (≈approximately 9 bar), short-duration extraction (25–30 seconds) produces a multiphase beverage containing emulsified lipids, suspended solids, dissolved gases, and concentrated volatiles. Pressure profiling has been demonstrated to reduce channelling and improve aromatic uniformity (Kowalski, 2025).

Pour-over and drip brewing: Gravity-driven percolation through a porous bed emphasises the importance of laminar flow, particle-size uniformity, filter geometry, and pouring technique. Pour-over methods (e.g., V60, Kalita) provide high flavour clarity and express terroir-driven nuances (Genovese, 2025).

French press: Immersion brewing with coarse grounds produces a high body and fuller mouthfeel due to the absence of filtration. Contact time (3–5 minutes) governs extraction balance, while suspended lipids and fines contribute to texture.

Cold brew: Extended immersion at low temperatures (12–24 h) reduces the extraction of organic acids and bitter CGA degradation products, resulting in smoother, lower-acid beverages. Hardtank® systems enhance the extraction of antioxidants and melanoidins (Kowalski, 2025).

Turkish coffee: Ultrafine grounds are simmered briefly, creating a suspension with high intensity, saturated mouthfeel, and distinct foam. The lack of filtration increases the extraction of lipids and suspended solids.

Each method emphasises different chemical fractions, illustrating the strong interplay between extraction physics and sensory chemistry. The main coffee brewing methods, together with their typical grind sizes, extraction mechanisms, chemical profiles, and recent findings, are summarised in Table 4.

6.4 Water chemistry and brewing

Water composition profoundly affects extraction. Calcium and magnesium ions act as flavour carriers, enhancing the extraction of aroma-active compounds, including chlorogenic acid derivatives and volatile aromatics. Bicarbonate buffers acidity; excessive alkalinity flattens brightness and reduces perceived complexity (Genovese, 2025). Speciality-coffee standards recommend:

- **Hardness**: 50–175 ppm
- **Alkalinity**: ~40 ppm
- **pH**: approximately neutral (≈7)

These parameters minimise undesirable mineral–flavour interactions and promote balanced extraction. Water chemistry is particularly important for high-altitude Arabica coffees, whose acidity and floral attributes are sensitive to bicarbonate levels.

6.5 Technological advances in brewing science

Modern brewing increasingly incorporates digital sensing technologies, enabling precise control and reproducibility. Flow-rate monitoring, thermal profiling, and refractometry are now standard tools in experimental coffee laboratories. Recent work demonstrates that integrating refractive index measurements with grind-size modelling can accurately predict extraction yield, supporting automated brewing systems and machine-learning-based optimisation (Méndez Harper and Hendon, 2024).

Extraction plays a central role in determining the balance of acidity, sweetness, bitterness, aroma, and aftertaste in coffee beverages. Organic acids primarily contribute to brightness and structural definition, while melanoidins formed during roasting influence body, colour, and antioxidant capacity. Lipid emulsions enhance mouthfeel and are particularly important for crema formation in espresso-style extractions. In addition, volatile compounds released during brewing are key contributors to overall aromatic complexity and sensory depth.

Table 4: Brewing methods and their extraction characteristics

Brewing method	Typical grind size	Contact time/ brew time	Extraction mechanism	Chemical profile	Strengths & weaknesses	Recent findings (2023–2025)
Espresso	Very fine (200–300 μm)	25–30 s	High pressure (≈9 bar); rapid extraction	High lipids, emulsions, volatile retention; strong body	Intense flavour & crema; prone to channelling	Pressure profiling increases uniformity and aroma complexity (Kowalski, 2025).
Pour-over (V60, Kalita)	Medium (500–700 μm)	2.5–4 min	Gravity-driven percolation	Clean cup, high clarity, bright acidity	Clarity; sensitive to pouring technique	Water chemistry significantly influences the perception of acidity (Genovese, 2025).
French Press	Coarse (800–1200 μm)	4–6 min	Immersion extraction	High body, more sediment, strong melanoidin presence	Rich body; lower clarity	Immersion boosts melanoidin-linked antioxidant activity (Martinez et al., 2022).
Cold Brew	Coarse (800–1200 μm)	12–24 h	Cold immersion	Low acidity, high sweetness; fewer bitter compounds	Smooth flavours; long preparation	Hardtank® method increases CGA extraction vs. traditional cold brew (Kowalski, 2025).
AeroPress	Fine-medium	1–2 min	Hybrid (immersion + pressure)	Balanced acidity, moderate body	Very consistent; sensitive to grind	Sensory analyses conducted in 2024–25 trials revealed a strong consumer preference (Fibrianto, 2025).
Moka Pot	Fine	1–2 min	Steam pressure percolation	High intensity, medium bitterness	Easy, rich taste; risk of over-extraction	Recent trials have shown higher phenolic extraction at temperatures exceeding 95°C (Uslu, 2025).
Turkish Coffee	Ultra-fine (150–200 μm)	Boil + short simmer	High solid suspension	Very high aroma intensity; thick mouthfeel	Maximum flavour saturation; sediment	Fine particulate suspension enhances lipid extraction (Fitri et al., 2025).

Thus, the sensory outcome reflects not only the intrinsic chemical composition of roasted beans but also the efficiency, uniformity, and selectivity of the brewing process.

7. Sensory evaluation and coffee quality

Sensory evaluation forms the essential link between the chemical composition of coffee and its human perception. It enables researchers, roasters, and producers to characterise flavour, aroma, mouthfeel, and aftertaste using systematic methodologies. While analytical chemistry identifies and quantifies chemical constituents, sensory science explains how these compounds are perceived, integrated, and interpreted. Together, these complementary approaches underpin modern frameworks for coffee quality assessment (Abdelwahab et al., 2025).

7.1 Importance of sensory science

Sensory evaluation is indispensable for connecting coffee chemistry with human perception. Recent work integrating sensory panel data with machine learning has demonstrated strong predictive accuracy for quality descriptors, such as fruitiness, sweetness, and body (Martinez et al., 2022; Fitri et al., 2025).

Modern quality assessment frameworks rely heavily on sensory science because sensory attributes are primary determinants of consumer acceptance and preference. Even subtle sensory differences often reflect underlying biochemical composition and variations in processing conditions, providing insight into quality differentiation. Sensory profiling also informs critical decisions related to roasting, blending, and post-harvest processing strategies. Moreover, sensory outcomes show strong correlations with environmental, genetic, and post-harvest variables, thereby supporting the scientific validation of terroir and product provenance. Consequently, sensory science functions not only as a tool for evaluating final beverage quality but also as an integral decision-making framework across the entire coffee value chain.

7.2 Core sensory attributes

Coffee's sensory experience emerges from the interaction of multiple perceptual modalities.

Aroma: Aroma is the most complex attribute, arising from hundreds of volatile compounds—including pyrazines, aldehydes, ketones, furans, and phenolic derivatives—formed during roasting. New 2025 profiling studies identify nitrogen-containing volatiles that contribute to floral and spice notes in lightly roasted *Arabica* (Fitri et al., 2025).

Taste: Coffee expresses four primary taste modalities:

- **Sweetness** — primarily from caramelised sugars and Maillard derivatives.
- **Acidity** — reflecting levels of citric, malic, and quinic acids, especially in light roasts.
- **Bitterness** — driven by caffeine, phenolics, and pyrolysis products.
- **Saltiness** — minimal but occasionally perceptible in mineral-rich waters.

Mouthfeel/Body: Mouthfeel reflects the presence of lipids, melanoidins, suspended solids, and polysaccharides. Espresso crema exemplifies the emulsification of lipids and CO_2 under pressure.

Aftertaste: The persistence and evolution of flavour following swallowing—depends on volatile retention, phenolic composition, and the balance of acidity and sweetness. High-quality coffees typically display long, clean, and coherent aftertastes.

7.3 Standardised cupping

Cupping remains the global standard for evaluating professional coffee. Research shows that consistent roast level (e.g., Agtron scale), grind uniformity, and mineral-adjusted brewing water significantly improve repeatability among trained cuppers (Genovese, 2025). A typical cupping protocol includes:

1. Grinding to a uniform medium-coarse particle size.
2. Assessing dry aroma.
3. Adding water at 93–96°C and steeping for four minutes.
4. Breaking the crust to evaluate wet aroma.
5. Slurping to aerate and distribute volatiles across retronasal pathways.
6. Scoring attributes such as flavour, acidity, sweetness, body, balance, and overall impression.

Cupping enables both qualitative and quantitative assessment, with scores above 80/100 indicating speciality-grade quality characterised by clarity, complexity, and absence of defects.

7.4 Coffee defects and sensory faults

Sensory analysis plays a crucial role in identifying faults originating from agronomic practices, post-harvest processing, or roasting. Recent defect-mapping studies have linked sensory faults with specific microbial metabolites and oxidation markers, improving consistency in defect identification (Uslu, 2025).

Common sensory defects in coffee include sour or quaker-like notes arising from immature or underdeveloped beans, as well as phenolic or medicinal flavours commonly associated with microbial contamination or poorly controlled fermentation processes. Mouldy or musty taints typically result from inadequate drying or improper storage conditions, whereas burnt or ashy flavours are indicative of over-roasting or scorching during thermal processing. Rubbery or earthy notes are frequently observed in poorly processed Robusta coffees. A clear understanding of the biochemical and processing origins of these defects enables producers to optimise processing parameters and implement corrective strategies, thereby improving overall coffee quality (Grandez-Alberca et al., 2025).

7.5 Instrumental analysis and sensory correlations

Instrumental methods, such as NMR, GC–MS, electronic noses, and HPLC, increasingly complement sensory panels. A 2024 NMR study successfully linked aromatic markers to sensory attributes, demonstrating strong correlations between

Table 5: Sensory attributes of coffee and their key chemical contributors

Sensory attribute	Primary chemical contributors	Chemical classes	Mechanisms/notes	Recent findings (2023–2025)
Aroma – Floral	Linalool, geraniol, benzyl alcohol	Terpenoids, alcohols	Formed during roasting and via fermentation precursors	Yeast-driven fermentation increases floral esters (Amorocho-Cruz et al., 2025).
Aroma – Fruity	Ethyl acetate, isoamyl acetate, phenethyl acetate	Esters	Microbial esterification during fermentation	Anaerobic and carbonic maceration significantly enhance ester levels (Lu, 2025).
Nutty/Roasted	Pyrazines	Nitrogen heterocycles	Thermally generated during Maillard reactions	High-altitude beans produce more pyrazine precursors (Fitri et al., 2025).
Caramel/Sweet	Furans, furanones, maltol	Thermal degradation products	Formed from sucrose breakdown during roasting	Higher sucrose → more caramel notes (Genovese, 2025).
Chocolate	2-methylpyrazine, 2,3-diethyl-5-methylpyrazine	Pyrazines	Core markers of medium-dark roasts	Optimal at mid-Maillard reaction stages.
Acidity (bright)	Citric, malic, phosphoric acids	Organic acids	Strongly influenced by origin and roasting	Light roasts retain the highest acidity; altitude has a positive correlation (Fitri et al., 2025).
Bitterness	Caffeine, phenylindanes, CGA-lactones	Alkaloids and phenolics	Formed during CGA degradation	Dark roasts ↑ phenylindanes → stronger bitterness (Uslu, 2025).
Body/Mouthfeel	Lipids, melanoidins, polysaccharides	Lipid fraction & Maillard polymers	Increase viscosity and thickness	Espresso's high pressure increases emulsified lipids (Genovese, 2025).
Astringency	Chlorogenic acids, polyphenols	Polyphenols	Bind salivary proteins	Robusta shows higher polyphenol-driven astringency (Bealu Girma, 2022).
Aftertaste	Volatile retention compounds; melanoidins	Melanoidins & volatiles	Persistence depends on roast profile and lipid content	Melanoidins also contribute to antioxidant capacity (Uslu, 2025).

instrumental and human evaluations (Martinez et al., 2022). Key instrumental contributions include:

- **GC–MS** — identification of volatile compounds underlying aroma.
- **HPLC** — quantification of caffeine, CGAs, and other non-volatile constituents.
- **Colourimetry** — objective roast-level assessment.
- **Rheology** — measurement of viscosity and mouthfeel in espresso and high-solids brews.

These tools enhance the precision of sensory interpretation, particularly for high-value speciality coffees or research applications.

7.6 Sensory training and calibration

Sensory performance depends on rigorous training. Programmes introduced in 2024–2025 incorporate digital flavour reference tools and cross-modal calibration exercises, improving descriptor accuracy and reducing inter-panel variability (Fitri et al., 2025).

Professional sensory training in coffee evaluation typically involves repeated exposure to coffees originating from diverse geographical regions and processed using different post-harvest methods. Trainees develop the ability to recognise specific volatile and flavour markers and undergo regular calibration with expert sensory panels to ensure alignment and reproducibility. Discriminative methods such as triangulation tests are commonly employed to enhance sensitivity to subtle sensory differences. In addition, structured sensory frameworks and reference tools, including the SCA Flavour Wheel and Le Nez du Café aroma kits, are systematically used to standardise vocabulary and perception. Such comprehensive training ensures that cuppers can consistently detect fine sensory variations and maintain reliable evaluations across sessions, panels, and professional settings. The major sensory attributes of coffee and their key chemical contributors, underlying mechanisms, and recent findings are summarised in Table 5.

8. Environmental and agronomic factors

Environmental and agronomic conditions have a profound influence on coffee plant performance, bean chemistry, and final sensory quality. Coffee is susceptible to temperature, altitude, rainfall patterns, soil characteristics, and biotic pressures, making its cultivation a dynamic interaction between plant genetics and environmental context. As climate change intensifies, understanding these factors has become indispensable to ensuring resilience and sustaining quality in global coffee production (Bilen et al., 2023).

8.1 Climate and temperature

Coffee production is strongly constrained by temperature. *Arabica* thrives at 15–24°C, with cool nights that slow fruit maturation and promote the accumulation of sugars and aromatic precursors (Grandez-Alberca et al., 2025). *Robusta*, by contrast, tolerates higher temperatures (24–30°C) and lowland environments due to its physiological robustness and higher caffeine and phenolic content (Bealu Girma, 2022).

Temperature stress induces measurable biochemical and physiological effects in coffee plants that directly influence bean quality and yield. Prolonged exposure to elevated temperatures suppresses lipid biosynthesis, leading to reduced body and diminished flavour complexity in the final beverage. High temperatures also accelerate fruit maturation and are associated with decreased concentrations of sucrose and chlorogenic acids, compounds that play key roles in sweetness, acidity, and antioxidant capacity (Abdelwahab et al., 2025). In addition, temperature fluctuations during the flowering period can disrupt anthesis and fruit set, resulting in reduced yields. Climate simulation models predict a substantial contraction of climatically suitable regions for *Coffea arabica* cultivation by 2050, with particularly pronounced impacts projected for East Africa and Central America (Bilen et al., 2023; Grandez-Alberca et al., 2025). These projections have intensified research efforts focused on the development of heat-tolerant cultivars, the use of shade-based microclimate buffering strategies, and the expansion of agroforestry systems to enhance climate resilience

8.2 Altitude

Altitude is a major determinant of coffee bean density, biochemical composition, and sensory expression. Higher elevations are characterised by lower mean temperatures, greater diurnal temperature variation, and increased solar radiation, creating environmental conditions that modulate key metabolic pathways in coffee plants. Under these conditions, fruit development proceeds more slowly, favouring the accumulation of sugars and organic acids that contribute to enhanced sweetness and structural acidity. Elevated altitudes are also associated with increased concentrations of chlorogenic acids, which are linked to brighter and more defined acidity in the cup (Abdelwahab et al., 2025). In addition, beans produced at higher elevations typically exhibit greater density, a physical attribute that improves heat transfer during roasting and promotes more uniform flavour development.

Transcriptomic studies reveal that altitude-driven temperature variation regulates the expression of genes related to carbohydrate and lipid metabolism, suggesting a molecular basis for altitude-associated quality differentiation (Reyes-Herrera et al., 2023). Current genetic and metabolomic work explores genotype–environment interactions to understand how altitude shapes metabolite formation (Medrano et al., 2025).

8.3 Rainfall and water availability

Optimal coffee production requires 1,200–2,000 mm annual rainfall, ideally distributed across distinct wet and dry seasons. Water availability influences flowering synchrony, fruit development, and microbial activity during post-harvest processing.

Rainfall patterns exert a strong influence on both coffee quality and post-harvest stability. Excessive rainfall during drying periods promotes mould growth and increases the risk of mycotoxin contamination, posing significant challenges for quality control and food safety. Conversely, drought stress reduces bean size, slows sugar accumulation, and may enhance bitterness through alterations in chlorogenic acid and caffeine biosynthetic pathways (Uslu, 2025). While irrigation can help stabilise yields under water-limited conditions, excessive water availability tends to dilute flavour intensity and disrupt the balance between acidity and sweetness. In response to climate-driven variability in rainfall, producers have increasingly adopted water-efficient processing technologies, such as mechanical demucilagers, to minimise environmental impact and improve consistency in post-harvest processing (Grandez-Alberca et al., 2025).

8.4 Soil composition and fertility

Soil characteristics, including pH, organic matter content, mineral availability, and microbial diversity, play central roles in plant nutrition, disease resistance, and the biochemical composition of coffee beans. Slightly acidic soils, typically within a pH range of 5.0 to 6.0, optimise nutrient availability and uptake by coffee plants. Nitrogen availability strongly influences vegetative growth and is closely linked to caffeine biosynthesis, while potassium contributes to carbohydrate translocation, sugar accumulation, and the maintenance of acidity balance in the developing beans. In addition, high levels of soil organic matter improve soil structure, enhance moisture retention, and support diverse microbial communities, collectively promoting plant health and contributing to improved bean quality.

Recent studies have demonstrated that the composition of the soil microbiome significantly influences nutrient absorption, root health, and plant resistance to pathogens, underscoring the importance of integrated soil fertility management for sustainable production (Uslu, 2025).

8.5 Shade management and agroforestry

Shade-grown and agroforestry systems moderate microclimatic extremes, enhance biodiversity, and improve ecosystem services. These systems:

- Reduce heat stress and buffer high-temperature events;
- Slow fruit maturation, augmenting sugar and trigonelline accumulation (Bealu Girma, 2022);
- Increase chlorogenic acid content and aromatic complexity;
- Improve soil health and carbon sequestration.

However, excessive shade can prolong fruit drying, increase disease incidence, and reduce yields. Effective shade management, therefore, requires balancing the benefits of microclimate against productivity demands. Recent ecological assessments underscore the value of multi-strata shade systems for long-term resilience (Bilen et al., 2023).

8.6 Pests and diseases

Biotic stressors are among the most significant threats to global coffee production. Major examples include:

Table 6: Environmental factors affecting coffee bean chemistry and sensory quality

Environmental factor	Impact on plant physiology	Effect on bean chemistry	Sensory consequences	Recent findings (2023–2025)
Temperature	Alters photosynthesis, stress hormones, and accelerates maturation	↓ sucrose, ↓ lipids, altered CGA ratios	↓ acidity, ↓ sweetness, flatter cup profile	Climate simulations indicate a significant decline in suitability for Arabica by 2050 (Bilen et al., 2023; Grandez-Alberca et al., 2025).
Altitude	Cooler temperatures slow maturation; increased bean density	↑ organic acids, ↑ sugars, ↑ aroma precursors	Brighter acidity, floral–fruit complexity	High altitude is correlated with elevated acidity and increased volatile abundance (Fitri et al., 2025).
Rainfall & Water Stress	Excess moisture → fungal risk; drought → stress metabolites	Irregular sugars, ↑ bitterness, ↑ CGA	Off-flavours under wet stress; sharper bitterness under drought	Water imbalance increases CGA accumulation (Uslu, 2025).
Soil Composition	Influences nutrient uptake, microbial diversity	K, N, and Mg deficiencies alter sucrose and lipid pathways	Weak body (low lipids), muted sweetness	Soil microbiome diversity improves nutrient-use efficiency (Uslu, 2025).
Shade/ Agroforestry	Reduces heat stress; modifies photosynthetic dynamics	More stable metabolic development; ↑ sugars	Smoother acidity, enhanced sweetness	Shade-grown coffee exhibits better resilience to heatwaves (Bilen et al., 2023).
Sun Exposure	Increases metabolic rate and stress	Faster CGA breakdown; lower lipid accumulation	More bitterness; thinner body	Excessive light reduces chlorogenic acid stability.
Pest Pressure (e.g., berry borer)	Damages cherries; alters stress metabolites	↑ phenolic off-flavour compounds	Astringency, woody or fermented notes	Expansion of pest habitats predicted with warming (Grandez-Alberca et al., 2025).
Fungal Pathogens (e.g., leaf rust)	Reduces plant energy; decreases sugar allocation	↓ carbohydrate accumulation in beans	Muted sweetness, reduced aroma complexity	New resistance-associated genes identified using 2024 genomics (Medrano et al., 2025).

- **Coffee leaf rust (*Hemileia vastatrix*)** — disrupts photosynthesis and dramatically reduces yield; *Arabica* is particularly susceptible (Bealu Girma, 2022).
- **Coffee berry disease (*Colletotrichum kahawae*)** — damages cherries and causes premature fruit drop.
- **Coffee berry borer (*Hypothenemus hampei*)** — burrows into cherries, reducing both yield and quality.

Climate change is increasing the prevalence, distribution, and severity of these pests and diseases (Bilen et al., 2023). Genomic analyses conducted from 2024 to 2025 have identified new resistance-associated loci in *Robusta* and wild *Arabica* populations, providing targets for breeding programmes (Medrano et al., 2025).

8.7 Sustainable agronomic practices

Sustainable agronomy integrates principles of regenerative agriculture, soil conservation, biodiversity enhancement, and resource-efficient production to support long-term productivity and quality in coffee systems. Recent innovations include the adoption of regenerative practices such as mulching, cover cropping, and compost-based fertilisation to improve soil structure and nutrient cycling, as well as precision irrigation strategies designed to optimise water use and mitigate drought-related risks. Intercropping systems are increasingly implemented to enhance ecosystem resilience and diversify farm income, while targeted pruning practices improve light distribution within the canopy and reduce disease pressure. In addition, selective harvesting aimed at ensuring uniform fruit ripeness is considered critical for achieving high sensory quality, particularly in washed coffee processing. Each of these practices modulates plant metabolic processes that ultimately shape the chemical composition and sensory attributes of coffee beans (Uslu, 2025). Consequently, a system-wide agronomic approach is essential to ensure both economic viability and excellence in coffee quality. The major environmental and agronomic factors influencing coffee bean chemistry and sensory quality are summarised in Table 6.

9. Coffee and health

Coffee is a complex matrix of bioactive compounds, and contemporary research continues to refine the understanding of its physiological effects. Recent epidemiological work highlights the benefits related to cardiometabolic health, neuroprotection, metabolic regulation, and mood, mediated by caffeine, chlorogenic acids (CGAs), diterpenes, and melanoidins (Shi, 2025; Li, 2025).

9.1 Bioactive compounds and mechanisms

Coffee contains several classes of bioactive constituents with distinct physiological roles:

Caffeine: Caffeine is the most studied component of coffee, acting primarily as an adenosine receptor antagonist. It enhances alertness, attention, reaction speed, and cognitive performance. New neurophysiological modelling suggests caffeine modulates dopaminergic and cholinergic pathways, supporting sustained attention and improved vigilance (Li, 2025).

Chlorogenic acids (CGAs): CGAs are the dominant polyphenols in green coffee and exhibit antioxidant, anti-inflammatory, and glucose-modulating activities. High-resolution chemical profiling reveals that lightly roasted Arabica retains the highest CGA concentrations, which correlate with enhanced antioxidant capacity (Fitri et al., 2025).

Diterpenes: cafestol and kahweol: These lipid-soluble compounds modulate liver enzyme activity and have been explored for potential anticancer and hepatoprotective effects. Diterpene content varies significantly by brewing method: unfiltered brews contain the highest levels, whereas paper filtration substantially reduces them.

Trigonelline: A precursor to nicotinic acid, trigonelline contributes to roasted aroma formation and exhibits emerging neuroprotective and antimicrobial properties (Medrano et al., 2025).

Melanoidins: Melanoidins—high molecular-weight compounds formed during the Maillard reaction—possess antioxidant activity and may function as dietary fibre, influencing gut microbiota composition (Abdelwahab et al., 2025).

Together, these compounds shape the biological footprint of coffee as both a stimulant and a functional food.

9.2 Cognitive and neurological health

Moderate coffee consumption has been linked with improved cognitive resilience, particularly in ageing populations. A 2025 longitudinal study demonstrated a reduced risk of mild cognitive impairment and improvements in executive function among individuals consuming 2–4 cups per day (Li, 2025). Neuroimaging data further associate habitual coffee intake with enhanced cerebral blood flow and reduced neuroinflammation.

Controlled intervention trials consistently demonstrate that caffeine intake is associated with improvements in cognitive performance, particularly in reaction time, sustained attention and vigilance, short-term memory, and perceived mental energy. These effects are primarily attributed to caffeine's antagonistic action on adenosine receptors, leading to enhanced neuronal excitability and increased neurotransmitter release.

Beyond caffeine, polyphenols and trigonelline may contribute additional neuroprotective effects. Epidemiological evidence supports an association between habitual coffee consumption and reduced incidence of neurodegenerative diseases such as Parkinson's and Alzheimer's disease (Abdelwahab et al., 2025).

9.3 Cardiovascular health

The relationship between coffee consumption and cardiovascular health is complex and modulated by both biochemical composition and individual genetic variability, particularly polymorphisms affecting caffeine metabolism such as variants of the CYP1A2 gene. Epidemiological evidence indicates that moderate coffee consumption, typically defined as three to five cups per day, is associated with a reduced risk of cardiovascular mortality. Chlorogenic acids (CGAs) have been shown to exert beneficial effects on vascular health by improving endothelial function and attenuating oxidative stress. In contrast, coffee diterpenes, notably cafestol and kahweol, may elevate low-density lipoprotein (LDL) cholesterol levels, especially in unfiltered brewing methods such as Scandinavian boiled coffee or French press preparations (Uslu, 2025). Filtered coffee, however, effectively removes the majority of these diterpenes, thereby substantially reducing lipid-related cardiovascular risk. Consequently, the brewing method represents a critical determinant of the cardiovascular effects associated with coffee consumption.

9.4 Metabolic health and diabetes

Strong epidemiological evidence supports an inverse association between coffee consumption and the risk of type 2 diabetes mellitus. Several biological mechanisms have been proposed to explain this protective effect. Chlorogenic acids (CGAs) may inhibit α-glucosidase activity, thereby reducing intestinal glucose absorption and postprandial glycaemic responses. In addition, coffee consumption has been linked to improved insulin sensitivity and antioxidant-mediated modulation of inflammatory pathways involved in metabolic dysregulation. Emerging evidence also suggests that coffee induces favourable alterations in gut microbiota composition, which may further contribute to improved glucose homeostasis. Notably, both caffeinated and decaffeinated coffee exhibit protective effects, indicating that non-caffeine polyphenolic compounds play a substantial role in mediating the metabolic benefits associated with coffee intake (Shi, 2025; Abdelwahab et al., 2025).

9.5 Liver health

Coffee consumption has been associated with reduced risk of liver fibrosis, cirrhosis, and hepatocellular carcinoma. Recent biochemical studies suggest that coffee-derived diterpenes activate detoxifying enzymes and reduce hepatic fat accumulation, providing a mechanistic explanation for observed epidemiological trends (Medrano et al., 2025). Benefits appear dose-dependent but are generally favourable within moderate intake levels.

9.6 Gut microbiome and gastrointestinal function

Coffee influences gut health through several mechanisms. Melanoidins and polyphenols function as prebiotic substrates, selectively stimulating the growth of beneficial microorganisms, such as *Bifidobacterium*. A 2024 cohort study showed that habitual coffee consumption increased microbial diversity and improved short-chain fatty acid production (Uslu, 2025).

Coffee also stimulates gastric motility, which may aid digestion but can exacerbate reflux symptoms in individuals who are sensitive to it.

9.7 Potential risks

Despite its well-documented benefits, coffee consumption may pose potential risks under specific physiological or clinical conditions. Individuals with heightened sensitivity to caffeine may experience adverse effects such as anxiety, insomnia, or tachycardia. During pregnancy, high caffeine intake—generally exceeding 200 mg per day—is discouraged because of potential risks to fetal development. Additionally,

Table 7: Summary of major health effects of coffee and underlying bioactive mechanisms

Health outcome/ domain	Key bioactive compounds	Proposed mechanisms	Evidence summary	Recent references
Cognitive performance & alertness	Caffeine, chlorogenic acids, and trigonelline	Adenosine receptor antagonism; modulation of dopaminergic and cholinergic signalling; improved cerebral blood flow	Improved attention, reaction time, and vigilance; protective association with cognitive decline in older adults	Li, 2025; Reddy et al., 2024
Mood, depression & anxiety	Caffeine, CGAs, polyphenols	Modulation of neurotransmitters (dopamine, serotonin); anti-inflammatory and antioxidant effects; HPA axis modulation	Lower risk of depression and suicidal ideation at moderate intake; mixed data for anxiety at high doses	Shi, 2025
Type 2 diabetes risk	CGAs, melanoidins, magnesium, polyphenols	Improved insulin sensitivity; delayed glucose absorption; modulation of gut microbiota and inflammatory pathways	Consistent inverse association between coffee intake and T2D risk; effect seen for both caffeinated and decaf	Emadi et al., 2025; Shi, 2025
Cardiovascular disease	CGAs, polyphenols, caffeine, diterpenes	Endothelial function improvement; antioxidant effects; modulation of lipid metabolism (cafestol, kahweol)	Moderate intake linked to ↓ CVD and stroke risk; unfiltered coffee can ↑ lower LDL via diterpenes	Uslu, 2025; Reddy et al., 2024
Liver health	Diterpenes (cafestol, kahweol), CGAs, melanoidins	Induction of detox enzymes; anti-fibrotic and anti-inflammatory effects; reduced hepatic fat accumulation	Lower risk of fibrosis, cirrhosis, and hepatocellular carcinoma among regular coffee drinkers	Medrano et al., 2025; Emadi et al., 2025
Antioxidant & anti-inflammatory status	CGAs, other polyphenols, melanoidins	Scavenging of reactive oxygen species; modulation of NF-κB and related inflammatory pathways	Higher antioxidant capacity in plasma; reduced markers of oxidative stress in intervention studies	Fitri et al., 2025; Uslu, 2025

(*Contd.*)

(*Contd.*)

Health outcome/ domain	Key bioactive compounds	Proposed mechanisms	Evidence summary	Recent references
Gut microbiome & GI function	Melanoidins, polyphenols, fibre from coffee, and by-products	Prebiotic effects: stimulation of beneficial bacteria; changes in SCFA profiles	Increased microbial diversity and favourable shifts in *Bifidobacterium*; mild laxative effect	Uslu, 2025; Tenea et al., 2025
Sleep & circadian rhythm	Caffeine	Adenosine antagonism; altered sleep latency and architecture	Improved alertness when timed early in the day; sleep disruption if consumed close to bedtime or in sensitive individuals	Shi, 2025; NHLBI, 2025
Potential risks (high intake / sensitive groups)	Caffeine, diterpenes	Anxiety, palpitations, GI discomfort; ↑ LDL with unfiltered coffee; effects in pregnancy	Risk mainly at high doses or specific brewing types; guidelines recommend <200 mg caffeine/day during pregnancy	Reddy et al., 2024; Emadi et al., 2025

excessive caffeine consumption has been associated with increased calcium excretion, although the long-term implications of this effect on bone health remain uncertain. Coffee may also interact with certain medications, including cardiovascular agents and anxiolytic drugs, potentially altering their pharmacokinetics or pharmacodynamic responses. Importantly, individual risk–benefit profiles vary considerably as a result of genetic variability in caffeine metabolism, particularly polymorphisms in the CYP1A2 gene.

9.8 Coffee as a functional food

With the growing interest in functional foods and nutraceuticals, research has increasingly focused on the health-promoting potential of coffee-derived components. Current areas of innovation include the development of chlorogenic acid (CGA)-enriched extracts and polyphenol-rich dietary supplements, as well as roasting strategies specifically designed to optimise the retention of antioxidant compounds. In addition, the valorisation of coffee by-products—such as cascara and silverskin—has gained considerable attention due to their high fibre content and abundance of bioactive compounds (Pyrzynska et al., 2024; Uslu, 2025). Advances in processing technologies, including antioxidant-enhanced cold-brew and intensified extraction approaches, further support the diversification of coffee-based functional products. Collectively, the broad spectrum of bioactive constituents in coffee firmly positions it within the expanding functional food and nutraceutical sectors. The major health outcomes associated with coffee consumption and their underlying bioactive mechanisms are summarised in Table 7.

10. Coffee industry, economics, and sustainability

The global coffee industry is one of the most complex and economically significant agro-industrial systems. Coffee is cultivated in more than seventy countries and supports the livelihoods of an estimated 125 million people, a majority of whom are smallholder farmers (Grandez-Alberca et al., 2025). As both an agricultural commodity and a culturally embedded beverage, coffee lies at the nexus of economics, sustainability, consumer trends, and technological innovation. Understanding these interconnected dimensions is essential for contextualising scientific research within real-world production and market systems.

10.1 Global production and economic structures

Coffee remains a cornerstone of global agricultural trade. Economic analyses highlight considerable price volatility driven by climate instability, supply chain disruptions, and speculative activity in futures markets (Grandez-Alberca et al., 2025). Approximately 70% of the global coffee supply is produced by smallholder farmers, who are disproportionately vulnerable to climate shocks, pest outbreaks, and fluctuating international prices.

Two species dominate the global supply—*Coffea arabica* and *C. canephora*. Latin America, led by Brazil and Colombia, remains the principal source of *Arabica*, while Vietnam and parts of Africa are major producers of *Robusta*.

Key structural characteristics of the global coffee economy include pronounced price volatility, largely driven by fluctuations in supply, climate variability, and shifts in international demand. The sector is also defined by asymmetric value chains, in which primary producers typically receive only a small fraction of the final retail price. Market segmentation further differentiates commodity-grade coffees from speciality-grade products, reflecting variations in quality standards, pricing mechanisms, and consumer expectations. In recent years, growing consumer demand for traceability, ethical sourcing, and superior sensory quality has increasingly shaped market dynamics and influenced production, certification, and marketing strategies across the coffee value chain.

These economic dynamics influence investment in processing technologies, agronomic innovation, and sustainability initiatives.

10.2 Speciality coffee and consumer behaviour

The speciality coffee sector has experienced rapid expansion, profoundly shaping both consumer expectations and industry practices. Speciality coffee places strong emphasis on terroir-driven flavour profiles, comprehensive traceability and transparency throughout the value chain, and high sensory quality, typically defined by cupping scores exceeding 80 points on the Specialty Coffee Association (SCA) scale. This segment is further characterised by artisanal roasting approaches, precision brewing techniques, and direct trade models that support long-term partnerships with producers. A recent consumer preference study reported an increased willingness to pay for coffees produced using controlled fermentation methods, reduced-water processing techniques, and verified ecological certifications (Genovese, 2025). In response to these market dynamics, scientific research has intensified in several key areas, including controlled microbial fermentation (Tran et al., 2016), genotype–environment interactions influencing flavour expression (Medrano et al., 2025), the identification of chemical markers associated with sensory quality (Abdelwahab et al., 2025), and the development of advanced brewing and extraction technologies (Méndez Harper and Hendon, 2024). Collectively, the speciality coffee movement functions not only as a market driver but also as a catalyst for scientific innovation.

10.3 Sustainability challenges

Climate change presents the most significant long-term challenge to global coffee production. Updated models project substantial contraction of *Arabica* habitat suitability by mid-century, accompanied by increased pest pressure and greater yield variability (Bilen et al., 2023; Grandez-Alberca et al., 2025).

In addition to environmental pressures, the coffee sector continues to face substantial economic challenges. Price instability remains a major concern, threatening income security for producers and increasing financial uncertainty. High production costs, particularly those associated with fertilisers, labour, and pest management, further constrain profitability. Limited access to finance restricts farmers' capacity to invest in improved technologies and sustainable practices, thereby reinforcing structural vulnerabilities within the sector. Moreover, coffee production remains highly susceptible to diseases, most notably coffee leaf rust

(*Hemileia vastatrix*), which continues to pose a serious threat to yield stability and long-term farm viability (Bealu Girma, 2022).

Given that smallholder farmers account for 70% of global coffee production, research in agronomy, breeding, and processing plays a crucial role in supporting resilience and improving livelihoods (Grandez-Alberca et al., 2025).

10.4 Certification and ethical sourcing

Certification schemes—including Fairtrade, Rainforest Alliance, Organic and UTZ—aim to promote sustainable production, improve labour conditions and provide price stability. However, recent reviews indicate that certification outcomes are highly context-dependent, with benefits varying according to farm size, geographic region, and market access (Grandez-Alberca et al., 2025). Significant sustainability pressures include:

Climate change: Increasing temperatures, erratic rainfall, and extreme weather events threaten both productivity and quality, particularly for climate-sensitive *Arabica* (Bilen et al., 2023).

Deforestation: Expansion of low-shade monocultures contributes to biodiversity loss and reduced ecological resilience.

Soil degradation: Continuous cultivation without adequate nutrient management leads to declining fertility and increased erosion.

Pests and diseases: Climate-driven changes in pest distribution intensify outbreaks of rust and the coffee berry borer.

Addressing these systemic challenges requires integrated strategies combining genetics, agroforestry, soil conservation, climate adaptation, and socio-economic policy.

10.5 Circular bioeconomy and by-product utilisation

Coffee production generates substantial quantities of by-products, including pulp, mucilage, husk, parchment, silverskin, and spent coffee grounds. Contemporary biorefinery research has increasingly focused on the valorisation of these materials into a wide range of value-added products, such as antioxidant- and polyphenol-rich extracts, dietary fibres and functional food ingredients, composts and organic fertilisers, biofuels and biogas, as well as biodegradable polymers and sustainable packaging materials (Uslu, 2025; Pyrzynska et al., 2024). The effective utilisation of coffee by-products not only mitigates environmental burdens associated with waste disposal but also supports the development of circular bioeconomy models, thereby aligning coffee industry practices with broader global sustainability objectives.

10.6 Technological innovation

Technological advancements are transforming multiple stages of coffee production and processing. Recent developments include:

- AI-driven roasting platforms enabling precision thermal profiling;

- Blockchain traceability systems ensuring transparency across supply chains;
- IoT-enabled fermentation tanks monitoring temperature, pH, and microbial dynamics;
- Infrared and ultrasonic drying technologies that enhance uniformity and reduce energy consumption;
- Hardtank® extraction systems, which improve extraction efficiency and antioxidant yield (Kowalski, 2025).

Simultaneously, valorisation of by-products—such as pulp for pectin extraction, silverskin as a fibre source, and spent grounds for biofuel or biopolymers—illustrates how technological innovation supports sustainability and economic diversification (Pyrzynska et al., 2024).

11. Conclusion

Coffee science integrates diverse disciplines—genomics, chemistry, sensory evaluation, processing technology, health research, and sustainability—into a cohesive understanding of how coffee quality is formed. Modern advances in genome sequencing, microbial ecology, and analytical chemistry have illuminated how environmental factors, post-harvest processes, and roasting dynamics shape the biochemical complexity of coffee beans.

Simultaneously, emerging research in nutrition and health has highlighted coffee's multifaceted physiological effects, reinforcing its role as both a cultural beverage and a functional food. Innovations in brewing and extraction continue to refine the flavour potential, while industry-wide interest in sustainability drives the exploration of agroforestry systems, climate-resilient cultivars, and by-product valorisation.

The global coffee sector now stands at the intersection of tradition and scientific innovation. As climate pressures intensify and consumer expectations evolve, interdisciplinary research will remain essential for supporting resilient production systems, improving quality, and advancing knowledge across the entire value chain—from plant genetics to sensory experience.

Coffee science has evolved into a rich and multidimensional field that integrates plant biology, chemistry, processing technology, sensory science, human health, and sustainability. As demonstrated throughout this chapter, coffee is not merely an agricultural commodity but a complex biochemical and socio-economic system shaped by interactions across the entire value chain—from genome to cup. The genetic foundations of *Coffea* species influence the biochemical composition long before the beans undergo fermentation, drying, and roasting. These biological traits, in turn, define the potential for flavour development, aroma profiles, and functional properties that emerge through thermochemical transformations during roasting and selective extraction during brewing.

The scientific character of coffee has been further shaped by its historical, cultural, and economic significance. The globalisation of coffee production and consumption has catalysed advances in agronomy, processing methods, and analytical technologies. Recent scientific progress—in genomics, metabolomics, microbial ecology, and data-driven sensory analysis—continues to deepen our understanding of quality formation and provide tools for producers, roasters, and researchers. At the same time, the intensification of climate pressures and market volatility underscores the urgent need for integrative, sustainable approaches that enhance resilience, reduce environmental footprints, and support smallholder livelihoods.

The intersection of tradition and innovation defines the current moment in coffee research. While long-standing practices such as washed, natural, and honey processing remain central to quality expression, emerging biotechnological techniques, precision roasting systems, and advanced brewing analytics offer new pathways for quality optimisation. Moreover, expanding evidence on the health effects of coffee's bioactive compounds has positioned coffee as both a culturally significant beverage and a potential functional food.

Ultimately, coffee science is most powerful when approached as a connected system rather than a set of isolated disciplines. Each component—from plant genetics to consumer experience—contributes to the final quality and cultural meaning of coffee. By presenting a scientifically grounded framework, this chapter aims to support deeper engagement with the specialised topics explored in subsequent chapters of this volume and to encourage ongoing interdisciplinary collaboration across the global coffee research community.

References

Abdelwahab, S. I., Elhassan Taha, M. M., Jerah, A. A., Aljahdali, I. A., Oraibi, B., Alfaifi, H. A. et al. (2024). Coffee arabica research (1932–2023): Performance, thematic evolution and mapping, global landscape, and emerging trends. *Heliyon*, 10(16), e36137. https://doi.org/10.1016/j.heliyon.2024.e36137

Amorocho-Cruz, C. M., Tarazona-Sánchez, P. A., and Ladino-Garzón, W. L. (2025). Effects of lactic acid bacteria and yeast inoculation on the sensory quality of Coffea arabica. *Coffee Science*, 20, e202361. https://doi.org/10.25186/.v20i.2361

Asadullah, M., Ikram, A., Sahni, O., Rasheed, M., Abdullah, Z., Menon, S. et al. (2025). Coffee beans chemical composition, processing and physiological nature: A comprehensive review. *European Food Research and Technology*, 251, 4135–4148. https://doi.org/10.1007/s00217-025-04893-w

Avallone, S., Guyot, B., Brillouet, J.-M., Olguin, E., and Guiraud, J.-P. (2001). Microbiological and biochemical study of coffee fermentation. *Current Microbiology*, 42(4), 252–256. https://doi.org/10.1007/s002840110213

Bealu, G. and Abrar, S. (2022). A review of coffee processing methods and their influence on aroma. *International Journal of Food Engineering and Technology*, 6(1), 7–16. https://doi.org/10.11648/j.ijfet

Bilen, C., El Chami, D., Mereu, V., Trabucco, A., Marras, S., and Spano, D. (2023). A systematic review on the impacts of climate change on coffee agrosystems. Plants, 12(1), 102. https://doi.org/10.3390/plants12010102.

Collazos-Chaves, E. F., Hidalgo-Troya, A., Hurtado-Gutiérrez, N. H., and Revelo-Romo, D. M. (2025). Microorganisms involved in the fermentation process of specialty coffees in Southwestern Colombia identified as potential starter cultures. *Revista Argentina de Microbiología.* https://doi.org/10.1016/j.ram.2025.09.009

Della Peruta, R., Mereu, V., Spano, D., Marras, S., Vezy, R., and Trabucco, A. (2025). Projecting trends of arabica coffee yield under climate change: A process-based modelling study at continental scale. *Agricultural Systems*, 227, 104353. https://doi.org/10.1016/j.agsy.2025.104353.

Emadi, R. C., and Kamangar, F. (2025). Coffee's impact on health and well-being. *Nutrients*, 17(15), 2558. https://doi.org/10.3390/nu17152558

Fitri, N., Nurhaliza, N., Anggraeni, E. S., Herawati, D., and Hunaefi, D. (2025). Chemical composition, antioxidant activity, and sensory profile of espresso-based Arabica coffees from different origins. *Beverage Plant Research*, 5, e008. https://doi.org/10.48130/bpr-0024-0040

Genovese, A., Caporaso, N., and Baiano, A. (2025). The impact of brewing methods on the quality of a cup of coffee. *Beverages*, 11(5), 125. https://doi.org/10.3390/beverages11050125.

Grandez-Alberca, M. A., Cotrina-Sanchez, A., Haro, N., Zabaleta-Santisteban, J. A., Silva-Melendez, T. B., Sanchez-Vega, J. A. et al. (2025). Agronomic and economic sustainability perspectives in coffee-based agroforestry system (Coffea arabica L.) in Rodríguez de Mendoza, Amazonas, Peru. *Trees, Forests and People*, 100924.https://doi.org/10.1016/j.tfp.2025.100924.

Harper, J. M., Bumbaugh, R. E., and Hendon, C. H. (2024). Strategies to mitigate electrostatic charging during coffee grinding. *IScience*, 27(9), 110639. https://doi.org/10.1016/j.isci.2024.110639

Kowalski, R., Kamil, B., Grażyna, K., Marzena,W., Artur, M., and Tomasz, B. (2025). High-efficiency coffee extraction technologies: Impact on chlorogenic acids and antioxidant potential Hardtank® technology in extraction of caffeine and chlorogenic acid from coffee: analysis of chemical composition and antioxidant properties of coffee beverages. *Frontiers in Chemistry*, 13, 1681137. | https://doi.org/10.3389/fchem.2025.1681137

Lee, L. W., Cheong, M. W., Curran, P., Yu, B., and Liu, S. Q. (2015). Coffee fermentation and flavor – An intricate and delicate relationship. *Food Chemistry*, 185, 182–191. https://doi.org/10.1016/j.foodchem.2015.03.124

Li, J., Yu, K., Bu, F. Li, P., and Hao, L. (2025). Exploring the impact of coffee consumption and caffeine intake on cognitive performance in older adults: A comprehensive analysis using NHANES data and gene correlation analysis. *Nutrition Journal*, 24, 102.https://doi.org/10.1186/s12937-025-01173-x

Lu, K., He, Y., Cao, Y., Zhu, T., Liu, Q., Liu, B. et al. (2025). The generation of coffee flavor compounds from fresh cherry to cup: Deciphering the contribution of microbiota metabolism. *Journal of Future Foods*. https://doi.org/10.1016/j.jfutfo.2024.10.011

Martinez, S. J., Batista, N. N., Bressani, A. P. P., Dias, D. R., and Schwan, R. F. (2022). Molecular, chemical, and sensory attributes fingerprinting of self-induced anaerobic fermented coffees from different altitudes and processing methods. *Foods*, 11(24), 3945. https://doi.org/10.3390/foods11243945.

Medrano, J. F., Cantu, D., Minio, A., Dreischer, C., Gibbons, T., Chin, J. et al. (2025). De novo whole-genome assembly and annotation of Coffea arabica var. Geisha, a high-quality coffee variety from the primary origin of coffee. *G3: Genes, Genomes, Genetics*, 15(1), jkae262. https://doi.org/10.1093/g3journal/jkae262

Pyrzynska, K. (2024). Useful extracts from coffee by-products: A brief review. *Separations*, 11(12), 334. https://doi.org/10.3390/separations11120334.

Reddy, V. S., Shiva, S., Manikantan, S., and Ramakrishna, S. (2024). Pharmacology of caffeine and its effects on the human body. *European Journal of Medicinal Chemistry Reports*, 10, 100138. https://doi.org/10.1016/j.ejmcr.2024.100138

Reyes-Herrera, D. F., Sánchez-Reinoso, A. D., Lombardini, L., and Restrepo-Díaz, H. (2023). Physiological responses of coffee (*Coffea arabica* L.) plants to biochar application under water deficit conditions. *Notulae Botanicae Horti Agrobotanici Cluj-Napoca*, 51(3), 12873. https://doi.org/10.15835/nbha51312873

Sasongko, D. P. (2025). A review of coffee kombucha brewing and health effects as a fermented beverage. *Jurnal Kebidanan Dan Kesehatan Tradisional*, 10(1), 53–65. https://doi.org/10.37341/jkkt.v10i1.590

Shi, Z., Luan, J., Zhang, Y., Wang, G., Mei, C., Chen, L. et al. (2025). Exploring the impact and mechanisms of coffee and its active ingredients on depression, anxiety, and sleep disorders. *Nutrients*, 17(19), 3037. https://doi.org/10.3390/nu17193037

Tenea, G. N., Cifuentes, V., Reyes, P., and Cevallos-Vallejos, M. (2025). Unveiling the microbial signatures of arabica coffee cherries: Insights into ripeness specific diversity, functional traits and implications for quality and safety. *Foods*, 14(4), 614. http://doi.org/10.3390/foods14040614

Tran, H. T., Lee, L. S., Furtado, A., Smyth, H., and Henry, R. J. (2016). Advances in genomics for the improvement of quality in coffee. *Journal of the Science of Food and Agriculture*, 96(10), 3300–3312. https://doi.org/10.1002/jsfa.7692

Uslu, N. (2025). Editorial: Coffee and coffee by-products: Innovative approaches fostering nutritional, sensory and chemical quality. *Frontiers in Nutrition*, 11, 1542109. https://doi.org/10.3389/fnut.2024.1542109

CHAPTER

2

Progress and Strategies for Conservation and Utilisation of *Coffea canephora* Genetic Diversity for Variety Development

Mildred Julian Nakanwagi[1,2,4*], Peter Wasswa[1], Pascal Musoli[2], Catherine Kiwuka[3], Penelope J. Bebeli[5], Elizabeth Balyejusa Kizito[4] and Godfrey Sseremba[2]

[1] Department of Crop Science and Horticulture, Makerere University, Kampala-Uganda
[2] National Coffee Research Institute, National Agricultural Research Organisation, Kituza-Mukono
[3] Plant Genetic Resource Center, National Agricultural Research Organisation, Entebbe-Uganda
[4] Faculty of Agricultural Sciences, Uganda Christian University, Mukono-Uganda
[5] Laboratory of Plant Breeding & Biometry, Department of Crop Science, Agricultural University of Athens, Iera Odos 75, 11855 Athens, Greece

This chapter presents an integrated synthesis of the genetic diversity, conservation, and cultivation of *Coffea canephora* (Robusta coffee), with a particular emphasis on Uganda's unique contributions to global coffee sustainability. It begins with an exploration of the taxonomy, origin, and evolutionary relationships within the genus Coffea, underscoring the importance of wild and underutilised species in climate resilience, including resistance to abiotic and biotic stressors. The geographic distribution and centres of diversity are contextualised through Africa's biodiversity hotspots, with Uganda highlighted as a key reservoir for Robusta germplasm. The chapter traces historical developments in Uganda's coffee sector, from colonial introductions and varietal selection to liberalisation and the rise of smallholder production. It critically examines the impacts of land tenure insecurity, ecological degradation, and shifting cultivation zones on production efficiency and biodiversity conservation. Studies on genetic structure spanning phenotypic characterisation, early isozyme analyses, microsatellite markers, and recent genome-wide approaches reveal regional adaptation patterns and the emergence of unique subgroups such as the Ugandan "Sub-group O." A focus is placed on conservation strategies encompassing ex situ and in situ efforts, germplasm documentation, field collection

*Corresponding author: mildredjuliannakanwagi@gmail.com

DOI: 10.1201/9781003689393-2

constraints, and community-based stewardship. Methodological challenges related to germplasm sampling, hybridisation, and administrative procedures are dissected to clarify their implications for breeding and genetic representativeness. Modern tools such as landscape genomics, species distribution modelling, and predictive analytics are discussed in relation to resilience breeding and habitat planning. Drawing on global perspectives, the chapter emphasises the need for coordinated international networks, participatory breeding models, and technological innovations, including bioinformatics and Artificial Intelligence, to safeguard coffee's genetic heritage. It concludes by identifying knowledge gaps in documentation, adaptive trait profiling, and farmer adoption, and offers strategic recommendations to strengthen Uganda's coffee sector through inclusive, science-based interventions. Together, these insights contribute to the sustainable management of *Coffea canephora* under changing climatic and socio-economic conditions. Additionally, strengthening national germplasm repositories, enhancing farmers' access to diverse planting materials, and integrating genetic conservation into agricultural development frameworks are critical steps toward preserving Uganda's role as a global reference point for Robusta coffee diversity and innovation.

1. Introduction

Coffee (*Coffea* spp.) is among the most economically significant agricultural commodities worldwide, because it sustains millions of livelihoods and contributes substantially to the Gross Domestic Product (GDP) of various tropical countries, including Uganda. Recent data (2023) indicate that global coffee consumption exceeds the 170 million 60-kg bags produced annually (ICO, 2023), with a notable increase in consumption in countries such as China and South Korea, which were previously non-traditional coffee markets. Urbanisation, lifestyle shifts, and the global proliferation of speciality coffee culture drive the increased demand and consumption. Coffee is no longer just a beverage but part of daily rituals, social ceremonies, and economic lifelines (Maspul, 2025; Poole et al., 2017). Two coffee species dominate commercial production: *Coffea arabica* (Arabica) and *Coffea canephora* (Robusta). Many other species, however, remain underutilised, despite offering critical traits for resilience to climate change, emerging pests, and evolving consumer preferences (Davis et al., 2011). While Ethiopia is acknowledged as the centre of origin for Arabica coffee (Lashermes et al., 2000), Uganda holds a unique and culturally rich position in the narrative of *C. canephora* as a centre of diversity. *C. canephora* has been embedded in Ugandan society for generations, transcending economic importance to serve cultural, social, and ceremonial roles (Thomas, 1935). As an example, in traditional Buganda culture, Robusta coffee was used to seal lifelong agreements, commemorate alliances, and extend hospitality, a practice that continues in marriage rites and community events today (Thomas, 1935).

Uganda holds a unique position in the global coffee landscape (Sseremba et al., 2025) as both a centre of diversity and a major producer of *C. canephora*. In 2024/25, Uganda exported approximately 7.93 million 60-kilogram bags of green coffee beans in the twelve months leading up to August 2025, valued at US$ 2.24 billion, with 85% of the beans being Robusta and 15% Arabica (MAAIF, 2025). The country lies within the East African biodiversity hotspot with varying agroecological conditions ranging from lowland forests to montane zones. This environment supports a broad range of genetic diversity in both wild and cultivated Robusta populations (Davis et al., 2006; Musoli et al., 2009). This diversity is not only taxonomically rich but also functionally significant, encompassing traits related to yield, drought tolerance, disease resistance, pest resistance, as well as cup and physical bean qualities that have become increasingly vital under shifting consumer demands and climatic regimes (Bunn et al., 2015; Kiwuka et al., 2021; Sseremba et al., 2025). Therefore, Uganda serves as a strategic reservoir of the Robusta coffee gene pool for breeding programs aiming at enhancing productivity, resilience, and market competitiveness. Unfortunately, the Uganda's coffee genetic heritage faces mounting threats from deployment of a small number of selected varieties on farms, changes in institutional and individual land-uses, changing policies on protected natural coffee habitats, deforestation, and climate variability, which are accelerating the erosion of cultivated and wild populations, as such reducing the genetic and connectivity across landscapes (Bunn et al., 2015; Davis et al., 2019). These are aggravated shifts in rainfall patterns, rising temperatures, and unpredictable weather events that are jeopardising traditional farming systems and will more likely render formerly suitable zones unproductive (Davis et al., 2023). The market-driven homogenization of cultivars and the limited institutional capacity for in situ conservation of cultivars and other genetic resources that are not in use at the time also pose challenges to the long-term sustainability of the coffee industry. Additionally, there are knowledge gaps regarding the composition, adaptation, and utilisation of Uganda's *C. canephora* germplasm (Ngugi and Aluka, 2019). These gaps limit strategic breeding and conservation efforts at a time when resilience is urgently needed. Addressing these risks requires a coordinated, multi-stakeholder approach that integrates understanding the genetic base of Ugandan and other *C. canephora* diversities, participatory breeding, policy support, and community-based conservation. This chapter provides a synthesis of the taxonomic and evolutionary foundations of *Coffea* diversity, traces the historical and modern trajectory of *C. canephora*, and presents a comprehensive analysis of genetic diversity, suggesting conservation strategies with a particular focus on Uganda. By integrating global perspectives with Uganda's unique ecological and institutional context, the chapter also offers actionable insights for breeding and policy development that position Uganda as a worldwide reference point for *C. canephora* genetic resource management.

2. Taxonomic and evolutionary bases of Coffea diversity

The genus *Coffea* belongs to the family Rubiaceae and comprises more than 130 species (Davis et al., 2011). This number of species has resulted from a progressive refinement of botanical understanding, starting with the initial classification by Carl Linnaeus in 1753, who described two species: *C. arabica* and *C. canephora.* Subsequently, other botanists expanded the genus by identifying several wild African species, emphasising morphological diversity and ecological adaptation. Modern advances in molecular phylogenetics and morphological analyses resulted in a critical revision expanding the genus to a catalogue of 103 species and seven intraspecific taxa Davis et al., 2006, 2011). This revision included the formal recognition of several previously overlooked species, such as *Coffea brassii*, *Coffea racemosa*, and *Coffea stenophylla*, among others. These species are predominantly evergreen shrubs or small trees distributed across tropical Africa, Madagascar, the Comoros, Mauritius, and various Indian Ocean islands, with a few introduced species present in Asia. Despite the species richness of the genus, global cultivation is dominated by just two species, *C. arabica* and *C. canephora*; *C. liberica* and *C. excelsa* (var. *dewevrei*) are being grown on a small scale, mainly for local consumption and small niche markets (Davis et al., 2022). Currently, other wild and underutilised species such as *C. racemosa*, *C. stenophylla*, and *C. eugenioides* are gaining attention for their unique agronomic and sensory traits, including low caffeine content, disease tolerance, and exceptional flavour profiles (Moat et al., 2019).

Geographically, Africa is the principal centre of origin and diversity for *Coffea*. Madagascar, with its isolated habitats, hosts over 60 endemic species (Davis et al., 2006), while continental Africa, particularly the Congo Basin and montane forests of East Africa, harbours the ancestral populations of the major cultivated types, *C. canephora* and *C. arabica*. (Table 1). *C. arabica* is an allotetraploid species ($2n = 4x = 44$), believed to have originated through natural hybridisation between *C. eugenioides* and *C. canephora* in Ethiopia's southwestern highlands approximately more than 1000 years ago (Lashermes et al., 1999; Scalabrin et al., 2020). This hybrid origin conferred self-fertility and a narrow genetic base, making Arabica highly vulnerable to disease and climate-related stresses in modern agriculture (Anthony

Table 1: Important cultivated and wild Coffea species, their ploidy level, country or region of origin, key traits, and cultivation status

Species	Ploidy level	Origin	Pollination type	Key traits	Cultivation status
C. arabica	Tetraploid	Ethiopia	Self-pollinating	Mild flavour, disease-prone	Widely cultivated
C. canephora	Diploid	Central & West Africa	Cross-pollinating	High yield, heat/disease tolerance	Widely cultivated
C. eugenioides	Diploid	East Africa	Self-pollinating	Low caffeine, Arabica parent	Wild
C. liberica	Diploid	West-Central Africa	Cross-pollinating	Tall trees, distinct flavour	Limited cultivation
C. stenophylla	Diploid	West Africa	Cross-pollinating	Heat-tolerant, aromatic flavour	Rediscovered, under study
C. racemosa	Diploid	Mozambique	Self-pollinating	Drought-tolerant, low caffeine	Wild/Niche markets

et al., 2002). In contrast, *C. canephora* (Robusta) is a diploid species (2n = 2x = 22), native to the lowland tropical forests of East, West, and Central Africa. It is cross-pollinating and allogamous, traits that maintain its high genetic diversity (Leroy et al., 2005). This diversity underpins its adaptive capacity to warmer and more humid climates and could be a strategic crop for climate adaptation.

Despite its taxonomic richness, cultivated coffee species represent a narrow subset of domesticated lineages. *C. arabica* particularly shows low intra-specific genetic variation because of its unique origin. On the other hand, wild populations of *C. canephora* and underutilised species retain significant genomic diversity that remains largely untapped in breeding programs (Depecker, 2023; Verleysen et al., 2023). A large proportion of wild *Coffea* species are highly localised in range, making them susceptible to threats such as habitat fragmentation, deforestation, and climatic shifts. Madagascar and the Congo Basin are particularly vulnerable hotspots, with over 60% of wild *Coffea* species estimated to be threatened with extinction (Davis et al., 2019). Similar destinies await the Ugandan forest as well. Preserving this diversity requires robust in situ conservation (e.g., forest protection) and ex situ strategies, such as germplasm repositories, seed banks, and field gene banks.

3. Historical development of Robusta Coffee production in Uganda

3.1 Colonial and post-independence developments

The global significance of *C. canephora* has evolved alongside shifting consumer demands and preferences, production dynamics, and national policy strategies. While Robusta is native to the lowland tropical rainforests of West, Central, and East Africa, including Uganda, its commercial cultivation gained prominence during colonial expansion. Before the colonisation of Uganda, a few trees of Robusta coffee were cultivated around homesteads to provide cherries and beans for use in traditional rituals, such as sealing lifelong agreements, commemorating alliances, and extending hospitality among some native tribes, including the Baganda. Although Robusta was traditionally grown by natives, Arabica coffee, first introduced by the colonial Government in 1900 (Thomas, 1944), overtook Robusta as a commercial crop because it was preferred due to its superior quality and higher global prices. Arabica coffee was propagated by seed, and seedlings were distributed to farmers in various parts of the country. However, leaf rust disease and poor adaptation of Arabica to lower-altitude areas forced farmers to adopt Robusta. Arabica has since remained a crop for areas above 1300 meters above sea level, especially on the slopes of Mt. Elgon, Mt. Rwenzori, the West Nile, and the southwestern highlands.

The devastation of Arabica coffee plantations in Uganda and Asia by leaf rust (*Hemileia vastatrix)* changed the global focus to Robusta (Cramer, 1957). This stimulated the colonial government in Uganda to develop the Robusta coffee industry (Maitland, 1926). Serious work on Robusta coffee began in 1916, when a range of coffee accessions from different sources were tested for their liquor content. Among these, tree No. 9, an erecta-type Toro accession, was identified as having superior yields and cup quality, as well as bold beans (Maitland, 1926). Half-sib seedlings from this tree were then raised to form seed gardens from which seed was distributed to farmers (Thomas, 1947). From 1920 to 1932, large quantities of seeds from the half-sib gardens of the Toro tree nine were distributed to farmers. However, the erecta progeny did not adapt well in most areas, perhaps due to inbreeding depression and poor root systems, and more so in the Sese Islands, where the soils were highly leached. Consequently, between 1932 and 1934, trees of the Nganda Robusta coffee type (characterised by flexible and spreading stems with small leaves) were selected from farms in the Sese Islands and on the mainland near Lake Victoria to provide seeds for new plantings (Thomas, 1947). From 1934 to the early 1970s, seeds from Nganda trees were distributed to farmers as the coffee industry progressed steadily (But et al., 1970), due to the high prices received by farmers in the 1950s (Thomas, 1947).

The variety improvement research, which began in 1956 and concluded in 1970 by releasing eight Robusta coffee clones that were high-yielding, resistant to leaf and red blister diseases, and had superior bean qualities (including a superior cup and large bean size), ran in parallel with the distribution of open-pollinated seedlings to farmers. In the 1980s and 1990s, farmers were supplied with rooted cuttings of these eight clones, three of which (1s/2, 1s/3, and 1s/6) were half-descendants of the Toro tree No. 9 identified in 1920, while five clones were of the Nganda/spreading type (Kibirige-Ssebunya et al., 1993). However, the distribution of these plants was limited due to a decline in the coffee trade between 1970 and 1990, largely driven by political turmoil, which led farmers to uproot both Arabica and Robusta trees. Recognising the strategic potential of the sector, the government initiated liberalisation reforms in 1991, which transferred export responsibilities from state-run institutions, such as the Coffee Marketing Board, to private sector actors (Ponte, 2001). This empowered farmers to negotiate competitive prices, which stimulated demand for planting materials of the eight improved coffee clones, and led to a quick recovery in production and exports, culminating in a record high of about 280,000 tons (4.7 million 60kg bags) exported in 1997 (ICO, 2006). As part of these reforms, coffee hulling machines were distributed to cooperatives to streamline post-harvest processing, while institutional support was bolstered by the creation of agencies such as the Uganda Coffee Development Authority (UCDA) and the Coffee Research Centre (COREC), now known as the National Coffee Research Institute (NaCORI). UCDA emerged as a cornerstone institution in sustaining national coffee productivity.

The emergence of Coffee Wilt Disease (CWD) in the 1990s, however, negatively affected Robusta coffee production. Still, it created an opening for the National Coffee Research Institute to search for new varieties, resulting in the production of 10 wilt-resistant varieties. These clones were selected for traits including early maturation, tolerance to CWD, and improved yield performance. The policy changes also led to the Government providing planting materials to farmers, which encouraged the mass planting of coffee and improved the deployment of extension services. To date, the Uganda coffee industry is undergoing a growing socio-economic transformation, with exports exceeding 8 million 60kg bags of green coffee beans, earning an income of over $2 billion, of which 75% goes to farmers (USDA, 2025). Its role as a major export commodity has been further accelerated by global climatic disruptions, most notably drought and frost episodes in Brazil and drought in Vietnam. The economic value generated by coffee has led to it becoming Uganda's leading cash crop, contributing 20-22% to the national Gross Domestic Product (GDP) (UCDA 2024). Nonetheless, the rapid expansion of private exporters has fragmented traceability systems and complicated quality assurance protocols. As of 2022, Uganda's coffee sector is home to over fifty registered entities. Ensuring long-term sustainability will require enforced inter-agency collaboration, expanded technical capacity, and innovative mechanisms such as price premiums for biodiversity conservation and quality enhancement, which could incentivise farmers to adopt environmentally sound and economically viable practices.

3.2 Uganda's market position in the global Robusta trade

Uganda's Robusta coffee has achieved international recognition for its distinct cup quality, attributed to cultivation at medium altitudes (typically above 1,000 meters above sea level) under less harsh environmental conditions (Sseremba et al., 2025). This positioning, combined with genetic diversity and demonstrated resistance to key diseases (Musoli et al., 2009), makes Ugandan Robusta particularly attractive for speciality and blend markets. Studies have shown its cup quality to be superior to that of many Robusta-producing regions, prompting favourable price evaluations across global commodity exchanges (Garavito et al., 2016). The species' remarkable ecological versatility allows it to thrive across Uganda's diverse landscapes, fostering widespread adoption among farmers. However, most of the production remains in the hands of smallholder farmers, often managing plots referred to as "shambas" with average land sizes ranging from 0.23 to 1.2 hectares (Baffes, 2006; Gwali et al., 2015). These farms typically interplant Robusta with banana (for shade and subsistence) and legumes (for soil fertility), reflecting integrated agroforestry systems suited for mixed-resource environments (Jassogne et al., 2013).

4. Emerging challenges and sustainable production strategies

4.1 Socioeconomic and policy constraints

Despite its adaptability, *C. canephora* production faces considerable threats, such as rapid urbanisation in traditional coffee-growing zones, leading to land-use conflicts and the displacement of plantations. Soil degradation and nutrient depletion, particularly in high-density farming zones, require costly inputs for restoration, yet resources are often inaccessible to smallholders (Kyalo et al., 2023). Land tenure insecurity, especially among farmers on rented plots, limits long-term investment, which in turn undermines sustainable production and leads to encroachment into biodiversity-rich areas. As large-scale producers expand into previously untouched regions, this disrupts native ecosystems and compromises evolutionary processes. Recent production shifts have successfully introduced *C. canephora* to northern Uganda (Alela et al., 2024), particularly at lower altitudes (~800 meters), where it demonstrates vigorous growth and promising yield potential. However, moving cultivation into new ecological contexts may affect the physical, chemical, and sensory properties of the beans, thereby altering market profiles and necessitating detailed agronomic and organoleptic evaluation (Belgis et al., 2023; Campuzano-Duque et al., 2021; Ngugi et al., 2016). Moreover, changing consumer behaviour adds another layer of complexity. While farmers seek to maximise yields and incomes, consumers increasingly prioritise cup quality and health attributes, such as reduced caffeine content and improved flavour. This is particularly relevant given Robusta's dominant role in instant coffee blends, which are under scrutiny for their stimulant properties. As health-conscious consumption trends grow, leveraging Uganda's intra-species diversity, especially variants with lower caffeine and superior flavour, could offer pathways to enhance both value addition and consumer satisfaction.

4.2 Agroforestry systems and integrated landscape management

C. canephora production sustainability lies in the potential of balancing intensive and traditional farming practices with environmental sustainability, social equity, and profitability (Poncet et al., 2024). In Uganda, more production meant expanding the production area. This contributed to the destruction of natural coffee habitats, such as forests (Maitima et al., 2009). For example, the germplasm sources in the gene bank can no longer be traced, and the areas appear to have already been inhabited due to the increasing population size, urbanisation, and other human activities, such as charcoal burning (Pender et al., 2004). Urbanisation has resulted in the displacement of indigenous peoples, loss of traditional livelihoods, and conflicts over land rights, which favour large-scale coffee farmers over small-scale farmers, who are the majority. From 1961 to 2021, there was no discernible relationship between yield and the expansion of land used for coffee production, even though this was done in an attempt to boost yield (FAO, 2023). This implies that the intensification of coffee production on small landholdings, accompanied by relatively low coffee production costs, makes the country fully competitive with other producers. Given the anticipated continuous land fragmentation in the country, increasing productivity remains a viable option to meet growing market demand through more sustainable practices, such as agroforestry (Gram et al., 2018). There have been initiatives to promote agroforestry, as it integrates coffee cultivation with native tree species, reforestation, and afforestation efforts. Agroforestry practices, such as the use of shade trees, are reported to increase land productivity by up to 75% through the trees' capacity to improve water filtration, control erosion by acting as windbreaks, and provide litter that decomposes (Gomez et al., 2010; Jonsson et al., 2015). Although coffee agroforestry practices appear to be a viable option for sustainable production, the increasing land fragmentation seems to be a persistent hindrance. The most recent agricultural survey indicated that the average land holding across the country lies between 2 and 3.2 acres (UBOS, 2020). This is attributed to the increasing population size, which indirectly downplays the benefits of agroforestry, as farmers prefer monocropping to utilise the small available space to optimise yields. Even with limited space, when farmers are asked to choose between cropping systems, as reported earlier, the majority prefer bananas and fruit trees as their shade trees (Gwali et al., 2015; Munyuli, 2011). This suggests that food security issues in Uganda may precede environmental sustainability concerns.

In other countries, such as Ethiopia, shade tree management has been widely adopted as an agroforestry practice. The country has had to invest in research and intensify extension services to enhance the adoption of region-specific tree species, including *Cordia Africana*, *Erythrina abyssinica*, *Albizia gummifera*, *Vernonia amygdalina*, and *Ficus sur* (Ehrenbergerová et al., 2021). The entry point for agroforestry adoption in Costa Rica has hinged on the fact that recommended tree species, such as *Inga densiflora*, have nitrogen-fixing capacities (Cerdán et al., 2012). With this in mind, they theorise that coffee production can be sustained with healthy soil. Some of the shade trees recommended in Uganda include *Albizia chinensis* and *Maesopsis eminii*, which are associated with coffee pests, such as black twig borers. As a result, farmers have resorted to cutting down such shade trees and using them for firewood. This highlights the need to investigate the negative social impacts of shade trees to make favourable recommendations. Furthermore, there has been a misconception that integrating shade trees directly affects coffee yield, which may not be true (Ehrenbergerová et al., 2021), hence the need to educate farmers more on the realistic benefits of optimum shading, which include high moisture retention, reducing carbon concentration, and providing alternatives to fuel sources through extension programs. Although agroforestry is a farming method, it also serves as a strategic framework for sustainable land use and ecological stewardship. Nevertheless, for its optimisation, it is essential to further align crop genetic resources with the environmental realities of each landscape.

5. Genetic diversity and ecological adaptation of *C. canephora* in Uganda

5.1 Evolution of genetic diversity assessment (morphology) in *Coffea* spp.

Locally adapted diversity is the biological engine that empowers resilient agroforestry systems. Understanding the genetic and ecological diversity within *Coffea* species is paramount to sustaining global coffee production, developing climate-resilient cultivars, and mitigating threats from pests and diseases. Historically, diversity assessments began with phenotypic characterisation, which, while accessible, was constrained by environmental influence and limited resolution. Breeding of *C. canephora* in Uganda has deep historical roots, beginning with farmer-based selection during the colonial period. Two major landraces, Nganda (characterised by a bushy phenotype) and Erecta (characterised by an upright architecture), were identified and propagated for their resilience and adaptability. In contrast, initial attempts to establish *C. arabica* in central Uganda faced disease, such as coffee leaf rust, and poor bean quality setbacks, resulting in its reintroduction to higher-altitude zones in the east and west, where climatic conditions proved more favourable (Baffes, 2006; Davis et al., 2023). By the early 1990s, breeding priorities shifted. Originally focused on high-yielding varieties, the program was forced to pivot after the outbreak of Coffee Wilt Disease (CWD) in 1993. The Coffee Research Centre (COREC) responded by intensifying selection for disease resistance, eventually releasing seven CWD-resistant Robusta lines by 2010. These cultivars showed promise for enhanced yield and resilience. Mass multiplication through tissue culture labs and nursery networks was initiated. Still, adoption rates remained suboptimal due to limited farmer awareness and training (Luzinda, 2018), inadequate seed multiplication capacity, inadequate mapping of optimal production zones given Uganda's diverse agroecological zones, significant intra-zone variation persistence, and cultural resistance to modern farming practices among smallholder communities.

Other classical diversity assessment approaches included morphological and agronomic descriptors. Traits such as leaf morphology, berry size, flowering habits, and plant stature formed the basis of early classification (IPGRI, 1996). However, these were subject to phenotypic plasticity, making it difficult to distinguish genetically similar individuals under variable cultivation conditions. Earlier studies on morphological diversity described *C. canephora* from different forest materials and correlated it with tolerance to heat, exposure to whole light, altitude, and leaf size (Thomas, 1935). More insight into genetic diversity was provided by Kibirige-Ssebunya (1993) using agronomic characteristics. The variations observed were crucial for selecting coffee varieties better suited to specific environmental conditions and production goals. A study by Montagnon in 2000 provided further relevance to the morphological diversity of *C. canephora* and highlighted unique morphological markers (Table 2).

A study conducted in Ghana by Akpertey et al. (2019) identified six key variables that contributed significantly to the observed morphological variation within Robusta coffee plants: height, stem diameter, span, number of laterals per tree, diameter of laterals, and the number of nodes per lateral. These findings indicate that specific morphological traits can serve as valuable markers for characterising and classifying Robusta coffee genotypes. However, these have not been verified among the largest proportion of the Ugandan population. Initially, there were two known morphological groups in Uganda, namely, the Erecta derived from seed, believed to have originated from Java, and the Nganda from the seed of native coffee. These groups were characterised by an erect plant habit, few main stems, and more rapid growth, while the Nganda were characterised by large stem size, many branches, and relatively smaller fruit size (Thomas, 1935). A comprehensive evaluation of 206 *C. canephora* genotypes was conducted to assess their organoleptic diversity, utilising a 10-point descriptive scale. The outcomes of this evaluation unveiled the presence of four groups that exhibited significant variations in fragrance, aroma, and flavour profiles. Group 1 encompassed genotypes that excelled in all organoleptic cup attributes. Meanwhile, Group 2 was distinguished by its higher acidity and a sweet mouthfeel among its genotypes. Group 3 included genotypes with superior characteristics in terms of fragrance, aroma, aftertaste, balance, and flavour. In contrast, Group 4 was associated with genotypes exhibiting inferior organoleptic cup attributes (Ngugi et al., 2016). This research provided insights into the unique morphological characteristics of Ugandan Robusta coffee varieties, contributing valuable knowledge to the global coffee community. However, it is essential to note that there is still much more to explore regarding the morphological diversity of *C. canephora*, of the current Ugandan populations.

Table 2: Phenotypic differences among Congolese and Guinean genetic groups based on accessions from wild and cultivated material

Trait	Congolese			Guinean	
	SG1	SG2			
	Cultivated	Cultivated	Wild	Cultivated	Wild
Leaf length (mm)	188	203	221	170	176
Leaf width (mm)	71	82	93	68	75
Acumen	Short to long	Long	Long	Long	Long
Petiole	Short to long	Medium to long	Long	Short to long	Medium to long
Domatia	Absent to clear visible	Slightly marked to clearly visible	Slightly marked to clearly visible	Slightly marked to clearly visible	Slightly marked to clearly visible
Peduncle	Average Court	Average court	Short to long	Short to long	Short to long
Disk size	Small	Small	Small	Large	Small
100 seeds weight (12% humidity)	14.2	15.4	9.5	10.0	9.0
Branching	Moderately strong	Low to medium	Low to medium	Moderately strong	Moderately strong
Phenology	Very late	Moderately late	Moderately late	Early	Early
Caffeine (% dry matter)	2.7	2.3	2.3	2.9	2.6
Resistance to drought	Good	Poor	Poor	Good	Good

Source: Adapted from Montagnon (2000)

5.2 Population structure and genetic richness using molecular assessment

Beyond morphology, cytogenetics, and karyotyping analyses provided insights into chromosomal architecture across *Coffea* species. Notably, *C. arabica* is the only naturally occurring allotetraploid ($2n = 4x = 44$), while *C. canephora* and others remain diploid ($2n = 2x = 22$). These studies were instrumental in unravelling evolutionary lineages and speciation events (Lashermes et al., 1999). Enzymatic assays (using isoenzyme markers) like esterase and peroxidase offered early biochemical differentiation. Although foundational, these markers exhibited limited polymorphism and tissue specificity, thereby constraining their application in breeding and taxonomy. The molecular era ushered in precision tools that transformed diversity assessment (Poncet et al., 2007). Highly polymorphic and codominant, SSR (Simple Sequence Repeats) revealed population structures in *C. canephora*, supported parentage analysis, and clarified species-level relationships (Aerts et al., 2013; Leroy et al., 2014). Though dominant in nature, Amplified Fragment Length Polymorphisms (AFLP) provided valuable insights into underexplored taxa and were used to detect genetic variability where SSRs were unavailable. The most abundant markers in plant genomes, Single Nucleotide Polymorphisms (SNPs), have underpinned studies in genome-wide association (GWAS) for traits like drought tolerance, caffeine metabolism, and cup quality (Denoeud et al., 2014). Genotyping by sequencing (GBS) uncovers thousands of SNPs, enabling analysis of genome-wide variation, given its efficiency for large-scale germplasm screening. Whole-genome sequencing (WGS) has mapped reference genomes for *C. arabica* and *C. canephora*, helping to identify genes linked to stress tolerance, flavour biosynthesis, and disease resistance (Charr et al., 2020; Hamon et al., 2015; Van Der Vossen et al., 2015).

5.3 Spatial diversity mapping and ecological stratification of *C. canephora* populations

Global perspective

Scientific efforts to unravel the genetic variation within this species have evolved over the past few decades, reflecting both technological advancements and an expanding geographic scope. Initial studies for assessing genetic variation based on enzyme activity in the 1980s employed isozyme analysis. These efforts by Berthaud and Charrier revealed two geographically distinct genetic clusters: the Guinean group, encompassing accessions from West Africa (Côte d'Ivoire and Guinea), and the Congolese group, which included populations from Central and East Africa (Berthaud, 1986). Progressing into the 1990s, Montagnon and colleagues identified additional subdivisions within the Congolese group, designating them as Sub-groups A and B (Montagnon et al., 1992). In 1999, the adoption of Restriction Fragment Length Polymorphism (RFLP), a molecular technique allowing for more precise genetic mapping, enabled differentiation of the Congolese populations into two new clusters: Sub-group B, predominantly from Eastern-Central Africa, and Sub-group C, from Western-Central regions (Dussert et al., 1999; Gomez et al., 2010). These findings illustrated a growing recognition of regional diversity.

The turn of the millennium marked a leap in analytical power with the use of SSRs, highly variable DNA markers well-suited for studying population structure and gene flow. Musoli et al. (2009) utilised SSRs to examine wild and feral populations across Uganda, uncovering a previously undocumented group labelled Sub-group O. This discovery not only reinforced Uganda's role as a genetic reservoir but also revealed the intricate in-country variation of *C. canephora*. Most recently, a study by Mérot-L'Anthoëne et al. (2019) expanded the genetic landscape further by differentiating two new groups, R and G, still using SSR markers and samples collected from the southern Democratic Republic of Congo and Angola. Collectively, these milestones illustrate a scientific journey facilitated by innovation and regional exploration, continually reshaping our understanding of robusta coffee's diversity. The expanding body of research into *C. canephora's* genetic diversity has resulted in the identification of eight distinct diversity groups, each associated with specific geographic regions across Africa (Figure 1). This geographical clustering is not coincidental; rather, it reflects *C. canephora's* natural distribution and adaptive divergence across ecological zones. The species spans from Guinea in West Africa, through the Central African Republic and the Democratic Republic of Congo in Central Africa, to Uganda in East Africa, and Angola in the south (Montagnon et al., 1998). Such regional segmentation has been consistently documented across studies, highlighting the influence of localised environmental pressures and historical migration on genetic differentiation (Davis et al., 2006; Garavito et al., 2016; Musoli et al., 2009a).

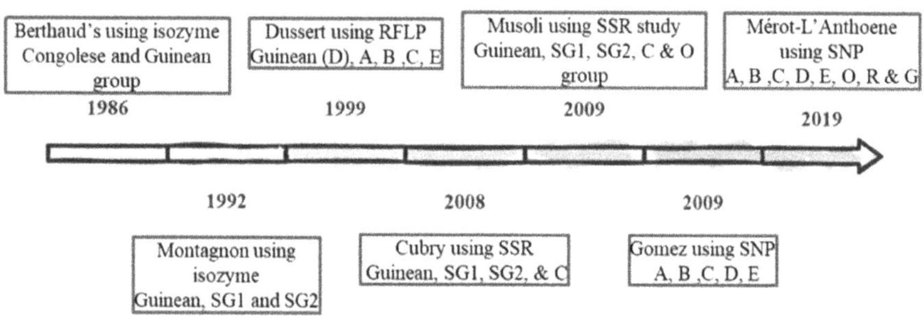

Figure 1: Summary of the different Robusta Coffee genetic clusters discovered over time using different markers

Ugandan context

Within Uganda, a particularly rich reservoir of genetic diversity has emerged. An intensive SSR-based investigation of wild and feral populations by Kiwuka et al. (2021) uncovered five genetic clusters within the Ugandan diversity group (O). These clusters represent populations from Zoka, Budongo, Itwara, Kibale, and a Southern-Central group, the latter considered ancestral to Uganda's domesticated robusta varieties cultivated today. Ugandan germplasm has contributed to three of the eight global diversity assessments, affirming its critical role in shaping our understanding of *C. canephora's* evolutionary history (Kiwuka et al., 2021; Mérot-L'Anthoëne et al., 2019; Musoli et al., 2009). These studies collectively emphasise a high degree of variability in wild populations across Africa. The observed distribution, from Guinea in the west to Uganda in the east, and from the Central African Republic in the north to Angola in the south, echoes a mosaic of climatic, edaphic, biotic, and abiotic influences that have guided natural selection and local adaptation (Montagnon et al., 1998).

5.4 Locally adapted Ugandan diversity of *Coffea canephora* and breeding implications

Uganda offers a variety of ecological zones that span altitudes from 1,000 to 1,600 meters above sea level, particularly around the Lake Victoria basin (Kiyingi and Gwali, 2012). Studies have demonstrated that *C. canephora* in Uganda has undergone regional adaptation to various climate and edaphic gradients. Wild populations found in tropical rainforests at an elevation of around 1,188 meters show site-specific traits shaped by environmental pressures. These adaptive processes are particularly evident in the country's south-central region, which has gained prominence as a cultivation hub due to its favourable climate and soil composition (Bunn et al., 2015). However, intra-regional differences in temperature, precipitation, and soil chemistry have led to heterogeneous yield performance, signalling the importance of micro-zonal analysis in breeding programs (Kiwuka et al., 2021). Over the past two decades, Uganda has contributed valuable insights into *C. canephora* diversity through targeted research efforts:

- A foundational study using 196 accessions from wild sites such as Itwara, Kibale, and Kalangala, as well as cultivated landraces including Nganda and Erecta, to assess potential resistance to Coffee Wilt Disease (CWD). As part of the study, Cubry et al. (2008) evaluated SSR marker transferability using material from Itwara and cultivated Nganda lines, supporting the reliability of molecular approaches in Ugandan Robusta germplasm. This research resulted in the identification of Sub-group O, a genetically distinct Ugandan lineage based on analysis of wild, feral, and cultivated populations (Musoli et al., 2009b). This discovery solidified Uganda's position as a reservoir of unique Robusta diversity.
- Ngugi and Aluka (2019) analysed 349 accessions, including 231 local landraces from 14 districts in western and central Uganda. Their study classified cultivated accessions into three main gene pools: Group 1: Kabarole, Bundibugyo, Hoima; Group 2: Kyenjojo, Kiboga, Kamuli; and Group 3: Kayunga, Mpigi, Mubende, Rakai, Mukono, Kamwenge.
- A subsequent study by Kiwuka et al. (2021) evaluated 295 accessions, including 191 wild samples from Zoka, Budongo, Itwara, Kibale, Malabigambo, Mabira, and Kalangala forests. Cultivated material from Kawanda and Kituza was also included, further enriching Uganda's germplasm profile.
- Prospection through funding from the EU under the ROBUST project, East Africa coffee.org (https://capacity4dev.europa.eu/projects/desira/info/robust_en), conducted a study between 2022 and 2025 using 1500 accessions. These included both wild (500) and cultivated accessions (1000) housed at Kituza. The wild samples were collected from ten Ugandan natural forests, including Mabira, Budongo, Kibale, Maramagambo, Malabigambo, Kalinzu, Semuliki, Bwindi Impenetrable, and Zoka forests.

Additionally, the Ugandan Robusta gene pool has also been enriched by exogenous introductions. Java-derived accessions planted in 1929 and 1933 were integrated into national breeding programs. These materials demonstrated good adaptability and, in some cases, enhanced self-fertility (Thomas, 1935), making them useful parents in clonal development. Despite these achievements, several methodological and conservation gaps hinder a comprehensive understanding of Uganda's *C. canephora* diversity. Limited forest material representation, as only 16 forest ecosystems out of 506 central forest reserves have been sampled, despite Uganda's vast natural vegetation diversity. This narrow scope may mask unrecognised diversity hotspots. Many studies rely on small sample sizes (fewer than 300 individual plants), which limits statistical power and reduces confidence in diversity extrapolations. Invasive flora has begun encroaching into native coffee habitats, posing a significant genetic risk to endemic populations through competition, altered pollination dynamics, and potential introgression (Davis et al., 2023). There have also been spatial gaps and redundancies, resulting from the overlap in collection zones and the lack of systematic grid sampling, which hinder full ecological coverage and reproducibility. To support sustainability and resilience, future initiatives should focus on expanding sampling efforts across underexplored forest ecosystems, augmenting sample sizes for robust genetic mapping, linking genotypes to climate and edaphic variables, enforcing invasive species control and forest protection, and integrating landscape genomics into conservation planning.

6. Germplasm sampling and characterisation framework

6.1 Institutional and administrative procedures

Uganda's rich biodiversity offers a wealth of genetic resources critical for crop improvement and conservation. Nevertheless, accessing this diversity, particularly through germplasm collection, requires navigating a complex web of legal, ecological, and administrative considerations. The Nagoya Protocol, adopted in 2010 by Uganda and entered into force in 2014, is a global agreement under the Convention on Biological Diversity (CBD) that ensures fair and equitable sharing of benefits arising from the use of genetic resources. Since ratifying the Nagoya Protocol in 2014, Uganda has committed to its principles of prior informed consent (PIC), mutually agreed terms (MAT), and benefit-sharing mechanisms. The country has been deliberate in aligning its national laws with international standards on access and benefit sharing (ABS). In compliance with the Nagoya protocol, the amended National Environment Act (2019) and Uganda's ABS regulations currently form the backbone of ethical germplasm collection, ensuring that both scientific inquiry and community rights are protected. Under this framework, researchers must secure formal authorisation from multiple institutions before entering the field.

The Uganda National Council for Science and Technology (UNCST) oversees national research compliance, while the National Forestry Authority (NFA) and Uganda Wildlife Authority (UWA) regulate access to forest reserves and protected areas. These agencies enforce protocols that safeguard ecological integrity and ensure that any biological material collected is done so with prior informed consent (PIC) and mutually agreed terms (MAT), as mandated by the Nagoya Protocol. Although these procedures are integral to ensuring ethical compliance with biodiversity conservation laws, access regulations, and institutional mandates, they also introduce substantial logistical complexity that can directly affect sampling efficiency and research outcomes. Before field deployment, researchers are required to secure formal authorisation from a set of regulatory institutions, including: the Uganda National Council for Science and Technology (UNCST), overseeing national research compliance, the National Forestry Authority (NFA), regulating forest access, use of biological resources, and environmental integrity, and the Uganda Wildlife Authority (UWA). For student researchers, the process begins with the approval of a proposal by an academic supervisor or an institutional review board. Once approved, the researcher submits documentation to UNCST for clearance. A critical prerequisite for UNCST approval is the presentation of forest access permits, obtained from NFA and UWA. This stage typically requires: an introductory letter from the hosting research institution, completion of formal documentation, including detailed study protocols, a Material Transfer Agreement (MTA) between collaborating institutions (when necessary), stipulating conditions for sample exchange and utilization, submission of a detailed list of research team members, clearly indicating nationals and international collaborators, definition of study objectives, sampling methods, target species, and specific forest sites to be visited and specification of the expedition timeline, beyond which sampling access is invalidated.

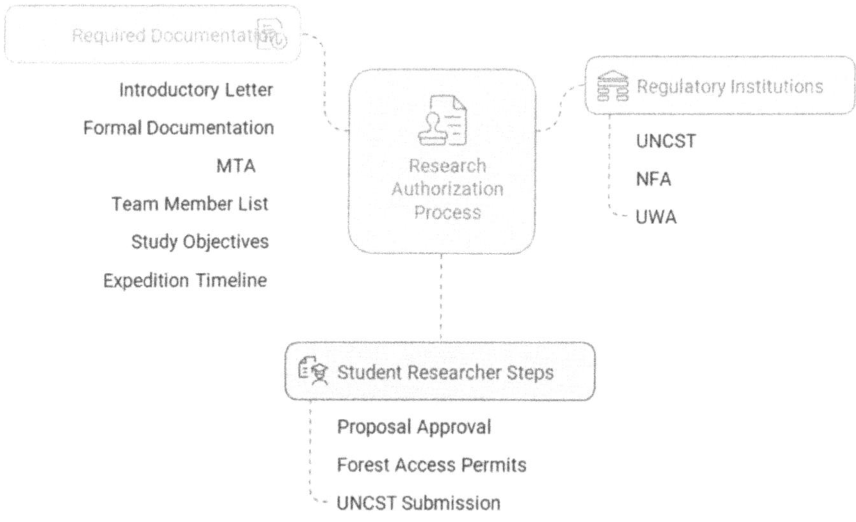

Figure 2: An illustration showing the key administrative steps and documents required before starting an expedition in Uganda

6.2 Perspectives on Pre-sampling authorization and ethical considerations

The procedural rigor is pronounced when collecting germplasm from Uganda's natural forest ecosystems, especially because these environments demand not only ecological sensitivity but also administrative precision. Researchers must navigate terrain that is both physically challenging and legally protected, often requiring months of preparation and coordination among multiple agencies. Further complications arise when custody is shared between different organisations/bodies, and the need to engage local communities whose cultural and ecological knowledge is integral to the process.

Although the authorisation process is designed to ensure ethical oversight and protect sensitive ecosystems, researchers frequently encounter operational delays and bottlenecks that can disrupt planned activities. These include inter-agency processing timelines that may not align, resulting in delayed clearance or incomplete documentation; Uncertain funding cycles, where delays in financial disbursement can result in permit expiration before field implementation; Strict access windows, where permits are valid only within stated timeframes, limiting flexibility in response to unforeseen logistical issues such as weather, staff availability, or forest conditions. Given these realities, researchers must build in extended timelines for regulatory approvals and budget for contingencies, balancing scientific objectives against institutional protocols. While such administrative rigour ensures national interests and biodiversity governance, the necessity for procedural flexibility becomes evident when sampling in dynamic, and often remote, forest landscapes. These administrative layers have direct implications for the quality and representativeness of germplasm collected. For example, a narrow sampling window may constrain researchers to easily accessible zones, resulting in spatial bias. Limited approval for sample types or species may exclude ecologically significant morphotypes from genetic evaluation. Time-constrained expeditions also reduce the likelihood of stratified ecological sampling, compromising insights into genotype-environment interactions. To mitigate these impacts, scientific and regulatory stakeholders must co-develop frameworks that promote timely access while safeguarding the natural resources that support robust, inclusive, and scientifically valid conservation strategies.

On the other hand, germplasm collection from farmer-managed fields presents a different set of dynamics. These fields, often located in agroforestry systems adjacent to forest margins, are shaped by generations of informal seed selection and environmental adaptation. Farmers often unknowingly maintain landraces with unique resistance traits, making their fields valuable sources of genetic diversity. The administrative protocol here must still adhere to the ABS protocols while incorporating local government structures and community engagement.

6.3 Site selection and ecological stratification in forest-based sites

The selection of forest sites for *C. canephora* germplasm collection is a critical preliminary step that influences the representativeness of sampling, ecological relevance, and the overall success of conservation efforts. In Uganda, site identification is primarily informed by historical botanical records, observed species presence, and logistical considerations such as accessibility. However, fewer than 20 forest environments have been systematically surveyed to date (Musoli et al., 2009b; Ngugi and Aluka, 2019), revealing a substantial gap in coverage across

Uganda's rich and varied landscape. While some forests have confirmed populations of *C. canephora*, researchers must exercise caution in relying exclusively on published records. The absence of documentation does not necessarily equate to the absence of a species. Wild coffee may persist in unsampled or overlooked microhabitats due to historical dispersal, vegetative regeneration, or environmental niche stability. Therefore, strategic site selection should incorporate exploratory objectives, enabling the discovery and documentation of previously unrecognised populations. Before field deployment, researchers should be encouraged to conduct informal reconnaissance investigations to confirm the presence or absence of *C. canephora*, characterize habitat features relevant to coffee growth, such as canopy cover, soil composition, altitude, and disturbance levels as well as engage with local communities and forest rangers who often possess practical knowledge of species occurrence, regeneration hotspots, and habitat dynamics. Such assessments refine the scope of sampling efforts and improve the ecological relevance of collected germplasm. It should be noted that forest governance in Uganda is administered under multiple custodial bodies, including NFA, UWA, and, in some cases, the Plant Genetic Resource Centre (PGRC). In some cases, forest areas are jointly administered, requiring researchers to liaise with both entities for access and compliance. Early communication with relevant authorities is essential not only to secure permission but also to verify the status of forest inventories, species occurrence, and allowable research activities.

6.4 Field-based sampling protocol: Case of 2023/2024 expedition

Purpose of the expedition

Despite the ecological significance of Uganda's forests as natural reservoirs of *C. canephora*, formal documentation of germplasm sampling procedures within these habitats remains limited in scientific literature. Most published studies reference generalised strategies such as randomised or transect-based sampling (Kiwuka et al., 2021) but fail to contextualise these methods within the environmental realities of Uganda's forest ecosystems. This section presents a documented protocol based on a 2023/2024 expedition led by the National Coffee Research Institute (NaCORI), designed to collect genetically and ecologically relevant samples to support ex situ conservation, genotyping efforts, and soil plant interaction studies. The expedition was structured to achieve the following specific objectives:

1. Conserve wild genetic resources by collecting stem cuttings, seeds (where applicable), and seedlings for ex situ storage and use in clonal breeding programs.
2. Generate genotypic data through leaf sample collection for DNA extraction and population-level diversity analysis.
3. Assess eco-geographic influences by collecting soil samples associated with sampled trees to examine potential edaphic drivers of genotype distribution and adaptation.

Sampling protocol and execution

The sampling took place in Semuliki National Park, targeting ecologically distinct areas within the lowland tropical forest, known to host wild *C. canephora*. Three forest zones were selected based on habitat accessibility and vegetation profiles, each separated by a minimum distance of 20 km to reduce spatial overlap. Depending on the forest size, the minimum distance between sampling sites was adjusted so that for some forests, the distance was more than 50 km apart. At each site, samples were collected as follows:

1. 60 trees per site, with individual trees spaced at ≥20 meters to minimise sampling redundancy.
2. Sampling initiated ≥1 km from the forest edge, reducing bias from transitional zones and edge effects.
3. Stem cuttings of ≥30 cm in length were harvested from healthy branches, immediately wrapped in moist cotton tissue, stored in water-containing buckets, and transferred into sealed polyethylene for transport.
4. Leaf tissue: Two mature leaves per tree were sterilised in ethanol, de-ribbed, and desiccated using silica gel in labelled Ziplock bags. Labels were codified by location (e.g., SNP-S1).
5. Soil samples were taken within the root zone (15–30 cm depth), labelled accordingly, cooled at –4°C initially, and then stored at –20°C post-survey.

In addition to biological samples, GPS coordinates, altitude (669–730 m), habitat descriptors, and tree-level characteristics, such as height and phenological stage, for each accession.

Sampling outcome and diversity insights

The expedition yielded a total of 60 individual coffee samples, representing both morphologically distinct and phenotypically familiar types. Two major morphotypes are identified, one characterised by large fruits, thick leaves, hardwood stems, and mature berries, potentially linked to long-lived genetic lineages. The other exhibits traits similar to those of ex situ cultivated clones, possibly indicating past feralization. Age stratification, with mature (>50 years) trees co-occurring with seedlings and juveniles (<3 years), suggests active regeneration and possible in situ conservation dynamics. Disease distribution, for example, in swampy clay soils, symptoms of black twig borer and brown eye spot were prominent, while coffee populations on clay loam soils showed resilience, pointing to possible soil-driven resistance mechanisms. Ecological adaptability, as evidenced by the observation that coffee trees thrive in both dry clay loam and waterlogged environments, indicates strong phenotypic plasticity and a potentially wide adaptive range of Ugandan *C. canephora*. This fieldwork demonstrated that standardised sampling designs must be ecologically flexible. While randomised sampling remains useful, forest accessibility, hydrological variation, and patchy distribution of coffee trees necessitated real-time methodological adjustments. Such adaptability ensured ecological breadth and minimised bias. Without comprehensive sampling, conservation efforts and genotype-environment modelling risk being based on incomplete data, thereby limiting breeding precision and germplasm utility.

7. *Coffea canephora* conservation imperatives in a changing climate

The increasingly volatile climate marked by rising temperatures, erratic precipitation, and prolonged drought episodes poses a fundamental threat to global coffee production. This is particularly true for *C. arabica*, whose narrow genetic base and dependence on stable montane climates render it highly susceptible to abiotic stresses (Bunn et al., 2015; Imbach et al., 2017). On the other hand, *C. canephora* and several wild relatives, including *C. racemosa* and *C. Stenophylla*, demonstrate greater tolerance to heat and drought, attributed to physiological traits such as efficient stomatal regulation and Osmoprotectant accumulation (Van Der Vossen et al., 2015). Harnessing the vast genetic diversity within *C. canephora* thus becomes a strategic imperative. Owing to its outcrossing nature, every seed represents a genetically distinct individual (Scalabrin et al., 2020), resulting in extensive variation across geographic, morphological, physiological, and genotypic dimensions. This diversity supports climate resilience, offering raw material for breeding programs that aim to develop robust cultivars with improved tolerance to environmental stressors, pests, and diseases (Bertrand et al., 2023).

7.1 In situ and ex situ conservation strategies

To preserve genetic variability in *Coffea* species, a blend of ex situ and in situ conservation techniques is recommended. Ex situ methods include field gene banks, in vitro slow-growth storage, cryopreservation, tissue culture, and seed banks (Engels and Ebert, 2021; Krishnan et al., 2015). Although field collections are widespread in developing countries like Uganda, they face challenges ranging from pest outbreaks and soil degradation to land availability and resource limitations (Dulloo et al., 2021; Y. Fu, 2017). The long-term sustainability of these collections is threatened by high inbreeding coefficients, genetic erosion due to limited population sizes, and inadequate documentation of genetic traits (Fu, 2015; Krishnan, 2011). The In Situ method involves protecting wild populations within their natural habitats. Studies emphasise the ecological importance of conserving *C. canephora* in native forest ecosystems across Uganda (Musoli et al., 2009; Kiwuka et al., 2021). However, existing conservation frameworks often prioritise wildlife and commercial timber trees, overlooking coffee species. For example, Zoka Forest, an important reservoir of wild *C. canephora*, faces imminent threats from encroachment and deforestation, endangering both genetic diversity and ecosystem health (Davis et al., 2023). To enhance genetic integrity, recent efforts involve collecting germplasm across broad climatic gradients in Uganda and other regions to capture adaptive traits shaped by long-term environmental exposure (Krishnan et al., 2018; Campuzano-Duque et al., 2021). These wild resources, while valuable, are increasingly at risk from habitat loss, invasive species, and climate change, underscoring the urgency of coordinated conservation action.

Introgressing alleles from wild relatives into cultivated coffee populations presents significant potential for climate-smart agriculture. Traits related to drought and heat tolerance, pest and disease resistance, and phenological plasticity have been identified in wild *Coffea* species and are now being targeted in pre-breeding pipelines (dos Santos et al., 2022; Mishra, 2019). Tools such as marker-assisted selection (MAS) and genomic selection (GS) allow breeders to integrate these traits while minimising undesired genetic linkage. Nonetheless, successful breeding hinges on the availability of well-characterised germplasm and the protection of wild populations that harbour untapped genetic resources (van der Vossen et al., 2015). Moreover, emerging technologies such as landscape genomics provide powerful insights into adaptation patterns by linking genomic variants to environmental gradients across geographic space (Dauphin et al., 2022). Applied to *Coffea*, this approach identifies loci associated with drought resilience or temperature tolerance, facilitating precision breeding and habitat conservation planning (Merot-L'anthoene et al., 2019). Coupled with species distribution modelling (SDM) and ecological niche modelling, landscape genomics enables forward-looking predictions of suitable coffee-growing regions under climate change scenarios (Läderach et al., 2017). These tools support informed germplasm collection, policy formation, and climate-smart interventions that ensure both the sustainability of the coffee sector and the preservation of biodiversity.

7.2 Utilisation of diversity for breeding

Hybridisation strategies, informed by ecological adaptation and genetic profiles, offer a pathway to enhance resilience and productivity. Breeding programs fundamentally rely on the exploitation of genetic variation. In *C. canephora*, where natural populations exhibit significant polymorphism due to allogamy and self-incompatibility (Berthaud and Charrier, 1985), artificial hybridisation offers a strategic pathway for harnessing and amplifying diversity. However, despite the biological feasibility, documented crossing designs in coffee remain scarce, particularly compared to other major perennial crops (Leroy et al., 1993). In Uganda, limited breeding documentation has constrained the reproducibility and scalability of promising crosses. Controlled hybridisation essential for targeted trait introgression has only recently gained momentum. National Coffee Research Institute (NaCORI) initiated reciprocal recurrent selection (RRS) and clonal breeding approaches to improve yield, cap quality, bean size, and climate resilience. Past efforts, which resulted in the release of ten CWD-resistant clones (Luzinda, 2018), seem to have exploited only the Southern-central genetic sub-group (Musoli et al., 2009a; Kiwuka et al., 2021). Amidst the increasing climate change effects, such as pests and diseases and drought, there is a need to explore inter-group hybridisation, thereby increasing chances of heterosis for favourable right combinations (Alkimim et al., 2021). What schemes have been applied to ensure diverse groups, how they lead to heterosis, and how they meet end-use needs (Montagnon et al., 2003)?

8. Conclusion and recommendation

8.1 Conclusion

Despite Robusta coffee's remarkable adaptability across ecological zones, the dominance of smallholder farmers presents significant challenges, including limited access to inputs, insecure land tenure, and degradation of fertile soils. The expansion of coffee cultivation into new regions has helped mitigate land scarcity, yet it raises concerns about deforestation and biodiversity loss. Agroforestry-based cultivation offers promising ecological benefits, but socio-cultural misconceptions and fragmented land ownership constrain adoption. Robusta coffee's genetic diversity, driven by its outcrossing reproductive system, offers immense potential for breeding climate-resilient and disease-tolerant cultivars. Wild accessions found across Uganda's forest ecosystems exhibit ecological and morphological variations that are indicative of adaptive genes crucial for resilience against abiotic stress and pest pressure. However, efforts to harness this diversity remain underdeveloped due to limited profiling of germplasm collections, fragmented research outputs, and low representation of wild gene pools in national breeding programs. To ensure long-term sustainability and sectoral transformation, genetic conservation must be integrated with practical breeding, responsive policy frameworks, and inclusive extension strategies. The convergence of technological innovation, ecological understanding, and participatory governance offers a pathway to enhance Uganda's coffee productivity while preserving its rich biodiversity.

8.2 Strategic recommendations

The following strategic actions are recommended:

- Prioritise the profiling and protection of ex situ collections while expanding in situ efforts to safeguard wild populations in Uganda's forest reserves.
- Invest in the characterisation of agroecological zones to ensure regionally tailored cultivar recommendations and maximise genotype-environment performance.
- Harmonise ongoing research efforts, standardise data collection, and establish open-access databases linking morphological, genotypic, and environmental metadata.
- Foster cooperation among research institutions, breeders, extension services, and farming communities to accelerate innovation uptake.
- The scattered nature of coffee germplasm across global centres of diversity necessitates harmonised conservation efforts. Initiatives like the Global Coffee Conservation Strategy (GCBS) exemplify integrated frameworks that unify ex situ and in situ conservation with breeding goals and climate resilience objectives (Bertrand et al., 2011; Anthony et al., 2007).
- Artificial intelligence (AI), machine learning, and bioinformatics platforms present transformative potential for trait prediction, data visualisation, and breeding optimisation (Raimondi et al., 2022). Such technologies can accelerate the identification of climate-resilient genotypes and streamline their deployment through marker-assisted and genomic selection.
- Community-based conservation approaches, centred around participatory breeding and traditional ecological knowledge, enhance on-farm biodiversity and ensure equitable benefit sharing (Bunn et al., 2015). Bridging technological advances with grassroots stewardship will strengthen coffee sustainability and socio-economic equity.

Acknowledgement

This initiative is funded at NaCORI by the European Union through CIRAD under a project 'Robusta coffee agroforestry to mitigate and adapt to climate change in Uganda'.

References

Aerts, R., Berecha, G., Gijbels, P., Hundera, K., Van Glabeke, S., Vandepitte, K. et al. (2013). Genetic variation and risks of introgression in the wild *Coffea arabica* gene pool in south-western Ethiopian montane rainforests. *Evolutionary Applications*, 6(2), 243–252. https://doi.org/10.1111/j.1752-4571.2012.00285.x

Akpertey, A., Anim-Kwapong, E., and Ofori, A. (2019). Assessment of genetic diversity in robusta coffee using morphological characters. *International Journal of Fruit Science*, 19(3), 276–299. https://doi.org/10.1080/15538362.2018.1502723

Alela, B., Katya Kule, E., Midamba, D. C., and Mugonola, B. (2024). Determinants of adoption of integrated soil fertility management practices among coffee producers in Mid-Northern Uganda. *Cogent Social Sciences*, 10(1), 2417807. https://doi.org/10.1080/23311886.2024.2417807

Alkimim, E. R., Caixeta, E. T., Sousa, T. V., Gois, I. B., Silva, F. L. da, Sakiyama, N. S. et al. (2021). Designing the best breeding strategy for *Coffea canephora*: Genetic evaluation of pure and hybrid individuals aiming to select for productivity and disease resistance traits. *PLOS ONE*, 16(12), e0260997. https://doi.org/10.1371/journal.pone.0260997

Aluka, P., and Ngugi, K. (2015). Sensory and organoleptic cup attributes of robusta coffee (*Coffea canephora* Pierre ex A. Froehner). *Journal of Agricultural Studies*, 4(1), 101. https://doi.org/10.5296/jas.v4i1.8789

Anthony, F., Astorga, C., Avendano, J., and Dulloo, E. (2007). Conservation of coffee genetic resources in the CATIE field genebank. Paris: Biodiversity International; IRD, 23–34. ISBN 978-92-9043-752-9

Baffes, J. (2006). Restructuring Uganda's coffee industry: Why going back to basics matters. *Development Policy Review*, 24(4), 413–436. https://doi.org/10.1111/j.1467-7679.2006.00332.x

Belgis, M., Zhafirah Arifin, T., Prameswari, D., Taruna, I., Choiron, M., Witono, Y. et al. (2023). Sensory profile on robusta coffee by rate-all-that-apply (RATA). *Pelita Perkebunan* (a Coffee and Cocoa Research Journal), 39(1). https://doi.org/10.22302/iccri.jur.pelitaperkebunan.v39i1.546

Berthaud, J. (1986). Les ressources genetiques pour l'amelioration des cafeires africains diploides (Vol. 188). ORSTOM.

Berthaud, J., and Charrier, A. (1985). Genetic resources of Coffea.

Bertrand, B., Davis, A. P., Maraval, I., Forestier, N., and Mieulet, D. (2023). Potential beverage quality of three wild coffee species (*Coffea brevipes*, *C. congensis* and *C. stenophylla*) and consideration of their agronomic use. *Journal of the Science of Food and Agriculture*, 103(7), 3602–3612. https://doi.org/10.1002/jsfa.12347

Bunn, C., Läderach, P., Ovalle Rivera, O., and Kirschke, D. (2015). A bitter cup: Climate change profile of global production of Arabica and Robusta coffee. *Climatic Change*, 129(1–2), 89–101. https://doi.org/10.1007/s10584-014-1306-x

Campuzano-Duque, L. F., Herrera, J. C., Ged, C., and Blair, M. W. (2021). Bases for the establishment of robusta coffee (*Coffea canephora*) as a new crop for Colombia. *Agronomy*, 11(12), 2550. https://doi.org/10.3390/agronomy11122550

Cerdán, C. R., Rebolledo, M. C., Soto, G., Rapidel, B., and Sinclair, F. L. (2012). Local knowledge of impacts of tree cover on ecosystem services in smallholder coffee production systems. *Agricultural Systems*, 110, 119–130. https://doi.org/10.1016/j.agsy.2012.03.014

Charr, J.-C., Garavito, A., Guyeux, C., Crouzillat, D., Descombes, P., Fournier, C. et al. (2020). Complex evolutionary history of coffees revealed by full plastid genomes and 28,800 nuclear SNP analyses, with particular emphasis on *Coffea canephora* (Robusta coffee). *Molecular Phylogenetics and Evolution*, 151, 106906. https://doi.org/10.1016/j.ympev.2020.106906

Cramer, P. J. S. (1957). A review of Literature of Coffee Research in Indonesia (Vol. 262). SIC Editorial, Inter American Institute of Agricultural Sciences.

Dauphin, B., Rellstab, C., Wüest, R. O., Karger, D. N., Holderegger, R., Gugerli, F. et al. (2022). Re-thinking the environment in landscape genomics. *Trends in Ecology & Evolution*. https://doi.org/10.1016/j.tree.2022.10.010

Davis, A. P., Govaerts, R., Bridson, D. M., and Stoffelen, P. (2006). An annotated taxonomic conspectus of the genus Coffea (Rubiaceae). *Botanical Journal of the Linnean Society*, 152(4), 465–512. https://doi.org/10.1111/j.1095-8339.2006.00584.x

Davis, A. P., Tosh, J., Ruch, N., and Fay, M. F. (2011). Growing coffee: Psilanthus (Rubiaceae) subsumed on the basis of molecular and morphological data; implications for the size, morphology, distribution and evolutionary history of Coffea. *Botanical Journal of the Linnean Society*, 167(4), 357–377.

Davis, A. P., Chadburn, H., Moat, J., O'Sullivan, R., Hargreaves, S., and Lughadha, E. N. (2019). High extinction risk for wild coffee species and implications for coffee sector sustainability. *Science Advances*, 5(1). https://doi.org/10.1126/sciadv.aav3473

Davis, A. P., Kiwuka, C., Faruk, A., Walubiri, M. J., and Kalema, J. (2022). The re-emergence of Liberica coffee as a major crop plant. *Nature Plants*, 8(12), Article 12. https://doi.org/10.1038/s41477-022-01309-5

Davis, A. P., Kiwuka, C., Faruk, A., Mulumba, J., and Kalema, J. (2023). A review of the indigenous coffee resources of Uganda and their potential for coffee sector sustainability and development. *Frontiers in Plant Science*, 13. https://www.frontiersin.org/articles/10.3389/fpls.2022.1057317

Denoeud, F., Carretero-Paulet, L., Dereeper, A., Droc, G., Guyot, R., Pietrella, M. et al. (2014). The coffee genome provides insight into the convergent evolution of caffeine biosynthesis. *Science*, 345(6201), 1181–1184. https://doi.org/10.1126/science.1255274

Depecker, J. (2023). Genetic diversity and structure in wild Robusta coffee (*Coffea canephora* A. Froehner) populations in Yangambi (DR Congo) and their relation to forest disturbance. *Heredity*. https://doi.org/10.1038/s41437-022-00588-0

dos Santos, T. B., da Silva Ferreira, M. F., Marques, I., Oliveira, S. C., Zaidan, I. R., Oliveira, M. G. et al. (2022). Current challenges and genomic advances towards the development resilient coffee genotypes to abiotic stresses. *In:* C. Kole (Ed.), Genomic Designing for Abiotic Stress Resistant Technical Crops (pp. 41–69). Springer International Publishing. https://doi.org/10.1007/978-3-031-05706-9_3

Dulloo, M. E., Solano, W., Dessauw, D., Astorga, C., and Guarino, L. (2021). A methodological approach for prioritization and rationalization of field genebank accessions of coffee genetic resources: A case study of CATIE International Coffee Collection, Costa Rica. *Frontiers in Sustainable Food Systems*, 5, 777415. https://doi.org/10.3389/fsufs.2021.777415

Dussert, S., Lashermes, P., Anthony, F., Montagon, C., Trouslot, P., Combes, M.-C. et al. (1999). Le caféier, *Coffea canephora*. *In:* Diversité génétique des plantes tropicales cultivées. CIRAD.

Ehrenbergerová, L., Klimková, M., Cano, Y. G., Habrová, H., Lvončík, S., Volařík, D. et al. (2021). Does shade impact coffee yield, tree trunk, and soil moisture on *Coffea canephora* plantations in Mondulkiri, Cambodia? *Sustainability*, 13(24), 13823. https://doi.org/10.3390/su132413823

Engels, J. M. M., and Ebert, A. W. (2021). A critical review of the current global ex situ conservation system for plant agrobiodiversity. I: History of the development of the global system in the context of the political/legal framework and its major conservation components. *Plants*, 10(8), 1557. https://doi.org/10.3390/plants10081557

Fu, Y. (2017). The vulnerability of plant genetic resources conserved ex situ. *Crop Science*, 57(5), 2314–2328. https://doi.org/10.2135/cropsci2017.01.0014

Fu, Y.-B. (2015). Understanding crop genetic diversity under modern plant breeding. *Theoretical and Applied Genetics*, 128(11), 2131–2142. https://doi.org/10.1007/s00122-015-2585-y

Garavito, A., Montagnon, C., Guyot, R., and Bertrand, B. (2016). Identification by the DArTseq method of the genetic origin of the *Coffea canephora* cultivated in Vietnam and Mexico. *BMC Plant Biology*, 16(1), 242. https://doi.org/10.1186/s12870-016-0933-y

Gomez, C., Batti, A., Le Pierres, D., Campa, C., Hamon, S., de Kochko, A. et al. (2010). Favourable habitats for Coffea inter-specific hybridization in central New Caledonia: Combined genetic and spatial analyses. *Journal of Applied Ecology*, 47(1), 85–95.

Gram, G., Vaast, P., Van Der Wolf, J., and Jassogne, L. (2018). Local tree knowledge can fast-track agroforestry recommendations for coffee smallholders along a climate gradient in Mount Elgon, Uganda. *Agroforestry Systems*, 92(6), 1625–1638. https://doi.org/10.1007/s10457-017-0111-8

Gwali, S., Agaba, H., Balitta, P., Hafashimana, D., Nkandu, J., Kuria, A. et al. (2015). Tree species diversity and abundance in coffee farms adjacent to areas of different disturbance histories in Mabira forest system, central Uganda. *International Journal of Biodiversity Science, Ecosystem Services & Management*, 11(4), 309–317. https://doi.org/10.1080/21513732.2015.1050607

Hamon, P., Hamon, S., Razafinarivo, N. J., Guyot, R., Siljak-Yakovlev, S., Couturon, E. et al. (2015). Coffea genome organization and evolution. *In:* Coffee in Health and Disease Prevention (pp. 29–37). Elsevier. https://doi.org/10.1016/B978-0-12-409517-5.00004-8

ICO (International Coffee Organization). (2006). Coffee production statistics. Available from http://www.ico.org

ICO. (2023). Coffee report and outlook. International Coffee Organisation. https://icocoffee.org/documents/cy2022-23/Coffee_Report_and_Outlook_April_2023_-_ICO.pdf

Imbach, P., Fung, E., Hannah, L., Navarro-Racines, C. E., Roubik, D. W., Ricketts, T. H. et al. (2017). Coupling of pollination services and coffee suitability under climate change. *Proceedings of the National Academy of Sciences*, 114(39), 10438–10442. https://doi.org/10.1073/pnas.1617940114

International Plant Genetic Resources Institute (IPGRI). (1996). Descriptors for coffee (Coffea spp. and Psilanthus spp.). International Plant Genetic Resources Institute 36 pp.

Jassogne, L., van Asten, P. J. A., Wanyama, I., and Baret, P. V. (2013). Perceptions and outlook on intercropping coffee with banana as an opportunity for smallholder coffee farmers in Uganda. *International Journal of Agricultural Sustainability*, 11(2), 144–158. https://doi.org/10.1080/14735903.2012.714576

Jonsson, M., Raphael, I. A., Ekbom, B., Kyamanywa, S., and Karungi, J. (2015). Contrasting effects of shade level and altitude on two important coffee pests. *Journal of Pest Science*, 88(2), 281–287. https://doi.org/10.1007/s10340-014-0615-1

Kibirige-Ssebunya, I., Nabasirye, M., Matovu, J., and Musoli, P. (1993). A comparison among various robusta coffee (*Coffea canephora* Pierre) clonal materials and their seedling progenies at different levels of nitrogen. *Uganda Journal of Agricultural Sciences*, 1(1), 5–12.

Kiwuka, C., Goudsmit, E., Tournebize, R., Aquino, S. O., Douma, J. C., Bellanger, L. et al. (2021). Genetic diversity of native and cultivated Ugandan Robusta coffee (*Coffea canephora* Pierre ex A. Froehner): Climate influences, breeding potential and diversity conservation. *PLOS ONE*, 16(2), e0245965. https://doi.org/10.1371/journal.pone.0245965

Kiyingi, I., and Gwali, S. (2012). Productivity and profitability of robusta coffee agroforestry systems in central Uganda. *Uganda Journal of Agricultural Sciences*, 13(1), 85–93.

Krishnan, S. (2011). Coffee biotechnology: Implications for crop improvement and germplasm conservation. *Acta Horticulturae*, 894, 33–44. https://doi.org/10.17660/ActaHortic.2011.894.2

Krishnan, S., Ranker, T. A., Davis, A. P., and Rakotomalala, J.-J. (2015). Current status of coffee genetic resources: Implications for conservation – Case study in Madagascar. *Acta Horticulturae*, 1101, 15–20. https://doi.org/10.17660/ActaHortic.2015.1101.3

Krishnan, S., Bramel, P., Horna, D., Lainoff, B., Montagnon, C., and Schilling, T. (2018). Development of a global conservation strategy for coffee genetic resources. *Acta Horticulturae*, (1205), 505–508. https://doi.org/10.17660/ActaHortic.2018.1205.62

Kyalo, G., Apunyo, P. C., Mwanjalolo, M., Luswata, C. K., Kawooya, R., & Niyibigira, E. I. (2023). Characterisation and mapping of soils in major coffee growing regions of Uganda. *Journal of Agricultural Science*, 16(1), 49. https://doi.org/10.5539/jas.v16n1p49

Läderach, P., Ramirez–Villegas, J., Navarro-Racines, C., Zelaya, C., Martinez–Valle, A., and Jarvis, A. (2017). Climate change adaptation of coffee production in space and time. *Climatic Change*, 141(1), 47–62. https://doi.org/10.1007/s10584-016-1788-9

Lashermes, P., Combes, M. C., Topart, P., Graziosi, G., Bertrand, B., and Anthony, F. (2000). Molecular breeding in coffee (*Coffea Arabica* L.). *In:* T. Sera, C. R. Soccol, A. Pandey, & S. Roussos (Eds), Coffee Biotechnology and Quality (pp. 101–112). Springer Netherlands. https://doi.org/10.1007/978-94-017-1068-8_7

Lashermes, P., Combes, M.-C., Robert, J., Trouslot, P., D'Hont, A., Anthony, F. et al. (1999). Molecular characterisation and origin of the *Coffea arabica* L. genome. *Molecular and General Genetics* (*MGG*), 261(2), 259–266. https://doi.org/10.1007/s004380050965

Leroy, T., De Bellis, F., Legnate, H., Musoli, P., Kalonji, A., Loor Solórzano, R. G. et al. (2014). Developing core collections to optimize the management and the exploitation of diversity of the coffee *Coffea canephora*. *Genetica*, 142(3), 185–199. https://doi.org/10.1007/s10709-014-9766-5

Leroy, T., Marraccini, P., Dufour, M., Montagnon, C., Lashermes, P., Sabau, X. et al. (2005). Construction and characterization of a *Coffea canephora* BAC library to study the organization of sucrose biosynthesis genes. *Theoretical and Applied Genetics*, 111, 1032–1041. https://doi.org/10.1007/s00122-005-0018-z

Leroy, T., Montagnon, C., Charrier, A., and Eskes, A. B. (1993). Reciprocal recurrent selection applied to *Coffea canephora* Pierre 1. Characterization and evaluation of breeding populations and value of intergroup hybrids. *Euphytica*, 67(1–2), 113–125.

Luzinda, H. (2018). Factors influencing adoption of improved robusta coffee technologies in Uganda. *Uganda Journal of Agricultural Sciences*, 18(1), 33–41. https://doi.org/10.4314/ujas.v18i1.3

Maitima, J. M., Mugatha, S. M., Reid, R. S., Gachimbi, L. N., Majule, A., Lyaruu, H. et al. (2009). The linkages between land use change, land degradation and biodiversity across East Africa. *African Journal of Environmental Science and Technology*, 3, 310–325.

Maitland, T. D. (1926). Coffea Robusta in Uganda. *In:* Uganda Protectorate, Department of Agriculture (ed.). Annual Conference of the Uganda Planters' Association, Government Press, Kampala, pp. 3–11.

Maspul, K. A. (2025). Coffee culture's role in community resilience: A case study of Buraidah Speciality Coffee. *Journal Health of Indonesian*, 3(01), 1–16. https://doi.org/10.58471/health.v3i01.144

Mérot-L'Anthoëne, V., Tournebize, R., Darracq, O., Rattina, V., Lepelley, M., Bellanger, L. et al. (2019). Development and evaluation of a genome-wide Coffee 8.5K SNP array and its application for high-density genetic mapping and for investigating the origin of *Coffea arabica* L. *Plant Biotechnology Journal*, 17(7), 1418–1430. https://doi.org/10.1111/pbi.13066

Mishra, M. K. (2019). Genetic resources and breeding of coffee (Coffea spp.). *In:* J. M. Al-Khayri, S. M. Jain, and D. V. Johnson (Eds), Advances in Plant Breeding Strategies: Nut and Beverage Crops (Volume 4, pp. 475–515). Springer International Publishing. https://doi.org/10.1007/978-3-030-23112-5_12

Moat, J., Gole, T. W., and Davis, A. P. (2019). Least concern to endangered: Applying climate change projections profoundly influences the extinction risk assessment for wild Arabica coffee. *Global Change Biology*, 25(2), 390–403. https://doi.org/10.1111/gcb.14341

Montagnon, C., Leroy, T., Cilas, C., and Charrier, A. (2003). Heritability of *Coffea canephora* yield estimated from several mating designs. *Euphytica*, 133(2), 209–218.

Montagnon, C., Leroy, T., and Eskes, A. (1998). Varietal improvement of *Coffea canephora*. 1: Criteria and breeding methods. *Plantations, Recherche, Développement*. https://agritrop.cirad.fr/390308/

Montagnon, C., Leroy, T., and Yapo, A. (1992). Genotypic and phenotypic diversity of some coffee groups (*Coffea canephora* Pierre) in the collections—Consequences on their use in breeding. *Cafe Cacao The*, 36(3), 187–198.

Munyuli, T. (2011). Farmers' perceptions of pollinators' importance in coffee production in Uganda. *Agricultural Sciences*, 02(03), 318–333. https://doi.org/10.4236/as.2011.23043

Musoli, P., Cubry, P., Aluka, P., Billot, C., Dufour, M., De Bellis, F. et al. (2009). Genetic differentiation of wild and cultivated populations: Diversity of *Coffea canephora* Pierre in Uganda. *Genome*, 52(7), 634–646. https://doi.org/10.1139/G09-037

Ngugi, K., Aluka, P., and Maina, D. (2016). Variation of mineral micronutrient elements in robusta coffee (*Coffea canephora* Pierre ex A. Froehner) as measured by energy dispersive X-ray fluorescence. *Annual Research & Review in Biology*, 9(2), 1–13. https://doi.org/10.9734/ARRB/2016/23068

Pender, J., Jagger, P., Nkonya, E., and Sserunkuuma, D. (2004). Development pathways and land management in Uganda. *World Development*, 32(5), 767–792. https://doi.org/10.1016/j.worlddev.2003.11.003

Poncet, V., Dufour, M., Hamon, P., Hamon, S., de Kochko, A., and Leroy, T. (2007). Development of genomic microsatellite markers in *Coffea canephora* and their transferability to other coffee species. *Genome*, 50(12), 1156–1161.

Poncet, V., Van Asten, P., Millet, C. P., Vaast, P., and Allinne, C. (2024). Which diversification trajectories make coffee farming more sustainable? *Current Opinion in Environmental Sustainability*, 68, 101432. https://doi.org/10.1016/j.cosust.2024.101432

Poole, R., Kennedy, O. J., Roderick, P., Fallowfield, J. A., Hayes, P. C., and Parkes, J. (2017). Coffee consumption and health: Umbrella review of meta-analyses of multiple health outcomes. *The BMJ*, 359, j5024. https://doi.org/10.1136/bmj.j5024

Scalabrin, S., Toniutti, L., Di Gaspero, G., Scaglione, D., Magris, G., Vidotto, M. et al. (2020). A single polyploidization event at the origin of the tetraploid genome of *Coffea arabica* is responsible for the extremely low genetic variation in wild and cultivated germplasm. *Scientific Reports*, 10(1), 4642. https://doi.org/10.1038/s41598-020-61216-7

Sseremba, G., Iyamulemye, E. N., Katwirenabo, G., Rweihangwe, D. R., Nanyonga, J., Mulindwa, R. et al. (2025). Regional flavor profiling of Ugandan coffee: Sensory characterization across agroecological zones. *Journal of Sensory Studies*, 40(5), e70077. https://doi.org/10.1111/joss.70077

Thomas, A. S. (1935). Types of robusta coffee and their selection in Uganda. *The East African Agricultural Journal*, 1(3), 193–198. https://doi.org/10.1080/03670074.1935.11663646

Thomas, A. S. (1944). The wild coffees of Uganda. *Empire Journal of Experimental Agriculture*, 45, 1–12.

Thomas, A. S. (1947). The cultivation and selection of Robusta coffee in Uganda. *The Empire Journal of Experimental Agriculture*, 15, 66–81.

Van Der Vossen, H., Bertrand, B., and Charrier, A. (2015). Next generation variety development for sustainable production of Arabica coffee (*Coffea arabica* L.): A review. *Euphytica*, 204(2), 243–256. https://doi.org/10.1007/s10681-015-1398-z

Verleysen, L., Bollen, R., Kambale, J.-L., Ebele, T., Katshela, B. N., Depecker, J. et al. (2023). Characterization of the genetic composition and establishment of a core collection for the INERA Robusta coffee (*Coffea canephora*) field genebank from the Democratic Republic of Congo. *Frontiers in Sustainable Food Systems*, 7. https://www.frontiersin.org/articles/10.3389/fsufs.2023.1239442

CHAPTER

3

Coffee Genomics: Mapping and Sequencing

Pengrui Luo[1,2,3]

[1] Yunnan Key Laboratory of Plant Diversity and Biogeography, Kunming Institute of Botany, Chinese Academy of Sciences, Kunming 650201, Yunnan, China
[2] Yunnan International Joint Laboratory for Biodiversity of Central Asia, Kunming Institute of Botany, Chinese Academy of Sciences, Kunming 650201, Yunnan, China
[3] University of Chinese Academy of Sciences, Beijing 100049, China

Coffee (*Coffea* spp.) is one of the most widely consumed beverages worldwide and a cornerstone agricultural commodity that sustains the livelihoods of millions of smallholder farmers, particularly in tropical regions. Over the past two decades, rapid advances in genomic technologies have fundamentally transformed our understanding of coffee biology, evolution, and improvement. High-throughput sequencing, long-read platforms (PacBio HiFi and Oxford Nanopore), chromosome conformation capture technologies (Hi-C), and sophisticated bioinformatics pipelines have enabled the generation of high-quality, chromosome-scale genome assemblies for several Coffea species. These resources have provided unprecedented insights into genome architecture, gene content, structural variation, and the evolutionary consequences of polyploidy, particularly in allotetraploid Coffea arabica.

This chapter provides a comprehensive overview of recent progress in coffee genomics, with a particular emphasis on genome architecture, genetic and physical mapping, sequencing strategies, and functional annotation. The genetic diversity and domestication history of Coffea species are discussed alongside the challenges posed by genome size variation and subgenome complexity. Advances in high-density SNP mapping, genotyping-by-sequencing, quantitative trait locus (QTL) analysis, and genome-wide association studies (GWAS) have greatly improved the resolution at which agronomically important traits—such as yield, cup quality, caffeine content, and resistance to pests and diseases—can be dissected. The integration of genetic

E-mail: luopengrui@mail.kib.ac.cn

DOI: 10.1201/9781003689393-3

linkage maps with physical genome assemblies has further strengthened trait localisation and comparative genomic analyses across Coffea species.

The chapter also highlights the implications of these genomic resources for modern coffee breeding. Marker-assisted selection, genomic selection, and machine learning–based predictive models are increasingly enabling data-driven breeding strategies that accelerate genetic gain in this perennial crop. Furthermore, the emergence of pan-genomic frameworks and the systematic inclusion of wild Coffea relatives offer powerful opportunities to broaden the narrow genetic base of cultivated Arabica coffee and enhance resilience to climate change.

Finally, key challenges and future directions are discussed, including haplotype-resolved genome assembly, structural variation analysis, ethical governance of genomic resources, and the integration of genomics with systems biology, phenomics, and environmental modelling. Together, these advances position coffee genomics as a critical foundation for sustainable crop improvement, supporting the development of climate-resilient, high-quality cultivars and safeguarding the long-term future of global coffee production.

1. Introduction and background

Coffee, produced from plants of the genus *Coffea* L., ranks among the most widely consumed beverages and globally traded agricultural commodities. It is essential to many tropical economies and provides a living for tens of millions of smallholder farmers. Coffee genomics has changed our knowledge of crop species over the past few decades. Advances in genomic technologies have allowed for precise characterisation of genetic diversity, gene function, and the genetic architecture underlying complex traits.

Genomic approaches enable increased resistance to pests and diseases, improve cup quality, enhance yield, and facilitate adaptation to climate change. *Coffea arabica* L. (Arabica) and *Coffea canephora* Pierre ex A. Froehner (Robusta) are the two species that dominate commercial production. Arabica coffee, which makes up about 60% of the world's coffee production, is prized for its flavour and fragrance. The remainder, robusta, is valued for its high caffeine content and resilience. For researchers and breeders, the genomic complexity of coffee, especially Arabica as an allopolyploid, offers both exceptional opportunities and challenges.

High-quality genome assemblies, dense genetic maps, genome-wide association studies (GWAS), and genomic selection pipelines in coffee have been made easier by the development of next-generation sequencing (NGS), long-read technologies, chromosome conformation capture (Hi-C), and sophisticated bioinformatics. Rapid breeding cycles, the study of wild germplasm, and integrative analyses integrating genomics, transcriptomics, and metabolomics are all made possible by these technological advancements.

A thorough review of coffee genomics is presented in this chapter, emphasising mapping and sequencing initiatives. The genome architecture and mapping techniques of coffee are examined in Section 2. Recent developments in genome assembly and sequencing technologies are discussed in Section 3. The field's current issues and prospects are covered in Section 4.

The goal is to highlight the significant scientific advancements made thus far and provide researchers, breeders, and stakeholders with a coherent summary of how genomics is influencing coffee improvement.

2. Genome architecture and mapping in coffee

2.1 Species diversity, domestication, and genome size variation

More than 130 species of coffee are known to exist, and they are indigenous to Madagascar, sub-Saharan Africa, and the Indian Ocean islands (Guyot et al., 2024). Their geographic distribution and major phylogenetic groupings are shown in Figure 1. However, almost all commercial production worldwide is accounted for by just two species: *C. arabica* (Arabica coffee) and *C. canephora* (Robusta coffee). Two diploid progenitors, *C. canephora* ($2n = 2x = 22$) and *C. eugenioides* S. Moore ($2n = 2x = 22$), spontaneously hybridised to form *C. arabica*, an allotetraploid ($2n = 4x = 44$). A narrow genetic base was established in cultivated *C. arabica* populations as a result of this hybridisation event, which is estimated to have occurred approximately 600,000 years ago. This has implications for genetic diversity, breeding, and genomic analyses (Dereeper et al., 2014; Salojärvi et al., 2024).

The genome size of each *Coffea* species varies significantly. Depending on the assembly and technique, the haploid genome of *C. arabica* ranges from 1.3 to 1.6 gigabases (Gb), whereas that of *C. canephora* is approximately 710 megabases (Mb). Because *C. arabica* is allotetraploid, it has two homeologous subgenomes, known as the E (from *C. eugenioides*) and C (from *C. canephora*) subgenomes. These subgenomes often complicate read alignment, variant calling, and gene annotation. Long-read sequencing and haplotype-resolved assemblies have recently helped alleviate these difficulties, revealing limited homoeologous recombination between subgenomes, expression dominance, and asymmetric gene loss (Rume et al., 2023; Scalabrin et al., 2020).

2.2 Genetic mapping in coffee

By offering a framework for linking phenotypic characteristics—like cup quality, caffeine content, yield, and disease resistance—with their underlying genomic loci, genetic mapping serves as the foundation for contemporary coffee genomics. It enables researchers to find quantitative trait loci (QTLs), track the inheritance of alleles across generations to come, and eventually support genomic and marker-assisted selection in breeding programs.

Low-throughput molecular markers such as restriction fragment length polymorphisms (RFLPs), amplified fragment length polymorphisms (AFLPs), and simple sequence repeats (SSRs) were used in the early stages of coffee genetic mapping. The early understanding of genome structure, inheritance patterns, and the genetic foundation of characteristics such as disease resistance, bean size, and caffeine biosynthesis was made possible by these early linkage maps. However, fine-scale mapping resolution was limited by their low transferability and limited marker density (Lashermes et al., 2000; Paillard et al., 1996).

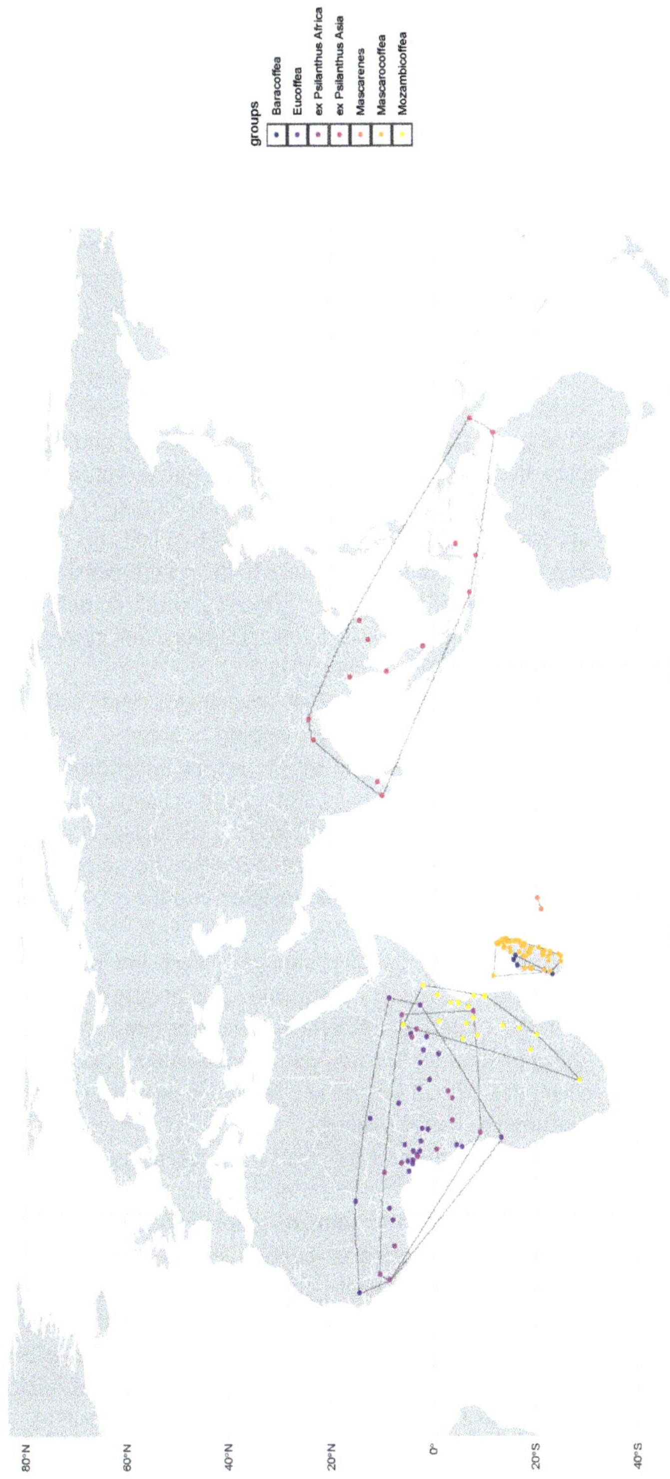

Figure 1: Distribution of wild *Coffea* species. The following phylogenetic groups are represented by coloured areas: Baracoffea, Eucoffea, the ex-Psilanthus genus from Africa, the ex-Psilanthus genus from Asia, Mascarenes Islands, Mascarocoffea, and Mozambicoffea (Guyot et al., 2024)

Coffee genetics has changed with the introduction of high-throughput genotyping and next-generation sequencing (NGS). The main tool for creating high-resolution maps is now single-nucleotide polymorphism (SNP) markers, which are widely dispersed throughout the genome. The Coffee 8.5K SNP array is one of the most important innovations. Merot-L'anthoene et al., 2019, comprising 8,580 verified SNPs obtained from the sequences of *C. arabica* and *C. canephora*. Near-saturated genome coverage is made possible by the fact that 82% of these SNPs are scoreable in mapping populations, with an average inter-marker distance of 54 kb. In addition to successfully mapping 945 segregating SNPs in *C. arabica* progeny, this array increased the marker density of the *C. canephora* genetic map by 1,307 SNPs. Additionally, the Coffee 8.5K array demonstrated high cross-species transferability, which makes it useful for researching germplasm diversity, population structure, and wild relatives.

In addition to SNP arrays, genotyping-by-sequencing (GBS) and RAD-seq technologies have enabled the detection of thousands of polymorphic loci at a lower cost, creating ultra-dense maps (Rasheed et al., 2017). Finding QTLs linked to drought tolerance, cup quality traits, and resistance to coffee leaf rust (*Hemileia vastatrix*) has been made possible, in large part, thanks to these high-density linkage maps. For example, the SH3 resistance locus from *C. liberica*, which confers durable resistance in Arabica cultivars, has been linked to linkage groups that correspond to QTLs governing leaf rust resistance (Ribas et al., 2011).

Chromosome rearrangements, homoeologous exchanges, and subgenome evolution have been clarified by the comparative mapping between *C. arabica*, *C. canephora*, and *C. eugenioides* made possible by the recent integration of high-density SNP data with whole-genome assemblies. Patterns of synteny, fractionation, and gene loss in these three species are shown in Figure 2 (Salojärvi et al., 2024). The development of molecular markers for subgenome-specific selection is an integrative analysis that also offers more profound insights into the evolution of the allopolyploid *C. arabica* genome.

To sum up, the evolution of coffee genetic mapping—from low-density SSR-based maps to high-resolution SNP and GBS platforms—has greatly improved our capacity to deconstruct complex traits and accelerated genetic progress. Multi-omics integration, which integrates phenomic, transcriptomic, metabolomic, and genomic data to develop predictive models of trait inheritance as sequencing costs continue to decline, is essential to the future of coffee genetics.

Together with advancements in machine learning and genomic selection, these integrative techniques will help next-generation breeding pipelines, enabling the creation of better, more resilient, and climate-adapted coffee varieties. All things considered, the functional annotation of Coffea genomes reveals how the diversification of this economically significant genus has been driven by genome duplication, transposable element dynamics, and lineage-specific evolution of gene families. Together with advancements in machine learning and genomic selection, these integrative techniques will help next-generation breeding pipelines, enabling the creation of better, more resilient, and climate-adapted coffee varieties.

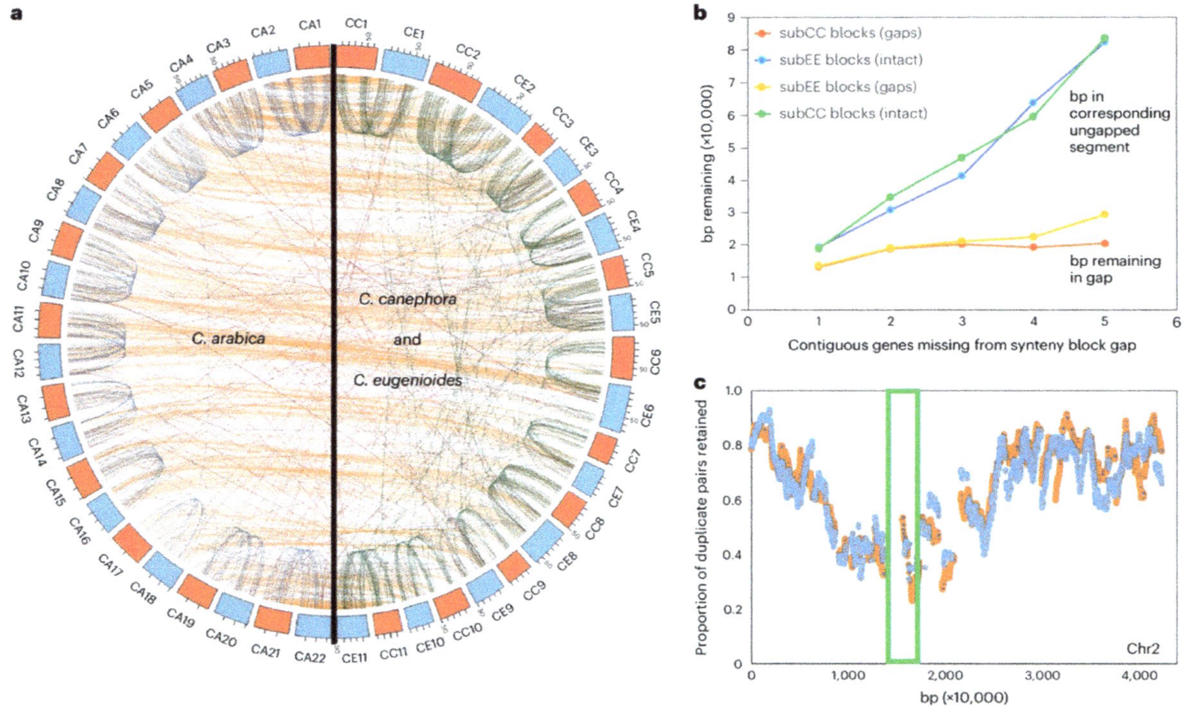

Figure 2: a: Corresponding syntenic blocks between CA subgenomes subCC (orange) and subEE (blue), and with the CC (orange) and CE (blue) genomes. b: The base pairs in intergenic DNA in synteny block gaps caused by fractionation in a subCC–subEE comparison, compared with the numbers of base pairs in homoeologous unfractionated regions, as a function of the number of consecutive genes deleted. c: Gene retention rates in synteny blocks plotted along subCC chromosome 2; subCC is plotted in orange and subEE in blue. The green box indicates the pericentromeric region. CA, *C. arabica*; CC, *C. canephora*; CE, *C. eugenioides* (Salojärvi et al., 2024)

2.3 Integrating genetic and physical maps

The integration of genetic linkage maps with physical genome assemblies represents a significant advancement in coffee genomics, providing a framework for connecting recombination-based genetic information to the genome's structural arrangement. Researchers can bridge the gap between molecular sequence data and phenotypic traits by ordering and orienting contigs into chromosome-scale pseudomolecules by aligning linkage markers to sequence-based scaffolds. In addition to improving the accuracy and complementarity of genome assemblies, this process enables the precise localisation of genes and quantitative trait loci (QTLs) underlying important agronomic and adaptive traits, including yield, disease resistance, and cup quality.

The resolution and integration potential of the early linkage maps of coffee were limited by the use of low-density markers, such as RAPDs, AFLPs, and SSRs. (De Kochko et al., 2010; Lashermes et al., 2000; Paillard et al., 1996). The chromosome-level assembly of *C. canephora* marked a major milestone by anchoring roughly 64% of the assembled sequences and 86% of annotated genes to 11 pseudochromosomes, radically altering the genomic landscape of coffee (Denoeud et al., 2014). The first comprehensive view of genome structure, recombination landscapes, and synteny with related Rubiaceae Juss. species was made possible by this integration, which was based on a dense map combining SSR and SNP markers. For comparative and evolutionary analyses within the *Coffea* genus, the reference genome of *C. canephora* became essential.

Later advancements in long-read sequencing and high-throughput genotyping technologies, including PacBio HiFi, Oxford Nanopore, and Hi-C, have improved the alignment of genetic and physical maps, resulting in chromosome-level assemblies of other *Coffea* species, most notably the subgenome-resolved assembly of *C. arabica* itself (Salojärvi et al., 2024).

In the allotetraploid *C. arabica* (2n = 4x = 44), derived from the hybridisation of *C. canephora* and *C. eugenioides*, integrating genetic and physical maps poses distinct challenges due to the high sequence similarity among homeologous chromosomes.

Recent chromosome-level assemblies of *C. arabica* and its diploid progenitors have facilitated subgenome partitioning and comparative analyses between the Cc (canephora-derived) and Ce (eugenioides-derived) components (Salojärvi et al. 2024). Although unique linkage maps for each subgenome have yet to be released, identified patterns of homoeologous exchange and gene retention suggest divergent evolutionary pathways. These results suggest potential allelic interactions and epistatic influences among subgenomes, which may contribute to intricate traits such as caffeine metabolism and flavour. Simultaneously, indications of biased recombination and subgenome divergence have begun to emerge (Combes et al., 2013); however, comprehensive epigenetic comparisons among subgenomes remain limited and represent a promising avenue for exploration.

Notably, a groundbreaking study recently employed an integrated sequencing approach (PacBio HiFi, Nanopore ultra-long, and Hi-C) to provide the most comprehensive near telomere-to-telomere (T2T) genome assembly for the allotetraploid *C. arabica* (cultivar 'Mundo Novo'). The independent evolution of

N-methyltransferases (NMTs), which are essential for the biosynthesis of caffeine, across various clades was made possible by this genomic resource, which enabled thorough functional and evolutionary analyses. In the Arabica subgenome, the researchers discovered and functionally described a copy-number-expanded UGT (UGT29 subfamily) that directly catalyses the one-step conversion of Rebaudioside A (Reb A) into the highly sought-after sweetener Reb M. This discovery creates a simplified, essential pathway for the synthetic biological production of Reb M in addition to offering a fundamental tool for metabolic research on *C. arabica* (Liu et al., 2025).

In conclusion, coffee genomics has evolved from disjointed sequences into logical, chromosome-scale reference frameworks through the integration of genetic and physical maps. The identification of genes associated with important agronomic traits is now accelerated by these integrated resources, which support pan-genomic analyses, QTL mapping, and genome-wide association studies (GWAS). In the future, combining multi-omics datasets—such as transcriptomic and pan-genomic—will provide a more comprehensive understanding of the coffee genome's architecture. In addition to enhancing efforts in gene editing and molecular breeding, such comprehensive frameworks will provide insight into the evolutionary processes that influence diversity and adaptation within the *Coffea* lineage.

2.4 Implications for breeding

The tools available for coffee breeding have been greatly accelerated by SNP arrays and high-density genetic maps, enabling genetic diversity assessments, quantitative trait locus (QTL) mapping, genome-wide association studies (GWAS), and marker-assisted selection (MAS). In perennial crops like coffee, where phenotyping is expensive and breeding cycles are drawn out, these molecular techniques are especially helpful.

Coffee breeding programs are increasingly able to establish marker-trait associations for key agronomic and quality traits, including disease resistance, yield stability, cup quality, bean size, and caffeine content. For example, researchers used the draft genome of *Coffea arabica* to identify 1,444 non-synonymous SNPs across 1,086 coding sequences that were statistically associated with caffeine content in a population of 232 genotypes (Tran et al., 2018).

The usefulness of SNP–trait links for accelerated breeding was demonstrated in another study, where GWAS in *C. arabica* identified SNPs significantly associated with seven of the eighteen agronomic traits assessed (e.g., plant height, yield components) (da Silva et al., 2024).

Disease resistance is another area where molecular breeding is showing promise. In a MAS program that targeted coffee leaf rust (CLR, *Hemileia vastatrix*) and coffee berry disease (CBD, *Colletotrichum kahawae*) in *C. arabica*, hybrids with a genetic gain of over 50% were produced by marker-assisted recurrent selection (L. M. Saavedra et al., 2023). Coffee yields are reduced by destructive phytonematodes known as *Meloidogyne* species. In a related effort, molecular markers linked to resistance to the nematode Meloidogyne exigua were found in coffee, showing that even root pest-resistant traits can be added to breeding pipelines using markers. These

SSR markers, confirmed in this population, provide a potential method for selecting nematode-resistant offspring in coffee breeding programs (Pereira et al., 2016).

Newer techniques, such as machine-learning-based prediction and genomic selection (GS), are gaining traction in the coffee industry, in addition to MAS and QTL mapping. Genomic selection (GS) offers a DNA-based alternative for the speedier selection of superior cultivars. Stacking Ensemble Learning (SEL) combines multiple models to potentially increase selection accuracy. A recent study demonstrated that stacking ensemble methods using more than 21,000 SNPs improved prediction accuracy for yield, fruit number, and pest incidence in *C. arabica*, indicating a data-driven approach to breeding. Taking everything into account, SEL presents GS with a viable strategy. By combining predictions from multiple models, SEL may enhance GS's predictive ability (PA) for complex traits (Nascimento et al., 2024).

In practice, the implications for breeders are substantial:

1. Parent-line selection becomes more precise, as favourable alleles can be tracked and selected early (seedling stage) via molecular screens rather than waiting for field performance (Nadeem et al., 2018).
2. It becomes possible to pyramid multiple advantageous alleles (e.g., combining disease-resistance loci and quality-trait QTLs), which decreases generation time and increases genetic gain per cycle (Laura Maritza Saavedra et al., 2023).
3. Breeders can monitor allele frequencies, prevent bottlenecks, and maintain broad gene pools by using SNP arrays for genetic diversity assessments. This is essential for crops with narrow genetic bases, such as Arabica (Salojärvi et al., 2024).
4. Additionally, marker-trait information enables precision introgression from diploid progenitors or wild relatives into elite lines, reducing linkage drag and accelerating variety development (Xu et al., 2017).

In conclusion, the paradigm shift from traditional phenotype-based selection, which is time-consuming and resource-intensive, to genotype-informed, accelerated breeding pipelines is being driven by modern molecular tools, which are revolutionising coffee breeding. The precise identification of alleles linked to important agronomic traits, such as disease resistance, yield stability, cup quality, and tolerance to drought or high temperatures, is made possible by high-density genetic maps, SNP arrays, and genome-wide association studies (GWAS). Breeders can also determine an individual's breeding value at an early developmental stage by combining genomic selection (GS) and predictive modelling, which shortens breeding cycles and increases genetic gain per generation.

With Arabica coffee's limited genetic base and vulnerability to biotic and abiotic stresses, these tools also make it easier to pyramid multiple advantageous alleles, allowing for the simultaneous improvement of multiple traits. Breeders can forecast genotype-by-environment interactions by combining molecular markers with environmental modelling and high-throughput phenotyping platforms. This aids in the creation of cultivars that are resilient to climate change.

Furthermore, by permitting accurate introgression and reducing linkage drag, molecular techniques improve the utilisation of genetic diversity from landraces and wild relatives. Allelic diversity conservation and knowledgeable germplasm collection management are further supported by marker-assisted and genomic-assisted breeding. When combined, these developments offer a framework for sustainable coffee production as well as more effective cultivar development, guaranteeing premium beans even in the face of shifting environmental conditions.

Overall, the use of high-resolution molecular tools in breeding programs is turning coffee improvement into a data-driven, predictive process. By combining environmental factors, phenotypic assessment, and genotypic data, superior *Coffea* cultivars with long-lasting disease resistance, optimal cup quality, and climate resilience can be produced more quickly.

3. Advances in sequencing technologies and genome assemblies

3.1 Technological evolution and assembly strategies

The past two decades have witnessed rapid advances in DNA sequencing technologies, transforming genome assembly from fragmented draft sequences to near-complete, chromosome-scale assemblies. DNA sequencing was first introduced in 1977, and next-generation sequencing technologies have been available only during the past decade. First-generation Sanger sequencing provided high-accuracy reads but was limited in throughput and cost. Next-generation sequencing (NGS) technologies can be broadly categorised into two approaches: short-read and long-read. Short-read methods offer high-accuracy and cost-effective data, making them ideal for large-scale population studies and the detection of genetic variants. Long-read sequencing, on the other hand, produces much longer reads, which are especially useful for de novo genome assembly and thorough full-length transcript characterisation (Goodwin et al., 2016; Mardis, 2017).

The advent of third-generation long-read technologies, such as Oxford Nanopore Technologies' (ONT) sequencing and Pacific Biosciences' (PacBio) HiFi reads, has revolutionised genome assembly and ushered in a new era of genomics. These platforms produce sequence reads that are frequently longer than tens of kilobases (some ONT reads can exceed a megabase), which represents a significant improvement over earlier short-read systems. The secret to their transformative power lies in their enormous read length. Long reads enable substantially more contiguous assemblies by covering vast genomic distances, leading to high contiguity, as indicated by metrics such as N50 or N50 contig. Importantly, they make it possible for researchers to successfully resolve complex genomic architecture, especially previously unresolvable regions like highly structured, polymorphic regions (highly heterozygous regions) and long, nearly identical repetitive elements (like transposable elements). As a result, long-read sequencing now frequently produces high-quality reference genomes at the chromosome level, thereby accelerating functional and comparative genomics research (Jain et al., 2018; Wenger et al., 2019).

By offering long-range genomic contacts that make it easier to order and orient contigs into chromosome-scale scaffolds, complementary scaffolding technologies—in particular, Hi-C chromatin conformation capture—have further improved the quality of assembly (Belton et al., 2012; Dudchenko et al., 2017). In order to achieve near telomere-to-telomere (T2T) completeness, hybrid assembly strategies now frequently combine short-reads for base accuracy, long-reads for contiguity, and Hi-C for chromosome-scale scaffolding. These methods have been extensively applied to the genomes of plants, animals, and microorganisms, revolutionising comparative studies, structural variant detection, and functional genomics.

Early coffee genome projects relied on short-read Illumina data supplemented by Sanger reads and BAC libraries. For example, the *C. canephora* project utilised ~30× Roche 454 reads, ~60× Illumina reads, and Sanger BAC-end reads to achieve a scaffold N50 of ~1.26 Mb (Denoeud et al., 2014). Later, hybrid assembly strategies emerged combining Pacific Biosciences (PacBio) long-reads or Oxford Nanopore (ONT) reads with Hi-C chromatin-capture scaffolding, which enables chromosome-scale contiguity. The recent Arabica assemblies (Geisha variety) employed both ONT and PacBio long-reads, as well as Hi-C, achieving a BUSCO completeness of ~97.7% (Medrano et al., 2025). The progression of technologies has significantly improved assembly contiguity, completeness and scaffold anchoring.

3.2 Reference genome assemblies in coffee

A major milestone in coffee genomics was the publication of the high-quality draft genome of *C. canephora* in 2014 (Denoeud et al., 2014). This diploid species (2n = 2x = 22) offered a comparatively straightforward model for assembly, anchoring a substantial portion of scaffolds to chromosomes and annotating ~25,574 protein-coding genes. The genome revealed several gene family expansions, including N-methyltransferases (NMTs) involved in caffeine biosynthesis and defence-related genes.

For *C. arabica*, the sequencing task is more complex owing to its allotetraploid nature. The first public Arabica (Red Bourbon variety) genome assembly, recorded at 1.51 Gb with 78,311 predicted genes, was made available pre-publication via World Coffee Research. A 2024 study presented chromosome-level assemblies of a di-haploid *C. arabica* accession and its diploid progenitors, enabling detailed sub-genome analyses (Salojärvi et al., 2024).

More recently, chromosome-level assemblies have been reported, including *C. arabica* var. *Geisha* was built using long-read sequencing (PacBio, ONT) and Hi-C scaffolding (e.g., Arabica var. Geisha, ~1.03 Gb, 47,062 gene loci, 97.7% BUSCO completeness) (Medrano et al., 2025).

Significant progress has been made this year in both sequencing technology and database establishment, laying a solid foundation for the utilisation of coffee genomes. The National Centre for Biotechnology Information (NCBI) provides a platform for storing and utilising genomic data, currently covering 11 chromosome-level genomes, including *C. arabica*, *C. canephora*, *C. eugenioides*, and *C. humblotiana*. Recently assembled *Coffea* genomes deposited in the NCBI database are summarised in Table 3.1.

3.3 Functional annotation and insights

High-quality genome assemblies form the foundation for comprehensive functional annotation, allowing accurate identification of genes, regulatory elements, and

Table 1: The assembled Coffea genome has been uploaded to the NCBI in recent years (https://www.ncbi.nlm.nih.gov/datasets/genome/?taxon=13442)

Assembly accession	Organism name	Release date	WGS project accession	Number of scaffolds
GCA_049114775.1	*C. arabica*	2025/3/28	JBLGCP01	967
GCA_041438355.1	*C.arabica*	2024/8/23	JBELAZ01	235
GCA_036785885.1	*C. arabica*	2024/2/23	JAZHGF01	133
GCF_036785885.1	*C. arabica*	2024/2/23	JAZHGF01	133
GCA_036785865.1	*C. canephora*	2024/2/23	JAZHGG01	3033
GCA_036785775.1	*C. arabica*	2024/2/23	JAZHSI01	8474
GCA_036785875.1	*C. eugenioides*	2024/2/23	JAZHGH01	117
GCA_030873655.1	*C. arabica*	2023/8/25	JARPVW01	1749
GCA_023065735.1	*C. humblotiana*	2022/4/22	JACYVW01	337
GCA_003713205.1	*C. eugenioides*	2018/11/8	RHJT01	3530
GCF_003713205.1	*C. eugenioides*	2018/11/8	RHJT01	3530
GCA_003713225.1	*C. arabica*	2018/11/8	RHJU01	2832
GCF_003713225.1	*C. arabica*	2018/11/8	RHJU01	2832
GCA_900059795.1	*C. canephora*	2014/8/21	CBUE02	13345

repetitive sequences. With the improvement of assembly contiguity and completeness, it has become possible to perform genome-wide synteny comparisons, characterise the repeat landscape, and conduct detailed gene family analyses across species (Campbell et al., 2013; Yandell & Ence, 2012).

In plant genomes, transposable elements (TEs) are major contributors to genome size variation and structural evolution. For instance, long terminal repeat (LTR) retrotransposons often constitute more than half of many angiosperm genomes and play crucial roles in genome plasticity and gene regulation (Bennetzen & Wang, 2014; Wicker et al., 2007). About 50% of the genome of Coffea canephora is made up of transposable elements, with LTR retrotransposons making up nearly 85% of this total (Denoeud et al., 2014). This type of repeat profiling provides information on genome expansion and contraction events, informing comparative studies between diploid and polyploid *Coffea* species.

Accurate gene model construction is made possible by functional annotation pipelines that combine RNA-seq evidence, ab initio prediction, and homology-based techniques (e.g., MAKER, BRAKER2, or AUGUSTUS) (Brůna et al., 2021; Stanke et al., 2006). These pipelines enable thorough and biologically consistent annotation of both coding and non-coding regions by combining extrinsic transcriptional and comparative evidence with intrinsic gene structure prediction.

Generally, transcriptome assemblies and protein homology data refine exon–intron boundaries and validate expressed genes. At the same time, ab initio algorithms predict gene models based on genomic sequence features, such as start/stop codons, splice sites, and codon usage bias. These integrative frameworks are now the industry standard for high-quality plant genome annotation, serving as a basis for subsequent analyses, such as pathway reconstruction, gene family evolution, and synteny inference (Campbell et al., 2013; Holt & Yandell, 2011).

In *C. canephora*, functional annotation identified approximately 25,000 protein-coding genes, consistent with other diploid angiosperms. The N-methyltransferase (NMT) gene family, which is essential for the biosynthesis of caffeine through the sequential methylation of xanthosine derivatives, has shown expansions in comparative genomic analyses (Denoeud et al., 2014). It's interesting to note that phylogenetic reconstruction of NMT homologs across the lineages of *Coffea* spp., *Camellia sinensis* (L.) Kuntze, and *Theobroma cacao* L. showed that caffeine biosynthesis evolved separately in each of these lineages, suggesting a polyphyletic origin for the pathway. The metabolic flexibility of secondary metabolism in plants is demonstrated by this convergent evolution, which also sheds light on the autonomous recruitment of related enzymatic functions under selective pressure for ecological adaptation and defence. Furthermore, the reconstruction of ancestral enzyme sequences suggests that the functional recruitment of pre-existing metabolic enzymes is most likely how the caffeine biosynthesis pathway emerged (Huang et al., 2016).

In the allotetraploid *C. arabica*, subgenome-aware annotation has further revealed the respective contributions of its progenitor species, *C. canephora* and *C. eugenioides*. Advanced read-partitioning algorithms and synteny analyses have enabled the discrimination of homologous chromosomes, clarifying how recombination, fractionation, and gene loss events have shaped the cultivated genome (Medrano et al., 2025; Salojärvi et al., 2024). For example, while strong synteny is maintained between diploid progenitors and *C. arabica* subgenomes, gene family diversification and subgenome-specific expansions reflect post-polyploidization adaptation and metabolic specialisation. These findings also provide a genomic framework for understanding key agronomic traits, including caffeine metabolism, flavour formation, and stress tolerance (Denoeud et al., 2014).

All things considered, the functional annotation of Coffea genomes reveals how the diversification of this economically significant genus has been driven by genome duplication, transposable element dynamics, and lineage-specific evolution of gene families. Future research integrating transcriptomic, metabolomic, and epigenomic data will help clarify the regulatory networks behind coffee's quality traits and adaptive evolution as annotation techniques and omics integration progress.

3.4 Implications for variation discovery and trait mapping

The availability of high-quality reference genomes has improved the accuracy and viability of resequencing and variation discovery. Whole-genome resequencing of 90 accessions in *C. arabica* revealed roughly 11.08 million variants, comprising 1.34 million insertions/deletions (InDels) and 9.74 million single-nucleotide polymorphisms (SNPs) (Mekbib et al., 2022).

After the coffee germplasm's genome was resequenced, a dense map of genetic polymorphisms—most notably SNPs and InDels—was created. The differentiation observed in functional gene networks supporting characteristics essential to coffee agriculture, including productivity, accumulation of secondary metabolites (such as caffeine), and defence mechanisms against pathogens and insects, is likely caused by these recently identified SNP markers. Together, the genetic variation insights and resequencing data obtained from the 90 accessions are essential tools for analysing the coffee gene pool and will greatly accelerate future genetic mapping and breeding initiatives.

For numerous subsequent genomic studies, these extensive compilations of genetic variation data (variant datasets) provide a crucial foundation for research. Detailed germplasm characterisation, in-depth investigation of population dynamics and demographic history, and genotype-to-phenotype association studies are important applications. It is crucial to be able to precisely map these sequence differences onto highly resolved reference assemblies at the chromosome level. This accuracy enables the fine-scale localisation of structural changes, haplotype structures, and quantitative trait loci (QTLs) that are directly associated with agronomic traits of commercial importance. This capability is a significant breakthrough that turns the initial identification of genetic polymorphisms into a thorough comprehension of their phenotypic effects and functional implications. Additionally, multi-omics association analyses are made possible by the deliberate combination of transcriptomics, metabolomics, and phenomics data with marker information derived from resequencing. Finding candidate genes, figuring out complex regulatory networks, and shedding light on the basic molecular mechanisms underlying complex traits all depend on this integrative approach. Even though research on coffee is still in its early stages, these combined approaches have shown great promise for the rapid genetic improvement of tropical and perennial crops due to their success with common agricultural species.

In summary, the transition to advanced genome-based breeding techniques, such as genomic selection and haplotype-focused breeding, is essentially made possible by the convergence of a high-fidelity reference genome with extensive variant data sets. In addition to accelerating cycles of crop improvement, the process of linking observable sequence variation to specific phenotypic differences also provides vital information for efforts to protect and optimise the genetic diversity found in *Coffea* species and their relatives.

4. The next horizon: Challenges and opportunities in coffee genomics

4.1 Polyploidy, assembly complexity, and haplotype resolution

C. arabica's allotetraploid genetic structure (2n = 4x = 44), which results from the hybridisation of *C. canephora* and *C. eugenioides*, presents ongoing and complex difficulties for genome assembly and analysis. Because of the high sequence similarity between its two sub-genomes (often referred to as A and B), it is technically challenging to distinguish homeologous chromosomes, phase allelic variants, and reconstruct the evolutionary trajectories of duplicated genes.

A fully phased *C. arabica* genome is still difficult to assemble and annotate, even with recent advances in chromatin conformation capture technologies (Hi-C, Omni-C) and long-read sequencing (e.g., PacBio HiFi, Oxford Nanopore). Understanding phenotypic traits requires knowledge of gene copy number variation, paralog-specific expression, and dosage effects, all of which are obscured during assembly by the collapse of numerous duplicated or homeologous loci. Because read misassignment can result in false SNPs or obscure true structural variants (SVs), the high degree of sequence identity among the sub-genomes further complicates variant calling.

Additionally, little is known about structural rearrangements within and between the two sub-genomes, such as inversions, translocations, and copy-number variants, which restricts our understanding of genome evolution and sub-genome dominance. Although they often represent consensus mosaics rather than haplotype-resolved genomes, current reference assemblies have provided a useful framework for gene annotation and comparative analyses.

The creation of a fully annotated, haplotype-resolved *C. arabica* pan-genome that captures allelic diversity across cultivars and wild accessions is a major future objective. Integrating ultra-long-read data, single-cell or trio-binning techniques, and sophisticated graph-based genome representations will probably be necessary to accomplish this. These resources will shed light on the genetic basis of significant agronomic traits, adaptation mechanisms, and the genetic basis of its superior quality and hybrid vigour, in addition to elucidating the degree of structural and sequence divergence between the A and B sub-genomes, thus answering the evolutionary story of *Coffea*.

4.2 Pan-genomics, wild relatives, and genetic resource utilisation

Future research in coffee genomics must increasingly focus on constructing pan-genomes—comprehensive genomic frameworks that integrate both core and dispensable genes across multiple Coffea species and cultivars. Unlike a single reference genome, a pan-genome can capture the full extent of structural and sequence diversity present in cultivated and wild coffee lineages. This approach is essential for understanding intraspecific variability, adaptive evolution, and the molecular basis of agronomic traits that underpin resilience and quality.

The genus *Coffea*, comprising over 130 recognised species distributed primarily across tropical Africa and Madagascar, represents a vast and largely untapped reservoir of genetic diversity. However, this number could increase to 141, considering the undescribed species in various collections (Guyot et al., 2024). Wild relatives, such as *C. eugenioides*, *C. canephora*, *C. congensis*, *C. racemosa*, and *C. liberica*, harbour alleles linked to drought and heat tolerance, disease resistance (including coffee leaf rust and coffee berry disease), and unique secondary metabolites that influence flavour and aroma. These genetic resources are crucial for breeding programs aiming to improve stress resilience and maintain yield stability in the face of climate change.

However, much of this diversity remains underutilised due to barriers in cross-compatibility, reproductive isolation, and limited genomic characterisation of wild taxa. Advances in comparative genomics, high-throughput resequencing, and graph-based genome representations now enable the systematic cataloguing of gene presence/absence variation, structural variants, and novel alleles across wild germplasm. Coupled with population genomics and environmental association analyses, these data can identify adaptive loci that have evolved in response to ecological gradients, such as temperature, precipitation, and soil chemistry (Tran et al., 2016).

Integrating wild germplasm into modern breeding pipelines will require genome-informed pre-breeding strategies, including marker-assisted introgression, genomic selection, and the use of bridge crosses to overcome reproductive barriers. Equally important are efforts to conserve and characterise germplasm, ensuring that wild *Coffea* populations—many of which face habitat loss—remain accessible for future research and breeding.

Ultimately, developing a *Coffea* pan-genome will provide a genomic roadmap for the rational utilisation of wild relatives, guiding the identification and deployment of beneficial alleles to enhance productivity, quality, and resilience. Such efforts will not only safeguard the genetic foundation of global coffee production but also expand our understanding of the evolutionary and ecological aspects of one of the world's most economically important crops.

4.3 Climate adaptation, sustainability imperatives, and ethical governance

The accelerating effects of climate change are putting the sustainability of the world's coffee production at greater risk, endangering the industry's socioeconomic and ecological underpinnings. Significant declines in yield and quality have been caused by rising temperatures, protracted droughts, erratic rainfall patterns, and the spread of pests and diseases, most notably *Hemileia vastatrix* (coffee leaf rust) and *Hypothenemus hampei* (coffee berry borer), especially in smallholder-dominated regions of Africa, Latin America, and Asia. Simultaneously, climate models predict significant changes in the zones suitable for cultivation, pushing production into higher altitudes and more delicate ecosystems, which will impact land use and biodiversity.

Advances in genomics and biotechnology have the potential to be revolutionary in this regard. The development of climate-resilient cultivars can be facilitated by identifying loci underlying disease resistance, drought resilience, and heat tolerance through the use of genomic selection, transcriptomic profiling, and genome-wide association studies (GWAS). However, a framework based on sustainability, inclusivity, and ethical governance is necessary to translate these scientific advancements into real benefits for growers.

Since smallholder farmers make up between 70 and 80 percent of the world's coffee production, the ethical application of genomics requires that new technologies benefit them. Many of these producers rely on traditional varieties that are not well-suited to new climatic stresses and operate within precarious socioeconomic systems.

To prevent the existing disparities in the global coffee value chain from worsening, fair benefit sharing, transparent data governance, and equitable access to genomic innovations are crucial.

The preservation of traditional landraces and wild genetic resources must remain a top priority. Long-term adaptive capacity depends on these genetic reservoirs, which are frequently preserved ex situ in germplasm collections or in situ within tropical forests. Therefore, genomic-assisted breeding ought to be used in conjunction with conservation measures that protect agro-biodiversity and the evolutionary potential of coffee species.

Understanding the cultural, economic, and environmental aspects of coffee production is another requirement for incorporating ethical governance principles. Policies should support participatory breeding initiatives, incorporate local knowledge, and align with global frameworks, such as the Nagoya Protocol on Access and Benefit-Sharing. The adoption of genomic technologies should also be accompanied by life-cycle and sustainability assessments to ensure they support social justice, environmental conservation, and climate mitigation.

In conclusion, although genomics offers powerful tools for adapting coffee to a rapidly changing world, sustainability imperatives must guide its application. A robust, diverse, and ethically run coffee industry, where scientific advancements promote environmental stewardship and human welfare, should be the ultimate objective.

4.4 Emerging methodologies: Advanced computation and systems biology

Beyond basic genomic resources, coffee genomics is undergoing a fundamental transformation and breeding capabilities are being accelerated due to the quick convergence of computational science and plant biology. Traditional analytical frameworks are no longer adequate for deriving significant biological insights from datasets that are becoming larger and more complex. Artificial intelligence (AI) and machine learning (ML) are becoming increasingly essential tools for data-driven discovery and decision-making in this rapidly evolving environment.

These methods can create predictive models that reveal intricate genotype–phenotype–environment relationships by combining data from high-resolution environmental and agronomic datasets with information from multi-omics layers, such as genomics, transcriptomics, proteomics, and metabolomics. These models are improving their ability to predict how plants will perform in various climatic conditions, identify potential genes associated with important agronomic or adaptive traits, and streamline breeding pipeline selection processes.

Furthermore, a more comprehensive view of the coffee plant as a dynamic, multiscale system is becoming possible thanks to systems biology frameworks. Researchers can mimic the effects of genetic perturbations or environmental stressors on physiological and developmental pathways by simulating regulatory networks, metabolic fluxes, and gene–environment interactions. Our understanding of coffee biology is enhanced by this integrative viewpoint, which also facilitates the logical design of cultivars with improved resilience and quality characteristics.

Technological developments in phenomics and remote sensing are further increasing research capacity. Real-time evaluation of plant growth, stress responses, and productivity across diverse landscapes is now possible thanks to sensor-based data collection, high-throughput imaging, and drone-assisted monitoring. These tools can provide quick, non-invasive phenotype characterisation when used in conjunction with AI algorithms, which is a crucial first step toward achieving precision agriculture in coffee.

Together, AI, ML, and systems biology signal a new paradigm for coffee science that uses computational intelligence for ethically guided, sustainable, and predictive crop improvement, overcoming the limitations of traditional experimentation. These technologies are likely to evolve over the next ten years, from research prototypes to essential elements of a comprehensive, data-driven breeding ecosystem for coffee species.

5. Conclusion: The path forward

Overall, advances in sequencing technology, gene mapping, and integrative omics research have profoundly transformed the field of coffee genomics. These advancements have provided researchers with unprecedented tools to explore the genetic diversity, functional gene variation, and evolutionary processes shaping coffee species. The resulting genomic insights not only deepen our understanding of the complex biology of coffee plants but also provide practical guidance for breeding programs aimed at improving coffee's resilience, yield, and quality. As climate change continues to alter the ecological environment of coffee production, integrating genomics into breeding and conservation strategies is crucial for protecting coffee's genetic heritage and ensuring its long-term sustainable development. Looking ahead, the next phase of coffee research will be characterised by a closer integration of high-resolution genomic data, computational and analytical innovations, and an ecological perspective. This integration will help translate scientific discoveries into practical resilience, enhancing the adaptability, development capacity, and ability to sustain livelihoods dependent on the global coffee industry.

References

Belton, J. M., McCord, R. P., Gibcus, J. H., Naumova, N., Zhan, Y., and Dekker, J. (2012). Hi-C: A comprehensive technique to capture the conformation of genomes. *Methods*, 58(3), 268–276. https://doi.org/10.1016/j.ymeth.2012.05.001

Bennetzen, J. L., and Wang, H. (2014). The contributions of transposable elements to the structure, function, and evolution of plant genomes. *Annu Rev Plant Biol*, 65, 505–530. https://doi.org/10.1146/annurev-arplant-050213-035811

Brůna, T., Hoff, K. J., Lomsadze, A., Stanke, M., and Borodovsky, M. (2021). BRAKER2: Automatic eukaryotic genome annotation with GeneMark-EP+ and AUGUSTUS supported by a protein database. *NAR Genom Bioinform*, 3(1), lqaa108. https://doi.org/10.1093/nargab/lqaa108

Campbell, M. S., Law, M., Holt, C., Stein, J. C., Moghe, G. D., Hufnagel, D. E. et al. (2013). MAKER-P: A tool kit for the rapid creation, management, and quality control of plant genome annotations. *Plant Physiology*, 164(2), 513-524. https://doi.org/10.1104/pp.113.230144

Combes, M.-C., Dereeper, A., Severac, D., Bertrand, B., and Lashermes, P. (2013). Contribution of subgenomes to the transcriptome and their intertwined regulation in the allopolyploid Coffea arabica grown at contrasted temperatures. *New Phytologist*, 200(1), 251–260. https://doi.org/10.1111/nph.12371

da Silva, R. A., Caixeta, E. T., Silva, L. F., Sousa, T. V., Barreiros, P., Oliveira, A. C. B. et al. (2024). Identification of SNP markers and candidate genes associated with major agronomic traits in Coffea arabica. *Plants* (Basel), 13(13). https://doi.org/10.3390/plants13131876

De Kochko, A., Akaffou, S., Andrade, A. C., Campa, C., Crouzillat, D., Guyot, R. et al. (2010). Advances in Coffea genomics. *In:* Advances in Botanical Research (Vol. 53, pp. 23–63). Elsevier.

Denoeud, F., Carretero-Paulet, L., Dereeper, A., Droc, G., Guyot, R., Pietrella, M. et al. (2014). The coffee genome provides insight into the convergent evolution of caffeine biosynthesis. *Science*, 345(6201), 1181–1184. https://doi.org/10.1126/science.1255274

Dereeper, A., Bocs, S., Rouard, M., Guignon, V., Ravel, S., Tranchant-Dubreuil, C. et al. (2014). The coffee genome hub: A resource for coffee genomes. *Nucleic Acids Research*, 43(D1), D1028–D1035. https://doi.org/10.1093/nar/gku1108

Dudchenko, O., Batra, S. S., Omer, A. D., Nyquist, S. K., Hoeger, M., Durand, N. C. et al. (2017). De novo assembly of the <i>Aedes aegypti</i> genome using Hi-C yields chromosome-length scaffolds. *Science*, 356(6333), 92–95. https://doi.org/doi:10.1126/science.aal3327

Goodwin, S., McPherson, J. D., and McCombie, W. R. (2016). Coming of age: Ten years of next-generation sequencing technologies. *Nature Reviews Genetics*, 17(6), 333–351. https://doi.org/10.1038/nrg.2016.49

Guyot, R., Gonzalez, L., and Bezandry, R. (2024). Wild coffea species: A modern genomic approach to unravel variations for future cultivated coffee improvement. *Proceedings*, 109(1), 23. https://www.mdpi.com/2504-3900/109/1/23

Holt, C., and Yandell, M. (2011). MAKER2: An annotation pipeline and genome-database management tool for second-generation genome projects. *BMC Bioinformatics*, 12(1), 491. https://doi.org/10.1186/1471-2105-12-491

Huang, R., O'Donnell, A. J., Barboline, J. J., and Barkman, T. J. (2016). Convergent evolution of caffeine in plants by co-option of exapted ancestral enzymes. *Proc Natl Acad Sci U S A*, 113(38), 10613–10618. https://doi.org/10.1073/pnas.1602575113

Jain, M., Koren, S., Miga, K. H., Quick, J., Rand, A. C., Sasani, T. A. et al. (2018). Nanopore sequencing and assembly of a human genome with ultra-long reads. *Nature Biotechnology*, 36(4), 338–345. https://doi.org/10.1038/nbt.4060

Lashermes, P., Andrzejewski, S., Bertrand, B., Combes, M. C., Dussert, S., Graziosi, G. et al. (2000). Molecular analysis of introgressive breeding in coffee (Coffea arabica L.). *Theoretical and Applied Genetics*, 100(1), 139–146. https://doi.org/10.1007/s001220050019

Liu, Y., Zong, H., Xing, Y., Jiao, X., Liu, Z., Niu, Y. et al. (2025). A near telomere-to-telomere genome assembly of Coffea arabica (Mundo Novo) provides insights into its secondary metabolism. *Molecular Ecology Resources*, 25(8), e70053. https://doi.org/https://doi.org/10.1111/1755-0998.70053

Mardis, E. R. (2017). DNA sequencing technologies: 2006–2016. *Nature Protocols*, 12(2), 213–218. https://doi.org/10.1038/nprot.2016.182

Medrano, J. F., Cantu, D., Minio, A., Dreischer, C., Gibbons, T., Chin, J. et al. (2025). De novo whole-genome assembly and annotation of Coffea arabica var. Geisha, a high-quality coffee variety from the primary origin of coffee. *G3 (Bethesda)*, 15(1). https://doi.org/10.1093/g3journal/jkae262

Mekbib, Y., Tesfaye, K., Dong, X., Saina, J. K., Hu, G.-W., and Wang, Q.-F. (2022). Whole-genome resequencing of Coffea arabica L. (Rubiaceae) genotypes identify SNP and unravels distinct groups showing a strong geographical pattern. *BMC Plant Biology*, 22(1), 69. https://doi.org/10.1186/s12870-022-03449-4

Merot-L'anthoene, V., Tournebize, R., Darracq, O., Rattina, V., Lepelley, M., Bellanger, L. et al. (2019). Development and evaluation of a genome-wide Coffee 8.5K SNP array and its application for high-density genetic mapping and for investigating the origin of Coffea arabica L. *Plant Biotechnol J*, 17(7), 1418–1430. https://doi.org/10.1111/pbi.13066

Nadeem, M. A., Nawaz, M. A., Shahid, M. Q., Doğan, Y., Comertpay, G., Yıldız, M. et al. (2018). DNA molecular markers in plant breeding: Current status and recent advancements in genomic selection and genome editing. *Biotechnology & Biotechnological Equipment*, 32(2), 261–285.

Nascimento, M., Nascimento, A. C. C., Azevedo, C. F., de Oliveira, A. C. B., Caixeta, E. T., and Jarquin, D. (2024). Enhancing genomic prediction with Stacking Ensemble Learning in Arabica Coffee. *Front Plant Sci*, 15, 1373318. https://doi.org/10.3389/fpls.2024.1373318

Paillard, M., Lashermes, P., and Pétiard, V. (1996). Construction of a molecular linkage map in coffee. *Theoretical and Applied Genetics*, 93(1), 41–47. https://doi.org/10.1007/BF00225725

Pereira, T. B., Setotaw, T. A., Santos, D. N., Mendes, A. N., Salgado, S. M., Carvalho, G. R., and Rezende, R. M. (2016). Identification of microsatellite markers in coffee associated with resistance to Meloidogyne exigua. *Genet Mol Res*, 15(3). https://doi.org/10.4238/gmr.15038054

Rasheed, A., Hao, Y., Xia, X., Khan, A., Xu, Y., Varshney, R. K., and He, Z. (2017). Crop breeding chips and genotyping platforms: Progress, challenges, and perspectives. *Molecular Plant*, 10(8), 1047–1064. https://doi.org/https://doi.org/10.1016/j.molp.2017.06.008

Ribas, A. F., Cenci, A., Combes, M.-C., Etienne, H., and Lashermes, P. (2011). Organization and molecular evolution of a disease-resistance gene cluster in coffee trees. *BMC Genomics*, 12(1), 240. https://doi.org/10.1186/1471-2164-12-240

Rume, G. d. C., de Oliveira, R. R., Ribeiro, T. H. C., and Chalfun-Júnior, A. (2023). Genome-wide and expression analyses of MADS-box genes in the tetraploid Coffea arabica L. and its diploid parental subgenomes. *Plant Gene*, 34, 100413. https://doi.org/https://doi.org/10.1016/j.plgene.2023.100413

Saavedra, L. M., Caixeta, E. T., Barka, G. D., Borém, A., Zambolim, L., Nascimento, M. et al. (2023). Marker-assisted recurrent selection for pyramiding leaf rust and coffee berry disease resistance alleles in Coffea arabica L. *Genes* (Basel), 14(1). https://doi.org/10.3390/genes14010189

Salojärvi, J., Rambani, A., Yu, Z., Guyot, R., Strickler, S., Lepelley, M. et al. (2024). The genome and population genomics of allopolyploid Coffea arabica reveal the diversification history of modern coffee cultivars. *Nature Genetics*, 56(4), 721–731. https://doi.org/10.1038/s41588-024-01695-w

Scalabrin, S., Toniutti, L., Di Gaspero, G., Scaglione, D., Magris, G., Vidotto, M. et al. (2020). A single polyploidization event at the origin of the tetraploid genome of Coffea arabica is responsible for the extremely low genetic variation in wild and cultivated germplasm. *Sci Rep*, 10(1), 4642. https://doi.org/10.1038/s41598-020-61216-7

Stanke, M., Keller, O., Gunduz, I., Hayes, A., Waack, S., and Morgenstern, B. (2006). AUGUSTUS: Ab initio prediction of alternative transcripts. *Nucleic Acids Res*, 34(Web Server issue), W435–439. https://doi.org/10.1093/nar/gkl200

Tran, H. T., Lee, L. S., Furtado, A., Smyth, H., and Henry, R. J. (2016). Advances in genomics for the improvement of quality in coffee. *Journal of the Science of Food and Agriculture*, 96(10), 3300–3312.

Tran, H. T. M., Ramaraj, T., Furtado, A., Lee, L. S., and Henry, R. J. (2018). Use of a draft genome of coffee (Coffea arabica) to identify SNPs associated with caffeine content. *Plant Biotechnol J*, 16(10), 1756–1766. https://doi.org/10.1111/pbi.12912

Wenger, A. M., Peluso, P., Rowell, W. J., Chang, P.-C., Hall, R. J., Concepcion, G. T. et al. (2019). Accurate circular consensus long-read sequencing improves variant detection and assembly of a human genome. *Nature Biotechnology*, 37(10), 1155–1162. https://doi.org/10.1038/s41587-019-0217-9

Wicker, T., Sabot, F., Hua-Van, A., Bennetzen, J. L., Capy, P., Chalhoub, B. et al. (2007). A unified classification system for eukaryotic transposable elements. *Nature Reviews Genetics*, 8(12), 973–982. https://doi.org/10.1038/nrg2165

Xu, Y., Li, P., Zou, C., Lu, Y., Xie, C., Zhang, X. et al. (2017). Enhancing genetic gain in the era of molecular breeding. *Journal of Experimental Botany*, 68(11), 2641–2666. https://doi.org/10.1093/jxb/erx135

Yandell, M., and Ence, D. (2012). A beginner's guide to eukaryotic genome annotation. *Nat Rev Genet*, 13(5), 329–342. https://doi.org/10.1038/nrg3174

CHAPTER

4

Coffee, Caffeine, and Human Genetics

Metin Çalışkan
Usak University, Faculty of Medicine, Department of Medical Biology, Usak, Turkey

Coffee is a globally consumed beverage with complex and sometimes contradictory associations with human health. Beyond caffeine, coffee contains numerous bioactive compounds whose concentrations vary by bean type, roasting, and preparation, making simple intake measures (e.g., cups per day) an imprecise proxy for biological exposure. This chapter synthesises evidence from human genetics, epidemiology, and multi-omics to explain why coffee-related health outcomes differ markedly across individuals, populations, and endpoints. We frame coffee exposure using three complementary constructs: intake (habitual consumption behavior), internal dose (biologically proximate measures such as circulating caffeine and metabolites), and sensitivity (physiological response at a given dose). Genetic variation in caffeine metabolism (notably CYP1A2 and AHR) and pharmacodynamics (e.g., ADORA2A) contributes to heterogeneity in sleep, anxiety, cardiovascular responses, and disease-related phenotypes, and underpins the "caffeine paradox," whereby higher genetically predicted intake may coincide with faster caffeine clearance rather than higher exposure. Genome-wide association studies reveal a polygenic architecture integrating metabolic, neurobehavioral, and sensory pathways, with important ancestry-specific loci that limit the portability of genetic instruments. Across neurological, cardiometabolic, renal, cancer, and pregnancy outcomes, Mendelian randomisation generally indicates small average causal effects of coffee intake, while highlighting meaningful context-dependent and subgroup-specific associations. Multi-omic signatures, including DNA methylation, transcriptomics, metabolomics, and microbiome profiles, offer more refined exposure markers and mechanistic insights but require careful control for confounding, particularly smoking. Collectively, the evidence supports a context-dependent model of coffee

E-mail: metin.caliskan@usak.edu.tr

DOI: 10.1201/9781003689393-4

biology rather than uniform benefit or harm. We conclude that precision nutrition approaches—integrating genetics, biomarkers, life stage, and clinical context—are better suited than one-size-fits-all recommendations to guide coffee consumption for specific health goals.

1. Introduction

Coffee is one of the most widely consumed beverages worldwide and an important dietary source of caffeine. Observational studies have reported associations between coffee consumption and diverse health outcomes, including cardiovascular disease, cancer, mental health, renal function, and mortality, sometimes with J- or U-shaped patterns in which moderate intake appears comparatively favorable for selected endpoints (Cornelis, 2019). A central challenge is that coffee is a complex matrix exposure: alongside caffeine, it contains chlorogenic acids, diterpenes, trigonelline, melanoidins, and other bioactives whose concentrations depend on bean type, roasting, and brewing method. Consequently, "cups per day" may represent materially different biochemical exposures across individuals and settings.

To improve conceptual clarity across the chapter, we distinguish three constructs. Intake refers to habitual consumption behavior (e.g., cups/day, caffeinated vs. decaffeinated, brewing method). Internal dose refers to biologically proximate exposure measures (e.g., circulating or urinary caffeine and coffee-derived metabolites). Sensitivity describes the physiological response at a given internal dose (e.g., sleep disruption, anxiety, or blood pressure changes). This framework helps explain why similar intake can yield different internal exposure profiles and why similar internal doses can yield different responses.

Human genetics provides an additional lens for explaining variability in both intake and response. Coffee consumption shows partial heritability, indicating that habitual intake is shaped not only by environment and culture but also by inherited predispositions (Vink et al., 2009). GWAS consistently implicate loci involved in xenobiotic metabolism, particularly the CYP1A1–CYP1A2 cluster and AHR, supporting links between habitual consumption and caffeine clearance biology (Cornelis et al., 2016; Sulem et al., 2011). These findings align with a "genetic titration" model in which individuals may adjust intake to achieve desired effects, and they also motivate the "caffeine paradox": variants associated with higher coffee consumption can coincide with faster caffeine metabolism, complicating simple interpretations of intake as dose (Giontella et al., 2023; Woolf et al., 2024). Pharmacodynamic variation further contributes to heterogeneity; for example, ADORA2A-related differences in adenosine signaling have been linked to variability in caffeine-related sleep and anxiety phenotypes (Childs et al., 2008; Retey et al., 2007). Genetic architecture and portability also vary by ancestry, with stronger effects of certain loci (e.g., 12q24) reported in East Asian populations, which has implications for polygenic scores and causal inference across diverse cohorts (Nakagawa-Senda et al., 2018).

Multi-omic studies have reported molecular signatures associated with coffee exposure, including DNA methylation and biomarker patterns, although interpretation requires careful attention to confounding—particularly smoking, which often correlates with coffee consumption (Karabegovic et al., 2021). Because conventional observational associations can be vulnerable to confounding, reverse causation, and exposure misclassification, Mendelian randomisation (MR) is increasingly used to strengthen causal reasoning, while recognizing potential limitations related to instrument validity and pleiotropy (Cornelis and Munafo, 2018; Pham et al., 2025).

This chapter synthesises genetic, epidemiological, and multi-omic evidence to clarify why coffee's health associations vary by exposure definition, individual biology, ancestry, coffee type, and endpoint. Section 2 reviews genetic determinants of caffeine metabolism and sensitivity, including CYP1A2, AHR, and ADORA2A. Section 3 summarises the broader genetic architecture of coffee consumption and implications for GWAS, polygenic scores, and MR. Sections 4–6 evaluate neurologic, cardiometabolic/renal, and cancer outcomes, respectively. Section 7 synthesises multi-omic signatures of coffee exposure, Section 8 addresses pregnancy and developmental considerations, and Section 9 discusses translational implications for precision nutrition and future research directions.

2. Genetic basis of caffeine metabolism and pharmacodynamics

This section delves into genetic variations that influence (i) the internal caffeine dosage through pharmacokinetics and (ii) the physiological response via pharmacodynamics. These factors collectively elucidate why coffee consumption is an inadequate proxy for biological exposure and why individuals experience varying effects on sleep, anxiety, blood pressure, and other outcomes at similar consumption levels (Cornelis et al., 2016; Zhou et al., 2009).

2.1 CYP1A2, AHR, and pharmacokinetic variation

Caffeine clearance is mainly governed by hepatic cytochrome P450 1A2 (CYP1A2), which metabolises most ingested caffeine, primarily converting it to paraxanthine (Cornelis et al., 2016; Zhou et al., 2009). As a result, genetic variations in CYP1A2 and its upstream regulation significantly contribute to the differences in internal caffeine exposure among individuals with similar intake levels. A well-researched functional variant is CYP1A2 rs762551 (−163C>A), which affects CYP1A2 inducibility. Individuals with the AA genotype are often categorised as "faster metabolisers," while those with the C-allele (AC/CC) generally exhibit reduced inducibility and slower clearance (Sachse et al., 1999). Slower clearance leads to prolonged caffeine exposure in the bloodstream, whereas faster clearance results in a quicker conversion to paraxanthine and a shorter caffeine half-life under similar conditions (Giersch et al., 2018; Palatini et al., 2009). Notably, CYP1A2 activity is influenced by environmental factors such as smoking and hormonal status, necessitating the consideration of genotype effects within a gene–environment context (Djordjevic et al., 2010). CYP1A2 expression is significantly regulated by the aryl hydrocarbon receptor (AHR), a ligand-activated transcription factor. GWAS meta-analyses consistently identify variants near AHR (e.g., rs4410790 and correlated signals) as strong predictors of habitual coffee consumption (Coffee and Caffeine Genetics Consortium et al., 2015; Cornelis et al., 2011; Sulem et al., 2011).

Mechanistically, AHR regulates CYP1A1/CYP1A2 expression through xenobiotic response elements in their shared promoter region, providing a plausible biological link between AHR variation, caffeine clearance capacity, and habitual intake (Sulem et al., 2011). This supports the interpretation that a significant portion of "coffee-consumption genetics" is attributable to the pharmacokinetic handling of caffeine rather than preference alone (Cornelis et al., 2016).

2.2 Pharmacodynamic targets: ADORA2A and phenotype-specific sensitivity

Pharmacokinetics explains the duration and intensity of caffeine's presence in the bloodstream, while pharmacodynamics describes how the nervous system responds to a specific internal dose. Caffeine primarily acts by non-selectively blocking adenosine receptors, particularly A1 and A2A, with the A2A receptor—encoded by ADORA2A—playing a vital role in arousal, anxiety, and sleep regulation (Fredholm et al., 1999; Rogers et al., 2010). Among the most extensively studied variants, ADORA2A rs5751876 shows endpoint-specific associations. In anxiety-related contexts, individuals with the rs5751876 T/T genotype report increased anxiety after caffeine consumption, including interactions with dopaminergic variants like DRD2 (Alsene et al., 2003; Childs et al., 2008). Conversely, controlled laboratory studies on sleep physiology reveal a different pattern: rs5751876 C/C carriers experience more pronounced caffeine-induced changes in NREM sleep EEG measures (e.g., increased beta power), while T/T carriers show minimal changes under similar dosing conditions (Retey et al., 2007). This distinction underscores a key principle throughout the chapter: "caffeine sensitivity" is not a singular trait, and genetic effects vary depending on the phenotype being measured (e.g., anxiety versus objective sleep physiology). Beyond ADORA2A, large-scale studies have identified additional loci potentially linked to reinforcement and reward processes in coffee-related behaviors, yet the most consistent signals for habitual intake are rooted in metabolism and clearance pathways; broader intake genetics is detailed in Section 3 (Coffee and Caffeine Genetics Consortium et al., 2015; Cornelis et al., 2016).

2.3 Integrating behavioral and metabolic genetics: why intake is not dose

Coffee consumption represents a form of partially self-regulated exposure. Genome-wide association studies (GWAS) indicate that alleles associated with increased coffee intake often correspond to faster caffeine metabolism, reflecting behavioral adjustments to maintain a desired psychoactive state (Coffee and Caffeine Genetics Consortium et al., 2015; Cornelis et al., 2016; Cornelis et al., 2011). This finding has two significant implications. First, self-reported coffee intake is an imprecise measure of internal caffeine dosage. Second, genetic instruments for "coffee intake" may partly reflect caffeine clearance and tolerance rather than a consistent increase in biological exposure, which is crucial when interpreting Mendelian randomisation analyses (Cornelis, 2014; Taylor et al., 2018). Several GWAS loci also link coffee consumption to broader metabolic biology. Variants in the genes GCKR and MLXIPL, which are essential for glucose and lipid metabolism, have been associated with coffee consumption, suggesting a shared biological pathway connecting energy sensing, metabolic traits, and stimulant use (Coffee and Caffeine Genetics Consortium et al., 2015; Zhong et al., 2019). Additional loci, such as ABCG2 and POR, may refine the pharmacokinetic landscape; for instance, POR facilitates electron transfer necessary for CYP enzymatic activity (Cornelis et al., 2016). Since many of these loci are related to cardiometabolic traits, they also represent potential sources of horizontal pleiotropy in causal inference frameworks, underscoring the need for sensitivity analyses and triangulation across exposure definitions.

2.4 Fast and slow metabolisers: implications for health and interpretation

Stratifying individuals based on CYP1A2-related clearance has been proposed as a strategy to address the varied findings observed across different studies. The central hypothesis suggests that prolonged caffeine exposure in individuals with slower metabolism rates may exacerbate adverse sympathomimetic effects, such as increased blood pressure. Conversely, faster clearance rates might mitigate caffeine-related risks, allowing non-caffeine constituents, like polyphenols, to predominate in net associations in certain contexts (Cornelis et al., 2006; Guest et al., 2018; Palatini et al., 2009). Effect modification is not always consistent. Large biobanks often show no effect, highlighting the sensitivity of interaction claims to phenotype definitions, intake measurement errors, and factors like smoking that affect CYP1A2 activity (Loftfield et al., 2018; Zhou and Hypponen, 2019). In the following sections of this chapter, metaboliser status (pharmacokinetics) and receptor-level sensitivity (pharmacodynamics) are explored as complementary dimensions that can influence outcomes in varying directions, depending on the endpoint and context. This framework is then applied to analyze disease outcomes, multi-omic signatures, and translational considerations within the realm of precision nutrition.

3. Genetic architecture of coffee consumption and related behavioral traits

Coffee consumption is a complex behavioral trait with moderate heritability, influenced by the interplay of pharmacological feedback, sensory perception, reinforcement learning, and socio-environmental factors. While often perceived as a lifestyle choice, genetic evidence suggests that both the inclination to drink coffee and the typical amount consumed are partially governed by a polygenic framework. This framework integrates the metabolic processing of caffeine with neurobehavioral and sensory pathways (Cornelis et al., 2016; Cornelis et al., 2011). This section synthesises evidence from heritability studies, genome-wide association studies (GWAS), cross-ancestry analyses, and sensory genetics, and then discusses the implications for downstream disease research and Mendelian randomisation (MR).

3.1 Heritability and phenotype definition

Research consistently highlights a genetic component in coffee consumption, although estimates vary across different cohorts, cultures, and definitions of the phenotype

(Vink et al., 2009). Notably, "coffee consumption" is not a singular phenotype; it encompasses beverage preference (coffee versus tea), frequency, quantity, the choice between caffeinated and decaffeinated options, and even high-intensity consumption that may suggest tolerance and dependence-like traits. Evidence indicates that heavier caffeine use patterns might be influenced by distinct genetic factors compared to moderate use; for instance, high-volume caffeine consumption has been reported to exhibit significantly higher heritability than typical intake in certain twin studies (Kendler et al., 2008). Heritability also appears to differ across various beverages, with some studies suggesting stronger genetic regulation for coffee preference compared to tea, highlighting the interaction between cultural availability, learned preferences, and biological predisposition (Luciano et al., 2005). These considerations necessitate methodological caution throughout this chapter: genetic signals and causal inferences will depend on how "coffee exposure" is operationalised.

3.2 Early GWAS and the core metabolic loci

The shift from classical twin studies to molecular genetics gained momentum with the advent of genome-wide association studies (GWAS), which identified reproducible loci associated with habitual coffee consumption. Initial GWAS conducted among populations of European ancestry consistently pinpointed two major loci that reached genome-wide significance: the CYP1A1–CYP1A2 region (15q24) and AHR (7p21) (Cornelis et al., 2011; Sulem et al., 2011). These findings are biologically consistent with the pharmacokinetic processing of caffeine, as discussed in Section 2: CYP1A2 is the primary hepatic enzyme responsible for caffeine metabolism, and AHR regulates the expression of CYP1A2. The recurrence of these loci across studies supports the interpretation that a significant portion of the genetics underlying coffee consumption reflects caffeine clearance capacity and behavioral adaptation to pharmacological feedback (Cornelis et al., 2016). Subsequent meta-analyses expanded the catalog of loci. A large consortium meta-analysis identified additional loci, including POR and ABCG2 (involved in xenobiotic metabolism/transport), as well as neurobehavioral and metabolic loci such as BDNF, SLC6A4, GCKR, and MLXIPL (Coffee and Caffeine Genetics Consortium et al., 2015; Cornelis, 2015). POR and ABCG2 further support a pharmacokinetic basis for consumption, while BDNF and SLC6A4 suggest the involvement of reward and mood-related neurobiology in habitual intake (Cornelis, 2015). Additional signals reported in some studies (e.g., NRCAM and ULK3) further link coffee consumption to neural development and addiction vulnerability pathways (Amin et al., 2012). Collectively, these findings support a polygenic model in which metabolic clearance, reinforcement, and preference-related processes jointly influence consumption behavior.

3.3 Cross-ancestry architecture and ancestry-specific loci

Cross-ancestry analyses reveal significant population differences in the genetic architecture of coffee consumption. While associations at CYP1A2 and AHR are consistently replicated, genome-wide association studies (GWAS) in East Asian cohorts consistently identify a notable signal at 12q24, implicating loci such as ALDH2 and HECTD4 (Kim et al., 2025; Nakagawa-Senda et al., 2018; Unohara et al.,

2025). These ancestry-specific effects have important biological and epidemiological implications: ALDH2 variation is closely linked to alcohol metabolism and drinking behavior, and the broader 12q24 region has been associated with multiple cardiometabolic traits. Consequently, genetic instruments and polygenic scores derived from European-ancestry GWAS may not be directly applicable to East Asian populations (and vice versa) due to differences in allele frequencies and ancestry-specific linkage disequilibrium structure (Nakagawa-Senda et al., 2018). This limits the portability of Mendelian randomisation (MR) instruments and the generalisability of gene–environment interaction findings across ancestries. Notably, HECTD4 variants such as rs2074356—common in East Asians but rare elsewhere—are associated not only with coffee intake but also with obesity- and metabolic syndrome–related traits, consistent with pleiotropic pathways that may differ from those typically observed in Western cohorts (Kim et al., 2025). Collectively, these findings highlight the need for ancestry-tailored instruments and risk scores to enhance behavioral prediction and causal inference in diverse populations.

3.4 Polygenic scores, genetic correlations, and common liability

With advancements in identifying genetic loci, polygenic scores (PGS) for coffee consumption have facilitated cross-trait analyses, situating coffee drinking within a broader behavioral and metabolic context. Recent studies have revealed genetic correlations between coffee intake and other substance use traits, such as smoking initiation and alcohol consumption, which align with a "common liability" framework. This framework may suggest shared neurobiological pathways that regulate stimulant seeking, reward learning, and habitual behavior (Thorpe et al., 2024; Yin et al., 2022). These shared liabilities have two significant implications. First, they elucidate why observational associations between coffee consumption and disease can be challenging to interpret: coffee intake often coincides with other health-related behaviors, and genetic overlap may reflect this clustering. Second, they highlight the risk of horizontal pleiotropy in Mendelian Randomisation (MR), particularly when coffee-associated loci also influence metabolic traits independently of coffee exposure. Consistent with this, variants linked to increased coffee consumption have been associated in some datasets with higher BMI or adverse lipid profiles, patterns that may reflect correlated behaviors (e.g., smoking) or broader metabolic biology rather than direct effects of coffee (Thorpe et al., 2024; Zhong et al., 2019). Further research using clustering frameworks suggests that heterogeneity within obesity phenotypes can modify the relationships between genetic profiles and dietary habits, emphasizing that cross-trait signals require careful interpretation rather than simplistic causal conclusions (Huang et al., 2022). Overall, PGS analyses reveal that coffee consumption is embedded within interconnected behavioral and metabolic networks, which aid in understanding co-occurrence but necessitate caution in causal inference.

3.5 Sensory pathways and learned preferences

An intriguing yet paradoxical aspect of coffee genetics lies in taste perception. Coffee is inherently bitter, a sensory trait that typically signals toxicity and would usually

Table 1: Key genetic determinants of caffeine metabolism, consumption behaviour, and sensitivity

Gene/Locus	Primary biological function	Phenotypic relevance	Key references	Maps primarily to (Intake/Internal dose/Sensitivity)
		Pharmacokinetics (Metabolism)		
CYP1A2 (15q24)	Encodes Cytochrome P450 1A2, the primary hepatic enzyme responsible for >95% of caffeine metabolism.	Determines "fast" (AA) vs. "slow" (AC/CC) metaboliser status; modifies risk for hypertension and myocardial infarction.	Cornelis et al., 2016; Sachse et al., 1999	Internal dose (PK clearance/half-life)
AHR (7p21)	Encodes Aryl Hydrocarbon Receptor, a transcription factor regulating *CYP1A2* expression.	Strongest GWAS signal for habitual coffee intake; regulates caffeine clearance capacity.	Sulem et al., 2011; Cornelis et al., 2011	Internal dose (PK regulation/clearance)
POR/ABCG2	Electron transfer for P450 enzymes (*POR*) and xenobiotic transport (*ABCG2*).	Associated with habitual consumption and caffeine metabolite ratios.	Coffee and Caffeine Genetics Consortium et al., 2015	Internal dose (PK network/transport)
Pharmacodynamics	(Sensitivity)			
ADORA2A (22q11)	Encodes Adenosine A2A Receptor, the primary target for caffeine's psychostimulant effects.	Modulates susceptibility to caffeine-induced anxiety and objective sleep disturbance (EEG beta power).	Retey et al., 2007; Childs et al., 2008	Sensitivity (PD response)
		Ancestry-Specific Loci		
12q24 (*ALDH2*, *HECTD4*)	Alcohol metabolism and ubiquitin-protein ligase pathways.	Primary determinant of coffee intake in East Asian populations; pleiotropic links to obesity and metabolic syndrome.	Nakagawa-Senda et al., 2018; Kim et al., 2025; Unohara et al., 2025	Intake (behavioral/ancestry-specific) + pleiotropy risk
		Sensory and Behavioral		
TAS2R Family	Bitter taste perception.	Paradoxical positive association: higher bitterness sensitivity correlates with higher coffee intake (conditioned preference).	Cornelis and van Dam, 2021; Ong et al., 2018	Intake (sensory preference)

deter consumption. However, genetic studies reveal that individuals genetically predisposed to perceive greater bitterness from caffeine—due to variations in bitter taste receptor genes like the TAS2R family—tend to consume more coffee. In contrast, variations linked to increased sensitivity to other bitter substances, such as quinine or PROP, are associated with lower consumption (Cornelis and van Dam, 2021; Ong et al., 2018). This phenomenon supports a "conditioned taste preference" model, where the psychoactive reinforcement of caffeine can override the natural aversion to bitterness, leading individuals to associate the specific taste profile of caffeine-containing drinks with desirable stimulatory effects (Cornelis and van Dam, 2021). Beyond taste, variations in olfactory receptor genes, such as those in the OR family noted in some GWAS, suggest that the multisensory integration of smell and taste may influence coffee preference and habitual consumption (Cornelis, 2015; Cornelis and van Dam, 2021). Thus, sensory genetics complements metabolic genetics by explaining why preference can increase despite the presence of bitterness.

3.6 Summary and implications for downstream disease studies

The genetic architecture of coffee consumption is characterised by its polygenic nature and mechanistic diversity, which includes (i) pharmacokinetic loci associated with caffeine clearance (e.g., CYP1A2, AHR, POR, ABCG2), (ii) neurobehavioral loci potentially linked to reinforcement and mood regulation (e.g., BDNF, SLC6A4, NRCAM), and (iii) sensory perception loci (e.g., TAS2R variants), with notable variations across different ancestries (Amin et al., 2012; Coffee and Caffeine Genetics Consortium et al., 2015; Cornelis et al., 2016; Nakagawa-Senda et al., 2018). These findings provide direct insights for Mendelian Randomisation (MR) and genetic epidemiology in subsequent sections. Utilizing coffee-intake variants as instruments can mitigate confounding from measured environmental factors; however, the validity of these instruments hinges on their specificity and sensitivity analyses, particularly given the pleiotropy risks from metabolic loci and ancestry-specific factors such as ALDH2 (Cornelis and Munafo, 2018). Therefore, interpreting MR findings in downstream disease sections requires considering what the underlying instruments likely represent—habitual intake, clearance capacity, reinforcement biology, or correlated behaviors—rather than presuming that "genetically predicted coffee consumption" constitutes a uniform biological exposure. This perspective also underscores the importance of refined exposure phenotypes (e.g., caffeinated vs. decaffeinated, preparation method, and biomarker-based measures) and multi-ancestry evidence as essential prerequisites for globally applicable and mechanistically credible nutritional conclusions. Key genetic loci involved in caffeine metabolism, consumption behaviour, and sensitivity are summarised in Table 1.

4. Coffee, human genetics, and neurological outcomes

Neurological phenotypes serve as an effective test case for genetically informed coffee research, as they include both (i) long-latency disease endpoints, such as neurodegeneration, and (ii) short-latency, pharmacodynamically proximal traits, like sleep and anxiety. Utilizing the previously established exposure framework (Sections 1–3), the primary question is not whether coffee is inherently "good" or "bad" for the brain. Rather, it concerns which neurological endpoints are plausibly influenced by coffee-related biology, under specific genetic backgrounds, and through particular exposure definitions (Cornelis and Munafo, 2018; Pham et al., 2025).

4.1 Parkinson's disease: susceptibility versus age at onset

In observational cohort studies, a higher intake of coffee, particularly caffeinated coffee, has consistently been associated with a reduced risk of Parkinson's disease (PD). This observation has spurred the development of mechanistic hypotheses centered on adenosine A2A receptor signaling within the basal ganglia circuitry (Nehlig, 2018). Gene–environment interaction studies suggest etiological heterogeneity. For example, the GRIN2A gene has been shown to modify the relationship between coffee consumption and PD, indicating a potential role for glutamatergic signaling in how caffeine-related exposures influence PD-relevant phenotypes (Hamza et al., 2011). Conversely, interactions involving caffeine-metabolism loci have been less consistently replicated, highlighting broader methodological challenges. These challenges include the fact that self-reported intake is an imperfect measure of internal caffeine exposure and that interaction estimates can be sensitive to phenotype definition, confounding structure, and measurement error (Cornelis and Munafo, 2018). Notably, the interpretation of coffee's neuroprotective role in PD has shifted from prevention to disease modification. While observational studies often suggest reduced PD susceptibility, recent Mendelian randomisation analyses using large genetic datasets indicate that genetically predicted higher coffee consumption is associated with a delayed age at onset of PD rather than a reduction in lifetime risk (Kuzovenkova et al., 2025; Senkevich et al., 2025). This distinction is consistent with a disease-modifier model where caffeine-mediated adenosine receptor antagonism delays clinical manifestation—by shifting symptom thresholds or attenuating early expression—without necessarily preventing the underlying neurodegenerative process.

4.2 Cognitive decline and dementia: Inconsistent causality signals

Compared to Parkinson's disease, research on coffee consumption and dementia shows greater methodological diversity. Observational studies often suggest a reduced risk of dementia or Alzheimer's disease associated with moderate coffee consumption. However, interpreting these associations is challenging due to the clustering of coffee intake with broader lifestyle patterns, frequent measurement errors, and the potential for reverse causality, such as prodromal cognitive changes affecting consumption habits (Cornelis and Munafo, 2018). Mendelian randomisation (MR) studies, which use genetic instruments to examine coffee consumption, have not provided substantial evidence supporting the hypothesis that coffee protects against dementia. Recent reviews frequently describe the evidence as ambiguous or inconsistent (Pham et al., 2025). Genetically informed analyses incorporating neuroimaging phenotypes offer additional insights but do not definitively resolve the causal question. Some MR studies have linked genetically predicted higher coffee consumption with less favorable structural brain measures, such as reduced gray matter volume. This may reflect aspects captured by intake instruments—such as clearance/titration behavior or correlated behavioral liabilities—rather than a sustained neuroprotective exposure effect (Cornelis and Munafo, 2018). These findings do not contradict the well-established acute alerting effects of caffeine; rather, they highlight that short-term cognitive performance, long-term neurodegeneration, and brain structural measures do not necessarily covary and may be influenced by distinct exposure constructs and confounding patterns. Another limitation pertains to phenotype specificity and the heterogeneity of background risk. The trajectories of dementia and its genetic architecture differ across baseline risk strata, suggesting that any effects related to coffee consumption, if they exist, may only be observable in specific subgroups that are often underrepresented in population-average MR designs (Acharya et al., 2025). For instance, genotype-stratified analyses indicate that associations may vary according to APOE status, with potentially distinct patterns among APOE ε4 carriers, underscoring the necessity for adequately powered studies focused on subgroups (Acharya et al., 2025). Overall, the most defensible conclusion remains cautious: current genetically informed evidence does not support a general assertion that coffee consumption prevents dementia. Future conclusions will likely depend on refined definitions of exposure (e.g., caffeinated versus decaffeinated coffee; biomarker-based internal dose), enhanced safeguards against reverse causality, and meticulous assessment of pleiotropy (Cornelis and Munafo, 2018; Pham et al., 2025).

4.3 Sleep and anxiety: Genetically stratified pharmacodynamic effects

The most reliable genetically informed indicators of caffeine consumption are found in neuropsychiatric phenotypes closely tied to pharmacodynamics, particularly sleep and anxiety. In this context, variations at the receptor level in adenosine signaling are directly relevant, aligning with the framework established in Section 2: genotype effects can be endpoint-specific, meaning that sleep sensitivity and anxiety sensitivity may not necessarily coincide within the same individual (Childs et al., 2008; Retey et al., 2007). Experimental sleep studies confirm that ADORA2A variation stratifies sensitivity to caffeine-induced changes in objective sleep physiology, while anxiety paradigms suggest overlapping but not identical patterns of association, including reported interactions with dopaminergic variants such as DRD2 (Childs et al., 2008; Retey et al., 2007). Practically, this indicates that genetically mediated susceptibility may alter thresholds—determining who experiences insomnia or anxiety at a given exposure—without necessitating significant effects on long-latency disease endpoints. This also explains why population-average studies may yield modest or inconsistent associations for broad psychiatric diagnoses: coffee-intake genetics encompasses multiple biological and behavioral components, and the most direct effects may manifest in acute or intermediate phenotypes rather than categorical disorders (Cornelis and Munafo, 2018; Pham et al., 2025).

4.4 Synthesis

Three integrative points emerge concerning neurological outcomes. Firstly, selecting appropriate endpoints is crucial: genetically informed evidence is most persuasive for phenotypes that closely match pharmacodynamic actions, such as sleep and anxiety, and potentially for symptom timing, like the age of onset in Parkinson's disease (PD). In contrast, evidence supporting broad prevention claims, such as the incidence of PD and dementia, often appears weaker or inconsistent (Kuzovenkova et al., 2025; Pham et al., 2025; Retey et al., 2007). Secondly, heterogeneity is anticipated: neurological effects, when present, are likely influenced by genetic background and the definition of "exposure," whether it pertains to intake, internal dose, or sensitivity (Cornelis and Munafo, 2018). Thirdly, these observations necessitate a consistent methodological standard throughout the chapter: neurological conclusions should be triangulated across various exposure definitions and study designs, and causal claims should be conservatively framed when instruments plausibly act as proxies for clearance/titration or broader behavioral liabilities rather than a uniform biological dose (Cornelis and Munafo, 2018).

5. Coffee, cardiometabolic disease, and kidney outcomes in a genetic framework

The cardiometabolic and renal outcomes are influenced by both the immediate hemodynamic effects of caffeine in coffee and its longer-term metabolic effects, which are attributed to its matrix components, such as polyphenols and other non-caffeine constituents. Genetic research highlights two consistent themes within this body of literature: (i) the average causal effects of "coffee intake" are generally modest, and (ii) variability is expected, as the same reported intake may lead to different internal exposures and physiological responses, depending on genetic background, ancestry, and the specific endpoint under consideration.

5.1 Caffeine metabolism, blood pressure, and myocardial infarction

A consistent finding in cardiovascular epidemiology is that CYP1A2-related caffeine metabolism influences risk at higher intake levels. In a classic case–control study, heavy coffee consumption was linked to an increased risk of nonfatal myocardial infarction, particularly among slow caffeine metabolisers, while fast metabolisers showed no significant increase and sometimes even a neutral-to-favorable pattern (Cornelis et al., 2006). Similarly, prospective studies have indicated that coffee intake is associated with a higher risk of hypertension, mainly among slow metabolisers (Palatini et al., 2009). Mechanistically, these patterns support the hypothesis that prolonged caffeine exposure may amplify sympathetic and pressor effects in susceptible individuals, whereas faster clearance might reduce these acute effects and highlight any beneficial components of the coffee matrix. Importantly, these interaction studies should be viewed as risk-stratification signals, rather than as evidence that coffee is uniformly harmful or beneficial. They illustrate that cardiovascular outcomes may reflect the balance between (i) caffeine persistence and (ii) non-caffeine constituents and lifestyle context. They also underscore why "cups/day" is an inadequate proxy for biological dose.

5.2 Mendelian randomisation and global cardiovascular risk

When the focus shifts from interaction to average causal effects using Mendelian randomisation (MR), the primary conclusion is that genetically predicted coffee consumption does not significantly cause ischemic cardiovascular disease in the general population (Nordestgaard and Nordestgaard, 2016; Yuan et al., 2021). MR studies on various heart-related conditions often reveal negligible or non-existent effects. For example, MR investigations into heart failure have not established a definitive causal relationship (van Oort et al., 2020). Genetically predicted coffee consumption is generally not associated with an overall risk of stroke, although some analyses suggest a weak or subtype-specific association, such as with small-vessel ischemic stroke, which remains uncertain (Qian et al., 2020). These findings support a model suggesting that coffee is unlikely to be a dominant causal factor for major cardiovascular events at the population level, while still allowing for meaningful heterogeneity within subgroups. Two additional considerations are relevant for interpretation. Firstly, MR instruments for "coffee intake" may partially act as proxies for metabolic traits and behavioral titration, potentially complicating causal interpretation. Secondly, the method of coffee preparation can alter cardiometabolic risk pathways: unfiltered coffee may elevate LDL cholesterol through diterpenes (cafestol and kahweol), potentially counteracting benefits in lipid-related pathways (Mendoza et al., 2023). Therefore, even if the average causal signal is minimal, actual cardiovascular profiles may still vary according to genotype, comorbidity (e.g., hypertension), and coffee type.

5.3 Obesity, metabolic syndrome, and type 2 diabetes

Observational epidemiological studies often indicate an inverse relationship between coffee consumption and conditions such as obesity, metabolic syndrome, and type 2 diabetes (T2D) when examining metabolic outcomes. However, analyses that incorporate genetic data provide a more nuanced perspective. Within a comprehensive Mendelian randomisation framework, genetically predicted coffee consumption does not establish a definitive causal link with obesity or T2D, suggesting that the observed benefits may partly arise from correlated behaviors and lifestyle clustering rather than coffee itself (Nordestgaard and Nordestgaard, 2016; Nordestgaard et al., 2015; Yuan et al., 2021). This does not negate the potential biological effects of coffee constituents; rather, it implies that the protective effects at the population level are likely modest and susceptible to confounding. Simultaneously, gene–environment interactions suggest that metabolic effects may be context-dependent. In the POUNDS LOST trial, individuals with a higher genetic predisposition to coffee consumption experienced more favorable changes in insulin resistance markers under specific dietary conditions, particularly within a low-fat dietary context (Han et al., 2017). Conversely, in genetically susceptible individuals, increased coffee consumption has been linked to worsened insulin resistance, indicating that coffee (or caffeine) may not be uniformly beneficial across different metabolic risk strata (Cornelis and Munafo, 2018). Evidence from East Asian cohorts further supports heterogeneity driven by ancestry-linked loci and metabolic background. In Korean studies, coffee consumption has shown differential associations with diabetes risk depending on diabetes-related variants (e.g., IGF2BP2, CDKAL1), and variants at 12q24 have been reported to modify associations with pre-diabetes/T2D in relation to black coffee consumption (Jin et al., 2020; Lee et al., 2015). Regarding obesity, variants such as HECTD4 rs2074356 have been reported to modify the coffee–obesity relationship, with the direction of risk depending on genotype (Kim et al., 2024). Finally, epigenetic signals (e.g., coffee-associated methylation changes near metabolic loci such as PHGDH) suggest plausible regulatory links between coffee exposure and hepatic lipid metabolism, although these remain mechanistic clues rather than direct evidence of clinical benefit (Karabegovic et al., 2021). Overall, the genetic evidence supports a "small average effect, large context" model: coffee is unlikely to serve as a stand-alone preventive measure for T2D or obesity, but it may interact with dietary composition, baseline risk, and ancestry-linked genetic architecture in ways that are significant for specific subgroups.

5.4 Non-alcoholic fatty liver disease and chronic liver disease

Liver outcomes are a major focus of observational research on coffee consumption, consistently indicating lower liver enzyme levels and a reduced prevalence of NAFLD and fibrosis among coffee drinkers. However, genetic causal inference remains uncertain. Mendelian randomisation (MR) analyses have not consistently demonstrated a definitive protective causal effect of genetically predicted coffee consumption on NAFLD risk (Zhang et al., 2021). One interpretation suggests that while coffee may have a modest impact, the primary causal factors influencing liver fat and progression are upstream metabolic determinants—particularly adiposity, insulin resistance, and related traits—making coffee's effects easily confounded or mediated. From a genetic perspective, the most defensible synthesis is twofold: (i) coffee may be compatible with liver health and could plausibly support hepatoprotective pathways, but (ii) the primary causal mechanisms for NAFLD and chronic liver disease are rooted in the metabolic architecture of obesity and insulin resistance, which may overshadow any coffee-specific contribution in MR analyses.

5.5 Kidney

Renal endpoints underscore the critical need for precise exposure definitions. Extensive genetic analyses indicate that genetically predicted higher coffee consumption might be associated with a reduced risk of chronic kidney disease (CKD) and more favorable estimated glomerular filtration rate (eGFR)-related outcomes, such as decreased albuminuria in specific contexts (Kennedy et al., 2020). However, genetically informed research differentiates between behavioral "coffee/caffeine intake" and internal caffeine exposure: while genetically predicted higher coffee consumption may correlate with enhanced renal function, genetically predicted elevated plasma caffeine levels might be linked to adverse renal profiles. This supports a model suggesting that caffeine accumulation, rather than intake itself, could be detrimental to susceptible individuals (Giontella et al., 2023). Pharmacogenetic evidence further suggests that increased coffee intake may be associated with albuminuria and hyperfiltration, particularly among slow metabolisers (Mahdavi et al., 2023). Beyond CYP1A2, kidney-related interactions appear to vary by ancestry and endpoint choice. In the Japanese J-MICC cohort, coffee intake demonstrated varied associations with eGFR and CKD prevalence across coffee-intake–related variants in AHR, HECTD4, and CYP1A2, with modest effect sizes in a cross-sectional framework (Unohara et al., 2025). Urinary metabolomics offers an additional perspective on exposure. In bidirectional Mendelian randomisation (MR) analyses of urinary metabolites, genetically instrumented higher urinary trigonelline and quinic acid concentrations—metabolites related to coffee—were linked to higher eGFR but also a higher urinary albumin-to-creatinine ratio (UACR), resulting in an internally discordant renal signature whose mechanism remains unclear (Valo et al., 2025). Notably, instrumental variants for these metabolites were mapped to the AHR locus and exhibited strong linkage disequilibrium with known coffee/caffeine intake variants, suggesting that part of the signal may reflect shared coffee-consumption biology and/or pleiotropy rather than a metabolite-specific causal pathway. Finally, for kidney stone disease, MR evidence supports a more straightforward protective relationship for coffee and caffeine (Yuan and Larsson, 2022), making nephrolithiasis one of the renal outcomes with comparatively stronger causal support.

5.6 Overall synthesis

The integration of extensive observational cohorts with modern genetic epidemiology has resulted in a complex—and sometimes conflicting—understanding of the health implications of regular coffee consumption. Coffee is a significant dietary source of bioactive compounds, including chlorogenic acids, diterpenes, and minerals like magnesium, which may plausibly influence metabolic and inflammatory pathways. However, translating these mechanistic possibilities into population-level disease prevention remains contentious. A primary tension exists between traditional observational epidemiology and causal inference frameworks. Observational studies have consistently reported inverse associations between coffee intake and various chronic outcomes, such as type 2 diabetes, cardiovascular mortality, and several site-specific cancers (Ding et al., 2014; Grosso et al., 2017). Nonetheless, these associations are highly susceptible to residual confounding by lifestyle factors—most notably smoking, diet quality, and physical activity—that often cluster with coffee consumption patterns. Mendelian randomisation (MR) offers a more rigorous test of causality and frequently results in less optimistic conclusions. Recent systematic syntheses of MR studies reveal frequent discrepancies between observational and genetic evidence for cardiometabolic and neoplastic outcomes, with many MR analyses failing to support a clear protective causal effect of genetically predicted coffee consumption (Nordestgaard and Nordestgaard, 2016; Pham et al., 2025; Yuan et al., 2021). This pattern aligns with the possibility that a significant portion of the observational "benefit" reflects confounding, reverse causality, or exposure misclassification rather than a direct therapeutic effect of coffee itself. Notable exceptions may exist for liver-related endpoints, where causal support has been comparatively stronger in certain parts of the literature (Kennedy et al., 2017). Overall, for many extra-hepatic conditions, coffee may function more reliably as a correlate of broader behavioral and metabolic profiles than as a stand-alone preventive intervention. Future research should prioritise triangulation across designs—integrating longitudinal phenotyping, improved exposure definitions (e.g., caffeine vs. non-caffeine constituents; biomarker-based internal dose), and well-powered multi-ancestry genetic instruments—to better distinguish coffee's biological effects from the environmental noise of human behavior.

6. Coffee, human genetics, and cancer outcomes

6.1 Overview: Observational patterns and genetic instruments

The relationship between coffee consumption and cancer risk has been extensively researched, yet determining causality remains complex. In observational epidemiology, coffee intake is associated with reduced risks for certain cancers, while for others, the risks are either increased or negligible. These associations often exhibit non-linear (J- or U-shaped) dose–response patterns, which are commonly interpreted as comparatively favorable outcomes at moderate intake, although residual confounding and exposure misclassification may contribute (Nordestgaard and Nordestgaard, 2016). Distinguishing these patterns from lifestyle habits can be challenging. Coffee consumption is linked to smoking, alcohol use, dietary habits, and other behaviors. Smoking, in particular, can confound results even after adjustments, complicating the isolation of coffee's specific effects (Loftfield et al., 2018). Genetic epidemiology, including Mendelian randomisation (MR), is increasingly used as a complementary tool in this field. Genome-wide association studies (GWAS) have identified consistent loci linked to coffee consumption, notably around CYP1A2 and AHR (Cornelis et al., 2011; Sulem et al., 2011). MR uses these loci as instruments to estimate effects that, under standard assumptions, are typically less susceptible to confounding than conventional observational estimates. However, careful consideration is needed to understand the biological implications of these instruments. Recent MR research indicates that a genetic predisposition to increased coffee consumption may correspond to reduced circulating caffeine levels, suggesting accelerated caffeine clearance. This distinction can meaningfully shape the expected direction of effects, depending on whether a cancer-related mechanism is primarily caffeine-driven or reflects non-caffeine components within the coffee matrix (Giontella et al., 2023; Woolf et al., 2024). In addition, several loci associated with coffee consumption are linked to broader metabolic pathways (e.g., glucose and lipid metabolism), raising the possibility of horizontal pleiotropy if variants affect cancer risk through pathways independent of coffee (Pham et al., 2025; Yuan and Larsson, 2020). Accordingly, observational evidence and genetically informed analyses should be interpreted together, with explicit attention to instrument biology and potential pleiotropy.

6.2 Hormonally sensitive cancers: Breast, endometrial, and ovarian

In the field of hormone-sensitive cancers, proposed mechanisms include the modulation of estrogen metabolism, insulin sensitivity, and inflammatory signaling. A particularly insightful line of evidence emerges from high-risk genetic subgroups, where gene–environment interactions may be more apparent than in population-average studies. Among BRCA1/2 mutation carriers, observational and pharmacogenetic studies have suggested that coffee consumption might be linked to a reduced risk of breast cancer in certain contexts. For instance, Nkondjock et al. (2006) found that coffee intake could be associated with a decreased risk of breast cancer among BRCA1 mutation carriers (Nkondjock et al., 2006). Kotsopoulos et al. (2007) further noted that a protective association was evident in BRCA1 carriers with the C allele of CYP1A2, supporting the hypothesis that slower caffeine clearance and/or prolonged exposure to coffee-derived bioactives may activate protective pathways during carcinogenesis (Kotsopoulos et al., 2007). These findings are best viewed as evidence of effect modification: the same intake may have different biological implications depending on metabolism-related genotype in a high-risk genetic background. In contrast, MR analyses in general-population datasets have not supported a clear causal protective effect of genetically predicted coffee consumption on breast cancer risk or on estrogen receptor–defined subtypes (Ellingjord-Dale et al., 2021). A coherent interpretation is that population-average MR estimates may not capture subgroup-specific effects (e.g., BRCA carriers), and that the genetic instruments used may not accurately map to the biologically relevant exposure dimension for hormonally mediated carcinogenesis. Regarding ovarian cancer, candidate gene interaction hypotheses have been proposed. Kotsopoulos et al. (2009) suggested that CYP19 polymorphisms could modify the coffee–ovarian cancer relationship, implying a potential role for steroidogenesis-related pathways (Kotsopoulos et al., 2009). However, MR analyses have not established a causal relationship between coffee consumption and epithelial ovarian cancer (Ong et al., 2018). For endometrial cancer, observational evidence often suggests that coffee consumption is associated with a lower risk, frequently interpreted through obesity and insulin-related mechanisms. Improved insulin sensitivity and increased sex hormone–binding globulin (SHBG) levels have been discussed as plausible protective pathways (Kunutsor et al., 2025; Zhou et al., 2015). Nevertheless, systematic syntheses emphasise that causal evidence remains unclear and that the apparent protective association may be partially explained by confounding, particularly by obesity and related metabolic factors (Pham et al., 2025). Overall, the most defensible conclusion for hormonally sensitive cancers is that any benefit, if present, is likely modest, context-dependent, and potentially most relevant in genetically defined high-risk groups rather than at the population average.

6.3 Digestive and hepatobiliary cancers

Among digestive and hepatobiliary outcomes, liver cancer—particularly hepatocellular carcinoma (HCC)—shows one of the more consistently reported inverse associations with coffee consumption. This pattern is biologically plausible, given that the liver is the primary site for metabolizing numerous coffee constituents and that coffee has been linked to anti-inflammatory and antifibrotic effects relevant to chronic liver disease progression. Mendelian randomisation (MR) analyses in some settings, including East Asian datasets, have reported that genetically predicted coffee consumption is associated with a lower risk of HCC (Deng et al., 2022), although interpretation remains contingent on instrument validity and pleiotropy.

Proposed mechanisms have included reductions in liver enzyme levels and potential effects on cirrhosis-related pathways, consistent with the etiological link between cirrhosis and HCC. Epigenetic studies further suggest that coffee consumption is associated with DNA methylation differences at loci involved in liver metabolism, such as PHGDH, providing mechanistic insights into how coffee-related exposures might intersect with hepatic lipid biology (Karabegovic et al., 2021). In the case of pancreatic cancer, historical observational concerns have largely been attributed to confounding factors, particularly smoking. More recent prospective studies and MR analyses have generally not supported a causal association between coffee consumption and pancreatic cancer risk. In an MR study incorporating European and East Asian datasets, Lu et al. (2024) reported no significant association between genetically predicted coffee consumption and pancreatic cancer (Lu et al., 2024). Esophageal cancer is a notable exception, as genetic analyses have reported a positive association in some datasets. Using UK Biobank data, Carter et al. (2022) found that genetically predicted higher coffee consumption was associated with an increased risk of esophageal cancer (reported OR = 2.79) (Carter et al., 2022; Pham et al., 2025). A key interpretive point is that this signal is plausibly driven by beverage temperature habits rather than the chemistry of coffee itself: very hot drinks can cause thermal injury to the esophageal mucosa and have been linked to carcinogenesis pathways. Accordingly, this association should be framed as a risk signal that is mechanistically compatible with hot beverage exposure and may reflect aspects of drinking behavior captured by instruments, rather than evidence that coffee constituents are intrinsically carcinogenic. For colorectal cancer, observational literature often suggests a protective association, but genetic evidence has generally not supported causality. He et al. (2024) reported no causal relationship for colorectal cancer in an MR framework (He et al., 2024). Collectively, these findings support an organ-specific model: liver outcomes show the strongest evidence consistent with benefit, whereas for several other digestive cancers, effects are neutral, inconsistent, or contingent on specific mechanisms such as beverage temperature.

6.4 Urologic and hematologic malignancies

Urologic cancers, including prostate, bladder, and kidney cancers, highlight a recurring theme in coffee–cancer research: the inconsistency between observational associations and genetic causality analyses. Observational studies often report an inverse relationship between coffee consumption and prostate cancer, particularly in advanced or fatal cases, frequently linked to effects on insulin metabolism and sex hormones. However, genetic epidemiology offers a different perspective. In a Mendelian randomisation (MR) study using data from the PRACTICAL consortium, Taylor et al. (2017) found no causal link between genetically predicted coffee consumption and the risk or progression of prostate cancer (Taylor et al., 2017). This discrepancy suggests that the observed protective effect might be due to confounding, selection effects, or exposure dimensions not fully captured by current instruments (e.g., preparation method, additives, or related health behaviors). Bladder cancer is especially susceptible to residual confounding because smoking, a major risk factor, is behaviorally linked to coffee consumption. Observational reports of increased bladder cancer risk among coffee drinkers are often attributed to incomplete control for smoking. Additionally, susceptibility may vary based on detoxification genetics. Covolo et al. (2008) found that variants in detoxification genes like GSTP1 could alter the relationship between coffee consumption and bladder cancer risk, with high consumption increasing risk in certain genotypes, independent of CYP1A2-mediated caffeine metabolism (Covolo et al., 2008). This underscores the broader point that coffee can interact with carcinogenesis through multiple pathways (detoxification and oxidative stress biology, not just caffeine kinetics). For kidney cancer, MR analyses generally show no causal link between genetically predicted coffee consumption and risk (Kunutsor et al., 2025; Rhee et al., 2021). Evidence for hematologic malignancies is relatively limited; however, recent MR work has identified a hypothesis-generating signal for multiple myeloma. Carter et al. (2022) reported that genetically predicted high coffee consumption was positively associated with multiple myeloma risk (reported OR = 1.30) (Carter et al., 2022; Pham et al., 2025). While the biological basis is not yet established, this finding encourages further exploration into immunomodulatory pathways and the bone marrow microenvironment as potential mediators.

6.5 Synthesis

The most defensible conclusion regarding cancer outcomes is the presence of heterogeneity. Coffee consumption is associated with improved liver health and does not appear to influence the risk of many common cancers. However, it may increase the risk of esophageal cancer, likely due to the consumption of very hot beverages rather than the coffee itself. Discrepancies between observational and genetic findings provide valuable insights, highlighting residual confounding factors, particularly smoking, potential non-linear dose–response relationships, and limitations in exposure definition when "coffee consumption" is represented by genetic variants that may reflect caffeine clearance, self-titration behavior, or broader metabolic pathways. These considerations underscore three methodological priorities. Firstly, research should more distinctly differentiate exposure constructs—such as self-reported intake versus biologically proximate measures like circulating caffeine metabolites and coffee-derived compounds—since etiological significance may reside in caffeine-independent constituents (e.g., polyphenols) or in internal dose rather than intake. Secondly, potential subgroup effects (e.g., in BRCA mutation carriers or strata defined by metabolic genotypes) should be explicitly framed as effect modification rather than generalised to population-average causality. Thirdly, given that genetic architecture and pleiotropy vary across ancestries, multi-ancestry evidence is crucial for both instrument validity and generalisable inference. Collectively, these points inform the subsequent section, which utilises epigenetic, transcriptomic, and broader multi-omic signatures to refine exposure measurement and elucidate mechanisms that intake-based genetics alone may not resolve—particularly when carcinogenesis is contingent upon exposures during etiologically critical windows. The divergence between associations observed in conventional epidemiology and causal inferences from Mendelian randomisation across major health outcomes is summarised in Table 2.

Table 2: Divergence between observational and genetic evidence across major health outcomes

Health outcome	Trend in observational epidemiology	Evidence from genetic epidemiology (Mendelian randomisation)	Synthesis and key insights	MR exposure instrument
Parkinson's Disease (PD)	Consistent protective association (reduced risk).	**Disease Modification:** Genetically predicted higher intake is causally linked to delayed *age-at-onset*, rather than reduced susceptibility.	Caffeine likely masks symptoms or delays clinical threshold rather than preventing neurodegeneration (Kuzovenkova et al., 2025; Senkevich et al., 2025).	Intake-based (coffee/caffeine consumption GWAS)
Cardiovascular Disease (CVD)	J-shaped curve; moderate intake associated with reduced mortality.	**Null/Neutral:** No clear causal protective effect for ischemic heart disease or stroke.	Benefits in observational studies likely stem from confounding; however, CYP1A2 slow metabolisers may face increased risk at high intake (Nordestgaard and Nordestgaard, 2016; Yuan et al., 2021).	Intake-based (coffee/caffeine consumption GWAS)
Type 2 Diabetes (T2D)	Strong inverse association (protective).	**Null/Inconsistent:** Genetic evidence generally fails to support causality.	Suggests observational protection is driven by lifestyle clustering or pleiotropy, though interactions with diet exist (Han et al., 2017; Pham et al., 2025).	Intake-based (coffee consumption GWAS)
Kidney Function (CKD)	Mixed; often protective.	**Discordant:** Genetically predicted *intake* is protective, but genetically predicted *plasma caffeine* is detrimental.	Highlights the "caffeine paradox": intake and internal dose may have opposing effects on renal physiology (Fujii et al., 2024; Giontella et al., 2023).	Both: Intake-based + biomarker-defined internal dose (plasma caffeine)
Liver Cancer (HCC) & NAFLD	Strong protective association.	**Supportive:** Evidence for causality is stronger than for other cancers, but still debated.	Biologically plausible due to caffeine's antifibrotic pathways; supported by some multi-ancestry MR data (Kennedy et al., 2017; Deng et al., 2022).	Primarily intake-based (multi-ancestry); limited biomarker MR
Esophageal Cancer	Mixed/Positive association.	**Risk Signal:** Genetically predicted intake associated with *increased* risk.	Likely mediated by thermal injury from consuming very hot beverages rather than coffee chemistry (Carter et al., 2022).	Intake-based (behavior/temperature may contribute)

7. Epigenetic, transcriptomic, and multi-omic signatures of coffee exposure

Genetic research offers insights into why individuals differ in their coffee consumption and caffeine response, but it does not directly explain how coffee-related exposures become biologically embedded in tissues over time. Multi-omics approaches address this gap by identifying molecular "fingerprints" of exposure—signals found in DNA methylation, gene expression, circulating metabolites, and the gut microbiome—that can (i) enhance exposure measurement beyond self-reporting and (ii) generate mechanistic hypotheses that complement Mendelian randomisation (MR) and disease association findings (Cornelis, 2015). The central interpretive principle is that omics signatures must be evaluated alongside key confounders (particularly smoking) and the distinction between intake behavior and internal dose.

7.1 DNA methylation and the coffee epigenome

Epigenome-wide association studies (EWAS) have consistently identified methylation differences associated with habitual coffee consumption. However, it is essential to meticulously control for confounding factors to ensure accurate interpretation. A comprehensive EWAS meta-analysis (n = 15,789) reported significant associations between coffee intake and hypomethylation at the AHRR and F2RL3 loci in peripheral blood (Karabegovic et al., 2021). Since these loci are also recognised as biomarkers of tobacco exposure, and coffee consumption often correlates with smoking in various populations, these findings should be cautiously interpreted as potentially indicative of residual confounding. In analyses adjusted for smoking, the AHRR signal was substantially attenuated, while a robust association persisted at PHGDH (cg14476101), suggesting that PHGDH methylation may serve as a more coffee-specific epigenetic marker (Karabegovic et al., 2021). Functionally, PHGDH is associated with metabolic regulation, and this finding has been posited as a plausible molecular link between coffee exposure and pathways relevant to hepatic lipid metabolism and fatty liver risk. Comparative epigenetic patterns across different beverages further emphasise that "coffee-like" exposures are not interchangeable. In sex-stratified analyses, tea consumption was associated with differential methylation at DNAJC16 and TTC17 in women but not in men, a pattern not observed for coffee (Ek et al., 2017). These results suggest that beverage-specific bioactives and hormonal contexts can influence epigenetic embedding, consistent with the heterogeneity observed in disease outcomes. Beyond locus-specific signals, emerging research connects coffee consumption to markers of biological aging. Coffee intake has been associated with a deceleration of epigenetic age acceleration using clocks such as GrimAge and FitAge, indicating a potential "geroprotective" molecular signature (Noroozi et al., 2024). While these findings remain observational and should be viewed as supportive molecular correlates rather than definitive causal evidence, they align with broader epidemiological reports that moderate coffee intake is often associated with reduced mortality risk.

7.2 Coffee and the blood transcriptome

Transcriptomic profiling provides a more immediate assessment of physiological states than methylation, illuminating pathways responsive to habitual exposure. In whole-blood gene expression analyses, high coffee consumption has been linked to differential expression of genes associated with metabolic and immune functions, although smoking consistently emerges as a significant confounder that can obscure expression patterns (B. Barnung et al., 2018). Notably, when analyses are restricted to non-smokers, the expression of POR (P450 oxidoreductase) shows a significant correlation with coffee consumption (B. Barnung et al., 2018). The biological plausibility of POR is evident, as it facilitates the electron transfer necessary for cytochrome P450 activity, thereby intersecting with caffeine metabolism. This convergence between expression data and GWAS findings implicating POR variants in coffee consumption supports the notion that the genetics of coffee consumption are partially observable at the transcriptional level through cis-regulatory effects (B. Barnung et al., 2018; Cornelis, 2015). A practical implication is that transcriptomic signals can help prioritise candidate pathways for subsequent mechanistic investigations, although their interpretation should remain cautious unless supported by replication, meticulous confounder control, and triangulation with genetic instruments.

7.3 Metabolomic fingerprints of coffee exposure

Among the various omics layers, metabolomics offers the most direct and objective markers of coffee exposure. Coffee consumption elevates circulating levels of exogenous or diet-derived metabolites, such as caffeine, paraxanthine, trigonelline, quinate (quinic acid), and hippuric acid, which are more reliable biomarkers than self-reported intake (Cornelis, 2015). Metabolomics also helps in operationalizing internal dose and distinguishing between caffeine-driven and non-caffeine components within the coffee matrix. A recurring interpretive challenge is that genetic predisposition to higher coffee intake may coincide with lower circulating caffeine levels due to accelerated metabolism. This necessitates caution when equating "genetically predicted coffee consumption" with "higher caffeine exposure" in associating metabolomic signatures with health outcomes (Giontella et al., 2023; Woolf et al., 2024). In this context, metabolomics can clarify which exposure dimension—intake behavior versus internal caffeine dose—most accurately correlates with a given outcome. Beyond exogenous markers, untargeted metabolomics suggests that coffee consumption can alter endogenous metabolic profiles, including changes in lipid-related pathways such as increased sphingomyelins and decreased acylcarnitines, consistent with hypotheses involving mitochondrial function and lipid oxidation (Cornelis, 2015). These signatures are mechanistically suggestive but require validation across cohorts and careful consideration of dietary and lifestyle contexts.

7.4 The gut microbiome and coffee-responsive taxa

The gut microbiome presents an additional avenue through which coffee may affect metabolic processes, particularly via polyphenols that reach the colon and undergo microbial transformation. Chlorogenic acids and related compounds are not completely absorbed in the small intestine and may be fermented by gut microbes into bioactive metabolites, such as short-chain fatty acids. Coffee consumption has been associated with increased microbial diversity and enhanced intestinal barrier function, changes that may help reduce metabolic endotoxemia and systemic inflammation, thereby intersecting with cardiometabolic and liver-related phenotypes (Zhang et al., 2022). Integrating microbiome signals with host epigenetic findings is an emerging direction; for instance, PHGDH methylation has been discussed as a potential node where microbial-metabolite signaling and hepatic lipid metabolism could converge (Karabegovic et al., 2021). Currently, microbiome evidence is best regarded as mechanistic plausibility that complements—rather than replaces—genetic and biomarker-based causal inference.

7.5 Integrating omics with human genetics

Integrative frameworks that merge omics readouts with human genetics have the potential to enhance causal reasoning and resolve apparent inconsistencies in observational findings. Multi-omic Mendelian Randomisation (MR) approaches can be employed to ascertain whether specific biomarkers, such as circulating caffeine or coffee-derived metabolites, mediate disease risk, and whether genetic instruments associated with intake reflect coffee constituents, metabolic rate, or broader behavioral liabilities. A pertinent example is renal physiology: MR analyses have revealed a divergence where genetically predicted higher coffee consumption behavior is linked to more favorable kidney function, while genetically predicted higher plasma caffeine levels are associated with a reduced estimated glomerular filtration rate (eGFR). This suggests that caffeine accumulation may pose a risk in susceptible contexts, whereas non-caffeine constituents or faster clearance may mitigate harm (Fujii et al., 2024; Giontella et al., 2023). A similar integrative approach has been applied in neuroimaging genetics, where the combination of neuroimaging phenotypes with transcriptomic annotation has connected coffee-intake genetic signals to brain expression patterns, such as CPLX3, thereby supporting mechanistic hypotheses regarding synaptic pathways underlying associations with structural brain measures (Kang et al., 2022). While these studies should be interpreted with caution, they underscore the importance of moving beyond intake-only models towards exposure biology that is measurable, tissue-informed, and mechanistically anchored. Overall, multi-omics evidence supports a multidimensional perspective of coffee exposure. Certain molecular signatures, such as AHRR hypomethylation, are highly susceptible to smoking confounding, whereas others, including PHGDH methylation, coffee-derived metabolite panels, and epigenetic aging signatures, may more plausibly reflect coffee-specific biology (Karabegovic et al., 2021; Noroozi et al., 2024). By integrating these layers with genetic instruments, future research can more effectively distinguish caffeine-driven effects from non-caffeine constituents, refine causal inference for specific endpoints, and enhance the interpretability and portability of findings across populations (Cornelis, 2015).

8. Coffee in pregnancy: Feto–maternal genetics and developmental outcomes

In perinatal medicine, coffee consumption during pregnancy raises significant concerns due to caffeine's ability to easily cross the placenta and the fetus's limited capacity to metabolise it. The physiological changes that occur throughout gestation further complicate exposure assessment, as caffeine's half-life is prolonged in the later trimesters compared to the non-pregnant state (Schellhas et al., 2023). Since the fetus shows minimal or no expression of CYP1A2, fetal caffeine clearance largely relies on maternal metabolism (Sasaki et al., 2017). Consequently, the maternal genotype—particularly along the CYP1A2–AHR axis discussed earlier in this chapter—can influence not only the duration of exposure but also the balance between caffeine and its downstream metabolites reaching the fetal–placental unit.

8.1 Maternal pharmacogenetics and fetal exposure

Maternal CYP1A2 activity is crucial in determining the internal caffeine dose during pregnancy. The unique physiological changes of pregnancy interact with inherited metabolic capacity: a slower clearance rate in late gestation can lead to prolonged fetal exposure even with moderate caffeine intake, while a faster metabolism may result in higher relative levels of metabolites like paraxanthine (Sasaki et al., 2017; Schellhas et al., 2023). This distinction is biologically significant, as metabolites may not merely serve as neutral intermediates. For instance, paraxanthine has been suggested to reduce the placental expression and activity of 11β-hydroxysteroid dehydrogenase-2 (11β-HSD2), potentially increasing fetal glucocorticoid exposure and restricting fetal growth (Sasaki et al., 2017). In this context, "metabolic efficiency" does not inherently imply a reduced fetal risk; rather, the critical factor may be the exposure profile (caffeine persistence versus metabolite predominance) shaped by maternal genotype in conjunction with pregnancy physiology.

8.2. Pregnancy loss, fetal growth, and timing of birth

Numerous epidemiological studies have linked increased caffeine consumption to adverse obstetric outcomes, particularly pregnancy loss and impaired fetal growth. However, the effect estimates are inconsistent and often affected by confounding and exposure misclassification. Some studies suggest that maternal genotype may influence these associations differently than adult cardiometabolic paradigms. For instance, high caffeine intake (≥300 mg/day) among non-smoking mothers with the CYP1A2 AA ("fast metaboliser") genotype was associated with lower birth weight (approximately 277 g less) and reduced head circumference (about 1 cm) compared to slow metabolisers, indicating that genotype-dependent exposure profiles may impact fetal growth (Sasaki et al., 2017). Supporting a metabolite-toxicity hypothesis, case-control studies on recurrent pregnancy loss reported a significantly increased risk among women consuming ≥300 mg/day who also had the fast-metaboliser genotype (Cornelis, 2014; Sasaki et al., 2017; Sata et al., 2005). Nonetheless, the literature remains inconsistent. As previous syntheses have summarised, some studies report no clear main effect of the CYP1A2 genotype on low birth weight or small-for-gestational-age outcomes, while others indicate an increased risk among fast metabolisers, highlighting endpoint- and context-specific heterogeneity (Cornelis, 2014). Furthermore, Mendelian randomisation analyses of pregnancy loss have not supported strong causal effects of genetically predicted coffee consumption, suggesting that at least part of the observational signal may be due to confounding (Kwok et al., 2016). Evidence regarding preterm birth and related complications is also mixed, with some studies suggesting a potential modification of neural tube defect risk by maternal CYP1A2 genotype under high caffeine exposure, though without a consistent pattern across study designs (Cornelis, 2014).

8.3 Offspring neurodevelopment, epigenetics, and overall assessment

Uncertainty persists regarding whether prenatal caffeine exposure has enduring neurodevelopmental effects, with epigenetic data offering a means to investigate potential biological embedding. In a comprehensive EWAS meta-analysis utilizing PACE consortium data (encompassing six European cohorts; n = 3,725), maternal caffeine consumption (from coffee, tea, and cola) did not reveal a strong or widespread cord-blood DNA methylation signature. Only a single CpG near PRRX1 (cg19370043) was associated with total caffeine intake, and a CpG near STAG1 (cg14591243) was linked to cola consumption (Schellhas et al., 2023). The weak and source-specific nature of these signals suggests either small true effects or residual confounding related to beverage choice (e.g., non-caffeine constituents or socioeconomic patterning). From a developmental outcomes perspective, fetal growth restriction—particularly reduced head circumference—has been proposed as a potential early-life marker relevant to later neurodevelopmental vulnerability, and genotype-stratified findings in high-exposure groups warrant attention (Sasaki et al., 2017). However, genetic causal inference again advises caution: within an MR framework, genetically predicted coffee consumption did not demonstrate a causal effect on childhood cognitive function when used as a negative-control outcome, arguing against significant direct neurodevelopmental effects attributable to coffee intake per se (Kwok et al., 2016). Collectively, the pregnancy literature supports a "non-uniform effect" model, wherein risk is most plausibly concentrated in higher-exposure strata and may depend on maternal metabolic genotype and co-occurring exposures (notably smoking). The most defensible synthesis is not that coffee or caffeine is uniformly harmful during pregnancy, but that internal exposure biology—shaped by pregnancy physiology and maternal pharmacogenetics—can create subgroups in which fetal growth and pregnancy-loss risks may be more detectable, while large-scale epigenetic and MR findings do not support broad population-wide developmental programming effects (Cornelis, 2014; Kwok et al., 2016; Sasaki et al., 2017; Schellhas et al., 2023).

9. Coffee in the context of precision nutrition

Precision nutrition is increasingly supplanting traditional "one-size-fits-all" dietary guidelines by incorporating genetic, phenotypic, and environmental data to customise recommendations for optimal health outcomes. Coffee serves as an ideal candidate for this approach due to its complex composition of bioactive compounds, such as caffeine, chlorogenic acids, and diterpenes, its widespread habitual consumption, and the significant inter-individual variability observed in both tolerance and health-related effects (Cornelis, 2014; Grosso et al., 2017). The preceding sections of this chapter outline three constructs crucial for effective translation: intake (behavioral consumption), internal dose (biologically proximate exposure), and sensitivity (pharmacodynamic responsiveness). Precision nutrition concerning coffee is most credible when recommendations are structured around these constructs rather than solely on intake.

9.1 Inter-individual variability: metabolism and sensitivity as distinct axes

Two complementary dimensions significantly influence the variability in individual responses to coffee consumption. Firstly, the rate at which caffeine is metabolised varies among individuals, largely due to genetic factors in the CYP1A2–AHR pathway (Cornelis et al., 2006; Sachse et al., 1999). Secondly, even with similar internal doses, caffeine sensitivity differs, with adenosine signaling—particularly ADORA2A—affecting variations in sleep disruption and anxiety-related responses (Alsene et al., 2003; Childs et al., 2008; Retey et al., 2007). These dimensions are distinct: a person may metabolise caffeine quickly yet remain highly sensitive to its effects, such as experiencing sleep disruption at low doses, or metabolise it slowly but exhibit relatively mild subjective stimulation. This distinction is crucial for personalised recommendations, as optimal coffee consumption advice depends on the clinical objective and endpoint. For instance, sleep disturbances are most directly linked to sensitivity and timing of intake, whereas blood pressure and certain cardiovascular endpoints may be more affected by prolonged internal exposure in susceptible contexts. In practice, precision nutrition should consider genotype as one layer of evidence, integrated with symptoms, comorbidities (e.g., hypertension, arrhythmia, anxiety, chronic kidney disease), life stage (e.g., pregnancy), and environmental modifiers of caffeine kinetics (e.g., smoking, medications, hormonal status).

9.2 From genetic markers to tailored coffee advice

A practical approach to translating coffee genetics into personalised recommendations involves employing nutrigenetic algorithms to assess the strength of evidence and prioritise contexts where genotype-based recommendations are most clinically actionable (De Caterina and El-Sohemy, 2016). A frequently cited example is the CYP1A2-mediated caffeine metabolism and cardiovascular responses at higher intake levels. Across observational and pharmacogenetic studies, heavy coffee consumption has been associated with increased risks of outcomes such as nonfatal myocardial infarction and hypertension, particularly among slower caffeine metabolisers, while faster metabolisers often exhibit neutral or less adverse associations at similar intake levels (Cornelis et al., 2006; Palatini et al., 2009). Similar patterns have been observed for renal endpoints, with heavy consumption linked to albuminuria and hyperfiltration primarily in slow metabolisers, suggesting that metaboliser status may be relevant beyond cardiometabolic phenotypes (Mahdavi et al., 2023). For translation, a key principle is that these findings support risk stratification rather than deterministic labeling. Genotype-informed advice should be communicated as probabilistic, endpoint-specific, and embedded in a clinical context. A practical framework is to (i) define the goal and endpoint (e.g., improve alertness without insomnia; manage blood pressure; minimise anxiety; ensure pregnancy safety), (ii) characterise the relevant phenotype and context (sleep quality, anxiety symptoms, hypertension/arrhythmia history, kidney disease, pregnancy, medication use), and (iii) use genotype—when available—to refine dose and timing thresholds, recognizing that coffee-intake measures and genetic proxies may index caffeine clearance and self-titration behavior rather than a uniform internal biological dose (Cornelis, 2014; De Caterina and El-Sohemy, 2016). Within this framework, individuals classified as slower metabolisers can be advised toward more conservative dosing and earlier timing, particularly when targeting outcomes plausibly worsened by prolonged caffeine exposure (e.g., uncontrolled hypertension or sleep disturbance). Conversely, faster metabolisers may tolerate moderate-to-higher intake with fewer hemodynamic consequences in some settings, although tolerance is phenotype-dependent and should not supersede symptom-guided counseling. For sensitivity-prone individuals (e.g., sleep disruption linked to ADORA2A), personalisation often centers on timing (earlier cut-off), dose reduction, or switching to lower-caffeine options rather than complete avoidance (Retey et al., 2007). Beyond single-variant approaches, the evidence increasingly supports moving toward polygenic and biomarker-informed personalisation. Coffee consumption is a polygenic trait integrating metabolic, reward, and sensory pathways (Cornelis, 2014; Pirastu et al., 2016), implying that more robust guidance will likely emerge from multi-locus risk models coupled with objective internal-dose markers to verify exposure. Finally, emerging findings suggest that "one-size-fits-all" advice may be particularly problematic for kidney function and micronutrient status. Stratified analyses indicate clinically meaningful genotype-by-intake interactions, with heavy coffee intake associated with renal risk markers only in genetically defined slow-metaboliser subgroups in some studies (Mahdavi et al., 2023). Precision nutrition models are also expanding to nutrient-related outcomes; for example, genotype-stratified analyses in metabolic syndrome suggest that CYP1A2 background may modify associations between coffee intake and vitamin B12/folate status, potentially revealing a therapeutic window for selected subgroups (Popa et al., 2025). Based on the integration of pharmacokinetic, pharmacodynamic, and ancestry-specific genetic evidence, a precision nutrition framework for coffee consumption is proposed (Table 3).

9.3 Personalised caffeine interventions: What do trials show?

Intervention studies are crucial for evaluating whether genotype-stratified guidance yields measurable benefits. Across various trials, the most compelling evidence of personalisation has emerged in acute performance and response phenotypes, while longer-term metabolic outcomes appear more context-dependent. In sports nutrition, randomised trial evidence supports genotype-dependent variability in ergogenic

Table 3: Precision nutrition framework for coffee consumption

Clinical profile/ Genotype	Mechanism	Potential risks with high intake	Tailored recommendation
Slow Metaboliser (*CYP1A2* AC or CC genotype)	Impaired caffeine clearance; prolonged systemic exposure.	Increased risk of nonfatal myocardial infarction, hypertension, and albuminuria (kidney dysfunction).	**"Moderate and Monitor":** Limit intake (e.g., <2 cups/ day). Caution is advised if a history of hypertension or renal compromise exists (Mahdavi et al., 2023; Palatini et al., 2009).
Fast Metaboliser (*CYP1A2* AA genotype)	Rapid caffeine clearance; accumulation of metabolites.	Generally lower cardiovascular risk; potential for fetal growth restriction in pregnancy due to metabolite toxicity.	**"Flexible / Ergogenic":** Higher tolerance for cardiovascular safety. Greater potential for acute ergogenic benefits in sports performance (Guest et al., 2018; Sasaki et al., 2017).
High Sensitivity (*ADORA2A* TT genotype)	Enhanced receptor-level response to adenosine blockade.	Anxiety, panic attacks, and sleep-onset insomnia even at low doses.	**"Symptom-Guided":** Strictly limit dosage and enforce early cut-off times (e.g., no caffeine after noon). Genotype affects sensitivity independent of metabolism rate (Retey et al., 2007).
Metabolic Syndrome Context	Interaction with nutrient status.	Potential B12/ Folate disruption or benefit depending on interaction.	**"Nutrient-Aware":** In CYP1A2 slow metabolisers, moderate intake may correlate with improved Vitamin B12 status, suggesting a specific therapeutic window (Popa et al., 2025).
Pregnancy	The fetus lacks CYP1A2 activity; reliance on maternal clearance.	Fetal growth restriction (lower birth weight), particularly in CYP1A2 fast metabolisers (metabolite hypothesis).	**"Precautionary":** Minimise intake regardless of maternal genotype due to fetal accumulation risks and potential metabolite toxicity (Schellhas et al., 2023; Sasaki et al., 2017).

response. For instance, in a trial administering caffeine at 4 mg/kg, performance benefits were significantly greater in CYP1A2 fast metabolisers, whereas slow metabolisers experienced reduced benefits or even impaired performance under the same protocol (Guest et al., 2018). However, not all studies replicate a strong genotype-by-caffeine interaction for physical performance, with some findings suggesting that genotype effects may be more consistent for cognitive endpoints (e.g., reaction time) than for endurance performance (Carswell et al., 2020). Variations in dose, timing, habitual caffeine use, outcome selection, and athletic population likely contribute to this heterogeneity (Southward et al., 2018). Therefore, the most defensible translational statement is that genotype-informed caffeine use may be most actionable for acute, proximal endpoints (performance, sleep, anxiety) rather than as a universal health strategy. For longer-term metabolic outcomes, evidence suggests interactions with the broader dietary environment. In the POUNDS LOST trial, a genetic risk score indicating a higher predisposition to coffee consumption was associated with more favorable changes in fasting insulin and insulin resistance indices under a low-fat dietary context compared to a high-fat context (Han et al., 2017). This finding illustrates a broader principle: for chronic outcomes, coffee-related biology may interact with macronutrient composition and baseline metabolic risk, making single-factor recommendations less applicable across dietary patterns. In precision nutrition terms, the "right" advice is unlikely to be coffee alone; it is coffee within a broader dietary and metabolic profile.

9.4 Future directions and ethical considerations

The next phase in coffee-related precision nutrition necessitates a transition from single-gene narratives to a "systems epidemiology" approach, integrating genetics with objective exposure markers and multi-omics readouts (Cornelis, 2015). Three methodological priorities are paramount. First, refining exposure biology is essential. Since coffee consumption does not consistently reflect internal caffeine dosage, future personalisation efforts should incorporate biomarkers such as caffeine–paraxanthine profiles and coffee-derived metabolites to differentiate intake, clearance, and internal exposure. This strategy directly addresses the caffeine paradox and enhances the interpretability of Mendelian randomisation and intervention studies (Giontella et al., 2023; Woolf et al., 2024). Second, it is crucial to expand genetic models beyond the most studied loci. While CYP1A2 and ADORA2A remain significant, genome-wide association studies (GWAS) have identified additional loci pertinent to coffee consumption and caffeine-related biology (e.g., AHR, PDSS2, and neurobehavioral loci), supporting the development of polygenic scores that may capture variability more effectively than any single SNP (Amin et al., 2012; Cornelis et al., 2011; Pirastu et al., 2016). Importantly, these models should be assessed against objective internal-dose markers to avoid conflating behavioral intake with biological exposure. Third, prioritizing multi-ancestry validity and equity is necessary. The genetic architecture of coffee consumption varies across ancestries, and tools developed in one population may not apply to another. Therefore, precision nutrition tools must be developed and validated across diverse cohorts to prevent exacerbating health disparities through inequitable predictive performance. Ethically, implementation must address the risks of overinterpretation and the realities of direct-to-consumer genetic testing. While genotype information can help tailor thresholds and expectations, it should not be presented as deterministic, nor should it replace clinical risk factors and symptoms. Clear communication of uncertainty, safeguards for privacy and consent, and avoidance of exaggerated health claims are imperative. The appropriate clinical approach is shared decision-making: combining genotype (when available) with phenotype (sleep quality, anxiety, blood pressure, kidney function, pregnancy status), environment (smoking, medications), and personal goals to determine a reasonable dose and timing strategy (De Caterina and El-Sohemy, 2016).

The evidence presented in this chapter collectively supports a context-dependent model of coffee biology, rather than a universal classification of coffee as inherently "good" or "bad." The benefits and drawbacks depend on specific parameters being assessed, such as intake versus biomarkers, the component responsible for the observed effects, like caffeine kinetics versus non-caffeine constituents, and the endpoint selected, such as acute symptoms versus long-term disease outcomes. Therefore, precision nutrition regarding coffee is best conceptualised as identifying the appropriate dose, timing, and preparation method tailored to the individual and endpoint, using genetics and multi-omics to enhance—rather than replace—rigorous causal inference and clinically informed judgment.

References

Acharya, V., Fan, K. H., Snitz, B. E., Ganguli, M., DeKosky, S. T., Lopez, O. L. et al. (2025). APOE genotype-stratified meta-analysis of cognitive decline reveals novel loci for language and global cognitive function in older adults. *Int J Mol Sci*, 26(14). https://doi.org/10.3390/ijms26146940

Alsene, K., Deckert, J., Sand, P., and de Wit, H. (2003). Association between A2a receptor gene polymorphisms and caffeine-induced anxiety. *Neuropsychopharmacology*, 28(9), 1694–1702. https://www.nature.com/articles/1300232

Amin, N., Byrne, E., Johnson, J., Chenevix-Trench, G., Walter, S., Nolte, I. M., et al. (2012). Genome-wide association analysis of coffee drinking suggests association with CYP1A1/CYP1A2 and NRCAM. *Mol Psychiatry*, 17(11), 1116–1129. https://doi.org/10.1038/mp.2011.101

B. Barnung, R., H. Nøst, T., Ulven, S. M., Skeie, G., and S. Olsen, K. (2018). Coffee consumption and whole-blood gene expression in the Norwegian women and cancer post-genome cohort. *Nutrients*, 10(8), 1047. https://doi.org/10.3390/nu10081047

Carswell, A. T., Howland, K., Martinez-Gonzalez, B., Baron, P., and Davison, G. (2020). The effect of caffeine on cognitive performance is influenced by CYP1A2 but not ADORA2A genotype, yet neither genotype affects exercise performance in healthy adults. *Eur J Appl Physiol*, 120(7), 1495–1508. https://doi.org/10.1007/s00421-020-04384-8

Carter, P., Yuan, S., Kar, S., Vithayathil, M., Mason, A. M., Burgess, S. et al. (2022). Coffee consumption and cancer risk: a Mendelian randomisation study. *Clin Nutr*, 41(10), 2113–2123. https://doi.org/10.1016/j.clnu.2022.08.019

Childs, E., Hohoff, C., Deckert, J., Xu, K., Badner, J., and de Wit, H. (2008). Association between ADORA2A and DRD2 polymorphisms and caffeine-induced anxiety. *Neuropsychopharmacology*, 33(12), 2791–2800. https://doi.org/10.1038/npp.2008.17

Coffee and Caffeine Genetics Consortium, Cornelis, M. C., Byrne, E. M., Esko, T., Nalls, M. A., Ganna, A. et al. (2015). Genome-wide meta-analysis identifies six novel loci associated with habitual coffee consumption. *Mol Psychiatry*, 20(5), 647–656. https://doi.org/10.1038/mp.2014.107

Cornelis, M. C. (2014). Gene-coffee interactions and health. *Current Nutrition Reports*, 3(3), 178–195. https://doi.org/10.1007/s13668-014-0087-1

Cornelis, M. C. (2015). Toward systems epidemiology of coffee and health. *Curr Opin Lipidol*, 26(1), 20–29. https://doi.org/10.1097/MOL.0000000000000143

Cornelis, M. C. (2019). The impact of caffeine and coffee on human health. *Nutrients*, 11(2). https://doi.org/10.3390/nu11020416

Cornelis, M. C., El-Sohemy, A., Kabagambe, E. K., and Campos, H. (2006). Coffee, CYP1A2 genotype, and risk of myocardial infarction. *Jama*, 295(10), 1135–1141. https://doi.org/10.1001/jama.295.10.1135

Cornelis, M. C., Monda, K. L., Yu, K., Paynter, N., Azzato, E. M., Bennett, S. N. et al. (2011). Genome-wide meta-analysis identifies regions on 7p21 (AHR) and 15q24 (CYP1A2) as determinants of habitual caffeine consumption. *PLoS Genet*, 7(4), e1002033. https://doi.org/10.1371/journal.pgen.1002033

Cornelis, M. C., Kacprowski, T., Menni, C., Gustafsson, S., Pivin, E., Adamski, J. et al. (2016). Genome-wide association study of caffeine metabolites provides new insights to caffeine metabolism and dietary caffeine-consumption behavior. *Hum Mol Genet*, 25(24), 5472–5482. https://doi.org/10.1093/hmg/ddw334

Cornelis, M. C., and Munafo, M. R. (2018). Mendelian randomization studies of coffee and caffeine consumption. *Nutrients*, 10(10). https://doi.org/10.3390/nu10101343

Cornelis, M. C., and van Dam, R. M. (2021). Genetic determinants of liking and intake of coffee and other bitter foods and beverages. *Sci Rep*, 11(1), 23845. https://doi.org/10.1038/s41598-021-03153-7

Covolo, L., Placidi, D., Gelatti, U., Carta, A., Scotto Di Carlo, A., Lodetti, P. et al. (2008). Bladder cancer, GSTs, NAT1, NAT2, SULT1A1, XRCC1, XRCC3, XPD genetic polymorphisms and coffee consumption: A case-control study. *Eur J Epidemiol*, 23(5), 355–362. https://doi.org/10.1007/s10654-008-9238-2

De Caterina, R., and El-Sohemy, A. (2016). Moving towards specific nutrigenetic recommendation algorithms: Caffeine, genetic variation and cardiovascular risk. *J Nutrigenet Nutrigenomics*, 9(2-4), 106–115. https://doi.org/10.1159/000446801

Deng, Y., Huang, J., and Wong, M. C. (2022). Associations between six dietary habits and risk of hepatocellular carcinoma: A Mendelian randomization study. *Hepatology Communications*, 6(8), 2147–2154. https://doi.org/10.1002/hep4.1960

Ding, M., Bhupathiraju, S. N., Satija, A., van Dam, R. M., and Hu, F. B. (2014). Long-term coffee consumption and risk of cardiovascular disease: A systematic review and a dose-response meta-analysis of prospective cohort studies. *Circulation*, 129(6), 643–659. https://doi.org/10.1161/CIRCULATIONAHA.113.005925

Djordjevic, N., Ghotbi, R., Jankovic, S., and Aklillu, E. (2010). Induction of CYP1A2 by heavy coffee consumption is associated with the CYP1A2-163C>A polymorphism. *Eur J Clin Pharmacol*, 66(7), 697–703. https://doi.org/10.1007/s00228-010-0823-4

Ek, W. E., Tobi, E. W., Ahsan, M., Lampa, E., Ponzi, E., Kyrtopoulos, S. A., Epigenome-Wide Association Study, C. et al. (2017). Tea and coffee consumption in relation to DNA methylation in four European cohorts. *Hum Mol Genet*, 26(16), 3221–3231. https://doi.org/10.1093/hmg/ddx194

Ellingjord-Dale, M., Papadimitriou, N., Katsoulis, M., Yee, C., Dimou, N., Gill, D. et al. (2021). Coffee consumption and risk of breast cancer: A Mendelian randomization study. *PLoS One*, 16(1), e0236904. https://doi.org/10.1371/journal.pone.0236904

Fredholm, B. B., Bättig, K., Holmén, J., Nehlig, A., and Zvartau, E. E. (1999). Actions of caffeine in the brain with special reference to factors that contribute to its widespread use. *Pharmacological Reviews*, 51(1), 83–133. https://doi.org/10.1016/S0031-6997(24)01396-6

Fujii, R., Nakatochi, M., and Del Greco, M. F. (2024). Coffee intake, plasma caffeine levels, and kidney function: two-sample mendelian randomization among east Asian and European ancestries. *Kidney Int Rep*, 9(4), 1083–1092. https://doi.org/10.1016/j.ekir.2024.01.024

Giersch, G. E. W., Boyett, J. C., Hargens, T. A., Luden, N. D., Saunders, M. J., Daley, H. et al. (2018). The effect of the CYP1A2 –163 C>A polymorphism on caffeine metabolism and subsequent cycling performance. *Journal of Caffeine and Adenosine Research*, 8(2), 65-70. https://doi.org/10.1089/caff.2017.0028

Giontella, A., de La Harpe, R., Cronje, H. T., Zagkos, L., Woolf, B., Larsson, S. C. et al. (2023). Caffeine intake, plasma caffeine level, and kidney function: A mendelian randomization study. *Nutrients*, 15(20). https://doi.org/10.3390/nu15204422

Grosso, G., Godos, J., Galvano, F., and Giovannucci, E. L. (2017). Coffee, caffeine, and health outcomes: An umbrella review. *Annu Rev Nutr*, 37(1), 131–156. https://doi.org/10.1146/annurev-nutr-071816-064941

Guest, N., Corey, P., Vescovi, J., and El-Sohemy, A. (2018). Caffeine, CYP1A2 genotype, and endurance performance in athletes. *Med Sci Sports Exerc*, 50(8), 1570–1578. https://doi.org/10.1249/MSS.0000000000001596

Hamza, T. H., Chen, H., Hill-Burns, E. M., Rhodes, S. L., Montimurro, J., Kay, D. M. et al. (2011). Genome-wide gene-environment study identifies glutamate receptor gene GRIN2A as a Parkinson's disease modifier gene via interaction with coffee. *PLoS Genet*, 7(8), e1002237. https://doi.org/10.1371/journal.pgen.1002237

Han, L., Ma, W., Sun, D., Heianza, Y., Wang, T., Zheng, Y. et al. (2017). Genetic variation of habitual coffee consumption and glycemic changes in response to weight-loss diet intervention: The Preventing Overweight Using Novel Dietary Strategies (POUNDS LOST) trial. *Am J Clin Nutr*, 106(5), 1321–1326. https://doi.org/10.3945/ajcn.117.156232

He, M., Huan, L., Wang, X., Fan, Y., and Huang, J. (2024). Nine dietary habits and risk of colorectal cancer: A Mendelian randomization study. *BMC Med Genomics*, 17(1), 21. https://doi.org/10.1186/s12920-023-01782-7

Huang, J., Huffman, J. E., Huang, Y., Do Valle, I., Assimes, T. L., Raghavan, S. et al. (2022). Genomics and phenomics of body mass index reveals a complex disease network. *Nat Commun*, 13(1), 7973. https://doi.org/10.1038/s41467-022-35553-2

Jin, T., Youn, J., Kim, A. N., Kang, M., Kim, K., Sung, J. et al. (2020). Interactions of habitual coffee consumption by genetic polymorphisms with the risk of prediabetes and Type 2 diabetes combined. *Nutrients*, 12(8). https://doi.org/10.3390/nu12082228

Kang, J., Jia, T., Jiao, Z., Shen, C., Xie, C., Cheng, W. et al. (2022). Increased brain volume from higher cereal and lower coffee intake: shared genetic determinants and impacts on cognition and metabolism. *Cereb Cortex*, 32(22), 5163–5174. https://doi.org/10.1093/cercor/bhac005

Karabegovic, I., Portilla-Fernandez, E., Li, Y., Ma, J., Maas, S. C. E., Sun, D. et al. (2021). Epigenome-wide association meta-analysis of DNA methylation with coffee and tea consumption. *Nat Commun*, 12(1), 2830. https://doi.org/10.1038/s41467-021-22752-6

Kendler, K. S., Schmitt, E., Aggen, S. H., and Prescott, C. A. (2008). Genetic and environmental influences on alcohol, caffeine, cannabis, and nicotine use from early adolescence to middle adulthood. *Arch Gen Psychiatry*, 65(6), 674–682. https://doi.org/10.1001/archpsyc.65.6.674

Kennedy, O. J., Roderick, P., Buchanan, R., Fallowfield, J. A., Hayes, P. C., and Parkes, J. (2017). Coffee, including caffeinated and decaffeinated coffee, and the risk of hepatocellular carcinoma: A systematic review and dose-response meta-analysis. *BMJ Open*, 7(5), e013739. https://doi.org/10.1136/bmjopen-2016-013739

Kennedy, O. J., Pirastu, N., Poole, R., Fallowfield, J. A., Hayes, P. C., Grzeszkowiak, E. J. et al. (2020). Coffee consumption and kidney function: A Mendelian randomization study. *Am J Kidney Dis*, 75(5), 753–761. https://doi.org/10.1053/j.ajkd.2019.08.025

Kim, M. J., Jin, H. S., and Eom, Y. B. (2024). Coffee consumption affects kidney function based on GCKR polymorphism in a Korean population. *Nutr Res*, 122, 92–100. https://doi.org/10.1016/j.nutres.2023.12.008

Kim, M. J., Jin, H. S., and Eom, Y. B. (2025). Relationship between HECTD4 gene variants, obesity, and coffee consumption. *Eur J Clin Nutr*, 79(3), 200–206. https://doi.org/10.1038/s41430-024-01541-6

Kotsopoulos, J., Ghadirian, P., El-Sohemy, A., Lynch, H. T., Snyder, C., Daly, M. et al. (2007). The CYP1A2 genotype modifies the association between coffee consumption and breast cancer risk among BRCA1 mutation carriers. *Cancer Epidemiol Biomarkers Prev*, 16(5), 912–916. https://doi.org/10.1158/1055-9965.EPI-06-1074

Kotsopoulos, J., Vitonis, A. F., Terry, K. L., De Vivo, I., Cramer, D. W., Hankinson, S. E., and Tworoger, S. S. (2009). Coffee intake, variants in genes involved in caffeine metabolism, and the risk of epithelial ovarian cancer. *Cancer Causes Control*, 20(3), 335–344. https://doi.org/10.1007/s10552-008-9247-1

Kunutsor, S. K., Lehoczki, A., and Laukkanen, J. A. (2025). Coffee consumption, cancer, and healthy aging: Epidemiological evidence and underlying mechanisms. *Geroscience*, 47(2), 1517–1555. https://doi.org/10.1007/s11357-024-01332-8

Kuzovenkova, D., Liu, L., Gan-Or, Z., and Senkevich, K. (2025). Coffee consumption is associated with later age-at-onset of Parkinson's disease. *Ann Clin Transl Neurol*, 12(8), 1706–1710. https://doi.org/10.1002/acn3.70076

Kwok, M. K., Leung, G. M., and Schooling, C. M. (2016). Habitual coffee consumption and risk of type 2 diabetes, ischemic heart disease, depression and Alzheimer's disease: A Mendelian randomization study. *Sci Rep*, 6(1), 36500. https://doi.org/10.1038/srep36500

Lee, J. K., Kim, K., Ahn, Y., Yang, M., and Lee, J. E. (2015). Habitual coffee intake, genetic polymorphisms, and type 2 diabetes. *Eur J Endocrinol*, 172(5), 595–601. https://doi.org/10.1530/EJE-14-0805

Loftfield, E., Cornelis, M. C., Caporaso, N., Yu, K., Sinha, R., and Freedman, N. (2018). Association of coffee drinking with mortality by genetic variation in caffeine metabolism: Findings from the UK biobank. *JAMA Intern Med*, 178(8), 1086–1097. https://doi.org/10.1001/jamainternmed.2018.2425

Lu, Y., Wang, P., Liu, H., Li, T., Wang, H., Jiang, D. et al. (2024). Coffee and risk of pancreatic cancer: Insights from two-sample and multivariable Mendelian randomization analyses. *Nutrients*, 16(21), 3723. https://doi.org/10.3390/nu16213723

Luciano, M., Kirk, K. M., Heath, A. C., and Martin, N. G. (2005). The genetics of tea and coffee drinking and preference for source of caffeine in a large community sample of Australian twins. *Addiction*, 100(10), 1510–1517. https://doi.org/10.1111/j.1360-0443.2005.01223.x

Mahdavi, S., Palatini, P., and El-Sohemy, A. (2023). CYP1A2 genetic variation, coffee intake, and kidney dysfunction. *JAMA Netw Open*, 6(1), e2247868. https://doi.org/10.1001/jamanetworkopen.2022.47868

Mendoza, M. F., Sulague, R. M., Posas-Mendoza, T., and Lavie, C. J. (2023). Impact of coffee consumption on cardiovascular health. *Ochsner J*, 23(2), 152–158. https://doi.org/10.31486/toj.22.0073

Nakagawa-Senda, H., Hachiya, T., Shimizu, A., Hosono, S., Oze, I., Watanabe, M. et al. (2018). A genome-wide association study in the Japanese population identifies the 12q24 locus for habitual coffee consumption: The J-MICC Study. *Sci Rep*, 8(1), 1493. https://doi.org/10.1038/s41598-018-19914-w

Nehlig, A. (2018). Interindividual differences in caffeine metabolism and factors driving caffeine consumption. *Pharmacol Rev*, 70(2), 384–411. https://doi.org/10.1124/pr.117.014407

Nkondjock, A., Ghadirian, P., Kotsopoulos, J., Lubinski, J., Lynch, H., Kim-Sing, C. et al. (2006). Coffee consumption and breast cancer risk among BRCA1 and BRCA2 mutation carriers. *Int J Cancer*, 118(1), 103–107. https://doi.org/10.1002/ijc.21296

Nordestgaard, A. T., and Nordestgaard, B. G. (2016). Coffee intake, cardiovascular disease and all-cause mortality: Observational and Mendelian randomization analyses in 95 000-223 000 individuals. *Int J Epidemiol*, 45(6), 1938–1952. https://doi.org/10.1093/ije/dyw325

Nordestgaard, A. T., Thomsen, M., and Nordestgaard, B. G. (2015). Coffee intake and risk of obesity, metabolic syndrome and type 2 diabetes: A Mendelian randomization study. *Int J Epidemiol*, 44(2), 551–565. https://doi.org/10.1093/ije/dyv083

Noroozi, R., Rudnicka, J., Pisarek, A., Wysocka, B., Masny, A., Boron, M. et al. (2024). Analysis of epigenetic clocks links yoga, sleep, education, reduced meat intake, coffee, and a SOCS2 gene variant to slower epigenetic aging. *Geroscience*, 46(2), 2583–2604. https://doi.org/10.1007/s11357-023-01029-4

Ong, J. S., Hwang, L. D., Zhong, V. W., An, J., Gharahkhani, P., Breslin, P. A. S. et al. (2018). Understanding the role of bitter taste perception in coffee, tea and alcohol consumption through Mendelian randomization. *Sci Rep*, 8(1), 16414. https://doi.org/10.1038/s41598-018-34713-z

Palatini, P., Ceolotto, G., Ragazzo, F., Dorigatti, F., Saladini, F., Papparella, I. et al. (2009). CYP1A2 genotype modifies the association between coffee intake and the risk of hypertension. *J Hypertens*, 27(8), 1594–1601. https://doi.org/10.1097/HJH.0b013e32832ba850

Pham, K., Kassaw, N. A., Mulugeta, A., Zhou, A., and Hypponen, E. (2025). Coffee and health outcomes: A systematic review of Mendelian randomisation studies. *Nutr Res Rev*, 38(2), 1006–1026. https://doi.org/10.1017/S0954422425100206

Pirastu, N., Kooyman, M., Robino, A., van der Spek, A., Navarini, L., Amin, N. et al. (2016). Non-additive genome-wide association scan reveals a new gene associated with habitual coffee consumption. *Sci Rep*, 6, 31590. https://doi.org/10.1038/srep31590

Popa, L. C., Abu-Awwad, A., Farcas, S. S., Abu-Awwad, S. A., and Andreescu, N. I. (2025). Interaction between CYP1A2-related caffeine metabolism and vitamin B12/Folate status in patients with metabolic syndrome: A novel biomarker axis. *Metabolites*, 15(7). https://doi.org/10.3390/metabo15070450

Qian, Y., Ye, D., Huang, H., Wu, D. J. H., Zhuang, Y., Jiang, X. et al. (2020). Coffee consumption and risk of stroke: A Mendelian randomization study. *Ann Neurol*, 87(4), 525–532. https://doi.org/10.1002/ana.25693

Retey, J. V., Adam, M., Khatami, R., Luhmann, U. F., Jung, H. H., Berger, W. et al. (2007). A genetic variation in the adenosine A2A receptor gene (ADORA2A) contributes to individual sensitivity to caffeine effects on sleep. *Clin Pharmacol Ther*, 81(5), 692–698. https://doi.org/10.1038/sj.clpt.6100102

Rhee, J., Loftfield, E., Freedman, N. D., Liao, L. M., Sinha, R., and Purdue, M. P. (2021). Coffee consumption and risk of renal cell carcinoma in the NIH-AARP Diet and Health Study. *Int J Epidemiol*, 50(5), 1473–1481. https://doi.org/10.1093/ije/dyab011

Rogers, P. J., Hohoff, C., Heatherley, S. V., Mullings, E. L., Maxfield, P. J., Evershed, R. P. et al. (2010). Association of the anxiogenic and alerting effects of caffeine with ADORA2A and ADORA1 polymorphisms and habitual level of caffeine consumption. *Neuropsychopharmacology*, 35(9), 1973–1983. https://doi.org/10.1038/npp.2010.71

Sachse, C., Brockmoller, J., Bauer, S., and Roots, I. (1999). Functional significance of a C-->A polymorphism in intron 1 of the cytochrome P450 CYP1A2 gene tested with caffeine. *Br J Clin Pharmacol*, 47(4), 445–449. https://doi.org/10.1046/j.1365-2125.1999.00898.x

Sasaki, S., Limpar, M., Sata, F., Kobayashi, S., and Kishi, R. (2017). Interaction between maternal caffeine intake during pregnancy and CYP1A2 C164A polymorphism affects infant birth size in the Hokkaido study. *Pediatr Res*, 82(1), 19–28. https://doi.org/10.1038/pr.2017.70

Sata, F., Yamada, H., Suzuki, K., Saijo, Y., Kato, E. H., Morikawa, M. et al. (2005). Caffeine intake, CYP1A2 polymorphism and the risk of recurrent pregnancy loss. *Mol Hum Reprod*, 11(5), 357–360. https://doi.org/10.1093/molehr/gah175

Schellhas, L., Monasso, G. S., Felix, J. F., Jaddoe, V. W., Huang, P., Fernandez-Barres, S. et al.

(2023). Maternal caffeine consumption during pregnancy and offspring cord blood DNA methylation: An epigenome-wide association study meta-analysis. *Epigenomics*, 15(22), 1179–1193. https://doi.org/10.2217/epi-2023-0263

Senkevich, K., Kuzovenkova, D., Liu, L., and Gan-Or, Z. (2025). Coffee consumption linked to delayed age at onset of Parkinson's disease. *Parkinsonism & Related Disorders*, 134. https://doi.org/10.1016/j.parkreldis.2025.107403

Southward, K., Rutherfurd-Markwick, K. J., and Ali, A. (2018). The effect of acute caffeine ingestion on endurance performance: A systematic review and meta-analysis. *Sports Med*, 48(8), 1913–1928. https://doi.org/10.1007/s40279-018-0939-8

Sulem, P., Gudbjartsson, D. F., Geller, F., Prokopenko, I., Feenstra, B., Aben, K. K. et al. (2011). Sequence variants at CYP1A1-CYP1A2 and AHR associate with coffee consumption. *Hum Mol Genet*, 20(10), 2071–2077. https://doi.org/10.1093/hmg/ddr086

Taylor, A. E., Davey Smith, G., and Munafo, M. R. (2018). Associations of coffee genetic risk scores with consumption of coffee, tea and other beverages in the UK biobank. *Addiction*, 113(1), 148–157. https://doi.org/10.1111/add.13975

Taylor, A. E., Martin, R. M., Geybels, M. S., Stanford, J. L., Shui, I., Eeles, R. et al. (2017). Investigating the possible causal role of coffee consumption with prostate cancer risk and progression using Mendelian randomization analysis. *Int J Cancer*, 140(2), 322–328. https://doi.org/10.1002/ijc.30462

Thorpe, H. H. A., Fontanillas, P., Pham, B. K., Meredith, J. J., Jennings, M. V., Courchesne-Krak, N. S. et al. (2024). Genome-wide association studies of coffee intake in UK/US participants of European ancestry uncover cohort-specific genetic associations. *Neuropsychopharmacology*, 49(10), 1609–1618. https://doi.org/10.1038/s41386-024-01870-x

Unohara, T., Fujii, R., Watanabe, T., Matsuura, A., Torii, Y., Kita, K. et al. (2025). Coffee intake, genetic variants, and chronic kidney disease: A cross-sectional analysis of the Japan Multi-Institutional Collaborative Cohort (J-MICC) study. *Eur J Nutr*, 64(7), 301. https://doi.org/10.1007/s00394-025-03819-2

Valo, E., Richmond, A., Mutter, S., Dahlstrom, E. H., Campbell, A., Porteous, D. J. et al. (2025). Genome-wide characterization of 54 urinary metabolites reveals molecular impact of kidney function. *Nat Commun*, 16(1), 325. https://doi.org/10.1038/s41467-024-55182-1

van Oort, S., Beulens, J. W. J., van Ballegooijen, A. J., Grobbee, D. E., and Larsson, S. C. (2020). Association of cardiovascular risk factors and lifestyle behaviors with hypertension: A Mendelian randomization study. *Hypertension*, 76(6), 1971–1979. https://doi.org/10.1161/HYPERTENSIONAHA.120.15761

Vink, J. M., Staphorsius, A. S., and Boomsma, D. I. (2009). A genetic analysis of coffee consumption in a sample of Dutch twins. *Twin Res Hum Genet*, 12(2), 127–131. https://doi.org/10.1375/twin.12.2.127

Woolf, B., Cronje, H. T., Zagkos, L., Larsson, S. C., Gill, D., and Burgess, S. (2024). Comparison of caffeine consumption behavior with plasma caffeine levels as exposure measures in drug-target Mendelian randomization. *Am J Epidemiol*, 193(12), 1776–1784. https://doi.org/10.1093/aje/kwae143

Yin, B., Wang, X., Huang, T., and Jia, J. (2022). Shared genetics and causality between decaffeinated coffee consumption and neuropsychiatric diseases: A large-scale genome-wide cross-trait analysis and Mendelian randomization analysis. *Front Psychiatry*, 13, 910432. https://doi.org/10.3389/fpsyt.2022.910432

Yuan, S., Carter, P., Mason, A. M., Burgess, S., and Larsson, S. C. (2021). Coffee consumption and cardiovascular diseases: A Mendelian randomization study. *Nutrients*, 13(7). https://doi.org/10.3390/nu13072218

Yuan, S., and Larsson, S. C. (2020). An atlas on risk factors for type 2 diabetes: A wide-angled Mendelian randomisation study. *Diabetologia*, 63(11), 2359–2371. https://doi.org/10.1007/s00125-020-05253-x

Yuan, S., and Larsson, S. C. (2022). Coffee and caffeine consumption and risk of kidney stones: A Mendelian randomization study. *Am J Kidney Dis*, 79(1), 9–14 e11. https://doi.org/10.1053/j.ajkd.2021.04.018

Zhang, Y., Liu, Z., Choudhury, T., Cornelis, M. C., and Liu, W. (2021). Habitual coffee intake and risk for nonalcoholic fatty liver disease: A two-sample Mendelian randomization study. *Eur J Nutr*, 60(4), 1761–1767. https://doi.org/10.1007/s00394-020-02369-z

Zhang, Z., Wang, M., Yuan, S., and Liu, X. (2022). Coffee consumption and risk of coronary artery disease. *Eur J Prev Cardiol*, 29(1), e29–e31. https://doi.org/10.1093/eurjpc/zwaa130

Zhong, V. W., Kuang, A., Danning, R. D., Kraft, P., van Dam, R. M., Chasman, D. I. et al. (2019). A genome-wide association study of bitter and sweet beverage consumption. *Hum Mol Genet*, 28(14), 2449–2457. https://doi.org/10.1093/hmg/ddz061

Zhou, A., and Hypponen, E. (2019). Long-term coffee consumption, caffeine metabolism genetics, and risk of cardiovascular disease: a prospective analysis of up to 347,077 individuals and 8368 cases. *Am J Clin Nutr*, 109(3), 509–516. https://doi.org/10.1093/ajcn/nqy297

Zhou, Q., Luo, M. L., Li, H., Li, M., and Zhou, J. G. (2015). Coffee consumption and risk of endometrial cancer: A dose-response meta-analysis of prospective cohort studies. *Sci Rep*, 5(1), 13410. https://doi.org/10.1038/srep13410

Zhou, S. F., Yang, L. P., Zhou, Z. W., Liu, Y. H., and Chan, E. (2009). Insights into the substrate specificity, inhibitors, regulation, and polymorphisms and the clinical impact of human cytochrome P450 1A2. *AAPS J*, 11(3), 481–494. https://doi.org/10.1208/s12248-009-9127-y

CHAPTER

5

Biochemical Composition of Coffee Beans

Betül Aydın[1], Gürkan Baytar[2*] and Serdal Öğüt[3]

[1] Department of Biology, Faculty of Science, Gazi University, Ankara, Türkiye

[2] Germencik State Hospital, Aydın, Türkiye

[3] Department of Community Nutrition, Aydin Adnan Menderes University, Aydın, Türkiye

Coffee is one of the most widely consumed beverages worldwide and represents a major agricultural commodity with substantial economic, cultural, and scientific significance. Beyond its sensory appeal, coffee beans constitute a highly complex biochemical matrix rich in macronutrients and bioactive compounds that contribute to both flavour development and health-related effects. This chapter provides a comprehensive overview of the biochemical composition of coffee beans, with a particular emphasis on carbohydrates, proteins, lipids, minerals, and water content, as well as bioactive and micronutrient constituents, including alkaloids, phenolic acids, flavonoids, diterpenes, and Maillard reaction–derived melanoidins. The biosynthesis, distribution, and transformation of key compounds—including caffeine, trigonelline, and chlorogenic acids—are discussed in relation to coffee species (*Coffea arabica* and *Coffea canephora*), genetic variability, environmental conditions, and post-harvest processing, especially roasting. In addition, the formation of volatile compounds and aroma precursors responsible for coffee's characteristic sensory profile is examined. The chapter also summarizes the major analytical techniques employed for the qualitative and quantitative determination of coffee constituents, including chromatographic and spectroscopic methods. By integrating biochemical composition with sensory attributes, processing effects, and documented physiological activities, this chapter aims to establish a foundational understanding of coffee beans as a functional biological resource, thereby supporting future research and applications in food science, nutrition, and health-related fields.

*Corresponding author: gurkan.baytar@saglik.gov.tr

DOI: 10.1201/9781003689393-5

1. Introduction

Coffee, derived primarily from the seeds (beans) of two species of coffee tree, *Coffea arabica* (Arabica) and *Coffea canephora* (Robusta), is one of the most widely consumed beverages globally. It also holds significant economic importance as a major agricultural commodity (Hanley et al., 2025). It is estimated that the global coffee market has an annual trade value exceeding $200 billion, with a production volume also estimated to exceed 10 million tons (Massrie, 2025). These figures reflect the sustainable demand observed within both traditional and emerging markets. Beyond the sensory experience associated with coffee consumption, the seeds of the coffee plant are notable for their high content of bioactive compounds. These compounds contribute to the distinctive flavour and aroma profiles of coffee and have been shown to possess significant health benefits (Todhanakasem et al., 2025). The biochemical composition of coffee beans encompasses a diverse array of macronutrients (carbohydrates, proteins, lipids, and minerals) and micronutrients/bioactive compounds (alkaloids such as caffeine and trigonelline, phenolic acids including chlorogenic acids, diterpenes like cafestol and kahweol, and others) (Saud and Salamatullah, 2021). These compounds not only underpin the sensory qualities of brewed coffee—through the formation of volatile aroma precursors during the roasting process via Maillard reactions and thermal degradation—but also exhibit physiological effects. Numerous studies have shown a positive correlation between moderate coffee consumption and several health benefits, including a reduced risk of chronic diseases, improved mental and physical performance, and increased longevity (Makiso et al., 2024; Emadi and Kamangar, 2025). Several studies have indicated that it may have the potential to reduce the risk of developing type 2 diabetes (Mohamed et al., 2024), cardiovascular diseases (Dewland et al., 2025), and certain types of cancer, including liver, prostate, and colorectal cancers (Kunutsor et al., 2025). The bioactive compounds in coffee, including caffeine, chlorogenic acid, and diterpenes, contribute to its antioxidant, anti-inflammatory, and anticarcinogenic properties, which are beneficial in preventing these diseases (Aborziza et al., 2024). The consumption of coffee has been shown to have neuroprotective effects, which may help attenuate the risk of neurodegenerative diseases such as Alzheimer's and Parkinson's diseases (Raeis-Abdollahi et al., 2024; Rodak et al., 2024). Furthermore, coffee consumption has been shown to improve physical performance by increasing fat oxidation and enhancing muscle endurance (Emadi and Kamangar, 2025). The aforementioned treatment has also been demonstrated to enhance cognitive acuity, stabilise mood, and promote alertness. These effects can contribute to the alleviation of depressive symptoms and the reduction of suicide risk (Shi Z. et al., 2025). The evidence suggests that moderate coffee consumption is associated with a reduced overall mortality rate, as well as a reduced rate of cause-specific mortality. This suggests a potential role for coffee in extending lifespan (Kobylińska et al., 2025). The protective effects of coffee against various diseases are a contributing factor to this association, as coffee drinkers tend to have lower rates of hospitalisation (Yan et al., 2025).

As demonstrated by the relevant literature, significant variations in the biochemical profiles of different coffee species are influenced by genetic factors, environmental conditions, and processing methods (Ali et al., 2022). The observed disparities are principally evident in the concentrations of key compounds, including caffeine, chlorogenic acids, trigonelline, and other specialized metabolites. It is essential to understand these variations to develop coffee-based products effectively, assess coffee quality, and improve its breeding. *Coffea arabica* and *Coffea canephora* (Robusta) are the most studied species. Robusta generally contains a higher concentration of caffeine, which contributes to a more pronounced bitterness. Conversely, Arabica is favoured for its superior aroma and milder flavour (Olechno et al., 2021a). Trigonelline, another significant alkaloid, exhibits considerable variation among coffee species and cultivars (Happyana et al., 2022).

This chapter provides a comprehensive examination of the composition of coffee beans, encompassing macronutrients, bioactive micronutrients, aroma precursors, influencing factors, and analytical methods. The objective of this preliminary chapter is to delineate the biochemical significance of coffee beans, summarise their general composition, and emphasise their economic value and health implications, thereby establishing the foundation for subsequent chapters.

2. General macronutrient composition

2.1 Carbohydrates

Naturally occurring carbohydrates, alongside nucleotides and amino acids, may be regarded as representing a third molecular alphabet of life (Dorst and Widmalm, 2024). Carbohydrates, essential macronutrients that serve as the primary source of energy for living organisms, are composed of carbon (C), hydrogen (H), and oxygen (O) atoms. They follow the general empirical formula $Cx(H_2O)n$ and contain a carbonyl functional group, which may exist as either an aldehyde or a ketone. Based on their chemical structure and degree of polymerization, carbohydrates are categorized into monosaccharides, disaccharides, oligosaccharides, and polysaccharides. Monosaccharides act as the fundamental units that can combine to form more complex carbohydrates such as disaccharides, oligosaccharides, and polysaccharides (Seeberger, 2022; H. Zhang and Su, 2025). Carbohydrates account for approximately 40–60% of the dry weight of coffee beans (Xu et al., 2025).

2.2 Monosaccharides

Monosaccharides are the simplest sugars and are important components of the diet. They cannot be broken down into smaller pieces. The components of larger saccharides, such as disaccharides, oligosaccharides, and polysaccharides, are converted into monosaccharides by digestive enzymes, providing the essential life energy (Seeberger, 2022). Monosaccharides are absorbed from the human small intestine (Qi and Tester, 2019). Monosaccharides are the simplest structural units of carbohydrates, which cannot be hydrolysed into smaller sugars. They are classified by the number of carbon atoms they contain as trioses (3 carbons), tetroses (4 carbons), pentoses (5 carbons), hexoses (6 carbons), and heptoses (7 carbons) (Seeberger, 2022).

They are also categorised based on their functional groups. Aldoses contain an aldehyde group (–CHO), whereas ketoses possess a ketone group (C=O). Among the six-carbon monosaccharides that play crucial roles in diet, health, and disease, the most significant ones include fructose, galactose, glucose, and mannose (Figure 1). Glucose and galactose, which possess an aldehyde group on the first carbon atom, are classified as aldohexoses. In contrast, fructose, containing a ketone group on the second carbon atom, is categorised as a ketohexose. Fructose is sweeter than both galactose and glucose. During the first pass through the intestine, fructose and galactose are almost completely metabolised in the liver. In muscle, erythrocytes, liver, and adipose tissues, glucose is metabolically converted into lactate, glycogen, and lipids (Qi and Tester, 2019). Monosaccharides possess multiple hydroxyl (–OH) groups, which confer polarity to their molecular structure. This polarity accounts for their pronounced solubility in water, rendering them distinctly hydrophilic compounds (Singh et al., 2021). Arabinose, a monosaccharide containing five carbon atoms, is classified as an aldose due to the presence of an aldehyde group at one terminal end of its structure (Asadullah et al., 2025).

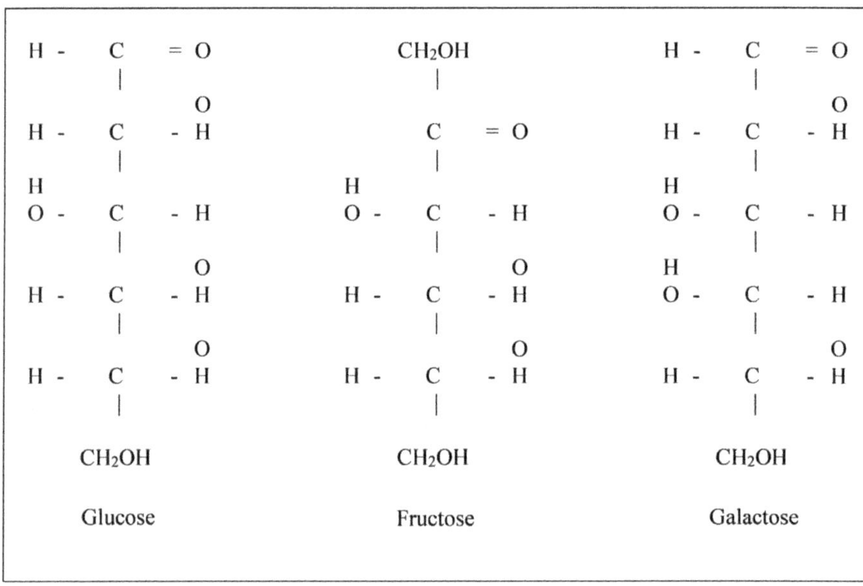

Figure 1: Linear structures of fructose, glucose, and galactose

2.3 Disaccharides

Disaccharides are formed through the combination of two monosaccharide units. During this process, the two monosaccharides are covalently linked by a glycosidic bond, established through the loss of a single water molecule (Pang and Zhang, 2021). A covalent bond is a type of chemical bond that involves the sharing of electron pairs between atoms to achieve stability. The linkage formed between two monosaccharides through this process is known as a glycosidic bond (Ordoñez-

Araque and Revelo-Vizuete, 2023). Maltose, lactose, and sucrose are the most common disaccharides (Pang and Zhang, 2021).

The most important disaccharide in nutrition is sucrose. It is known as table sugar and is mostly processed from sugar cane or beet. Sucrose is formed by the combination of the anomeric carbon 1 of glucose and the anomeric carbon 2 of fructose (β(2→1) bond). It is a widely consumed sugar and is also used as a food additive. It is a water-soluble disaccharide with a high sweetness ratio. Lactose, known as milk sugar, is formed by the anomeric carbon 1 of D-galactose bonding with the anomeric carbon 4 of D-glucose, forming a β (1 → 4) bond. It is a disaccharide with a low sweetness ratio. It is broken down into glucose and galactose by the action of the lactase enzyme. Maltose is formed by the linkage of two glucose molecules through the 1st and 4th carbon atoms. It is hydrolyzed by the enzyme maltase into two glucose units. This hydrolysis process takes place in the intestines (Ordoñez-Araque and Revelo-Vizuete, 2023).

2.4 Polysaccharides

Polysaccharides, also known as glycans, are biological macromolecules formed by the linkage of hundreds of monosaccharide units through glycosidic bonds. These compounds serve as storage carbohydrates within cells, accumulating in the form of granules. In plants, the main storage polysaccharide is starch, whereas in animals it is glycogen. Because both starch and glycogen are composed solely of glucose residues, they are collectively referred to as glucans. Most polysaccharides appear as white, amorphous, and tasteless substances. Natural, non-toxic polysaccharides have a wide range of applications due to their biological activities and diverse industrial uses (Pang and Zhang, 2021).

Polysaccharides are divided into three main groups based on their sources: plant polysaccharides, animal polysaccharides, and microbial polysaccharides. Most polysaccharides in nature are derived from plant polysaccharides. The most important of these are cellulose, mannan, arabinogalactan, and starch (Xu et al., 2025). Cellulose is formed by the combination of many β-glucose molecules. They are linked by β-1,4-glycosidic bonds. It is a water-insoluble and indigestible polysaccharide. It is a food source for beneficial bacteria. Mannans are macromolecules composed of numerous mannose monosaccharide units linked together through glycosidic bonds. These polysaccharides are enzymatically degraded by mannanases. Based on their carbohydrate composition, mannans are classified into four main categories: homomannan, glucomannan, galactomannan, and galactoglucomannan. Mannan, found in seeds, provides cell wall rigidity. It is also found in coffee beans. Galactomannan, which consists of mannose and galactose monosaccharides, is found dominantly at 69% in coffee beans (Voiniciuc, 2022). Arabinogalactan polysaccharides contain 6 C galactose and 5 C arabinose monosaccharides (Asadullah et al., 2025).

Animal polysaccharides are generally bound to proteins. The most well-known are heparin, hyaluronic acid, chondroitin sulfate, and chitin. Microbial polysaccharides can be synthesised by a variety of microorganisms, including bacteria, fungi, and algae. Among these biopolymers, xanthan gum, pullulan, and gellan gum are recognised as the most prominent examples (Xu et al., 2025).

3. Proteins and amino acids

The basic units of proteins are smaller amino acids. Amino acids combine to form long chains known as polypeptides, which in turn form large and complex molecules called proteins. These chains of amino acids, linked together by peptide bonds, fold into specific three-dimensional conformations, thereby acquiring the structural organisation necessary for the formation of functional proteins (Figure 2) (Rehlund et al., 2025).

There are hundreds of amino acids found in nature; however, only 20 of them are incorporated into the structure of proteins. Each amino acid has a central alpha carbon atom to which four different groups are attached: an amino group ($-NH_2$), a carboxyl group (–COOH), a hydrogen atom (–H), and a variable side chain known as the R group. The R group is the distinguishing group. The R group determines whether amino acids are polar, nonpolar, acidic, or basic. The $-NH_2$ of one amino acid is covalently bonded to the carboxyl group of another amino acid through a bond called a peptide bond (Yi et al., 2025).

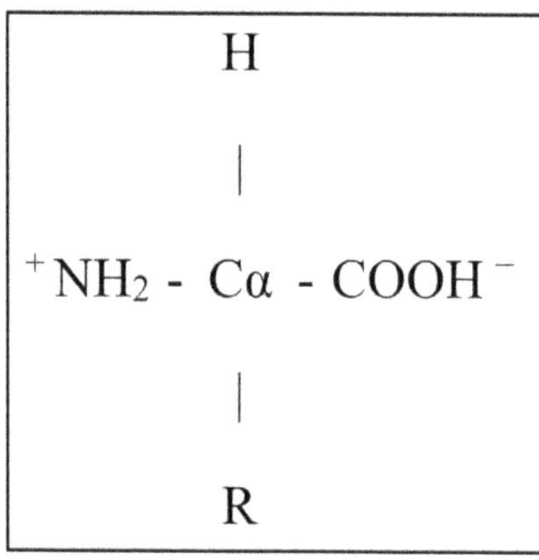

Figure 2: Amino acid structure

- Polar (Uncharged) Side Chain Amino Acids; Serine (Ser, S), Threonine (Thr, T), Cysteine (Cys, C), Asparagine (Asn, N), Glutamine (Gln, Q), Tyrosine (Tyr, Y)
- Amino Acids with Nonpolar Side Chains are hydrophobic; Glycine (Gly, G), Alanine (Ala, A), Valine (Val, V), Leucine (Leu, L), Isoleucine (Ile, I), Proline (Pro, P), Methionine (Met, M), Phenylalanine (Phe, F), Tryptophan (Trp, W)
- Amino acids with acidic side chains have a carboxyl group (–COOH) in their side chains; They are negatively charged at physiological pH. Aspartic Acid (Asp, D), Glutamic Acid (Glu, E)
- Basic Amino Acids with Basic Side Chains; These amino acids possess side chains containing basic functional groups such as amino groups ($-NH_2$). At physiological pH, their side chains are positively charged due to protonation of these basic groups. Lysine (Lys, K), Arginine (Arg, R), Histidine (His, H) (Yi et al., 2025).

L-Glutamic acid and L-Aspartic acid are used as flavour enhancers. They also have a similar effect on coffee flavour (Murkovic and Derler, 2006; Zimmermann et al., 2025).

Fresh coffee beans (unroasted) contain protein fractions, including globulin, glutelin, and albumin. Globulin (11S legumin-type globulins) is the most abundant and important storage protein in coffee beans (Bhattarai et al., 2022; Toribio-Nava et al., 2024). Glutelin is the most abundant storage protein fraction after globulins. It is soluble in dilute acid or base solutions. Albumin is the third most abundant protein fraction after the other protein fractions. It is water-soluble and metabolically active.

Roasting coffee beans takes place at high temperatures. This denatures the proteins. These denatured proteins undergo a Maillard reaction with sugars, which creates the coffee's characteristic brown colour, aroma, and flavour. The Maillard reaction occurs between the free amino groups of amino acids and reducing sugars or lipid oxidation products. The non-enzymatic reaction forms early volatile products and medium- to high-molecular-weight polymers. Maillard reactions cause coffee beans to brown and develop aromas. Pyrrole, imidazole, and pyrazine are formed as a result of Maillard reactions. These are heterocyclic aromatic compounds that play a critical role in aroma (Lee et al., 2024). Pyrazins, which contain two nitrogen atoms, provide the various aromas that define coffee's character. Pyrrole, which contains one nitrogen atom, provides coffee with earthy, hayy, and malty aromas. Imidazole, which has two nitrogen atoms, imparts bitter, salty, and metallic aromas to coffee (Moon et al., 2025).

The Maillard reaction proceeds through a series of interconnected stages that collectively drive the formation of colour, aroma, and flavour in thermally processed foods such as coffee. The initial stage, condensation, involves a reaction between the free amino group ($–NH_2$) of an amino acid and the carbonyl group (–C=O) of a reducing sugar, leading to the release of a water molecule (H_2O) and the formation of a Schiff base, also known as an N-glycosylamine; this step is considered the first and irreversible phase of the Maillard reaction. In the subsequent reorganisation stage, the unstable Schiff base undergoes molecular rearrangement to form the more stable Amadori product, which serves as a key intermediate and flavour precursor. This is followed by the disintegration stage, in which Amadori products degrade via different pathways depending on environmental conditions such as pH. Under acidic conditions, the deoxosone pathway favours the formation of furan derivatives, including furfural and 5-hydroxymethylfurfural, which impart sweet, caramel-like, berry-like, and occasionally woody aroma notes. Under basic conditions, alternative degradation routes lead to the formation of compounds such as furaneol, contributing primarily to caramel and fruity aromas. In the Strecker degradation stage, amino acids react with dicarbonyl compounds, producing Strecker aldehydes that are crucial contributors to characteristic roasted and malty flavours. Finally, during the condensation and polymerisation stage, high-molecular-weight nitrogen-containing polymers known as melanoidins are formed, which are responsible for the characteristic brown colouration of coffee and also contribute to its antioxidant properties (Xiong et al., 2024).

4. Lipids

4.1 Fatty acids

Lipids, a heterogeneous group of hydrophobic organic molecules, are composed of fatty acids with a carboxylic acid group (–COOH) at their end. These fatty acids, which have a hydrocarbon structure of 4–28 carbons long, also have a methyl group (CH_3) at their other end. They are also the precursors of complex lipids. The methyl group is hydrophobic, while the carboxyl group is hydrophilic. Fatty acids are classified in many ways. However, classification according to saturation and unsaturation is important.

Unsaturated fatty acids contain one or more double bonds. Monounsaturated fatty acids (MUFAs) contain one double bond. Olive oil is one of the MUFAs. Polyunsaturated fatty acids (PUFAs) contain two or more double bonds. Both MUFAs and PUFAs are liquid at room temperature. Linoleic acid is among the PUFAs. Linoleic acid is found in coffee beans and is prone to oxidation due to its high number of double bonds. Oxidation can cause bitterness and the development of undesirable aromas. This plays a role in the staling of coffee and limits its shelf life (A. Alsanad et al., 2021).

Saturated fatty acids have no double bonds between their carbon atoms. The chain is "saturated" with hydrogen, making it solid at room temperature. Palmitic acid is also a saturated fatty acid. Therefore, the palmitic acid found in coffee beans is more stable against oxidation (Vu et al., 2021).

4.2 Triglycerides

Triglycerides are the storage form of energy in both plants and animals. A glycerol molecule is bonded to three fatty acids to form a triglyceride. Triglycerides constitute a significant portion of the lipids in coffee beans. The amount of unsaturated fatty acids, such as linoleic acid, increases their susceptibility to oxidation. Light and heat are other factors that affect oxidation (Fukagawa et al., 2017).

4.3 Sterols

It consists of four interconnected carbon rings with a hydrocarbon tail. It can occur in nature in free form or esterified with fatty acids. β-Sitosterol, Stigmasterol, and Campesterol are the primary sterols found in coffee beans and are listed in order of most abundant to least abundant. Sterols such as Δ5-avenasterol, campestanol, 24-methylenecholesterol, sitostanol, 7-stigmasterol, 7-avenasterol, and clerosterol are also present. Sterols are susceptible to oxidation due to the double bonds in their structures. It occurs in the form of autoxidation and thermal oxidation. During the roasting process, temperatures above 150°C cause thermal oxidation (Caño-Carrillo et al., 2025).

5. Minerals and water content

Mineral and water content are among the fundamental components of coffee beans. Water, which makes up approximately 12% of the coffee bean, directly affects its shelf life, controls enzymatic activity, inhibits the growth of mold and bacteria, and preserves flavour precursors. The amount of water present in coffee beans before roasting decreases to approximately 1% after roasting. Water is also important for Maillard reactions, flavour development, and the physical transformation of coffee beans (Theurillat et al., 2024).

Coffee beans contain about 4% of the minerals in their dry form. Potassium is the most abundant element in coffee beans. Potassium is followed by magnesium, calcium, phosphorus, sodium, manganese, and trace elements (Olechno et al., 2021b).

6. Bioactive and micronutrient compounds

Coffee beans have a reputation that is not solely attributed to their macronutrient composition; the beverage's sensory characteristics and potential health benefits are also significantly impacted by the presence of a rich array of bioactive micronutrients and secondary metabolites. The presence of these compounds, which vary in concentration according to species, genetic factors, environmental conditions, and processing methods, includes alkaloids, phenolic acids, flavonoids, terpenoids (primarily diterpenes), vitamins, and roasting-derived polymers such as melanoidins (Figure 3) (Makiso et al., 2024). It is evident from a range of in vitro, animal, and epidemiological studies that bioactive compounds in coffee exert multifaceted physiological effects, including antioxidant, anti-inflammatory, neuroprotective, antidiabetic, and anticarcinogenic activities (Machado et al., 2024).

The following subsections detail the major classes of these compounds, their typical concentrations in green and roasted beans, biosynthetic pathways where known, and documented health implications.

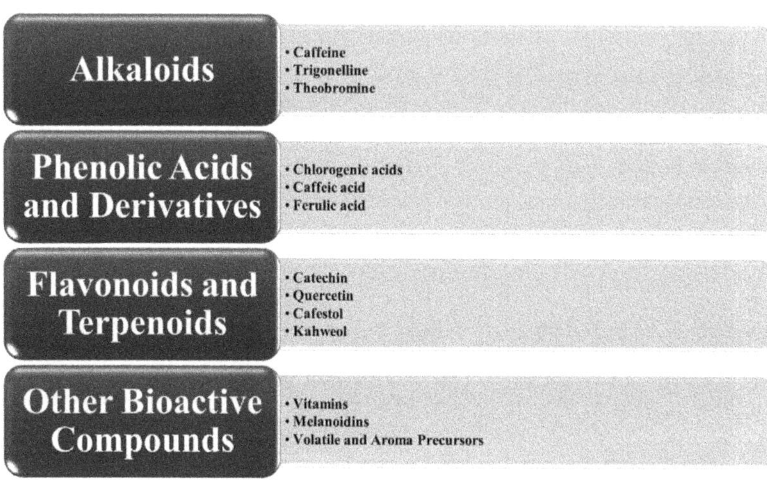

Figure 3: Some examples of bioactive and micronutrient compounds in coffee

6.1 Alkaloids

Alkaloids represent a significant class of nitrogen-containing secondary metabolites present in coffee beans, consisting predominantly of purine (methylxanthine) and pyridine derivatives (Ashihara, 2006; Ashihara, 2016). The predominant alkaloids are caffeine (1,3,7-trimethylxanthine, typically 1–2% dry weight in green beans) (Langi et al., 2023; Sutrisno et al., 2023) and trigonelline (N-methylnicotinic acid, 0.5–1% dry weight), with minor amounts of theobromine (3,7-dimethylxanthine) and trace levels of theophylline (Mehari et al., 2016). These chemical compounds have been shown to contribute to the bitterness of coffee, exert a significant influence on the formation of aroma precursors during the process of roasting, and to exert significant physiological effects in consumers, including central nervous system stimulation, cardiovascular modulation, and the potential to offer neuroprotective benefits (dePaula and Farah, 2019).

Caffeine

Caffeine (1,3,7-trimethylxanthine) is the most abundant and best-studied purine alkaloid found in coffee beans, accounting for approximately 0.9–1.3% of the dry weight of green *Coffea arabica* beans and 1.5–2.5% in *Coffea canephora* (Robusta). Some varieties have been found to contain up to 4% caffeine (Ky et al., 2001). It is the principal agent responsible for the bitterness perceived in coffee beverages and acts as a natural defence compound against herbivores and pathogens in the coffee plant. The process of caffeine biosynthesis in coffee is primarily located in the young leaves, pericarp, and developing endosperm (seeds). The pathway begins with xanthosine, which undergoes methylation to form 7-methylxanthosine. This reaction is catalyzed by xanthosine methyltransferase (XMT) (Uefuji et al., 2003). The ribose moiety is removed from 7-methylxanthosine, resulting in 7-methylxanthine. This step is facilitated by a nucleosidase enzyme. 7-methylxanthine is then methylated to form theobromine, a reaction catalyzed by 7-methylxanthine methyltransferase (MXMT). Theobromine undergoes a final methylation to produce caffeine, catalyzed by 3,7-dimethylxanthine methyltransferase (DXMT) (Ogita et al., 2004) (Figure 4). N-Methyltransferases (NMTs) are crucial for the methylation steps in caffeine biosynthesis. The three main NMTs involved are XMT, MXMT, and DXMT, each responsible for a specific methylation step in the pathway (Zhou et al., 2020). Biosynthesis is developmentally regulated, with the highest activity in immature fruits, declining as beans mature. Environmental factors, such as light intensity and altitude, can influence enzyme expression and the final caffeine content (Perrois et al., 2015).

Figure 4: The major biosynthetic pathway of caffeine from xanthosine

The transformation of caffeine during coffee processing involves several stages, each affecting the caffeine content and structure in different ways. The primary processes include roasting, decaffeination, and microbial fermentation, each contributing uniquely to the caffeine profile in coffee. These transformations are crucial as they influence the final quality and characteristics of the coffee consumed. It is imperative to recognise the pivotal role that the roasting process plays in determining the caffeine content of a coffee. As the temperature of the roasting process increases, the caffeine content of coffee beans has been observed to decrease. However, the structural integrity of caffeine is not compromised during the roasting process, as evidenced by Gas Chromatography-Mass Spectrometry (GC-MS) analysis (Suaniti et al., 2022). Different roasting levels (light, medium, dark) also affect caffeine content, but the variations are not solely dependent on the roasting level (Muzykiewicz-Szymańska et al., 2021). Decaffeination is a process that is designed to remove caffeine from green coffee beans. It is typically carried out using solvents such as water, ethyl acetate, or supercritical carbon dioxide. The objective of this process is to produce coffee with a reduced caffeine content while maintaining flavour quality. The decaffeination process is a method of extracting caffeine from coffee beans or tea leaves into an aqueous solution. The caffeine is then removed from this solution using activated charcoal or other similar adsorbents. The selection of solvent and method has been shown to significantly influence the efficiency of caffeine removal and the preservation of coffee's aromatic compounds (Shofinita et al., 2024). Microbial fermentation represents a further method by which the caffeine content of coffee beans can be reduced. Specific microorganisms, including Lactobacillus casei and Saccharomyces cerevisiae, have been found to degrade caffeine into other compounds, such as dimethylxanthine and methylxanthine. This transformation is characterised by the conversion of caffeine into paraxanthine and, subsequently, into methylxanthine. These reactions are catalysed by distinct microbial enzymes, which follow different degradation pathways (Shen et al., 2024; Ran et al., 2025).

Caffeine exerts a range of physiological effects on the human body. As a non-selective antagonist for adenosine receptors, caffeine contributes to greater wakefulness and a reduced experience of fatigue (Song et al., 2023). It has the potential to enhance cognitive functioning, particularly in activities that require prolonged focus and alertness (Ricupero and Ritter, 2024). The ingestion of caffeine is correlated with both advantageous and detrimental effects on cardiovascular health. It possesses the capacity to enhance cardiovascular endurance and exhibits properties that are protective to the heart (Bjelıca et al., 2023). Nevertheless, elevated dosages may result in an augmented heart rate (tachycardia) and heightened blood pressure, thereby posing potential hazards for individuals with pre-existing cardiovascular ailments (Turnbull et al., 2017). Caffeine increases the release of catecholamines, which can enhance metabolic rate and promote lipolysis, leading to increased free fatty acids in the bloodstream (Virgili et al., 2023). Improvements in glucose metabolism and a reduction in the risk of type 2 diabetes are possible, although the influence of genetic factors must be taken into account when interpreting these outcomes (Song et al., 2023). Evidence has demonstrated its capacity to induce bronchodilation, a property that may prove advantageous for individuals afflicted with respiratory ailments, such as asthma (Reddy et al., 2024). Caffeine has been demonstrated to possess neuroprotective properties, which may serve to reduce the risk of neurodegenerative diseases, including Alzheimer's and Parkinson's (M Yelanchezian et al., 2022). Additionally, evidence has been found to demonstrate anticarcinogenic activity; however, further research is required to fully comprehend the underlying mechanisms (Tej and Nayak, 2018). Nevertheless, the effects of caffeine are complex and variable, influenced by factors such as dosage, individual sensitivity, and genetic disposition. The ingestion of elevated quantities of caffeine has the potential to induce insomnia, anxiety, and nervousness (Liu et al., 2024). Genetic polymorphisms can increase the risk of adverse effects, including hypertension and cardiovascular complications (Fulton et al., 2018). The ingestion of caffeine has been demonstrated to be a factor that may increase the likelihood of developing osteoporosis and experiencing complications during pregnancy (Olchowik et al., 2011).

Trigonelline

Trigonelline (N-methylnicotinic acid, also known as 1-methylpyridinium-3-carboxylate) is a pyridine alkaloid. It is the second most abundant alkaloid in coffee beans, after caffeine, typically comprising 1–3% of the dry weight of coffee beans, with variations observed based on the coffee species and growing conditions (Atlabachew et al., 2021; Di Gioacchino et al., 2024).

The biosynthesis of trigonelline in coffee plants involves a complex pathway that converts nicotinic acid into the aforementioned alkaloid, which is found in significant quantities in coffee seeds. The metabolism of nicotinic acid and its byproducts depends on this mechanism, which is a part of the larger pyridine nucleotide cycle. The primary biosynthetic pathway for trigonelline in coffee plants is as follows: NAD → nicotinamide mononucleotide → nicotinamide riboside → nicotinamide → nicotinic acid → trigonelline. This pathway demonstrates the conversion of nicotinic acid, a pivotal intermediate, into trigonelline through methylation processes (Figure 5). The pyridine nucleotide cycle (PNC) is integral to this process, whereby nicotinamide is converted to nicotinic acid. This, in turn, forms nicotinic acid mononucleotide (NaMN) and nicotinic acid adenine dinucleotide (NaAD), ultimately regenerating NAD (Zheng et al., 2004). Trigonelline synthase activity has been detected in coffee plants, although the specific gene encoding this enzyme has not yet been cloned. It has been established that this enzyme is responsible for the methylation of nicotinic acid, thereby forming trigonelline (Ashihara et al., 2015). Trigonelline has been detected in all parts of the coffee plant; however, it is found in particularly high quantities in the seeds. Its presence has also been observed in young leaves and fruits, where it accumulates during the developmental process (Zheng et al., 2004).

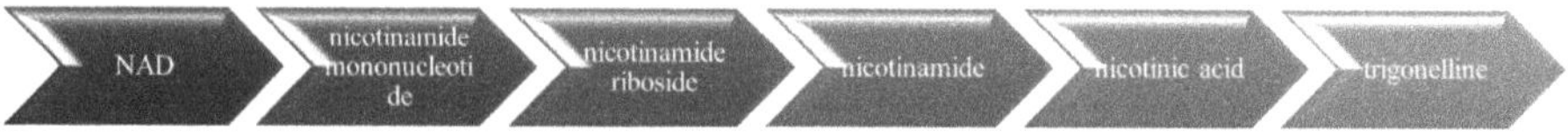

Figure 5: The major biosynthetic pathway of trigonelline from NAD

Trigonelline undergoes various transformations during coffee processing, particularly during the roasting process. During the roasting process, trigonelline undergoes thermal degradation, resulting in the conversion into nicotinic acid and other compounds, including N-methylpyridinium ions and methyl nicotinate (Stadler et al., 2002). The extent of the roasting process has been shown to have a substantial impact on the rate of trigonelline degradation, with a greater loss of this compound being observed in Robusta coffee compared to Arabica. The degradation process has been linked to the formation of nicotinic acid, which in turn impacts the nutritional profile of roasted coffee (Jeszka-Skowron et al., 2020).

Trigonelline exhibits a wide range of physiological effects contributing to its potential therapeutic applications. The substance exerts its influence on a range of molecular targets and pathways, thereby impacting metabolic, inflammatory, and oxidative processes. Trigonelline has been demonstrated to promote glucose and lipid homeostasis, making it beneficial for the treatment of diabetes and related metabolic conditions. The process under investigation can enhance the regeneration of β-cells, as well as insulin secretion. Additionally, it has been demonstrated to modulate enzymes involved in glucose metabolism (Zhou et al., 2012; Zhou et al., 2013). It has been shown that the substance being studied affects white adipocyte browning, which increases lipid metabolism and promotes lipid breakdown. This suggests the possibility of anti-obesity effects (Choi et al., 2021). Trigonelline has been shown to possess significant anti-inflammatory properties. These properties result from the inhibition of pro-inflammatory factor release and the modulation of pathways such as NF-κB and MAPK, which play a crucial role in inflammatory responses (Zhang et al., 2021). The substance has also been shown to act as an antioxidant, thereby reducing oxidative stress by blocking detrimental pathways, such as the Nrf2 pathway, in cases where autophagy is impaired, and enhancing stress resistance in models like *C. elegans* (Nguyen et al., 2024). Trigonelline prevents high-glucose-induced endothelial-to-mesenchymal transition, oxidative stress, and mitochondrial dysfunction, thereby maintaining endothelial function and angiogenic activity (Peerapen et al., 2024).

Theobromine

Theobromine is an intermediate in the primary caffeine biosynthetic pathway in coffee plants (Figure 4). Theobromine concentrations are typically low in both *C. arabica* and *C. canephora*, with some reports indicating slightly higher levels in certain Arabica varieties. However, these concentrations are negligible with respect to the contribution of coffee to the diet in general (Jeszka-Skowron et al., 2022). Theobromine demonstrates comparable heat stability to caffeine, exhibiting minimal degradation during the roasting process. Nonetheless, it has been demonstrated that darker roasts may exhibit slight reductions as a consequence of thermal effects or migration (Hermund et al., 2025).

Theobromine has been demonstrated to enhance cognitive function and provide neuroprotection, particularly in conditions related to dementia and Alzheimer's disease. The substance under discussion has been shown to facilitate synaptic transmission and normalise synaptic plasticity by antagonising adenosine receptors. This, in turn, has been demonstrated to help prevent neuronal loss and long-term potentiation deficits in the hippocampus (Valada et al., 2022). Furthermore, it has been demonstrated to upregulate brain-derived neurotrophic factor (BDNF) and facilitate motor learning in mice, thus indicating its role in supporting neuronal functions and learning processes (Yoneda et al., 2017). Theobromine has been shown to have a beneficial effect on cardiovascular health, with studies demonstrating that it can improve lipid profiles and lower blood pressure. It has been hypothesised that the phenomenon under investigation exerts a favourable influence on inflammatory markers and vascular function. Nevertheless, the validity of these findings remains to be confirmed by further research in the form of long-term human studies (Sharifi-Zahabi et al., 2023). It has been demonstrated to stimulate thermogenesis by inducing the browning of white adipose tissue and activating brown adipocytes, thereby enhancing both lipid metabolism and energy expenditure. It can thus be hypothesised that the substance has the potential to act as an agent in the management of obesity and related metabolic disorders (Jang et al., 2018). It has been demonstrated that theobromine has significant anti-inflammatory properties, which are beneficial in reducing the low-grade chronic inflammation associated with obesity (Tuğal Aslan and Göktaş, 2025).

6.2 Phenolic acids and derivatives

Phenolic acids and their derivatives constitute the most abundant class of polyphenols in coffee beans, making coffee one of the richest dietary sources of these compounds. The significance of these compounds in coffee in terms of its health benefits and sensory attributes is a subject of considerable interest. Chlorogenic acids (CGAs) are the predominant group found in the coffee sample. Free phenolic acids, including caffeic acid, ferulic acid, and minor concentrations of sinapic and p-coumaric acids, are also present (Risso et al., 2007).

Chlorogenic acids (CGAs)

Chlorogenic acids (CGAs) are the predominant phenolic compounds in coffee beans, representing the major subclass of phenolic acids and derivatives. According to the International Union of Pure and Applied Chemistry (IUPAC) nomenclature, the term "chlorogenic acid" specifically refers to 5-O-caffeoylquinic acid (5-CQA). However, it is commonly used more broadly to encompass a family of esters formed between quinic acid and one or more trans-hydroxycinnamoyl moieties (caffeic, ferulic, or p-coumaric acids). The main isomers include 3-CQA, 4-CQA, and 5-CQA (mono-caffeoylquinic acids), as well as di-caffeoylquinic acids (3,4-diCQA, 3,5-diCQA, 4,5-diCQA), feruloylquinic acids (FQAs), and p-coumaroylquinic acids (p-CoQAs) (Upadhyay and Mohan Rao, 2013; Naveed et al., 2018). The total content of CGA in green coffee beans ranges from 5–8% dry weight in *Coffea arabica* to 7–12% in *C. canephora* (Robusta), with 5-CQA typically being the most abundant (Colomban et al., 2020; Pimpley et al., 2020; Rojas-González et al., 2022).

CGA biosynthesis in coffee has been observed to occur primarily in young leaves and developing fruits via the shikimate-phenylpropanoid pathway. The pathway commences with the deamination of phenylalanine by phenylalanine ammonia-lyase (PAL) to form trans-cinnamic acid. This is followed by a series of hydroxylations and methylations catalysed by cinnamate 4-hydroxylase (C4H), 4-coumarate: CoA ligase (4CL), and other enzymes, resulting in the production of caffeoyl-CoA, feruloyl-CoA, and p-coumaroyl-CoA. The key esterification step involves hydroxycinnamoyl-CoA: quinate hydroxycinnamoyl transferase (HQT), which transfers the acyl group to quinic acid (a product of the shikimate pathway) to form mono-CQAs. Derivatives of di- and tri-acyl groups are formed through the process of further acylations, which may involve the involvement of hydroxycinnamoyl-CoA: shikimate/quinate transferase (HCT) intermediates (Koshiro et al., 2007; Lallemand et al., 2012; Clifford et al., 2017).

Chlorogenic acids (CGAs) undergo substantial transformations during coffee processing, particularly during roasting. These transformations are critical, as they affect the chemical composition, flavour, and potential health benefits of coffee. During the process of roasting, there is a degradation of CGAs into volatile components, resulting in the formation of quinic acid and its epimers, as well as chlorogenic acid lactones. It is evident that this process also results in alterations to the ratios between different CGA isomers and dicaffeoylquinic acid isomers (Dawidowicz and Typek, 2017).

Chlorogenic acids (CGAs) have been the subject of extensive research due to their diverse physiological effects. It has been hypothesised that CGA, which has been demonstrated to regulate glucose and lipid metabolism in vivo and improve insulin sensitivity, may prevent and treat diabetes mellitus, obesity, and metabolic syndrome (Meng et al., 2013; Santana-Gálvez et al., 2017). CGAs possess neuroprotective properties, with the potential to benefit neurodegenerative disorders and diabetic peripheral neuropathy by targeting neuroreceptors and ion channels (Heitman and Ingram, 2017). They may also reduce the carcinogenic potential of other compounds, such as caffeic acid, by regulating prooxidant activities and preventing DNA damage (Li et al., 2023). It has been demonstrated that CGA supplementation has been associated with a reduction in malondialdehyde (MDA) levels and an increase in testosterone in cases of diabetes, thus indicating their role in mitigating oxidative stress and improving hormonal balance (Sofyanita et al., 2025).

Caffeic acid (CA)

Caffeic acid (3,4-dihydroxycinnamic acid) is a hydroxycinnamic acid and a key phenolic compound present in coffee beans. The occurrence of this substance is predominantly in bound form as the acyl moiety in chlorogenic acids (CGAs), contributing significantly to the total phenolic content. However, it is present in free form at much lower concentrations (Birková et al., 2020). CA is biosynthesised from the amino acids tyrosine or phenylalanine and is often present in foods as chlorogenic acid, an ester of caffeic acid and quinic acid (Espíndola et al., 2019). Caffeic acid is released from CGAs during the roasting process via hydrolysis, with levels peaking in medium roasts before undergoing partial degradation or incorporation into melanoidins in darker roasts. It has been demonstrated that the substance in question contributes to the development of astringency, as well as to the oxidation process, which in turn leads to the evolution of colour and aroma (Dawidowicz and Typek, 2017).

Caffeic acid exhibits a range of physiological effects that contribute to its potential health benefits. Caffeic acid has been demonstrated to exhibit significant antioxidant activity by scavenging reactive oxygen species (ROS) and chelating metal ions, thereby helping to prevent oxidative damage. It is essential to note that this activity plays a crucial role in reducing lipid peroxidation and preserving cellular components from oxidative stress (Genaro-Mattos et al., 2015). In addition, the substance exhibits anti-inflammatory properties by inhibiting the production of pro-inflammatory cytokines and enzymes, such as cyclooxygenase-2 (COX-2), through the suppression of signalling pathways, such as NF-κB and AP-1 (Kang et al., 2009; Shin et al., 2017). In the nervous system, caffeic acid has been observed to reduce oxidative stress and microglial activation in the hippocampus, which are associated with neurological disorders. This finding indicates the possible positive effects on brain health and the potential for the prevention of neurodegenerative diseases (Koga et al., 2019). Caffeic acid has been identified as a contributing factor to metabolic health through its role in reducing lipid accumulation and ROS production in adipocytes. This property has the potential to assist in the management of obesity and related metabolic disorders (Dantas et al., 2018). In addition, it has been demonstrated to exhibit vasorelaxant effects with the capacity to reduce blood pressure and to protect against cardiovascular diseases. The aforementioned effects are mediated through a range of mechanisms, including increased endothelial nitric oxide secretion and modulation of calcium and potassium channels (Silva and Lopes, 2020).

Ferulic acid

Ferulic acid (4-hydroxy-3-methoxycinnamic acid) is a derivative of caffeic acid and is present in coffee in both free and conjugated forms, often as part of chlorogenic acids (CGAs), where it is ester-bonded to quinic acids (Saud and Salamatullah, 2021). Free ferulic acid is present in low concentrations in green beans, but its concentration increases during the roasting process due to partial hydrolysis of feruloylquinic acids (FQAs) and other esters. Ferulic acid is recognised for its potent antioxidant properties, which help neutralise free radicals and reduce oxidative stress. It is imperative to recognise the significance of this factor to avert potential cellular damage and ensure the preservation of general well-being (Kumar et al., 2025). The substance exerts anti-inflammatory effects by regulating inflammatory cytokines and modulating signalling pathways such as nuclear factor-kB (NF-κB) and mitogen-activated protein kinase (MAPK). These signalling pathways are involved in inflammatory responses (Liu et al., 2025). As demonstrated by clinical trials, ferulic acid supplementation has been shown to have a significant effect on inflammatory markers, such as hs-CRP and TNF-α. This suggests it may alleviate inflammation-related conditions. Ferulic acid has demonstrated encouraging results in the enhancement of cardiovascular health, achieved by means of improved lipid profiles and reduced oxidative stress. The decrease in total cholesterol, LDL-C, and triglycerides, whilst the increase in HDL-C, is of critical importance in the prevention of cardiovascular disease (Bumrungpert et al., 2018). Ferulic acid has been identified as a promising agent in the management of metabolic syndrome. It has been observed to exert several beneficial effects, including improved glucose and lipid metabolism, reduced blood pressure, and modulation of inflammatory pathways (Jacobo-Velázquez, 2025).

6.3 Flavonoids and terpenoids

Flavonoids and terpenoids represent important classes of secondary metabolites in coffee beans, contributing to antioxidant capacity, sensory properties, and potential health effects, though present at lower concentrations compared to phenolic acids or alkaloids. Flavonoids in coffee primarily include flavan-3-ols such as (+)-catechin and (-)-epicatechin, as well as flavonols like quercetin, kaempferol, and their glycosides (rutin, quercetin-3-rutinoside) at trace to minor levels (Zengin et al., 2020; Mediani et al., 2023). Flavonoids in coffee contribute to its antioxidant capacity, which is associated with various health benefits, including protection against cardiovascular diseases and neurodegenerative disorders (Lu et al., 2013; Barreca et al., 2021). Terpenoids in coffee are primarily composed of ent-kaurane diterpenes, specifically cafestol and kahweol, with cafestol being more prevalent in both Arabica and Robusta, while kahweol is almost exclusively found in Arabica (Silva et al., 2023). Terpenoids in coffee offer a range of health benefits due to their diverse bioactive properties. Cafestol and kahweol have been shown to induce apoptosis in tumor cells and inhibit angiogenesis, a process crucial for tumor growth and metastasis (Ren et al., 2019). Kahweol has been demonstrated to reduce the production of pro-inflammatory cytokines and chemokines in human keratinocytes, which has been shown to be beneficial in the management of inflammatory skin conditions, such as atopic dermatitis (Kwon et al., 2024). In studies utilising *Caenorhabditis elegans*, kahweol has been demonstrated to prolong lifespan through the insulin/insulin-like growth factor-1 and AMP-activated protein kinase signalling pathways. This suggests that kahweol may promote longevity by modulating vital metabolic pathways (Cho et al., 2023). It has been demonstrated that Kahweol has the capacity to reduce neuroinflammation and to improve outcomes in traumatic brain injury models. This outcome is realised through the attenuation of pro-inflammatory cytokine secretion and the reduction of microglia/macrophage activation, which are pivotal factors in the process of neuroinflammation (Lee et al., 2019). It has been demonstrated that Kahweol protects against acetaminophen-induced hepatotoxicity by inhibiting oxidative stress, reducing hepatocyte death, and suppressing inflammation. This finding suggests the potential use of the substance as a therapeutic agent in managing hepatic injuries (Kim et al., 2022). Cafestol has been found to inhibit the production of prostaglandin E2 (PGE2), a key mediator in inflammatory responses, by targeting the extracellular signal-regulated kinase (ERK) pathway. This inhibition has been demonstrated to reduce the expression of cyclooxygenase-2 (COX-2) in lipopolysaccharide-activated macrophages, thereby highlighting its potential as an anti-inflammatory agent (Shen et al., 2010). Cafestol has demonstrated potential in enhancing glucose metabolism. In experimental models of the disease, the substance has been demonstrated to increase insulin secretion and enhance glucose uptake in muscle cells. These findings suggest that it may have therapeutic potential in the treatment of type 2 diabetes (Mellbye et al., 2017).

6.4 Melanoidins

Coffee melanoidins are defined as complex, high-molecular-weight compounds formed during the Maillard reaction between reducing sugars, amino acids/proteins, and phenolic compounds (particularly chlorogenic acids), a process that occurs during the roasting of coffee beans. It has been determined that these compounds constitute up to 25% of the dry weight in roasted coffee beans, with the content increasing with the degree of roast. They thus represent a major component of coffee brew (approximately 10–30% of soluble solids in medium to dark roasts), thus making coffee one of the primary dietary sources of melanoidins (Borrelli et al., 2002; Böhm et al., 2023). Melanoidins are structurally heterogeneous and consist of a complex skeleton incorporating polysaccharides (galactomannans and arabinogalactans), proteins, and phenolics. These components significantly contribute to the colour, body, and aroma of brewed coffee (Moreira et al., 2012). The significance of these compounds lies not only in their contribution to the sensory qualities of coffee, such as colour and flavour, but also in their potential health benefits. Coffee melanoidins have been demonstrated to possess a variety of biological activities, including antioxidant, antimicrobial, anticariogenic, anti-inflammatory, antihypertensive, and antiglycative properties (Langner and Rzeski, 2014; Antonietti et al., 2022). Melanoidins have been shown to possess antimicrobial properties, which have been demonstrated to inhibit the growth of pathogenic bacteria and to promote the growth of beneficial gut microbiota, such as Bifidobacterium and Lactobacillus (Morales et al., 2012; Mesías and Delgado-Andrade, 2017). Whilst the health benefits of melanoidins are encouraging, it is important to acknowledge that the majority of the current evidence originates from in vitro and animal studies. The evidence from human studies is currently limited, and further research is needed to confirm these benefits in human populations.

7. Volatile and aroma precursors

Volatile compounds have a significant influence on sensory quality, roasting, and fermentation processes. Volatile compounds are associated with olfactory perception in coffee. Aroma precursors are formed through complex interactions that occur during the roasting process, which depends on the chemical composition of the coffee beans. The complex interaction of over 800 volatile compounds gives coffee its distinctive aroma. Volatile compounds are classified as furans, pyrazines, phenols, aldehydes, ketones, esters, sulfur compounds, alcohols, acids, and terpenes (Ji et al., 2025). Furfural, 5-methylfurfural, furfuryl alcohol, and furaneol are classified as furans. Sugar degradation (Maillard reactions) results in caramel, sweet, candy-like, and nutty aromas (Rusinek et al., 2022). 2-methylpyrazine, 2,3-dimethylpyrazine, 2-ethyl-3-methylpyrazine, and trimethylpyrazine are classified as pyrazines. After the amino acids and sugars react, they provide nutty, earthy, roasted, and chocolatey aromas (Yu et al., 2021). Guaiacol and 4-vinylguaiacol are classified as phenolic compounds. Lignin degradation during roasting provides spicy, smoky, fermentation aromas (Perdani et al., 2019). Acetaldehyde, propanal, and 2-butanone are classified as aldehydes and ketones. They provide sharp, fruity aromas (Dippong et al., 2022). Ethyl acetate, methyl formate, and ethyl butyrate, formed from the interaction of alcohols and acids, are classified as esters. They provide fruity and sweet aromas resulting from the esterification reactions of alcohols and acids (Moeenfard et al., 2015). 2-furfurylthiol and methanethiol are classified as sulfur compounds. They provide roasted, burnt, and delicious aromas as a result of the decomposition of sulfur-containing amino acids (Flament, 2002). 2-methyl-1-butanol, 2-phenylethanol, and furfuryl alcohol are classified as alcohols. Fermentation and Maillard reaction intermediates contribute to the development of sweet, floral, and alcoholic aromas (Zhang et al., 2022). Acetic acid, formic acid, and butanoic acid are classified as acids. Carbohydrate and lipid degradation result in sour and pungent aromas. Limonene and linalool are classified as terpenes. They provide citrus, floral, and herbal aromas (Góngora et al., 2020). Maltol, sotolon, and kahweoli are classified as pyrones and lactones. They provide creamy and caramelised aromas (Farah vd., 2005). Aromas are affected by roasting temperature and duration. Terpenes come to the fore with light roasting (180–200°C), while pyrazines and furans come to the fore with medium roasting (200–220°C). Phenolic compounds increase with dark roasting (220–240°C) (Flores-Chávez et al., 2022).

8. Factors affecting biochemical composition

The biochemical composition of coffee beans is influenced by a complex interplay of genetic, environmental, agronomic, and post-harvest factors. These factors collectively determine the concentrations of macronutrients, bioactive compounds, and aroma precursors. The resultant impact of these factors on sensory quality and health-related properties is significant. The genetic and varietal distinctions observed between *Coffea arabica* and *C. canephora* (Robusta), in addition to intraspecific variations, are responsible for significant disparities. For instance, comparing regional landraces from Ethiopia with those from Brazil reveals notable differences. Similarly, Ethiopian coffee varieties exhibit a considerable variation in chlorogenic acid and caffeine content, the specifics of which depend on the particular variety in question (Tolessa et al., 2019). A study of local coffee cultivars in Saudi Arabia has revealed significant differences in the concentrations of caffeine, trigonelline, and chlorogenic acid among different accessions. These results indicate that genetic diversity is a key factor in the study of these compounds (Oteef et al., 2024). It has been demonstrated that altitude exerts a deleterious effect on the content of caffeine and chlorogenic acids in coffee (Tolessa et al., 2017; Worku et al., 2018). The controlled irrigation and application of nitrogen, phosphorus, and potassium fertilisers have shown to enhance the accumulation of beneficial biochemical compounds in coffee beans. For instance, the application of potassium has been demonstrated to enhance the levels of lipids and chlorogenic acids. Conversely, the supplementation of nitrogen has been shown to increase caffeine levels (Vinecky et al., 2017). The method of processing coffee beans has been shown to have a significant effect on their biochemical composition. The application of washing methods has been shown to increase the content of chlorogenic acid and caffeine. Conversely, the utilisation of dry methods has been shown to enhance sucrose levels (Tolessa et al., 2019). Conventional roasting techniques can significantly reduce the total phenolic content, trigonelline, and chlorogenic acid in coffee beans, thereby substantially affecting their flavour profile (Mengesha et al., 2024). As previously indicated, the factors cited above are of paramount significance in determining the biochemical composition of coffee. However, it is also crucial to consider the potential interactions between these variables. It is essential to understand these complex interactions to develop effective strategies that ensure the preservation and improvement of coffee quality in the face of changing environmental conditions.

9. Analytical methods for determination

Coffee is composed of a complex structure containing hundreds of chemical compounds, including macromolecules. Techniques like spectroscopy, electrochemistry, and chromatography are used to quantify coffee content (dos Santos and Boffo, 2021).

9.1 Chromatographic techniques

9.1.1 High-performance liquid chromatography (HPLC)

The amounts of caffeine, chlorogenic acids, and phenolic compounds are determined. The extract is prepared and then pretreated with solid-phase extraction (SPE), and the eluent (mobile phase) is prepared accordingly. The prepared eluent is placed in the HPLC device. Sample injection and chromatographic separation are performed. Various detectors are used to identify the compounds and convert them into signals (La Pera et al., 2008).

UV-visible detector: Light absorption measurement in the ultraviolet (UV, 190-380 nm) or visible light (380-780 nm) range is used to measure the quantities of chemicals (caffeine and phenolics) in the sample. The light absorbed by the molecule is converted into an electrical signal by the detector. Caffeine exhibits maximum absorption at 272 nm. The measurement is made quantitatively. This method can be disadvantageous if other compounds exhibit absorption at the same wavelength (Kalita et al., 2025).

Fluorescence detectors are more sensitive than UV-visible techniques, as certain chemical compounds emit light at different wavelengths when excited by light of a specific wavelength. Phenolic compounds, trigonelline, and aromatic amines found in coffee fluorescence (Suhandy et al., 2025).

A Diode-Array Detector (DAD) can monitor multiple wavelengths (UV, 190–780 nm) simultaneously in the UV-visible range. Processing and analysis times are longer than those with UV-V, but multiple compounds can be monitored simultaneously (Seccia et al., 2025).

Liquid Chromatography (LC–MS/MS) combines HPLC with mass spectrometry (MS) in a method where compounds are ionised after chromatographic separation. The mass-to-charge ratio (m/z) of the ions is measured. The MS/MS system provides structural information by fragmenting the ions. All compounds found in coffee can be measured (Fukushima et al., 2025).

9.1.2 Gas Chromatography (GC)

The volatile compounds responsible for the aroma of coffee are analysed. Gas chromatography separates the compounds. Mass spectrometry determines the chemical structure and molecular mass of each compound. This technique provides high sensitivity for volatile compounds and is used to determine the flavour profiles of complex mixtures such as coffee aroma. During sample preparation, headspace sampling, solid-phase microextraction (SPME), or steam distillation techniques can effectively isolate aroma compounds. Headspace sampling minimises contamination by collecting volatile compounds in the vapour phase above the coffee sample. Solid-phase microextraction (SPME), which provides high sensitivity and reproducibility, uses a coated fiber to adsorb volatile compounds directly from the sample cavity. Steam distillation helps concentrate volatile compounds for subsequent GC analysis by co-distilling the volatiles with water vapour. After the preparation stage, Flame Ionisation Detector (FID), Mass Spectrometry (MS), and Thermal Conductivity Detector (TCD) can be used to measure volatile compounds (Bordiga et al., 2025; Shi X. et al., 2025).

9.2 Spectroscopic and other methods

Spectroscopic methods are widely used to determine the chemical structure and quality of coffee. Due to their high accuracy and sensitivity, they provide quantitative and qualitative information about the molecular structure and components of coffee.

9.2.1 Nuclear magnetic resonance (NMR) spectroscopy

NMR is important for the detailed chemical characterisation of coffee. Both proton (^{1}H) and carbon (^{13}C) NMR techniques play a role in determining the structure of the components. It allows the identification of substances present in even very small amounts in coffee extracts. The levels of diterpenes, such as kahweol and cafestol, can be used to distinguish between species. Hundreds of metabolites can be identified and quantified simultaneously. It can be used in various areas, including food safety, food control, sensory quality, geographic origin, and food and beverage fingerprinting. It can also be used to assess roasting degree and detect potential adulteration by monitoring chemical changes. Possible adulteration includes the addition of other plants. Mixed plants can be detected (Borém et al., 2023; Trimigno et al., 2015).

9.2.2 Fourier transform infrared spectroscopy (FTIR)

FTIR is used to quickly identify chemical bonds and functional groups in foods. It measures the vibrational frequencies of organic functional and chemical bonds in coffee. The vibrations occur at characteristic frequencies depending on the bond type and atomic masses. It is a fast, inexpensive, and small sample-requiring method. The characteristic vibration bands of carbonyl, hydroxyl, and aromatic compounds enable the determination of quality control, roasting degree, and moisture content in coffee samples. It allows the detection and identification of different coffee beans. It measures the absorbance of infrared light in a sample. Each functional group produces a characteristic absorption band in the spectrum. FTIR spectroscopy is a technique that combines the mathematics of the Michelson interferometer and the Fourier Transform (Li et al., 2025).

10. Conclusion and Future Perspectives

The biochemical composition of coffee beans has been shown to contain a complex matrix of macronutrients, bioactive micronutrients, and processing-derived compounds. Collectively, these elements underpin both the coffee's exceptional sensory attributes and its increasingly well-documented health-promoting properties. The complex chemical composition of coffee encompasses a multitude of bioactive compounds, including carbohydrates, proteins, lipids, and various alkaloids such as caffeine and trigonelline. Phenolic acids, predominantly comprising chlorogenic acids and their derivatives, along with flavonoids, diterpenes (cafestol and kahweol), and Maillard-derived melanoidins, are also present. This diverse array of bioactive compounds contributes to the unique stimulant, antioxidant, anti-inflammatory, and metabolic and neurological pathway-modulating properties of coffee. The presence of these compounds, which exhibit notable concentrations depending on the species (*Coffea arabica* vs. *C. canephora*), genetic variation, environmental conditions, agronomic practices, and post-harvest processing, contributes to the observed associations between moderate coffee consumption and reduced risks of major chronic diseases, including type 2 diabetes, cardiovascular disease, certain cancers, neurodegenerative disorders, and liver conditions. The industrial significance of this composition extends beyond the quality of beverages to emerging applications in functional foods, nutraceuticals, cosmetics, and pharmaceutical development. In these areas, coffee-derived bioactives are being explored for their antioxidant, anti-aging, and therapeutic potential. In the future, several significant research gaps and opportunities require attention. The repercussions of climate change on the biochemistry of coffee are of particular concern, given the projected increases in temperature, shifts in precipitation patterns, and the advent of pests that have not previously been observed. These factors pose a substantial threat to the bioactive content, yield, and overall quality of coffee. This necessitates the undertaking of studies on adaptive genotypes and resilient agronomic strategies. Progress in omics technologies, including genomics, metabolomics, and transcriptomics, presents significant opportunities for clarifying biosynthetic pathways, pinpointing essential regulatory genes, and expediting breeding initiatives aimed at improving nutritional profiles or minimising undesirable substances, such as acrylamide formation. A more profound understanding of compound bioavailability, gut microbiota interactions (particularly with melanoidins and polyphenols), and personalised responses based on genetic polymorphisms in metabolism will refine dietary recommendations. The development of low-impact processing methods that preserve bioactives while minimising waste, the valorisation of by-products (e.g., silverskin and spent grounds as sources of antioxidants and fibre), and the promotion of agroecological systems that enhance soil health and biodiversity will be essential for ensuring the long-term viability of coffee production and its continued contribution to human health and well-being.

References

A. Alsanad, M., Sassine, Y. N., El Sebaaly, Z., and Abou Fayssal, S. (2021). Spent coffee grounds influence on Pleurotus ostreatus production, composition, fatty acid profile, and lignocellulose biodegradation capacity. *CyTA-Journal of Food*, 19(1), 11–20. https://doi.org/10.1080/19476337.2020.1845243

Aborziza, M., Amalia, R., Zuhrotun, A., Ikram, N. K. K., Novitasari, D., and Muchtaridi, M. (2024). Coffee bean and its chemical constituent caffeine and chlorogenic acid as promising chemoprevention agents: Updated biological studies against cancer cells. *Molecules*, 29(14), 3302. https://doi.org/10.3390/molecules29143302

Ali, A., Zahid, H. F., Cottrell, J. J., and Dunshea, F. R. (2022). A comparative study for nutritional and phytochemical profiling of Coffea arabica (*C. arabica*) from different origins and their antioxidant potential and molecular docking. *Molecules*, 27(16), 5126. https://doi.org/10.3390/molecules27165126

Antonietti, S., Silva, A. M., Simões, C., Almeida, D., Félix, L. M., Papetti, A., & Nunes, F. M. (2022). Chemical composition and potential biological activity of melanoidins from instant soluble coffee and instant soluble barley: A comparative study. *Frontiers in Nutrition*, 9, 825584. | https://doi.org/10.3389/fnut.2022.825584

Asadullah, M., Ikram, A., Sahni, O., Rasheed, M., Abdullah, Z., Menon, S. V., ... & Kinki, A. B. (2025). Coffee beans chemical composition, processing and physiological nature: A comprehensive review. *European Food Research and Technology*, 1–14. https://doi.org/10.1007/s00217-025-04893-w

Ashihara, H. (2006). Metabolism of alkaloids in coffee plants. *Brazilian Journal of Plant Physiology*, 18, 1–8. https://doi.org/10.1590/S1677-04202006000100001

Ashihara, H. (2016). Biosynthetic pathways of purine and pyridine alkaloids in coffee plants. *Natural Product Communications*, 11(7), 1934578X1601100742. https://doi.org/10.1177/1934578X160110074

Ashihara, H., Ludwig, I. A., Katahira, R., Yokota, T., Fujimura, T., and Crozier, A. (2015). Trigonelline and related nicotinic acid metabolites: Occurrence, biosynthesis, taxonomic considerations, and their roles in planta and in human health. *Phytochemistry Reviews*, 14(5), 765–798. https://doi.org/10.1007/s11101-014-9375-z

Atlabachew, M., Abebe, A., Alemneh Wubieneh, T., and Tefera Habtemariam, Y. (2021). Rapid and simultaneous determination of trigonelline, caffeine, and chlorogenic acid in green coffee bean extract. *Food Science & Nutrition*, 9(9), 5028–5035. https://doi.org/10.1002/fsn3.2456

Barreca, D., Trombetta, D., Smeriglio, A., Mandalari, G., Romeo, O., Felice, M. R., ... and Nabavi, S. M. (2021). Food flavonols: Nutraceuticals with complex health benefits and functionalities. *Trends in Food Science & Technology*, 117, 194–204. https://doi.org/10.1016/j.tifs.2021.03.030

Bhattarai, R. R., Al-Ali, H., and Johnson, S. K. (2022). Extraction, isolation and nutritional quality of coffee protein. *Foods*, 11(20), 3244. https://doi.org/10.3390/foods11203244

Birková, A., Hubková, B., Bolerázska, B., Mareková, M., & Čižmárová, B. (2020). Caffeic acid: A brief overview of its presence, metabolism, and bioactivity. *Bioactive Compounds in Health and Disease*, 3(4), 74-81. Online ISSN: 2574-0334; Print ISSN: 2769-2426.

Bjelıca, B., Aksovıć, N., Zelenović, M., Pržulj, R., and Radenkovic, O. (2023). Effect of caffeine on the cardiovascular system and performance: A systematic review. *Turkish Journal of Kinesiology*, 9(2), 125–132.

Böhm, W., Zinke, L., Rehle, A. K., and Henle, T. (2023). Role of proteins in the formation of melanoidins during coffee roasting. *Journal of Agricultural and Food Chemistry*, 71(47), 18499–18509. https://doi.org/10.31459/turkjkin.1263307

Bordiga, M., Disca, V., Manfredi, M., Barberis, E., Carrà, F., Navarini, L., ... and Arlorio, M. (2025). Fingerprinting of green Arabica Coffee Volatile Organic Compounds (VOCs): HS-GC-IMS Versus GC× GC-MS. *International Journal of Food Science*, 2025(1), 1302823. https://doi.org/10.1155/ijfo/1302823

Borém, F. M., de Abreu, G. F., Ferreira, A. G., da Silva Santos, M., da Conceição Alves, T., and de Carvalho Alves, A. P. (2023). 1H NMR spectroscopy applied to identify chemical aging markers in green coffee (Coffea arabica L.). *Food Chemistry*, 405, 134667. https://doi.org/10.1016/j.foodchem.2022.134667

Borrelli, R. C., Visconti, A., Mennella, C., Anese, M., & Fogliano, V. (2002). Chemical characterization and antioxidant properties of coffee melanoidins. *Journal of Agricultural and Food Chemistry*, 50(22), 6527-6533. https://doi.org/10.1021/jf025686o

Bumrungpert, A., Lilitchan, S., Tuntipopipat, S., Tirawanchai, N., and Komindr, S. (2018). Ferulic acid supplementation improves lipid profiles, oxidative stress, and inflammatory status in hyperlipidemic subjects: A randomized, double-blind, placebo-controlled clinical trial. *Nutrients*, 10(6), 713. https://doi.org/10.3390/nu10060713

Caño-Carrillo, I., Batteau, M., Bruna, L., Gilbert-López, B., García-Reyes, J. F., Chatel, G., and Faure, K. (2025). Determination of free and derivative terpenoids and sterols in spent coffee grounds by two-dimensional liquid chromatography-supercritical fluid chromatography coupled to high-resolution mass spectrometry. *Food Chemistry*, 144853. https://doi.org/10.1016/j.foodchem.2025.144853

Cho, J., and Park, Y. (2023). Kahweol, a coffee diterpene, increases lifespan via insulin/insulin-like growth factor-1 and AMP-activated protein kinase signaling pathways in Caenorhabditis elegans. *Current Research in Food Science*, 7, 100618. https://doi.org/10.1016/j.crfs.2023.100618

Choi, M., Mukherjee, S., and Yun, J. W. (2021). Trigonelline induces browning in 3T3-L1 white adipocytes. *Phytotherapy Research*, 35(2), 1113–1124. https://doi.org/10.1002/ptr.6892

Clifford, M. N., Jaganath, I. B., Ludwig, I. A., and Crozier, A. (2017). Chlorogenic acids and the acyl-quinic acids: Discovery, biosynthesis, bioavailability and bioactivity. *Natural Product Reports*, 34(12), 1391–1421.

Colomban, S., Guercia, E., and Navarini, L. (2020). Validation of a rapid ultra-high-performance liquid chromatography–tandem mass spectrometry method for quantification of chlorogenic acids in roasted coffee. *Journal of Mass Spectrometry*, 55(11), e4634. https://doi.org/10.1039/C7NP00030H

Dantas, M. B., Sampaio, T. L., Menezes, R. D., Ferreira, J. M., Melo, T. S., Meneses, R. R. C., ... and Queiroz, M. D. (2018). Caffeic acid reduces lipid accumulation and reactive oxygen species production in adipocytes. *African Journal of Pharmacy and Pharmacology*, 12(20), 263–268. 10.5897/AJPP2018.4937

Dawidowicz, A. L., and Typek, R. (2017). Transformation of chlorogenic acids during the coffee beans roasting process. *European Food Research and Technology*, 243(3), 379–390. https://doi.org/10.1007/s00217-016-2751-8

dePaula, J., and Farah, A. (2019). Caffeine consumption through coffee: Content in the beverage, metabolism, health benefits and risks. *Beverages*, 5(2), 37. https://doi.org/10.3390/beverages5020037

Dewland, T. A., van Dam, R. M., & Marcus, G. M. (2025). Coffee and cardiovascular disease. *European Heart Journal*, 46(36), 3546–3554. https://doi.org/10.1093/eurheartj/ehaf421

Di Gioacchino, M., Ricci, M. A., and Bruni, F. (2024). Science in a cup of coffee: A structural study of a trigonelline aqueous solution. *Journal of Molecular Liquids*, 396, 123972. https://doi.org/10.1016/j.molliq.2024.123972

Dippong, T., Dan, M., Kovacs, M. H., Kovacs, E. D., Levei, E. A., and Cadar, O. (2022). Analysis of volatile compounds, composition, and thermal behavior of coffee beans according to variety and roasting intensity. *Foods*, 11(19), 3146. https://doi.org/10.3390/foods11193146

Dorst, K. M., and Widmalm, G. (2024). Conformational preferences at the glycosidic linkage of saccharides in solution as deduced from NMR experiments and MD simulations: Comparison to crystal structures. *Chemistry – A European Journal*, 30(15), e202304047. https://doi.org/10.1002/chem.202304047

dos Santos, H. D., and Boffo, E. F. (2021). Coffee beyond the cup: Analytical techniques used in chemical composition research—A review. *European Food Research and Technology*, 247(4), 749–775. https://doi.org/10.1007/s00217-020-03679-6

Emadi, R. C., and Kamangar, F. (2025). Coffee's impact on health and well-being. *Nutrients*, 17(15), 2558.

Espíndola, K. M. M., Ferreira, R. G., Narvaez, L. E. M., Rosario, A. C. R. S., Da Silva, A. H. M., Silva, A. G. B., ... and Monteiro, M. C. (2019). Chemical and pharmacological aspects of caffeic acid and its activity in hepatocarcinoma. *Frontiers in Oncology*, 9, 541. https://doi.org/10.3390/nu17152558

Farah, A., de Paulis, T., Trugo, L. C., and Martin, P. R. (2005). Effect of roasting on the formation of chlorogenic acid lactones in coffee. *Journal of Agricultural and Food Chemistry*, 53(5), 1505–1513.

Flament, I. (2002). Coffee Flavor Chemistry. 424. https://books.google.com/books/about/Coffee_Flavor_Chemistry.html?hl=tr&id=NQi1LYJxFvUC

Flores-Chávez, B., Garza-López, P. M., Angel-Cuapio, A., Hernández-León, S., Espinosa-Zaragoza, S., Villafaña-Rivera, F. J., ... and Espitia-López, J. (2022). Analysis verbal sensory methods in the evaluation of organic coffee with different roasting intensities. *International Journal of Food Properties*, 25(1), 708–716. https://doi.org/10.1080/10942912.2022.2057532

Fukagawa, S., Haramizu, S., Sasaoka, S., Yasuda, Y., Tsujimura, H., and Murase, T. (2017). Coffee polyphenols extracted from green coffee beans improve skin properties and microcirculatory function. *Bioscience, Biotechnology, and Biochemistry*, 81(9), 1814–1822. https://doi.org/10.1080/09168451.2017.1345614

Fukushima, Y., Yamamoto, K., Machida, K., Kotani, A., and Hakamata, H. (2025). A simple LC–MS method for the determination of free fatty acids in coffee beans and cow milk. *Food Analytical Methods*, 1–13.

Fulton, J. L., Dinas, P. C., Carrillo, A. E., Edsall, J. R., Ryan, E. J., & Ryan, E. J. (2018). Impact of genetic variability on physiological responses to caffeine in humans: A systematic review. *Nutrients*, 10(10), 1373. https://doi.org/10.1007/s12161-025-02765-3

Genaro-Mattos, T. C., Maurício, Â. Q., Rettori, D., Alonso, A., and Hermes-Lima, M. (2015). Antioxidant activity of caffeic acid against iron-induced free radical generation—A chemical approach. *PLoS One*, 10(6), e0129963. https://doi.org/10.1371/journal.pone.0142402

Góngora, C. E., Tapias, J., Jaramillo, J., Medina, R., Gonzalez, S., Casanova, H., ... and Benavides, P. (2020). Evaluation of terpene-volatile compounds repellent to the coffee berry borer, Hypothenemus hampei (Ferrari) (Coleoptera: Curculionidae). *Journal of Chemical Ecology*, 46(9), 881–890. https://doi.org/10.1007/s10886-020-01202-5

Hanley, A., Liu, W. H., Semrau, F. O., and Görlich, D. (2025). Climbing the coffee global value chain—Three decades of trade and upgrading. *Journal of International Development*. https://doi.org/10.1002/jid.70033. https://doi.org/10.1002/jid.70033

Happyana, N., Syah, Y. M., and Hakim, E. H. (2022). Discrimination of metabolite profiles of Gayo Roasted Arabica and Robusta Coffees. *Molekul*, 17(1), 98-106. https://doi.org/10.20884/1.jm.2022.17.1.5603

Heitman, E., & Ingram, D. K. (2017). Cognitive and neuroprotective effects of chlorogenic acid. *Nutritional Neuroscience*, 20(1), 32–39. https://doi.org/10.1179/1476830514Y.0000000146

Hermund, D. B., Larsen, L. K., Trangbæk, S. R., Madsen, Q. K. R. M. T., Sørensen, A. D. M., Kaya, J., and Jacobsen, C. (2025). Fate of flavonoids and theobromine in cocoa beans during roasting: Effect of time and temperature. *Journal of the American Oil Chemists' Society*, 102(1), 35–45. https://doi.org/10.1002/aocs.12853

Jacobo-Velázquez, D. A. (2025). Ferulic acid: Mechanistic insights and multifaceted applications in metabolic syndrome, food preservation, and cosmetics. *Molecules*, 30(18), 3716. https://doi.org/10.3390/molecules30183716

Jang, M. H., Kang, N. H., Mukherjee, S., and Yun, J. W. (2018). Theobromine, a methylxanthine in cocoa bean, stimulates thermogenesis by inducing white fat browning and activating brown adipocytes. *Biotechnology and Bioprocess Engineering*, 23(6), 617–626. https://doi.org/10.1007/s12257-018-0434-y

Jeszka-Skowron, M., Frankowski, R., and Zgoła-Grześkowiak, A. (2020). Comparison of methylxantines, trigonelline, nicotinic acid and nicotinamide contents in brews of green and processed Arabica and Robusta coffee beans – Influence of steaming, decaffeination and roasting processes on coffee beans. *LWT*, 125, 109344. https://doi.org/10.1016/j.lwt.2020.109344

Jeszka-Skowron, M., Frankowski, R., Zgoła-Grześkowiak, A., and Płatkiewicz, J. (2022). Comprehensive analysis of metabolites in brews prepared from naturally and technologically treated coffee beans. *Antioxidants*, 12(1), 95. https://doi.org/10.3390/antiox12010095

Ji, C., Liu, C., Wang, R., Ruan, M., Yao, Z., Wan, H., ... and Ye, Q. (2025). The science behind vegetable aromas: Types, synthesis, and influencing factors of volatile compounds. *Plant Signaling & Behavior*, 20(1), 2527958. https://doi.org/10.1080/15592324.2025.2527958

Kalita, L., Sinha, S. K., Deka, B., Dasgupta, D., Deka, B., Patowary, P., ... and Ali, R. (2025). Caffeine analysis through the lens of green chemistry: A review of sustainable techniques and emerging trends. *Critical Reviews in Analytical Chemistry*, 1–15. https://doi.org/10.1080/10408347.2025.2562364

Kang, N. J., Lee, K. W., Shin, B. J., Jung, S. K., Hwang, M. K., Bode, A. M., ... and Dong, Z. (2009). Caffeic acid, a phenolic phytochemical in coffee, directly inhibits Fyn kinase activity and UVB-induced COX-2 expression. *Carcinogenesis*, 30(2), 321–330. https://doi.org/10.1093/carcin/bgn282

Kim, J. Y., Leem, J., and Kim, G. M. (2022). Kahweol protects against Acetaminophen-induced hepatotoxicity in mice through inhibiting oxidative stress, hepatocyte death, and inflammation. *BioMed Research International*, 2022(1), 8121124. https://doi.org/10.1155/2022/8121124

Kobylińska, Z., Biesiadecki, M., Kuna, E., Galiniak, S., and Mołoń, M. (2025). Coffee as a source of antioxidants and an elixir of youth. *Antioxidants*, 14(3), 285. https://doi.org/10.3390/antiox14030285

Koga, M., Nakagawa, S., Kato, A., and Kusumi, I. (2019). Caffeic acid reduces oxidative stress and microglial activation in the mouse hippocampus. *Tissue and Cell*, 60, 14–20. https://doi.org/10.1016/j.tice.2019.07.006

Koshiro, Y., Jackson, M. C., Katahira, R., Wang, M. L., Nagai, C., and Ashihara, H. (2007). Biosynthesis of chlorogenic acids in growing and ripening fruits of Coffea arabica and Coffea canephora plants. *Zeitschrift für Naturforschung C*, 62(9-10), 731–742.

Kumar, M., Kaushik, D., Shubham, S., Kumar, A., Kumar, V., Oz, E., ... and Oz, F. (2025). Ferulic acid: Extraction, estimation, bioactivity and applications for human health and food. *Journal of the Science of Food and Agriculture*, 105(8), 4168–4177. https://doi.org/10.1002/jsfa.13931

Kunutsor, S. K., Lehoczki, A., and Laukkanen, J. A. (2025). Coffee consumption, cancer, and healthy aging: Epidemiological evidence and underlying mechanisms. *GeroScience*, 47(2), 1517–1555. https://doi.org/10.1007/s11357-024-01332-8

Kwon, Y. J., Kwon, H. H., Leem, J., and Jang, Y. Y. (2024). Kahweol Inhibits pro-inflammatory cytokines and chemokines in tumor necrosis factor-α/interferon-γ-stimulated human keratinocyte HaCaT cells. *Current Issues in Molecular Biology*, 46(4), 3470–3483. https://doi.org/10.3390/cimb46040218

Ky, C. L., Louarn, J., Dussert, S., Guyot, B., Hamon, S., & Noirot, M. (2001). Caffeine, trigonelline, chlorogenic acids and sucrose diversity in wild Coffea arabica L. and C. canephora P. accessions. *Food Chemistry*, 75(2), 223–230. https://doi.org/10.1016/S0308-8146(01)00204-7

La Pera, L., Avellone, G., Lo Turco, V., Di Bella, G., Agozzino, P., and Dugo, G. (2008). Influence of roasting and different brewing processes on the ochratoxin A content in coffee determined by high-performance liquid chromatography-fluorescence detection (HPLC-FLD). *Food Additives and Contaminants*, 25(10), 1257–1263. https://doi.org/10.1080/02652030802112627

Lallemand, L. A., Zubieta, C., Lee, S. G., Wang, Y., Acajjaoui, S., Timmins, J., ... and McCarthy, A. A. (2012). A structural basis for the biosynthesis of the major chlorogenic acids found in coffee. *Plant Physiology*, 160(1), 249–260. https://doi.org/10.1104/pp.112.202051

Langi, T. M., Paat, F. J., Kusuma, S. D., Oessoe, Y. Y., Liwu, S. L., Mamuaja, C. F., ... and Mamarimbing, R. (2023). The effect of arabica and robusta coffee blends on caffeine content, acidity and organoleptic properties of instant coffee. *Journal of Agriculture*, 2(02). https://doi.org/10.47709/joa.v2i02.2806

Langner, E., and Rzeski, W. (2014). Biological properties of melanoidins: A review. *International Journal of Food Properties*, 17(2), 344–353. https://doi.org/10.1080/10942912.2011.631253

Lee, H. F., Lin, J. S., and Chang, C. F. (2019). Acute kahweol treatment attenuates traumatic brain injury neuroinflammation and functional deficits. *Nutrients*, 11(10), 2301. https://doi.org/10.3390/nu11102301

Lee, S., Choi, E., and Lee, K. G. (2024). Kinetic modelling of Maillard reaction products and protein content during roasting of coffee beans. *LWT*, 211, 116950. https://doi.org/10.1016/j.lwt.2024.116950

Li, X., Yang, L., Hao, M., Song, T., He, Y., Yang, M., and Zhang, J. (2023). Chlorogenic acid as an indispensible partner of caffeic acid in coffee via selective regulation of prooxidative actions of caffeic acid. *Food Research International*, 173, 113482. https://doi.org/10.1016/j.foodres.2023.113482

Li, Z., Gao, Z., Li, C., Yan, J., Hu, Y., Fan, F., ... and Tian, H. (2025). Rapid discrimination of different primary processing Arabica coffee beans using FT-IR and machine learning. *Food Research International*, 205, 115979. https://doi.org/10.1016/j.foodres.2025.115979

Liu, C., Wang, L., Zhang, C., Hu, Z., Tang, J., Xue, J., and Lu, W. (2024). Caffeine intake and anxiety: A meta-analysis. *Frontiers in Psychology*, 15, 1270246. https://doi.org/10.3389/fpsyg.2024.1270246

Liu, J., Guan, Y., Yang, L., Fang, H., Sun, H., Sun, Y., ... and Wang, X. (2025). Ferulic acid as an anti-inflammatory agent: Insights into molecular mechanisms, pharmacokinetics and applications. *Pharmaceuticals*, 18(6), 912. https://doi.org/10.3390/ph18060912

Lu, M. F., Xiao, Z. T., and Zhang, H. Y. (2013). Where do health benefits of flavonoids come from? Insights from flavonoid targets and their evolutionary history. *Biochemical and Biophysical Research Communications*, 434(4), 701-704. https://doi.org/10.1016/j.bbrc.2013.04.035

M Yelanchezian, Y. M., Waldvogel, H. J., Faull, R. L., and Kwakowsky, A. (2022). Neuroprotective effect of caffeine in Alzheimer's disease. *Molecules*, 27(12), 3737. https://doi.org/10.3390/molecules27123737

Machado, F., Coimbra, M. A., Castillo, M. D. D., and Coreta-Gomes, F. (2024). Mechanisms of action of coffee bioactive compounds – A key to unveil the coffee paradox. *Critical Reviews in Food Science and Nutrition*, 64(28), 10164–10186. https://doi.org/10.1080/10408398.2023.2221734

Makiso, M. U., Tola, Y. B., Ogah, O., and Endale, F. L. (2024). Bioactive compounds in coffee and their role in lowering the risk of major public health consequences: A review. *Food Science & Nutrition*, 12(2), 734–764. https://doi.org/10.1002/fsn3.3848

Massrie, K. D. (2025). Why is the price of coffee rising globally? Future prospects for Ethiopian coffee. *Frontiers in Sustainable Food Systems*, 9, 1545168. https://doi.org/10.3389/fsufs.2025.1545168

Mediani, A., Kamal, N., Lee, S. Y., Abas, F., and Farag, M. A. (2023). Green extraction methods for isolation of bioactive substances from coffee seed and spent. *Separation & Purification Reviews*, 52(1), 24–42. https://doi.org/10.1080/15422119.2022.2027444

Mehari, B., Redi-Abshiro, M., Chandravanshi, B. S., Atlabachew, M., Combrinck, S., and McCrindle, R. (2016). Simultaneous determination of alkaloids in green coffee beans from Ethiopia: Chemometric evaluation of geographical origin. *Food Analytical Methods*, 9(6), 1627–1637. https://doi.org/10.1007/s12161-015-0340-2

Mellbye, F. B., Jeppesen, P. B., Shokouh, P., Laustsen, C., Hermansen, K., and Gregersen, S. (2017). Cafestol, a bioactive substance in coffee, has antidiabetic properties in KKAy mice. *Journal of Natural Products*, 80(8), 2353–2359. https://doi.org/10.1021/acs.jnatprod.7b00395

Meng, S., Cao, J., Feng, Q., Peng, J., and Hu, Y. (2013). Roles of chlorogenic acid on regulating glucose and lipids metabolism: A review. *Evidence-Based Complementary and Alternative Medicine*, 2013(1), 801457. https://doi.org/10.1155/2013/801457

Mengesha, D., Retta, N., Woldemariam, H. W., and Getachew, P. (2024). Changes in biochemical composition of Ethiopian Coffee arabica with growing region and traditional roasting. *Frontiers in Nutrition*, 11, 1390515.

Mesías, M., and Delgado-Andrade, C. (2017). Melanoidins as a potential functional food ingredient. *Current Opinion in Food Science*, 14, 37–42. https://doi.org/10.3389/fnut.2024.1390515

Moeenfard, M., Silva, J. A., Borges, N., Santos, A., and Alves, A. (2015). Quantification of diterpenes and their palmitate esters in coffee brews by HPLC-DAD. *International Journal of Food Properties*, 18(10), 2284–2299. https://doi.org/10.1080/10942912.2014.933351

Mohamed, A. I., Erukainure, O. L., Salau, V. F., and Islam, M. S. (2024). Impact of coffee and its bioactive compounds on the risks of type 2 diabetes and its complications: A comprehensive review. *Diabetes & Metabolic Syndrome: Clinical Research & Reviews*, 18(7), 103075. https://doi.org/10.1016/j.dsx.2024.103075

Moon, S. A., Wongsakul, S., Kitazawa, H., & Saengrayap, R. (2025). Impact of roasting and storage conditions on the shelf stability of Thai Arabica Coffee. *Journal of Agriculture and Food Research*, 102060.

Morales, F. J., Somoza, V., and Fogliano, V. (2012). Physiological relevance of dietary melanoidins. *Amino Acids*, 42(4), 1097–1109. https://doi.org/10.1016/j.jafr.2025.102060

Moreira, A. S., Nunes, F. M., Domingues, M. R., and Coimbra, M. A. (2012). Coffee melanoidins: Structures, mechanisms of formation and potential health impacts. *Food & Function*, 3(9), 903–915. https://doi.org/10.1039/C2FO30048F

Murkovic, M., & Derler, K. (2006). Analysis of amino acids and carbohydrates in green coffee. *Journal of Biochemical and Biophysical Methods*, 69(1-2), 25–32. https://doi.org/10.1016/j.jbbm.2006.02.001

Muzykiewicz-Szymańska, A., Nowak, A., Wira, D., and Klimowicz, A. (2021). The effect of brewing process parameters on antioxidant activity and caffeine content in infusions of roasted and unroasted Arabica coffee beans originated from different countries. *Molecules*, 26(12), 3681. https://doi.org/10.3390/molecules26123681

Naveed, M., Hejazi, V., Abbas, M., Kamboh, A. A., Khan, G. J., Shumzaid, M., ... and XiaoHui, Z. (2018). Chlorogenic acid (CGA): A pharmacological review and call for further research. *Biomedicine & Pharmacotherapy*, 97, 67–74. https://doi.org/10.1016/j.biopha.2017.10.064

Nguyen, V., Taine, E. G., Meng, D., Cui, T., and Tan, W. (2024). Pharmacological activities, therapeutic effects, and mechanistic actions of trigonelline. *International Journal of Molecular Sciences*, 25(6), 3385. https://doi.org/10.3390/ijms25063385

Ogita, S., Uefuji, H., Morimoto, M., and Sano, H. (2004). Application of RNAi to confirm theobromine as the major intermediate for caffeine biosynthesis in coffee plants with potential for construction of decaffeinated varieties. *Plant Molecular Biology*, 54(6), 931–941. https://doi.org/10.1007/s11103-004-0393-x

Olchowik, G., Chadaj-Polberg, E., Tomaszewski, M., Polberg, M., and Tomaszewska, M. (2011). The influence of caffeine on the biomechanical properties of bone tissue during pregnancy in a population of rats. *Folia Histochemica et Cytobiologica*, 49(3), 504–511. 10.5603/FHC.2011.0071

Olechno, E., Puścion-Jakubik, A., Socha, K., and Zujko, M. E. (2021a). Coffee brews: Are they a source of macroelements in human nutrition? *Foods*, 10(6), 1328. https://doi.org/10.3390/foods10061328

Olechno, E., Puścion-Jakubik, A., Zujko, M. E., and Socha, K. (2021b). Influence of various factors on caffeine content in coffee brews. *Foods*, 10(6), 1208. https://doi.org/10.3390/foods10061208

Ordoñez-Araque, R., and Revelo-Vizuete, B. (2023). Perspective chapter: Sugar and its impact on health. *In:* Combating Malnutrition through Sustainable Approaches. https://doi.org/10.5772/intechopen.104454

Oteef, M. D., Khemira, H., Tounekti, T., Mahdhi, M., AlFaifi, Z., Sharma, M., ... and AlSolami, W. (2024). Biochemical composition of the beans of local coffee cultivars from southwestern Saudi Arabia. *Journal of Food Composition and Analysis*, 129, 106089. https://doi.org/10.1016/j.jfca.2024.106089

Pang, J., and Zhang, F. (2021). Carbohydrates. *In:* Essentials of Food Chemistry. https://doi.org/10.1007/978-981-16-0610-6_4

Peerapen, P., Boonmark, W., Chantarasaka, S., and Thongboonkerd, V. (2024). Trigonelline prevents high-glucose-induced endothelial-to-mesenchymal transition, oxidative stress, mitochondrial dysfunction, and impaired angiogenic activity in human endothelial EA.hy926 cells. *Biomedicine & Pharmacotherapy*, 179, 117320. https://doi.org/10.1016/j.biopha.2024.117320

Perdani, C. G., and Pranowo, D. (2019, February). Total phenols content of green coffee (Coffea arabica and Coffea canephora) in East Java. *In:* IOP Conference Series: Earth and Environmental Science (Vol. 230, No. 1, p. 012093). IOP Publishing. 10.1088/1755-1315/230/1/012093

Perrois, C., Strickler, S. R., Mathieu, G., Lepelley, M., Bedon, L., Michaux, S., ... and Privat, I. (2015). Differential regulation of caffeine metabolism in Coffea arabica (Arabica) and Coffea canephora (Robusta). *Planta*, 241(1), 179–191. https://doi.org/10.1007/s00425-014-2170-7

Pimpley, V., Patil, S., Srinivasan, K., Desai, N., and Murthy, P. S. (2020). The chemistry of chlorogenic acid from green coffee and its role in attenuation of obesity and diabetes. *Preparative Biochemistry & Biotechnology*, 50(10), 969–978.

Qi, X., and Tester, R. F. (2019). Fructose, galactose and glucose – In health and disease. *Clinical Nutrition ESPEN*, 33, 18–28. https://doi.org/10.1016/j.clnesp.2019.07.004

Raeis-Abdollahi, E., Raise-Abdullahi, P., Rashidy-Pour, A., Meamar, M., and Askari, H. (2024). Coffee's protective mechanisms against neurodegeneration. *Progress in Brain Research*, 288, 167–200. https://doi.org/10.1016/bs.pbr.2024.06.017

Ran, L. X., Wei, X. Y., Ren, E. F., Qin, J. F., Rasheed, U., and Chen, G. L. (2025). Application of microbial fermentation in caffeine degradation and flavor modulation of coffee beans. *Foods*, 14(15), 2606. https://doi.org/10.3390/foods14152606

Reddy, V. S., Shiva, S., Manikantan, S., and Ramakrishna, S. (2024). Pharmacology of caffeine and its effects on the human body. *European Journal of Medicinal Chemistry Reports*, 10, 100138. https://doi.org/10.1016/j.ejmcr.2024.100138

Rehlund, A. S. G., Canoy, T. S., Nielsen, D. S., Lund, M. N., and Poojary, M. M. (2025). Protein hydrolysis for amino acid analysis revisited. *Food Chemistry*, 146224. https://doi.org/10.1016/j.foodchem.2025.146224

Ren, Y., Wang, C., Xu, J., and Wang, S. (2019). Cafestol and kahweol: A review on their bioactivities and pharmacological properties. *International Journal of Molecular Sciences*, 20(17), 4238. https://doi.org/10.3390/ijms20174238

Ricupero, S., and Ritter, F. E. (2024). Caffeine and cognition: A cognitive architecture-based review. *Theoretical Issues in Ergonomics Science*, 25(6), 655–679. https://doi.org/10.1080/1463922X.2024.2323547

Risso, É. M., Péres, R. G., and Amaya-Farfan, J. (2007). Determination of phenolic acids in coffee by micellar electrokinetic chromatography. *Food Chemistry*, 105(4), 1578–1582. https://doi.org/10.1016/j.foodchem.2007.03.028

Rodak, K., Bęben, D., Birska, M., Siwiela, O., Kokot, I., Moreira, H., ... & Kratz, E. M. (2024). Evaluating the neuroprotective potential of caffeinated coffee in the context of aluminum-induced neurotoxicity: Insights from a PC12 cell culture model. *Antioxidants*, 13(3), 342. https://doi.org/10.3390/antiox13030342

Rojas-González, A., Figueroa-Hernández, C. Y., González-Rios, O., Suárez-Quiroz, M. L., González-Amaro, R. M., Hernández-Estrada, Z. J., and Rayas-Duarte, P. (2022). Coffee chlorogenic acids incorporation for bioactivity enhancement of foods: A review. *Molecules*, 27(11), 3400. https://doi.org/10.3390/molecules27113400

Rusinek, R., Dobrzański Jr, B., Oniszczuk, A., Gawrysiak-Witulska, M., Siger, A., Karami, H., ... and Gancarz, M. (2022). How to identify roast defects in coffee beans based on the volatile compound profile. *Molecules*, 27(23), 8530. https://doi.org/10.3390/molecules27238530

Santana-Gálvez, J., Cisneros-Zevallos, L., and Jacobo-Velázquez, D. A. (2017). Chlorogenic acid: Recent advances on its dual role as a food additive and a nutraceutical against metabolic syndrome. *Molecules*, 22(3), 358. https://doi.org/10.3390/molecules22030358

Saud, S., and Salamatullah, A. M. (2021). Relationship between the chemical composition and the biological functions of coffee. *Molecules*, 26(24), 7634. https://doi.org/10.3390/molecules26247634

Seccia, S., Albrizio, S., and Dini, I. (2025). SPE-HPLC-DAD dosage of seven neonicotinoids in green coffee. *Molecules*, 30(9), 1930. https://doi.org/10.3390/molecules30091930

Seeberger, P. H. (2022). Monosaccharide diversity. *In:* Içinde A. Varki et al. (Eds), Essentials of Glycobiology (4. bs). Cold Spring Harbor Laboratory Press. https://doi.org/10.1101/GLYCOBIOLOGY.4E.2

Sharifi-Zahabi, E., Hajizadeh-Sharafabad, F., Nachvak, S. M., Mirzaian, S., Darbandi, S., and Shidfar, F. (2023). A comprehensive insight into the molecular effect of theobromine on cardiovascular-related risk factors: A systematic review of in vitro and in vivo studies. *Phytotherapy Research*, 37(9), 3765–3779. https://doi.org/10.1002/ptr.7916

Shen, T., Lee, J., Lee, E., Kim, S. H., Kim, T. W., and Cho, J. Y. (2010). Cafestol, a coffee-specific diterpene, is a novel extracellular signal-regulated kinase inhibitor with AP-1-targeted inhibition of prostaglandin E2 production in lipopolysaccharide-activated macrophages. *Biological and Pharmaceutical Bulletin*, 33(1), 128–132. https://doi.org/10.1248/bpb.33.128

Shen, X., Wang, Q., Wang, H., Fang, G., Li, Y., Zhang, J., and Liu, K. (2024). Microbial characteristics and functions in coffee fermentation: A review. *Fermentation*, 11(1), 5. https://doi.org/10.3390/fermentation11010005

Shi, X., Li, Y., Huang, D., Chen, S., and Zhu, S. (2025). Characterization and discrimination of volatile compounds in roasted Arabica coffee beans from different origins by combining GC-TOFMS, GC-IMS, and GC-E-Nose. *Food Chemistry*, 481, 144079. https://doi.org/10.1016/j.foodchem.2025.144079

Shi, Z., Luan, J., Zhang, Y., Wang, G., Mei, C., Chen, L., ... and Cheng, J. (2025). Exploring the impact and mechanisms of coffee and its active ingredients on depression, anxiety, and sleep disorders. *Nutrients*, 17(19), 3037.

Shin, H. S., Satsu, H., Bae, M. J., Totsuka, M., and Shimizu, M. (2017). Catechol groups enable reactive oxygen species scavenging-mediated suppression of PKD-NFkappaB-IL-8 signaling pathway by chlorogenic and caffeic acids in human intestinal cells. *Nutrients*, 9(2), 165. https://doi.org/10.3390/nu17193037

Shofinita, D., Lestari, D., Purwadi, R., Sumampouw, G. A., Gunawan, K. C., Ambarwati, S. A., ... and Tjahjadi, J. T. (2024). Effects of different decaffeination methods on caffeine contents, physicochemical, and sensory properties of coffee. *International Journal of Food Engineering*, 20(8), 561–581.

Silva, H., and Lopes, N. M. F. (2020). Cardiovascular effects of caffeic acid and its derivatives: A comprehensive review. *Frontiers in Physiology*, 11, 595516. https://doi.org/10.3389/fphys.2020.595516

Silva, M. A. E., Brand, A. L. M., Novaes, F. J. M., and Rezende, C. M. (2023). Cafestol, kahweol and their acylated derivatives: Antitumor potential, pharmacokinetics, and chemopreventive profile. *Food Reviews International*, 39(9), 7048–7080. https://doi.org/10.1080/87559129.2022.2141776

Singh, V., Indoria, S., Jisha, K. J., and Gardas, R. L. (2021). Structure and solubility of polysaccharides. *In:* Içinde, A. A. K. et al. (Eds.), Polysaccharides: Properties and applications (1. bs, ss. 325–336). John Wiley & Sons, Ltd. https://doi.org/10.1002/9781119711414.ch15

Sofyanita, E. N., Juniarto, A. Z., Suwondo, A., and Nugroho, H. (2025). The effect of chlorogenic acid content in coffee can reduce Malondialdehyde (MDA) and increase testosterone hormone levels in DM conditions. *Gema Lingkungan Kesehatan*, 23(1), 20–26. https://doi.org/10.36568/gelinkes.v23i1.163

Song, X., Kirtipal, N., Lee, S., Malý, P., and Bharadwaj, S. (2023). Current therapeutic targets and multifaceted physiological impacts of caffeine. *Phytotherapy Research*, 37(12), 5558–5598. https://doi.org/10.1002/ptr.8000

Stadler, R. H., Varga, N., Hau, J., Vera, F. A., and Welti, D. H. (2002). Alkylpyridiniums. 1. Formation in model systems via thermal degradation of trigonelline. *Journal of Agricultural and Food Chemistry*, 50(5), 1192–1199. https://doi.org/10.1021/jf011234k

Suaniti, N. M., Saraswati, A. A. S. D., and Putra, A. A. B. (2022). Analisis Kafein Dalam Kopi Arabika (Coffea arabica L.) Pada Berbagai Suhu Penyangraian Dengan Metode Spektrofotometer Uv-Vis Dan Gc-Ms. *Jurnal Kimia (Journal of Chemistry)*, 16(1), 115–21. https://doi.org/10.24843/JCHEM.2022.v16.i01.p15

Suhandy, D., Yulia, M., Widodo, S., Naito, H., Al Riza, D. F., and Rahman, A. (2025). Identification of Lampung fine Robusta peaberry green coffee beans with different fermentation methods using portable front-face fluorescence spectroscopy. *Discover Applied Sciences*, 7(5), 1–13. https://doi.org/10.1007/s42452-025-06820-w

Sutrisno, E., Nurhayati, A., and Apriyanto, M. (2023). Physical and chemical properties of robusta coffee beans at different temperatures and roasting times. *Open Global Scientific Journal*, 2(01), 25–36.

Tej, G. N. V. C., & Nayak, P. K. (2018). Mechanistic considerations in chemotherapeutic activity of caffeine. *Biomedicine & Pharmacotherapy*, 105, 312–319.

Theurillat, V., Dubois, M., XueFeng, D., Pellegrino, G., and Lamberti, G. (2024). Coffee and mineral oil hydrocarbons: potential dietary intake. *Food Additives & Contaminants: Part A*, 41(11), 1516–1520. https://doi.org/10.1080/19440049.2024.2397805

Todhanakasem, T., Van Tai, N., Kunyanee, K., and Pitinidhipat, N. (2025). Physicochemical characteristics and metabolite content of roasted arabica coffee in relation to consumer preference. *LWT*, 217, 117438. https://doi.org/10.1016/j.lwt.2025.117438

Tolessa, K., D'heer, J., Duchateau, L., and Boeckx, P. (2017). Influence of growing altitude, shade and harvest period on quality and biochemical composition of Ethiopian specialty coffee. *Journal of the Science of Food and Agriculture*, 97(9), 2849–2857. https://doi.org/10.1002/jsfa.8114

Tolessa, K., Alemayehu, D., Belew, D., and Boeckx, P. (2019). Biochemical composition of Ethiopian coffees (Coffea arabica L.) as influenced by variety and postharvest processing methods. *African Journal of Food Science*, 13(2), 48–56. https://doi.org/10.5897/AJFS2018.1770

Toribio-Nava, E., Estrada-Girón, Y., Velázquez-Juárez, G., López-Mercado, J., Delgado-Saucedo, J. I., Castro-García, P. B., ... and Puebla-Pérez, A. M. (2024). Chan Seeds (Hyptis suaveolens L.) storage proteins: Isolation, characterization, and cytotoxic effect on three cancer cell lines. *Separations*, 11(12), 350. https://doi.org/10.3390/separations11120350

Trimigno, A., Marincola, F. C., Dellarosa, N., Picone, G., and Laghi, L. (2015). Definition of food quality by NMR-based foodomics. *Current Opinion in Food Science*, 4, 99–104. https://doi.org/10.1016/j.cofs.2015.06.008

Tuğal Aslan, D., and Göktaş, Z. (2025). The therapeutic potential of theobromine in obesity: A comprehensive review. *Nutrition Reviews*, 83(5), 859–868. https://doi.org/10.1093/nutrit/nuae122

Turnbull, D., Rodricks, J. V., Mariano, G. F., and Chowdhury, F. (2017). Caffeine and cardiovascular health. *Regulatory Toxicology and Pharmacology*, 89, 165–185. https://doi.org/10.1016/j.yrtph.2017.07.025

Uefuji, H., Ogita, S., Yamaguchi, Y., Koizumi, N., and Sano, H. (2003). Molecular cloning and functional characterization of three distinct N-methyltransferases involved in the caffeine biosynthetic pathway in coffee plants. *Plant Physiology*, 132(1), 372–380. https://doi.org/10.1104/pp.102.019679

Upadhyay, R., and Mohan Rao, L. J. (2013). An outlook on chlorogenic acids—Occurrence, chemistry, technology, and biological activities. *Critical Reviews in Food Science and Nutrition*, 53(9), 968–984. https://doi.org/10.1080/10408398.2011.576319

Valada, P., Alçada-Morais, S., Cunha, R. A., and Lopes, J. P. (2022). Theobromine targets adenosine receptors to control hippocampal neuronal function and damage. *International Journal of Molecular Sciences*, 23(18), 10510. https://doi.org/10.3390/ijms231810510

Vinecky, F., Davrieux, F., Mera, A. C., Alves, G. S. C., Lavagnini, G., Leroy, T., ... and Andrade, A. C. (2017). Controlled irrigation and nitrogen, phosphorous and potassium fertilization affect the biochemical composition and quality of Arabica coffee beans. *The Journal of Agricultural Science*, 155(6), 902–918.

Virgili, J., Motitis, P., Julal, G., Mavrommatis, Y., and Pilic, L. (2023). The impact of genetic variability on the relationship between caffeine and cardiometabolic outcomes: A systematic review. *Nutrition Bulletin*, 48(1), 28–42. https://doi.org/10.1111/nbu.12606

Voiniciuc, C. (2022). Modern mannan: A hemicellulose's journey. *New Phytologist*, 234(4), 1175–1184. https://doi.org/10.1111/nph.18091

Vu, D. C., Vu, Q. T., Huynh, L., Lin, C. H., Alvarez, S., Vo, X. T., and Nguyen, T. H. (2021). Evaluation of fatty acids, phenolics and bioactivities of spent coffee grounds prepared from Vietnamese coffee. *International Journal of Food Properties*, 24(1), 1548–1558. https://doi.org/10.1080/10942912.2021.1977657

Worku, M., De Meulenaer, B., Duchateau, L., and Boeckx, P. (2018). Effect of altitude on biochemical composition and quality of green arabica coffee beans can be affected by shade and postharvest processing method. *Food Research International*, 105, 278–285. https://doi.org/10.1016/j.foodres.2017.11.016

Xiong, K., Li, M. M., Chen, Y. Q., Hu, Y. M., and Jin, W. (2024). Formation and reduction of toxic compounds derived from the Maillard Reaction during the thermal processing of different food matrices. *Journal of Food Protection*, 87(9), 100338. https://doi.org/10.1016/j.jfp.2024.100338

Xu, B. W., Li, S. S., Ding, W. L., Zhang, C., Rehman, M. U., Tareen, M. F., ... and Huang, S. C. (2025). From structure to function: A comprehensive overview of polysaccharide roles and applications. *Food Frontiers*, 6(1), 15–39. https://doi.org/10.1002/fft2.490

Yan, G., Dai, X., Yan, Y., Yan, J., Tian, W., and Jiang, R. (2025). Association between coffee consumption and life expectancy: A prospective cohort study from NHANES 2001–2018. *Public Health Nutrition*, 28(1), e141. https://doi.org/10.1017/S1368980025100888

Yi, Z., Ma, X., Tong, Q., Ma, L., Tan, Y., Liu, D., ... and Li, X. (2025). A library of polyphenol-amino acid condensates for high-throughput continuous flow production of nanomedicines with ultra-high drug loading. *Advanced Materials*, 37(15), 2417534. https://doi.org/10.1002/adma.202417534

Yoneda, M., Sugimoto, N., Katakura, M., Matsuzaki, K., Tanigami, H., Yachie, A., ... and Shido, O. (2017). Theobromine up-regulates cerebral brain-derived neurotrophic factor and facilitates motor learning in mice. *The Journal of Nutritional Biochemistry*, 39, 110–116. https://doi.org/10.1016/j.jnutbio.2016.10.002

Yu, H., Zhang, R., Yang, F., Xie, Y., Guo, Y., Yao, W., & Zhou, W. (2021). Control strategies of pyrazines generation from Maillard reaction. *Trends in Food Science & Technology*, 112, 795–807. https://doi.org/10.1016/j.tifs.2021.04.028

Zengin, G., Sinan, K. I., Mahomoodally, M. F., Angeloni, S., Mustafa, A. M., Vittori, S., ... and Caprioli, G. (2020). Chemical composition, antioxidant and enzyme inhibitory properties of different extracts obtained from spent coffee ground and coffee silverskin. *Foods*, 9(6), 713. https://doi.org/10.3390/foods9060713

Zhang, H., and Su, Q. (2025). Low-FODMAP diet for irritable bowel syndrome: Insights from microbiome. *Nutrients*, 17(3), 544. https://doi.org/10.3390/nu17030544

Zhang, K., Cheng, J., Hong, Q., Dong, W., Chen, X., Wu, G., and Zhang, Z. (2022). Identification of changes in the volatile compounds of robusta coffee beans during drying based on HS-SPME/GC-MS and E-nose analyses with the aid of chemometrics. *LWT*, 161, 113317. https://doi.org/10.1016/j.lwt.2022.113317

Zhang, W., Zhang, Y., Chen, S., Zhang, H., Yuan, M., Xiao, L., ... and Xu, H. (2021). Trigonelline, an alkaloid from Leonurus japonicus Houtt., suppresses mast cell activation and OVA-induced allergic asthma. *Frontiers in Pharmacology*, 12, 687970. https://doi.org/10.3389/fphar.2021.687970

Zheng, X. Q., Nagai, C., and Ashihara, H. (2004). Pyridine nucleotide cycle and trigonelline (N-methylnicotinic acid) synthesis in developing leaves and fruits of Coffea arabica. *Physiologia Plantarum*, 122(4), 404–411. https://doi.org/10.1111/j.1399-3054.2004.00422.x

Zhou, J., Chan, L., and Zhou, S. (2012). Trigonelline: A plant alkaloid with therapeutic potential for diabetes and central nervous system disease. *Current Medicinal Chemistry*, 19(21), 3523–3531. https://doi.org/10.2174/092986712801323171

Zhou, J., Zhou, S., and Zeng, S. (2013). Experimental diabetes treated with trigonelline: Effect on β cell and pancreatic oxidative parameters. *Fundamental & Clinical Pharmacology*, 27(3), 279–287. https://doi.org/10.1111/j.1472-8206.2011.01022.x

Zhou, M. Z., Yan, C. Y., Zeng, Z., Luo, L., Zeng, W., and Huang, Y. H. (2020). N-methyltransferases of caffeine biosynthetic pathway in plants. *Journal of Agricultural and Food Chemistry*, 68(52), 15359–15372. ttps://doi.org/10.1021/acs.jafc.0c06167

Zimmermann, Y. C., Schwarz, S., and Lachenmeier, D. W. (2025). Post-roast maturation and coffee quality: A sensory perspective on freshness. *European Food Research and Technology*, 251(11), 3839–3845. https://doi.org/10.1007/s00217-025-04873-0

CHAPTER

6

Chromatographic Techniques in Coffee Science: Analysis of Bioactive Compounds and Quality Control

Abidin Gümrükçüoğlu[1] **and Alper Durmaz**[2*]

[1] Artvin Çoruh University, Medicinal-Aromatic Plants Application and Research Center, Artvin, Türkiye

[2] Artvin Çoruh University, Ali Nihat Gökyiğit Botanical Garden Application and Research Center, Artvin, Türkiye

Coffee is a complex matrix containing a broad range of bioactive compounds, including caffeine, chlorogenic acids, trigonelline, and diterpenes, which contribute not only to its sensory properties but also to its recognised physiological benefits. The chemical variability among *Coffea arabica* and *Coffea canephora* cultivars is influenced by genetic, environmental, and processing factors, necessitating the use of advanced analytical techniques for precise characterisation and quality control. Chromatographic methods, especially High-Performance Liquid Chromatography (HPLC), Gas Chromatography–Mass Spectrometry (GC–MS), and Liquid Chromatography–Tandem Mass Spectrometry (LC–MS/MS), have become indispensable tools for profiling coffee's bioactive constituents. Recent advances in multivariate statistical analysis, such as Principal Component Analysis (PCA) and Hierarchical Cluster Analysis (HCA), have further enhanced the ability to classify coffee varieties and origins based on chromatographic fingerprints.

This chapter provides a comprehensive overview of chromatographic applications in coffee science, emphasising sample preparation techniques, chromatographic method optimisation, and data integration for authenticity and quality assessment. The integration of chemometric tools and green analytical methodologies is also discussed as an emerging approach toward sustainable and efficient analysis. By combining the expertise of phytochemistry and analytical chemistry, this chapter highlights how chromatographic technologies support both scientific understanding and industrial quality assurance in modern coffee research.

*Corresponding author: a.alperdurmaz@gmail.com

DOI: 10.1201/9781003689393-6

1. Introduction: Coffee as a functional beverage

Coffee, one of the most widely consumed beverages worldwide, has evolved from a cultural and economic commodity to a scientifically recognised source of bioactive nutrients. The beverage is derived from the roasted seeds of Coffea species, primarily *Coffea arabica* and *Coffea canephora*, which together dominate global production and trade (Jeszka-Skowron et al., 2015). The chemical composition of coffee encompasses numerous classes of compounds, including caffeine, chlorogenic acids, trigonelline, diterpenes, and various phenolics, that determine its flavour, aroma, and biological effects (Nuhu, 2014). The diversity of these phytochemicals reflects both botanical origin and environmental conditions such as altitude, soil composition, and post-harvest treatment (Grumezescu and Holban, 2019).

From a nutritional and pharmacological perspective, these compounds exhibit antioxidant, anti-inflammatory, and neuroprotective properties, supporting the categorisation of coffee as a functional food. Caffeine acts as a central nervous system stimulant and a performance enhancer, while chlorogenic acids contribute to antioxidant activity and the regulation of lipid metabolism. Trigonelline and diterpenes, including cafestol and kahweol, further modulate metabolic and hepatic functions (Jeszka-Skowron et al., 2015; Nuhu, 2014).

Given the complexity of coffee chemistry, the application of robust analytical methodologies has become essential for qualitative and quantitative assessment. Chromatographic techniques—particularly High-Performance Liquid Chromatography (HPLC), Gas Chromatography (GC), and advanced hybrid systems such as Liquid Chromatography–Mass Spectrometry (LC–MS)—enable the detection, identification, and quantification of bioactive constituents with high precision and selectivity (Nascimento et al., 2020). These analytical approaches are complemented by chemometric and multivariate tools, such as Principal Component Analysis (PCA) and Partial Least Squares (PLS), which allow discrimination of coffee samples based on their chemical fingerprints.

Moreover, integrating chromatographic data with spectroscopic and sensory analyses provides a multidimensional understanding of coffee quality and authenticity (Grumezescu and Holban, 2019). This convergence of analytical chemistry and phytochemical profiling represents a critical step toward standardisation, ensuring both consumer safety and the preservation of coffee's unique functional attributes.

Plant metabolites occur as complex mixtures of numerous compounds exhibiting a wide range of polarity and hydrophobicity (Romanik et al., 2007). Within plant cells, these compounds can be broadly classified into low-polarity constituents, such as waxes and terpenoids; moderately polar constituents, including lipids, phenolic compounds, and relatively non-polar alkaloids; and highly polar constituents, such as polar glycosides, polar alkaloids, sugars, peptides, and proteins (Romanik et al., 2007; Awuchi, 2019; Hussein and El-Anssary, 2018). Among these groups, polyphenols are of particular interest due to their well-documented anti-inflammatory, antioxidant, and anticancer properties, which render them highly valuable for applications in the food, pharmaceutical, and cosmetic industries (Hussein and El-Anssary, 2018; Taha et al., 2019; Yenduri et al., 2025).

The recovery of bioactive compounds from plant cells requires appropriate sample preparation steps and carefully designed extraction strategies (Romanik et al., 2007; Zhang et al., 2020). The structural diversity of polyphenols, combined with their susceptibility to thermal degradation, presents significant challenges in maintaining biological activity during the extraction process (Yenduri et al., 2025; Zhang et al., 2020). Consequently, an effective extraction protocol should aim to maximise the recovery of target compounds while preserving thermolabile metabolites and prioritising sustainable and environmentally friendly approaches (Azwanida, 2015; Yenduri et al., 2025). Achieving this balance is essential not only for analytical reliability but also for environmental responsibility.

Extraction techniques enable the isolation of therapeutic and functional compounds from plant matrices by optimising key parameters, including solvent selection, temperature, pressure, extraction time, and solid-to-liquid ratio (Azwanida, 2015; Meinhart et al., 2010; Zhang et al., 2020). In recent years, technological advancements have facilitated a transition from conventional extraction methods—such as maceration, infusion, and decoction—to advanced techniques, including microwave-assisted extraction, ultrasound-assisted extraction, and supercritical fluid extraction (Ajila et al., 2010; Azwanida, 2015; Vinatoru et al., 2017; Ziyatdinova et al., 2013). These advanced approaches have significantly improved extraction efficiency and selectivity while reducing processing time and solvent consumption. Nevertheless, each technique possesses inherent advantages and limitations, and the selection of an appropriate method depends on multiple factors, including the chemical nature of the target compounds, the characteristics of the plant matrix, analytical objectives, and economic considerations (Ajila et al., 2010; Vinatoru et al., 2017).

These general extraction principles and methodological considerations become particularly critical for plant-derived products of high economic importance. In this context, coffee represents one of the most illustrative examples, given its extensive global consumption and wide range of industrial applications. Coffee is no longer regarded solely as a beverage but as a strategic agricultural commodity that shapes socioeconomic structures in tropical regions and influences billions of consumers worldwide. The production and international trade of coffee constitute a substantial proportion of export revenues for many developing countries, underscoring its central role in the global agricultural economy (United States Department of Agriculture, 2019).

Ensuring coffee quality is critical not only during production but also in terms of the isolation and analytical evaluation of bioactive compounds from plant material. Coffee quality depends not only on the genetic background but also on several stages throughout the production chain. In the initial phase, farmers select the cultivation area, identify the appropriate coffee species, and prepare the soil. During cultivation, practices such as irrigation, fertilisation, and disease control play a decisive role in shaping bean quality. Harvest timing is equally important because harvesting at an inappropriate stage of maturity can prevent beans from reaching their expected quality attributes. Post-harvest operations, including drying, fermentation, and thermal processing, further determine the chemical composition and characteristic properties of coffee (Alliance for Coffee Excellence, 2019).

After processing, coffee is stored under conditions that strongly influence the preservation of its initial quality attributes. During storage, changes in the chemical profile are largely unavoidable, and factors such as temperature, relative humidity, and the gaseous environment significantly influence the duration for which quality can be maintained (Alliance for Coffee Excellence, 2019). Degradation processes may reduce the levels of alkaloids, phenolic compounds, and aroma-related constituents, ultimately leading to a measurable decline in overall coffee quality (Brazilian Association of Coffee Industries, 2004).

Following these stages, the bioactive compound profile and overall quality of coffee can be accurately evaluated only through advanced analytical methods coupled with optimised extraction protocols. A range of analytical approaches is employed to gain a deeper understanding of coffee bioactives. Because earlier analytical procedures are often labour-intensive and time-consuming, they have progressively been replaced by more efficient modern technologies (Brazilian Association of Coffee Industries, 2018). In sample preparation, protocols based on phase distribution principles and matrix-supported extraction approaches are frequently applied, depending on the target analytes (Gomez-Ruiz et al., 2007). The resulting extracts can then be characterised using high-resolution liquid chromatography systems capable of resolving complex mixtures, while complementary platforms such as infrared spectroscopy, mass spectrometry, and nuclear magnetic resonance provide additional structural and compositional information (Brazilian Association of Coffee Industries, 2020; Motilva et al., 2013).

In particular, the complex polyphenolic profile of coffee and the susceptibility of certain constituents to thermal degradation necessitate careful selection of solvent systems and extraction temperatures in line with general extraction principles. More recently, microextraction strategies and techniques based on gas/liquefied-gas systems under controlled conditions have been increasingly adopted in research laboratories to enhance sensitivity and reduce solvent consumption (Motilva et al., 2013; Shrivas and Wu, 2007).

2. Coffee production

Global coffee production reached approximately 168.71 million 60-kg bags during the 2019/2020 harvest season. The majority of this production was supplied by *Coffea arabica* and *Coffea canephora* (Robusta), with estimated outputs of 96.22 million and 72.50 million bags, respectively. Brazil, Vietnam, and Colombia collectively accounted for nearly 70% of global coffee production, representing a marked increase compared to their combined share of approximately 60% a decade earlier (International Coffee Organization, 2020). Brazil remained the world's largest coffee producer, with a total production of about 57 million bags, including roughly 37 million bags of Arabica and 20 million bags of Robusta. Vietnam recorded a production increase of approximately 4.4%, reaching 31.2 million bags, while Colombia produced around 14.1 million bags, maintaining its position as the second-largest Arabica producer worldwide (International Coffee Organization, 2020).

Colombia's sustained production growth has been partly attributed to the renovation of coffee plantations and the adoption of rust-resistant coffee varieties, which have contributed to improved yields and crop stability (United States

Department of Agriculture, 2019). In Brazil, approximately 2.1 million hectares are dedicated to coffee cultivation. Favourable tropical climatic conditions, adequate annual rainfall, and average temperatures ranging from 18 to 26 °C provide optimal conditions for coffee production (United States Department of Agriculture, 2019). Minas Gerais is the leading coffee-producing state in Brazil, accounting for approximately 49.8% of the national output, equivalent to 24.55 million bags. Other central producing states include Espírito Santo (13.50 million bags) and São Paulo (4.30 million bags). The state of Bahia ranks fourth, producing nearly 3 million bags from an estimated cultivation area of 130,000 hectares (United States Department of Agriculture, 2019).

Within Bahia, the Chapada Diamantina region is particularly notable due to its favourable soil characteristics and climatic conditions, which are suitable for high-quality coffee production. Coffee originating from this region has received international recognition through various speciality coffee competitions, reflecting its distinctive sensory and quality attributes (Alliance for Coffee Excellence, 2019).

During the 2019/2020 marketing year, Brazil exported approximately 26.95 million bags of coffee. Of this total, about 10.2% consisted of processed coffee products, while the remainder comprised mainly green Arabica (21.28 million bags) and green Robusta (2.94 million bags) coffee beans (International Coffee Organization, 2020). Global coffee consumption increased to approximately 169.34 million bags, representing an annual rise of around 1.24 million bags. This imbalance between production and consumption contributed to market pressures that favoured rising coffee prices (International Coffee Organization, 2020). Between October 2019 and March 2020, Brazil supplied nearly 20% of the European Union's soluble coffee demand. In addition, coffees originating from Brazil and Colombia were among the most heavily imported by consumers in the United States (53.6%) and Japan (46.0%), underscoring their dominant position in the international coffee trade (International Coffee Organization, 2018).

3. Coffee quality classification

In Brazil, coffee purity and quality certification has been formally implemented through the *Programa de Qualidade do Café* (PQC), launched in 2004 by the Brazilian Association of Coffee Industries (ABIC). This program evaluates a wide range of quality parameters, including aroma, body, acidity, bitterness, astringency, and the presence of defective beans, such as green, black, burnt, and broken beans (Brazilian Association of Coffee Industries, 2004). The PQC grading system is based on a numerical scale ranging from 0 to 10, with a minimum acceptable beverage quality score of 5. In addition to sensory parameters, optical microscopy is employed to assess coffee purity and detect foreign materials and defects (Brazilian Association of Coffee Industries, 2004).

Sensory evaluation of coffee in Brazil is further conducted through the officially regulated "cup test," administered by the Ministry of Agriculture, Livestock, and Food Supply. This standardised procedure provides an additional framework for assessing beverage quality and consistency within national coffee production and trade systems (Ministry of Agriculture, Livestock, and Food Supply, 2019).

The classification of green coffee beans is primarily based on defect count, sieve size, and cup quality. The Brazil/New York Green Coffee Classification System is regarded as more sensitive than the Speciality Coffee Association of America (SCAA) method, although it is also more time-consuming. Brazilian legislation limits the proportion of foreign matter in green coffee beans to a maximum of 1% by weight, reflecting strict regulatory control over both export and domestic market quality (Brazilian Ministry of Agriculture, 2019; Coffee Research Institute, 2020).

Intrinsic defects include beans that deviate from healthy coffee beans, such as those that are black, burnt, or immature. Additionally, extrinsic defects may occur due to insect damage, excessive drying, or improper processing practices, resulting in broken or physically damaged beans (Ministry of Agriculture, Livestock, and Food Supply, 2019). According to ABIC quality standards, Traditional and Extra Strong coffee categories may contain up to 20% defective beans by weight, whereas this proportion is reduced to 10% for Superior coffees. In contrast, Gourmet coffee classifications do not permit the presence of faulty beans under any circumstances (Brazilian Association of Coffee Industries, 2018).

International coffee certification and standardisation are primarily governed by standards developed by the International Organisation for Standardisation (ISO). The ISO Technical Committee ISO/TC 34/SC 15 encompasses the entire coffee production chain, from green coffee to final consumption, including sampling procedures, analytical and testing methods, product specifications, terminology, and packaging requirements (International Organisation for Standardisation, 2020). Standards such as ISO 3509:2005 define commonly used coffee-related terminology, while ISO 10470:2004 establishes five major defect categories for green coffee. Additionally, ISO 4149:2005 outlines methods for detecting foreign matter and defects through visual and olfactory inspection (International Organisation for Standardisation, 2020).

For roasted coffee, the determination of moisture content is standardised under ISO 11294:1994 and ISO 11817:1994. Instant coffee products are subject to specific analytical requirements defined in ISO 3726:1983 and ISO 20938:2008. Caffeine content is typically determined using ISO 20481:2008, which describes High-Performance Liquid Chromatography as the reference analytical method, as well as ISO/AWI 24386 for related applications (International Organisation for Standardisation, 2020).

Although ISO standards are widely adopted, some coffee-producing countries, including Angola, Colombia, and Cuba, apply country-specific classification and certification systems alongside or in place of international standards, reflecting regional production practices and regulatory frameworks (International Coffee Organization, 2019).

4. Phenolic acids and terpenoids: Coffee composition and biological effects

Consumer demand for authentic, high-quality coffee products has motivated extensive research worldwide to elucidate the chemical composition of coffee and to quantitatively determine its constituent compounds. Among these constituents, caffeine is the most widely recognised compound due to its pronounced stimulatory effects on the central nervous system (Debastiani et al., 2019). Caffeine is also present in other beverages such as tea and energy drinks and is widely used in pharmaceutical formulations (Weldegebreal et al., 2017). As a result, caffeine is considered the most frequently consumed psychoactive substance globally. In addition to caffeine, trigonelline and formic acid are well-known constituents of coffee, and their influence on the organoleptic properties of the final product necessitates their accurate determination in coffee matrices (Del Campo et al., 2010).

Coffee is particularly rich in phenolic compounds. The major phenolic acids identified in coffee include *p*-hydroxybenzoic acid, vanillic acid, *p*-coumaric acid, ferulic acid, and chlorogenic acid (CGA). Most of these compounds occur as derivatives such as caffeic acid, caffeoylquinic acids, dicaffeoylquinic acids, 3-*O*-feruloylquinic acid, and mixed feruloyl–caffeoyl quinic acid esters. Chlorogenic acid is the predominant phenolic acid in coffee and has been reported to exhibit hypolipidemic, antioxidant, and antibacterial activities (Zhang et al., 2016; Martins et al., 2014). Several studies have demonstrated that chlorogenic acid reduces total cholesterol, triglyceride, and low-density lipoprotein (LDL) levels while increasing high-density lipoprotein (HDL) concentrations. At the cellular level, chlorogenic acid protects against DNA damage (Zhang et al., 2016) and has been shown to reverse corticosteroid-induced autophagy and apoptosis in PC12 cells (Martins et al., 2014). Furthermore, chlorogenic acid exhibits antibacterial activity against microorganisms, such as Pseudomonas aeruginosa, and contributes to the maintenance of intestinal barrier integrity (Zhang et al., 2016).

During coffee roasting, chlorogenic acid undergoes thermal degradation, leading to the formation of bitter phenolic derivatives and various aromatic compounds. Additionally, chlorogenic acid participates in melanoidin-forming reactions with proteins, contributing to the development of coffee's colour and influencing its acidic characteristics (Martins et al., 2014).

Coffee is also a rich source of terpenoid compounds. The diterpenoid fraction of coffee mainly consists of *p*-kaurane-type diterpenes and coffeol-related structures. Among these, coffeol and 16-*O*-methylcoffeol are the most abundant compounds. Notably, 16-*O*-methylcoffeol has been proposed as a chemotaxonomic marker for coffee species and was first identified in small coffee beans using high-field 600 MHz nuclear magnetic resonance spectroscopy and liquid chromatography–mass spectrometry techniques (Chu et al., 2016). In recent years, several novel diterpenoid glycosides, including four pairs of mascarosides, one pair of villanovane I derivatives, and seven pairs of kaurane diterpenoid glycosides, have been isolated and structurally characterised from coffee beans (Chu et al., 2016; Iwamoto et al., 2019).

Biological activity studies have shown that most of these newly identified diterpenoid glycosides do not exhibit inhibitory effects against tumour cell lines such as HL60, A549, SMMC-7721, MCF-7, and SW480. However, specific compounds, including caffarolides C, D, and F, were found to activate platelet aggregation in vitro (Chu et al., 2016; Wang et al., 2018). In addition, conuri diterpenes isolated from sun-dried coffee beans were reported to inhibit lipopolysaccharide-induced mediator production in macrophages (Iwamoto et al., 2019).

Further investigations into the biological activities of coffee terpenoids have focused on compounds such as cafestol, kahweol, and their derivatives. Cafestol acetate and kahweol exhibit dose-dependent inhibitory effects on human prostate cancer cells (Iwamoto et al., 2019). Cafestol has also been shown to suppress the proliferation of breast cancer cells and induce apoptosis via a caspase-3-dependent pathway (Oh et al., 2018). Moreover, studies examining the effects of caffeine on leukaemia cell lines, including NB4, K562, HL60, and KG1, have revealed that HL60 and KG1 cells exhibit the highest cytotoxic responses (Lima et al., 2017).

Beyond its cytotoxic effects, caffeine plays a role in metabolic regulation by stimulating insulin secretion and enhancing glucose metabolism. Caffeine increased insulin secretion by approximately 12% and 16% at concentrations of 10^{-8} mol/L and 10^{-6} mol/L, respectively, while prolonged exposure resulted in increases of up to 34% and 68% (Mellbye et al., 2015). Additionally, caffeine enhanced glucose uptake by approximately 8% in human skeletal muscle cells, underscoring its contribution to metabolic processes (Seo et al., 2018).

5. Determination of volatile compounds in coffee

At the beginning of the twentieth century, knowledge regarding the aromatic constituents of coffee was minimal. Over time, however, a variety of compounds—including methylamine, ammonia, trimethylamine, pyrrole, pyridine, acetone, formic acid, valeric acid, furfural, and furfuryl alcohol—were identified using classical chemical methods. During the 1920s, with financial support from the Swiss government, 29 new volatile compounds were discovered in coffee (Morton and Macleod, 1986). Among the most frequently detected compounds during this period were furfuryl mercaptan, α-diketones, and alkylpyrazines. Between 1929 and 1962, an additional 27 compounds were identified, while the application of gas capillary chromatography between 1963 and 1982 led to the identification of more than 600 volatile compounds associated with roasted coffee aroma. These studies enabled the classification of the major chemical groups responsible for the characteristic aroma of roasted coffee (Morton and Macleod, 1986). The principal aroma constituents of coffee were further investigated by Ky et al. (2001).

In another study, the aroma profiles of espresso coffees obtained from different botanical varieties and roasting conditions were examined (Maeztu et al., 2001). Volatile compounds were extracted using headspace techniques and subsequently analyzed by gas chromatography–mass spectrometry (GC–MS). A total of 77 volatile compounds were identified, among which 13 were determined to have the greatest impact on coffee aroma (Maeztu et al., 2001). Chromatographic data showed strong agreement with sensory flavor profiles. For instance, aldehydes were associated with fruity notes, diketones with buttery aromas, and pyrazines with burnt and earthy characteristics. Similar investigations were conducted using Arabica coffee from Colombia and Robusta coffee from Indonesia (Bücking and Steinhart, 2002). These studies employed gas chromatography coupled with flame ionisation detection (FID), mass spectrometry, olfactometry, and headspace techniques, revealing approximately 30 key volatile compounds responsible for coffee aroma. Additionally, the influence of milk addition on coffee aroma was evaluated, demonstrating that milk reduced aroma intensity while generally preserving the overall aroma profile (Bücking and Steinhart, 2002).

In recent years, advanced analytical techniques such as solid-phase microextraction (SPME) and comprehensive two-dimensional gas chromatography (GC×GC-FID) have been increasingly applied to the analysis of coffee volatiles. In a study by Ryan et al. (2004), comprehensive two-dimensional gas chromatography employing a polar first column and a non-polar second column enabled the identification of 44 volatile compounds. Time-of-flight mass spectrometry (TOF-MS) was also utilised to enhance compound detection and identification. The identified aroma constituents included ketones, thiols, furans, phenols, and pyrroles (Ryan et al., 2004).

A subsequent study in 2017 established correlations between the chemical composition of coffee volatiles and their sensory attributes (Bressanello et al., 2017). Chemical analyses were performed on ground coffee samples using headspace solid-phase microextraction (HS-SPME) followed by GC–MS analysis. The resulting data were consistent with cupping (sensory evaluation) results, highlighting the applicability of these analytical approaches for quality control purposes.

The effects of different roasting temperature profiles on volatile aroma compounds associated with roasted coffee defects—such as light, burnt, dark, baked, and underdeveloped roasts—have also been investigated (Yang et al., 2016). In these studies, HS-SPME and GC–MS techniques were employed to characterise volatile profiles. Volatile thiols were identified as some of the most influential compounds contributing to coffee flavour due to their extremely low odour thresholds, which allow them to exert strong sensory effects even at trace concentrations. Thiols are primarily formed during coffee roasting and play a crucial role in determining the aroma of coffee. Gas chromatography coupled with mass spectrometry (GC–MS) is commonly used for thiol analysis, with low-temperature injection conditions applied to minimise thermal degradation of these highly reactive compounds (Czerny et al., 1999; Dulsat-Serra et al., 2016).

6. Determination of non-volatile compounds in coffee

6.1 Simultaneous determination of caffeine, theobromine, and theophylline

Methylxanthines, including caffeine, theobromine, and theophylline, are naturally present in beverages such as coffee, tea, cocoa, and mate. Their quantitative determination has been widely performed using high-performance liquid chromatography (HPLC). HPLC has simultaneously determined these compounds coupled with ultraviolet detection at a wavelength of 273 nm, achieving a detection limit as low as 1×10^{-11} g (Bispo et al., 2002). In the same study, caffeine content was analysed not only in beverages such as coffee but also in human urine samples following consumption, highlighting its relevance in dietary exposure assessment.

Caffeine intake at high concentrations has been associated with increased gastric acid secretion, impaired renal function, and adverse effects on the central nervous system. In contrast, theobromine and theophylline are widely used in the treatment of asthma, and their adverse health effects are generally considered less restrictive than those of caffeine (Bispo et al., 2002). In another study, the caffeine content in instant coffee samples—including Africafe, Nescafé, and Dormans—was quantified using HPLC. The results indicated caffeine levels of 1.64% in Dormans, 3.42% in Africafe, and 3.12% in Nescafé instant coffee products (Wanyika et al., 2010).

6.2 Determination of coffee diterpenes (kahweol and cafestol)

Coffee contains diterpene alcohols such as cafestol and kahweol, whose biological effects have been extensively investigated. These diterpenes are partially preserved during the roasting process, although they may undergo dehydration and structural transformation into related compounds (Roos et al., 1997). Cafestol is present in both *Coffea arabica* and *Coffea canephora* (Robusta), whereas kahweol has been detected exclusively in Arabica coffee.

Coffee diterpenes have been reported to exhibit protective effects against toxic compounds such as aflatoxin B_1 and possess anticancer, antioxidant, and hepatoprotective properties (Cavin et al., 2002; Kim et al., 2009). However, cafestol has also been linked to elevated serum cholesterol levels, raising concerns about excessive intake. The brewing method plays a critical role in determining diterpene concentrations in the final beverage; for example, espresso-type coffees may contain five to ten times higher diterpene levels compared to filtered coffee (Urgert et al., 1995). In addition to their physiological relevance, the cafestol and kahweol profiles are commonly used as chemotaxonomic markers to distinguish between Arabica and Robusta coffee varieties.

6.3 Determination of sugars in coffee

Low-molecular-weight sugars in coffee beans vary depending on the processing method. Arabica coffee beans subjected to wet processing have been reported to contain lower levels of fructose and glucose, whereas beans processed by dry methods exhibit higher concentrations of these sugars (Knopp et al., 2006). Sucrose is the most abundant sugar in coffee beans and occurs at lower levels in Robusta coffee compared to Arabica varieties. In contrast, fructose content is generally higher in Robusta coffee, while glucose levels remain relatively similar between the two species.

The sugar composition of coffee is also influenced by geographical origin and post-harvest processing conditions. Ion-exchange chromatography has been employed for the determination of low-molecular-weight sugars using sodium hydroxide as the eluent and amperometric detection. This analytical approach has proven to be a valuable tool for characterising coffee processing technologies and distinguishing between different production methods (Mabrok and Deatheroge, 1956).

6.4 Determination of amino acids in coffee

Amino acids present in coffee have been determined using classical ion-exchange chromatography and HPLC techniques, either after derivatisation or directly with amperometric detection (Nakhmedov, 1984; Bytof et al., 2005). Studies have identified amino acids such as glutamic acid, aspartic acid, serine, histidine, glycine, and arginine in green coffee beans, with concentrations ranging from 4.4 mg/100 g to 1075 mg/100 g. Dry processing methods typically yield higher amino acid contents compared to wet processing techniques.

In addition to proteinogenic amino acids, coffee contains non-protein amino acids such as γ-aminobutyric acid (GABA). GABA has been associated with various health benefits, including the improvement of stress-related insomnia, reduction of cardiovascular disease risk, and delayed ageing processes (Bytof et al., 2005; Granvogl and Schieberle, 2007). GABA is produced through the α-decarboxylation of glutamic acid and has been found at higher levels in coffee beans obtained through dry processing methods.

7. Extraction methods and techniques

7.1 Pre-treatment procedures

7.1.1 Sample collection and initial handling

In analytical chemistry, the representativeness of the sample constitutes a fundamental prerequisite for obtaining accurate, reproducible, and comparable results (Romanik et al., 2007). A representative sample is defined as one whose composition, handling, processing, and storage conditions adequately reflect the average composition of the bulk material under investigation (Romanik et al., 2007). In the analysis of plant-derived matrices, achieving sample representativeness is particularly challenging due to the intrinsic biological variability among individual plants within the same species or cultivar (Romanik et al., 2007; Namieśnik, 2000).

To reduce sampling bias arising from this variability, composite sampling strategies are commonly recommended. Accordingly, plant material is collected from a confined sampling area, typically on the order of several square centimetres, and a sufficient number of individual plants—generally a dozen or more—are combined to form a representative composite sample (Romanik et al., 2007). This approach improves the statistical representativeness of the sample and enhances the reliability of subsequent analytical determinations.

Following collection, adhering soil residues and surface contaminants are removed from the plant material. Samples are initially rinsed with deionised water to eliminate loosely bound particulate matter (Romanik et al., 2007). When required, additional washing steps may be applied using chelating agents such as ethylenediaminetetraacetic acid (EDTA), dilute hydrochloric acid solutions, or, in specific cases, organic solvents (Namieśnik, 2000; Zhang et al., 2020). These pre-treatment procedures are implemented to minimise matrix-related interferences, thereby improving analytical accuracy and ensuring that the obtained results are transferable to other comparable plant material samples (Romanik et al., 2007; Namieśnik, 2000).

7.1.2 Drying methods

Chemical analyses of natural plant samples are generally performed on dried material rather than fresh matrices (Romanik et al., 2007). Drying enables the expression of analyte concentrations on a dry-weight basis, substantially reducing analytical complications associated with high intrinsic moisture content (Romanik et al., 2007). It should be emphasised, however, that drying does not result in the complete removal of water; the term "dry mass" commonly refers to plant material retaining residual moisture, typically ranging from a few per cent up to approximately 10% (Romanik et al., 2007).

The water content of plant material is influenced by multiple factors, including species, developmental stage, environmental growth conditions, and post-harvest storage practices (Romanik et al., 2007; Asrate and Ali, 2025). The drying process is governed by mass transfer mechanisms involving the removal of water molecules from plant tissues (Asrate and Ali, 2025). The primary objective of drying is to reduce water potential and lower water activity to levels that inhibit enzymatic reactions and microbial growth (Asrate and Ali, 2025). Water activity is a critical parameter representing the fraction of free, unbound water within the matrix. It plays a decisive role in determining the chemical and microbiological stability of plant materials (Asrate and Ali, 2025).

7.1.3 Hot air drying

Hot air drying is among the most frequently employed drying techniques in both laboratory-scale and industrial applications due to its simplicity, reproducibility, and cost-effectiveness (Romanik et al., 2007; Asrate and Ali, 2025). In this method, plant material is exposed to a controlled stream of heated air, facilitating the diffusion-driven removal of moisture from the cellular structure to the surrounding environment (Asrate and Ali, 2025).

Under laboratory conditions, hot air drying is commonly conducted in ventilated ovens at temperatures of around 70 °C or in systems providing controlled convective airflow (Romanik et al., 2007; Asrate and Ali, 2025). Nevertheless, recent studies have demonstrated that effective drying can be achieved at considerably lower temperatures, typically between 40 °C and 50 °C, without compromising drying efficiency (Asrate and Ali, 2025). The use of reduced drying temperatures is particularly critical when target analytes exhibit volatility or susceptibility to sublimation, as elevated temperatures may result in analyte loss or chemical degradation, thereby adversely affecting analytical accuracy (Romanik et al., 2007).

7.1.4 Vacuum drying

Laboratory vacuum ovens are widely used for drying plant materials that contain non-volatile and non-sublimable constituents, and are considered adequate for sample preparation before chemical analysis (Romanik et al., 2007). In vacuum drying, the diffusion of water molecules from the material is accelerated by reducing the ambient pressure surrounding the sample (Romanik et al., 2007). The reduction in pressure significantly lowers the boiling point of water, thereby facilitating moisture removal at reduced temperatures (Romanik et al., 2007).

This technique is particularly suitable for drying of thermally sensitive compounds, as effective dehydration can be achieved at relatively low temperatures, typically in the range of 40–60 °C (Romanik et al., 2007). By minimising thermal stress, vacuum drying reduces the risk of chemical degradation and preserves the integrity of heat-labile analytes during sample preparation.

7.1.5 Freeze-drying (Lyophilisation)

Freeze-drying, also known as lyophilisation, involves the initial freezing of the material, typically at temperatures between −20 and −40 °C, followed by the removal of water through sublimation under reduced pressure. In this process, water is directly converted from the solid phase to the gaseous phase, bypassing the liquid state.

Freeze-drying is considered the most suitable drying technique for materials containing highly thermolabile compounds, as it minimises both thermal and oxidative degradation. Although the process is relatively slow, typically requiring 12–48 h to complete, the resulting dried material retains its original structural, chemical, and morphological characteristics to a high degree (Romanik et al., 2007). Consequently, freeze-drying is frequently employed in analytical workflows where the preservation of native compound profiles is crucial.

7.1.6 Grinding and homogenisation

The subsequent step in preparing raw plant material for chemical analysis involves grinding and homogenisation, which are essential for achieving sample uniformity and enhancing extraction efficiency (Romanik et al., 2007; Namieśnik et al., 2003). The choice of grinding technique depends primarily on the physical properties of the material, including its hardness, fibrous structure, and moisture content.

In the case of raw materials containing volatile constituents, such as essential oils, any increase in temperature during grinding must be strictly avoided. Therefore, grinding should be performed in small batches to minimise frictional heat and prevent the loss of volatile compounds (Szumny et al., 2010). In laboratory practice, homogenisation is commonly carried out using ceramic or agate mortars and pestles, which allow gentle size reduction while limiting heat generation.

Although various mechanical homogenisers are available and widely used, their application may result in localised overheating of the sample matrix, even when external cooling is applied. Such localised temperature increases can lead to the thermal degradation of heat-sensitive analytes, thereby compromising analytical accuracy (Romanik et al., 2007). For this reason, the selection of grinding and homogenisation techniques must be carefully matched to the physicochemical properties of the target analytes.

8. Extraction methods

8.1 Extraction principles and general approaches

As a general rule, the isolation of analytes from plant matrices is carried out using extraction/leaching procedures with a sequence of solvents applied in order of increasing polarity (Romanik et al., 2007; Jin et al., 2011; Nyiredy, 2004). By employing different solvents, it is possible to obtain extracts enriched in different classes of analytes depending on their physicochemical properties (Romanik et al., 2007; Jin et al., 2011; Nyiredy, 2004). During extraction, a portion of target compounds may remain associated with cell walls or organelles due to adsorption phenomena, which can influence the composition of the final extract (Romanik et al., 2007). Therefore, to relate the metabolite composition measured in the final extract to the actual content in the original tissue or organism, it is necessary to determine recovery values (Romanik et al., 2007).

The major extraction techniques applied to coffee beans, coffee oil, and spent coffee grounds, together with their operating principles, advantages, limitations, and target compound classes, are summarised in Table 1.

8.2 Maceration (Cold soaking extraction)

Maceration is defined as keeping plant material in contact with a solvent at room temperature or under low-temperature conditions for a predetermined period (Cujić et al., 2016). This technique is particularly suitable for isolating thermolabile compounds that may be degraded by heat (Hofler et al., 1996). The procedure is simple and low-cost: milled plant material is left in the selected solvent for a defined duration, commonly 3–7 days (Cujić et al., 2016). Historically, maceration has been widely used in traditional pharmacology and in the preparation of plant-based remedies (Hofler et al., 1996). The principal mechanism is diffusion-driven transfer of analytes from plant tissues into the solvent phase (Cujić et al., 2016). In the extraction of polyphenols, optimisation of maceration conditions has been reported to yield high extraction efficiencies in dried plant matrices (Cujić et al., 2016).

Advantages include minimal equipment requirements and improved preservation of heat-sensitive compounds (Hofler et al., 1996). Limitations include long extraction times (3–7 days), low extraction rate, the need for relatively high solvent volumes, and a higher risk of contamination during prolonged contact (Cujić et al., 2016).

8.3 Infusion (Hot water extraction)

Infusion is the process of steeping plant material in hot or boiling water to extract polyphenols and other soluble constituents (Cittan et al., 2018; Mahmudati et al., 2020; Perera et al., 2017). It is among the most common methods used in the household preparation of teas and herbal beverages (Cittan et al., 2018; Mahmudati et al., 2020; Perera et al., 2017). Operationally, plant material is added to hot/boiling water and maintained for a defined period (typically 5–15 min), after which the liquid phase is separated (Cittan et al., 2018; Mahmudati et al., 2020; Perera et al., 2017). The extraction mechanism relies on hot water to disrupt cell wall structures and facilitate the diffusion of analytes into the aqueous phase (Sharif et al., 2014). Because water is a polar solvent, it is particularly effective for extracting polar phenolic constituents (Cittan et al., 2018; Mahmudati et al., 2020). Key parameters typically considered for optimisation include plant species, temperature (commonly 80–100 °C), steeping time, and plant-to-water ratio (Mahmudati et al., 2020; Perera et al., 2017).

In ginger (*Zingiber officinale*), antioxidant activity has been evaluated using infusion and compared with decoction, yielding comparable outcomes (Mahmudati et al., 2020). For *Syzygium cumini* fruit, infusion and decoction have likewise been compared for isolating antidiabetic compounds (Perera et al., 2017). Additionally, the phenolic profile and antioxidant capacity of *Tilia cordata* fruit have been successfully determined using conventional hot-water infusion (Cittan et al., 2018).

Table 1: Major extraction techniques used for coffee and coffee-derived matrices

Technique	Principle	Typical operating conditions	Target compounds	Main advantages	Main limitations	References
Mechanical pressing (screw/hydraulic)	Physical pressure is applied to release oil	35–45 °C; yield ~6–10% (green beans), ~10–15% (roasted)	Lipids, diterpenes	Solvent-free, environmentally friendly, aroma preserved	Low selectivity, moderate yield	Ribeiro et al., 2024; Clarke and Macrae, 1985
Soxhlet extraction	Continuous solvent reflux and percolation	4–16 h; petroleum ether; laboratory reference method	Oils, diterpenoids	Standardised, reproducible	Long extraction time, high solvent and energy consumption	Ribeiro et al., 2024; Gloess et al., 2013
Microwave-assisted extraction (MAE)	Volumetric heating by microwave radiation	~10 min; 10–30 mL solvent; ~45 °C	Oils, bioactive compounds	Short processing time, reduced solvent use	Equipment cost, careful optimisation required	Ribeiro et al., 2024; Pandhi and Kumar, 2025
Ultrasound-assisted extraction (UAE)	Acoustic cavitation enhances mass transfer	20–40 kHz; 150–300 W	Polyphenols, alkaloids	Mild temperatures, improved extraction kinetics	Post-extraction filtration needed	Vinatoru et al., 2017; Yenduri et al., 2025
Ultrasound + microwave (UMAE)	Synergistic acoustic and electromagnetic energy	Highest reported yield (~20.53%)	Oils and bioactives	Very high extraction efficiency	Process complexity, higher capital cost	Pandhi and Kumar, 2025; Ribeiro et al., 2024
Supercritical CO_2 extraction (SFE)	Solvation above the critical point of CO_2	>31.1 °C; 70–90 °C, 327–373 bar	Lipids, diterpenes, aroma compounds	Green technology, no solvent residues	High initial investment	Ribeiro et al., 2024; Andrade et al., 2012

Pressurised liquid extraction (PLE)	Liquid solvents at elevated temperature and pressure	High pressure and temperature	Polyphenols, phytosterols	High efficiency, reduced extraction time	Specialised equipment	Ribeiro et al., 2024
Extrusion + Soxhlet (hybrid)	Mechanical pretreatment enhances matrix accessibility	68 °C, 60 rpm extrusion followed by Soxhlet	Coffee oil	Substantially improved yield vs. Soxhlet alone	Additional processing step	Ribeiro et al., 2024
Spent coffee grounds (SCG): aqueous/ ultrasound	Water-based extraction with ultrasound support	Ultrasound + water	Chlorogenic acid, caffeine	Low environmental impact, waste valorisation	Matrix-dependent efficiency	Bouhzam et al., 2023; Conde and Mussatto, 2016
Maceration	Passive diffusion over extended periods	Hours to days at ambient temperature	Secondary metabolites	Low cost, no energy input	Very long extraction time, low yield	Pandhi and Kumar, 2025

Advantages include simplicity, rapid operation (5–15 min), low cost, and minimal equipment requirements (Cittan et al., 2018; Sharif et al., 2014). Limitations include the possibility of thermal degradation and losses of heat-sensitive metabolites (Mahmudati et al., 2020).

8.4 Decoction (Boiling extraction)

Decoction involves boiling plant material in water at or near the boiling point for an extended period, typically 10–30 min (Mahmudati et al., 2020; Perera et al., 2017). Compared with infusion, decoction employs longer heating times and sustained high-temperature exposure (Mahmudati et al., 2020; Perera et al., 2017). Historically, it has been preferred in traditional medicine for extracting bioactive compounds from hard plant materials, such as roots, barks, and other rigid tissues (Mahmudati et al., 2020; Perera et al., 2017). The underlying mechanism is associated with prolonged exposure to hot water, which enables deeper solvent penetration into compact matrices and facilitates the release of compounds that may not be efficiently recovered by infusion (Mahmudati et al., 2020; Perera et al., 2017). Studies on ginger have demonstrated that temperature and time parameters can significantly influence both total phenolic content and antioxidant activity (Mahmudati et al., 2020). Similarly, in *Syzygium cumini*, antidiabetic compounds have been successfully obtained via decoction and compared with infusion-based preparations (Perera et al., 2017).

Advantages include improved extraction from hard matrices using simple equipment (Mahmudati et al., 2020; Perera et al., 2017). Limitations include longer processing times, higher energy requirements, and an elevated risk of thermal degradation (Mahmudati et al., 2020; Perera et al., 2017).

8.5 Liquid–liquid extraction

Liquid–liquid extraction is a technique based on the selective transfer of analytes between two immiscible solvent phases to achieve enrichment in one phase (Sharif et al., 2014; Zhang et al., 2005). Typically, an aqueous phase is combined with an organic phase, depending on the partition behaviour of the target compounds (Sharif et al., 2014; Zhang et al., 2005). In tea decaffeination, liquid–liquid extraction has been used to demonstrate that infusion time and temperature critically affect the distribution of analytes between phases (Sharif et al., 2014). In another application, extraction techniques were optimised and compared for sanguinarine and chelerythrine from *Macleaya cordata* fruits, where liquid–liquid extraction was reported to provide optimal outcomes under the tested conditions (Zhang et al., 2005).

Advantages include selectivity, operational speed, and controllability (Sharif et al., 2014; Zhang et al., 2005). Limitations include reliance on organic solvents and potential emulsion formation, which can complicate phase separation (Sharif et al., 2014).

8.6 Soxhlet extraction

Soxhlet extraction is among the oldest classical approaches for isolating metabolites from solid natural matrices and is still commonly used as a reference technique in comparative extraction studies (Romanik et al., 2007; Hofler et al., 1995; Alara et al.,

2018; Chin et al., 2013; Rezvankhah et al., 2019). In this technique, the solid sample is placed in a thimble and repeatedly washed with a condensed solvent; the extract accumulates and is periodically siphoned back into the boiling flask, allowing for the continuous recycling of solvent through the matrix (Romanik et al., 2007; Hofler et al., 1996).

A major limitation is the long extraction time (commonly 6–48 h in many practical applications), which increases solvent use and energy demand and can promote thermal degradation of heat-labile compounds due to prolonged heating (Romanik et al., 2007; Alara et al., 2018; Rezvankhah et al., 2019). The Soxhlet method has been successfully applied for recovering phenolic fractions from plant leaves, such as *Vernonia cinerea*, producing extracts with measurable antioxidant potential (Alara et al., 2018). It has also been used to compare the extraction performance of tea polyphenols and alkaloids with direct extraction approaches (Chin et al., 2013). In oil-rich matrices, the Soxhlet method has been frequently used as a benchmark when evaluating alternative intensified techniques, such as microwave-assisted or pressurised approaches (Rezvankhah et al., 2019; Hu et al., 2021).

8.7 Accelerated solvent extraction (ASE)

Accelerated solvent extraction (ASE) (often discussed under pressurised/accelerated extraction platforms) operates under elevated temperature and pressure to keep solvents in the liquid phase and enhance mass transfer from solid matrices (Romanik et al., 2007; Hofler et al., 1996; Rahmalia et al., 2015; Brachet et al., 2001). Under these conditions, solvent viscosity decreases and diffusivity increases, improving matrix penetration and accelerating solute desorption and diffusion (Romanik et al., 2007; Rahmalia et al., 2015). ASE is typically characterised by relatively short extraction cycles compared with classical approaches, contributing to reduced processing time and often lower solvent consumption (Romanik et al., 2007; Hofler et al., 1996).

ASE has been utilised across various plant-based matrices for the recovery of bioactive compounds and for method optimisation, often supported by experimental design strategies to maximise recovery and reproducibility (Brachet et al., 2001; Jentzer et al., 2015). Examples include optimisation of accelerated extraction for glycosidic targets (Jentzer et al., 2015) and comparative/optimisation workflows applied to complex plant matrices (Brachet et al., 2001). Comprehensive identification of bioactives from avocado peel has also been reported using high-resolution LC-based chemical profiling workflows that rely on efficient extraction and sample preparation strategies (Figueroa et al., 2018).

8.8 Ultrasonic-assisted extraction (UAE)

Ultrasound-assisted extraction (UAE) uses acoustic energy (ultrasound) to intensify extraction by promoting mechanical disruption and improving mass transfer in the solid–liquid system (Vinatoru et al., 2017; Śliwiński, 1993; Romanik et al., 2007). The key mechanism is acoustic cavitation, which involves bubble formation and collapse, generating micro-jets and localised shear forces that can erode plant tissues and enhance solute release (Vinatoru et al., 2017; Śliwiński, 1993). As a result, the

UAE frequently increases extraction efficiency and/or decreases extraction time relative to purely diffusion-controlled conventional methods (Vinatoru et al., 2017; Romanik et al., 2007).

UAE performance depends strongly on operating variables such as frequency, power, temperature, time, and solvent composition, and these factors are commonly optimised to target specific phytochemical classes (Vinatoru et al., 2017; Machado et al., 2019; González-Centeno et al., 2014; Wang et al., 2013). The UAE has been widely applied for polyphenol recovery from multiple plant matrices, including pomegranate peel (Machado et al., 2019), grape pomace (González-Centeno et al., 2014), and other botanicals, where response surface approaches have been used to identify optimal acoustic conditions (Wang et al., 2013). Analytical applications include the ultrasound-assisted preparation of phenolic compounds in strawberries (Herrera and Luque-de-Castro, 2004) and the ultrasound-enabled recovery/assessment of phenolics and antioxidants from diverse fruit/vegetable matrices (Altemimi et al., 2016).

Combined approaches have also been explored, such as enzyme- and ultrasound-assisted extraction of polyphenols from *Citrus sinensis* peel (Nishad et al., 2019) and ultrasound-assisted extraction of phenolic compounds/antioxidants/anthocyanins from grape seeds (Ghafoor et al., 2009). For *Tilia cordata*, the influence of extraction approaches on the quantification of flavonoids and phenolic acids has been investigated, supporting the relevance of UAE for phenolic profiling in medicinal plants (Oniszczuk and Podgórski, 2015). A practical advantage is that UAE can be conducted at mild temperatures, supporting the preservation of thermolabile compounds, while a common downstream requirement is filtration/clarification due to suspended particulates (Romanik et al., 2007; Vinatoru et al., 2017).

8.9 Microwave-assisted extraction (MAE)

Microwave-assisted extraction (MAE) employs microwave energy (commonly 2.45 GHz) to heat polar solvents and moisture-containing matrices through dielectric mechanisms, enabling rapid heating and accelerated extraction kinetics (Kaufmann and Christen, 2002; Routray and Orsat, 2012; Bagade and Patil, 2019). The extent of microwave heating depends on the dielectric properties of the solvent–matrix system, and polar solvents typically couple strongly with microwave energy (Kaufmann and Christen, 2002; Routray and Orsat, 2012). MAE is frequently described as a time-saving approach and may reduce solvent consumption compared with prolonged conventional extraction under otherwise comparable conditions (Bagade and Patil, 2019; Kaufmann and Christen, 2002).

MAE has been optimised for bioactive enrichment in aqueous green tea extracts (Bilge and Aksoylu Özbek, 2020). It has been applied in comparative studies against Soxhlet extraction, including seed oil extraction contexts (Hu et al., 2021). MAE has also been used for recovering carotenoids from carrot juice processing waste using oil-based solvent systems (Elik et al., 2020) and for extracting *Curcuma longa* oil with compositional/antioxidant evaluation and comparison with conventional Soxhlet extraction (Fernández-Marín et al., 2021). Additionally, MAE optimisation has been reported for ergosterol recovery from Agaricus bisporus by-products (Heleno et al., 2016) and polyphenolic content/antioxidant activity optimisation in medicinal plants (Ismail-Suhaimy et al., 2021), while MAE has also been evaluated for bioactive recovery from eggplant peel by-products (Doulabi et al., 2020). These studies collectively support MAE as an efficient alternative extraction platform for plant-derived compounds under optimised conditions (Bagade and Patil, 2019; Kaufmann and Christen, 2002).

8.10 Pressurised hot water extraction (PHWE) and subcritical water extraction

Pressurised hot water extraction (PHWE) uses water at temperatures above its atmospheric boiling point while maintaining pressure to keep the solvent in the liquid phase, thereby altering water's physicochemical properties (Matshediso et al., 2015; Nuapia et al., 2018; Bursać Kovačević et al., 2018; Vergara-Salinas et al., 2012). Under these conditions, the dielectric constant of water decreases, increasing its ability to solubilise less-polar organic compounds relative to ambient water (Matshediso et al., 2015; Nuapia et al., 2018). Because water is non-toxic and recyclable, PHWE is frequently framed as a green extraction strategy (Matshediso et al., 2015; Bursać Kovačević et al., 2018).

PHWE has been developed and optimised for extracting essential compounds from Moringa oleifera leaves (Matshediso et al., 2015) and for the selective recovery of vitamin C and phenolic compounds with model-based optimisation approaches (Nuapia et al., 2018). It has also been applied as a green recovery approach for steviol glycosides from *Stevia rebaudiana* leaves (Bursać Kovačević et al., 2018). Additionally, operational variables (temperature/time) have been demonstrated to affect the polyphenolic content and antioxidant activity in PHWE of thyme matrices (Vergara-Salinas et al., 2012). Subcritical water extraction has been used for the recovery of phenolic/antioxidant constituents from pistachio hulls (Erşan et al., 2018) and implemented with natural deep eutectic solvents for sustainable phenolic extraction from winemaking by-products (Loarce et al., 2020).

8.11 Supercritical fluid extraction (SFE)

Supercritical fluid extraction (SFE) uses solvents above their critical temperature and pressure, where they exhibit gas-like diffusivity and liquid-like solvating power, enhancing mass transfer and extraction efficiency (Uwineza et al., 2020; Kassama et al., 2008). Carbon dioxide is the most widely used supercritical solvent due to its low critical temperature and pressure, making it suitable for thermolabile constituents and enabling relatively mild thermal processing (Uwineza et al., 2020; Kassama et al., 2008). Because CO_2 is relatively non-polar, it is well-suited for non-polar and weakly polar compounds, while co-solvents (modifiers) are commonly used to enhance the extraction of more polar analytes (Rosas-Quina and Mejía-Nova, 2021; Hassim et al., 2021).

Reported applications include ultrahigh-pressure supercritical extraction and chromatography of seed lipids from *Moringa* species (Belo et al., 2019), alkaloid extraction with co-solvent, and comparisons to conventional methods (Rosas-Quina and Mejía-Nova, 2021), and phenolic extraction from potato peels (de Andrade Lima et al., 2021). High-pressure/supercritical technologies have been utilised for phytocannabinoid fractionation from hemp residues (Vági et al., 2020) and for the recovery of bioactive compounds from radish leaves (Goyeneche et al., 2018). Scale-up strategies using ethanol–water modified CO_2 have also been reported (Hassim et al., 2021). SFE has been extensively optimised for lycopene recovery from tomato skin and related by-products, including the impact of operational parameters on yield and antioxidant activity (Kassama et al., 2008; Yi et al., 2009; Pellicanò et al., 2020). Related high-pressure assisted extraction strategies have also been used to enhance antioxidant and antityrosinase activities in longan pericarp extracts (Nagendra et al., 2010).

8.12 Enzyme-assisted extraction (EAE)

Enzyme-assisted extraction employs enzymes (e.g., cellulases, pectinases, proteases) to degrade or loosen cell-wall polymers, facilitating the controlled release of intracellular bioactives under mild conditions. This approach distinguishes it from purely physicochemical extraction routes (Puri et al., 2012; Liu et al., 2016). Because plant cell walls are composed primarily of cellulose, pectins, hemicelluloses, and associated proteins, enzymatic hydrolysis can increase permeability and enhance the extraction of phenolic and other bioactive compounds (Puri et al., 2012; Liu et al., 2016). EAE conditions are typically optimised to meet enzyme requirements (temperature/pH), enabling efficient extraction while minimising thermal stress (Puri et al., 2012).

Applications include enzyme-assisted extraction of non-extractable polyphenols from sweet cherry pomace (Domínguez-Rodríguez et al., 2021), phenolic recovery from citrus peels (Liu et al., 2016), and lycopene recovery from tomato processing waste (Zuorro et al., 2011). Integrated microwave- and enzyme-assisted extraction has been reported for the recovery of phenolics from olive pomace (Macedo et al., 2021). Enzyme-assisted extraction has also been employed for the recovery of bioactive compounds from brown seaweeds (Habeebullah et al., 2019) and for the extraction of glycyrrhizic acid from liquorice roots using combined processing strategies (Giahi et al., 2021). Bioactivity-oriented applications include phenolic-enriched wine lees extracts (López-Fernández-Sobrino et al., 2021) and the recovery of polyphenols from green yerba mate (Heemann et al., 2019).

8.13 Combination (hybrid) extraction methods

Hybrid extraction approaches combine two or more techniques to overcome limitations of single methods and to increase yield, selectivity, and/or sustainability (Vinatoru et al., 2017; Macedo et al., 2021; Yang et al., 2017; Anh et al., 2021). For example, integrated microwave- and enzyme-assisted extraction has been utilised for phenolic recovery from olive pomace, offering an intensified yet environmentally friendly workflow (Macedo et al., 2021). Enzyme–ultrasound combinations have been reported to enhance the extraction efficiency of plant polysaccharides, such as pectin, from sisal waste (Yang et al., 2017). In citrus by-products, enzyme- and ultrasound-assisted extraction conditions for flavonoid recovery from pomelo peel residues have been optimised using response surface methodology (Anh et al., 2021). Comparative investigations have also assessed enzyme–ultrasound strategies for polyphenol recovery from *Citrus sinensis* peel (Nishad et al., 2019). These examples underscore the growing importance of hybrid platforms for the efficient recovery of plant-derived high-value compounds (Vinatoru et al., 2017; Macedo et al., 2021).

9. Various technical approaches used to obtain coffee extracts

Coffee (*Coffea arabica* and *C. canephora*) is among the most widely consumed beverages worldwide, and the recovery of its bioactive constituents represents a key aspect of valorising this botanical resource (Pandhi and Kumar, 2025; Bouhzam et al., 2023; Gloess et al., 2013; Ribeiro et al., 2024). During extraction, process variables such as thermal input, pressure, solvent type, extraction time, and solid-to-liquid ratio directly shape the chemical composition and functional properties of the final product (Osorio-Tobon, 2020; Kurzrock and Speer, 2001). Recent research further indicates that careful control of these factors can enable the development of more efficient and selective procedures that target desired compounds more effectively (Osorio-Tobon, 2020; Kurzrock and Speer, 2001).

9.1 Mechanical pressing

One of the most basic and environmentally compatible approaches for obtaining coffee oil is mechanical pressing (Ribeiro et al., 2024). In practice, pressing systems are mainly implemented as screw presses and hydraulic presses, where applied mechanical force releases the oil fraction from the coffee matrix (Ribeiro et al., 2024). In screw-type systems, the rotational motion of a metal screw generates compressive stress on beans, and processing capacities on the order of kilograms per hour have been reported for such configurations (Ribeiro et al., 2024).

A principal advantage of mechanical pressing is the absence of added chemical solvents, which supports its environmental suitability and facilitates its acceptance for "natural" product categories (Ribeiro et al., 2024). The obtained oil may better retain its original sensory properties (taste and aroma profile), which contributes to its industrial relevance (Ribeiro et al., 2024). Nevertheless, when green beans are processed, reported oil yields are relatively limited (approximately 6–10%), whereas roasted beans may provide higher yields (about 10–15%); however, process-associated heating (reported around 35–45 °C) may compromise certain bioactive constituents (Clarke and Macrae, 1985).

In a parametric evaluation conducted on green Arabica material, variables such as particle size, screw speed, outlet nozzle geometry, and thermal pretreatment were found to influence the amount of extracted oil (Ribeiro et al., 2024). Under optimised conditions, cafestol and kahweol levels were reported to be in the ranges of 37–47 mg/g oil and 13–17 mg/g oil, respectively, and serotonin-like alkaloids were also identified among the measured constituents (Ribeiro et al., 2024). The physical integrity of beans influenced oil recovery, with intact beans yielding about 10.8% oil and damaged beans yielding about 9.9% oil, supporting the view that mechanical pressing behaves as a largely non-selective recovery process at the matrix level (Ribeiro et al., 2024).

Overall, mechanical pressing remains widely used at an industrial scale because it is practical and environmentally favourable. However, it can be limited in yield and/or purity compared with more advanced technologies (Ribeiro et al., 2024). Despite these constraints, the food and pharmaceutical sectors continue to consider this technique attractive for high-quality products due to the absence of solvent-derived residues (Ribeiro et al., 2024).

9.2 Soxhlet-based extraction

Under laboratory conditions, Soxhlet extraction is commonly regarded as a standard approach for recovering coffee oil, and it is reported to be recognised in the context of standardised analytical procedures (Ribeiro et al., 2024). In this technique, an organic solvent (e.g., petroleum ether) is refluxed and cycled through the sample for extended periods (typically 4–16 hours), allowing repeated contact between the solvent and the solid matrix (Ribeiro et al., 2024). Operationally, the sample is placed in a thimble; the solvent is heated to a vapour, condensed in a condenser, and drips onto the sample. Meanwhile, the solvent containing solubilised compounds returns to the boiling flask after siphoning, thereby sustaining the extraction cycle (Ribeiro et al., 2024).

In a comparative evaluation of multiple extraction strategies, nine methods were assessed, and the Soxhlet method yielded an extraction efficiency of approximately 15.07%. In contrast, the combined application of ultrasound and microwave energy was reported to provide a higher yield (20.53%) under the tested conditions (Pandhi and Kumar, 2025). In another study focused on coffee brewing/extraction practices, the Soxhlet method was selected as a reference method for comparative purposes (Gloess et al., 2013).

Using green Arabica beans and hexane as the solvent, an 8-hour Soxhlet protocol was reported to produce approximately 11–15% oil, while roasted beans resulted in slightly lower yields (about 10–12%) (Ribeiro et al., 2024). Soxhlet extraction has also been described as particularly effective for the recovery of diterpenoids (Ribeiro et al., 2024). Specifically, the cafestol and kahweol values were reported as 3.84 g/kg and 4.76 g/kg, respectively, in extracts from green beans. After roasting, these values decreased to 3.28 g/kg and 3.98 g/kg, respectively (Ribeiro et al., 2024).

Although Soxhlet extraction offers practicality and can deliver good yields, it also has significant disadvantages, including a long processing time (4–16 h), high consumption of organic solvent, potential thermal degradation of sensitive compounds, and an unfavourable environmental profile compared to greener alternatives (Ribeiro et al., 2024). Consequently, it is generally not favoured for large-scale industrial production despite its utility as a benchmark method (Ribeiro et al., 2024).

9.3 Microwave energy-based extraction

The use of microwave radiation in extraction is considered an alternative approach that can improve processing speed while supporting more environmentally compatible workflows through reduced solvent and time demand (Ribeiro et al., 2024). When ultrasound and microwave energy are combined (UMAE), the highest extraction yield among the compared methods (20.53%) has been reported. This performance is attributed to matrix disruption by acoustic effects coupled with enhanced penetration and heating efficiency during microwave treatment (Pandhi and Kumar, 2025).

For green Arabica, Soxhlet extraction reportedly yielded approximately 7.5–9.5%, whereas microwave-assisted extraction (MAE) alone resulted in yields of about 5.8–7.6% under the evaluated conditions (Ribeiro et al., 2024). An optimum MAE condition was reported at 45 °C with a 10-min processing time (Ribeiro et al., 2024). Under these parameters, MAE was reported to increase the recovery of cafestol and kahweol to a level substantially higher than Soxhlet (Ribeiro et al., 2024). The main advantages of MAE include markedly shorter extraction times (on the order of minutes) and reduced solvent volumes (reported to be around 10–30 mL). In contrast, limitations include challenges in maintaining consistent outcomes and the capital cost of equipment (Ribeiro et al., 2024).

9.4 Supercritical carbon dioxide extraction

Supercritical fluid extraction leverages the solvent properties that emerge when temperature and pressure exceed the critical point, at which the fluid exhibits both gas-like mass transfer and liquid-like solvating power, thereby enabling efficient extraction behaviour (Ribeiro et al., 2024). In coffee oil production, supercritical CO_2 is widely considered because it is readily available, non-toxic, and has relatively low critical parameters (approximately 31.1 °C and 72.8 bar) that allow processing under comparatively mild conditions (Ribeiro et al., 2024).

For green coffee, a maximum extraction yield of 17.65% was reported at 90 °C and 373 bar under conditions corresponding to a CO_2 density of 0.77 g/mL (Ribeiro et al., 2024). Regarding diterpenoids, more favourable outcomes were reported at 70 °C and 327 bar, suggesting that optimum conditions may vary depending on whether maximising total oil yield or targeting diterpenoid enrichment is prioritised (Ribeiro et al., 2024; Andrade et al., 2012).

Because pure CO_2 is non-polar, it has limited capacity to extract polar compounds, and co-solvents such as ethanol or methanol are therefore applied to increase solubility for more polar constituents (Ribeiro et al., 2024). In this context, the addition of 2.9% ethanol to supercritical CO_2 was reported to improve solubility by 63% compared with pure CO_2, and diterpenoid (cafestol and kahweol) levels were found to be 4–5 times higher than those obtained through mechanical pressing (Ribeiro et al., 2024).

Key advantages of this technique include minimal or no requirement for organic solvents, rapid processing, recoverability/recyclability of CO_2, and improved preservation of heat-sensitive compounds (Ribeiro et al., 2024; Andrade et al., 2012). The main barriers remain high system cost and technical challenges associated with scale-up and large-scale implementation (Ribeiro et al., 2024). Despite these limitations, interest continues to grow in the food and pharmaceutical sectors where high purity and quality targets can justify the higher investment (Ribeiro et al., 2024).

9.5 Recovery of bioactive compounds from coffee processing wastes

Granular residues generated after coffee beverage preparation, commonly referred to as spent coffee grounds (SCG), represent an opportunity for valorisation through the recovery of bioactive compounds (Bouhzam et al., 2023; Conde and Mussatto, 2016). Different strategies for recovering polyphenolic compounds and alkaloids from SCG have been compared, including the application of ultrasonic energy, simple water-based vortex treatment, and the use of synthetic solvents (Bouhzam et al., 2023).

The combination of ultrasound with an aqueous medium produced the highest outcome for chlorogenic acid, reported at 1.15 mg/g, whereas extraction using only water yielded 0.827 mg/g; in contrast, the synthetic-solvent approach resulted in 0.116 mg/g, corresponding to an approximately ten-fold difference relative to the best-performing condition (Bouhzam et al., 2023). For caffeine, the ultrasound + water approach yielded 0.972 mg/g (Bouhzam et al., 2023).

From a life cycle assessment (LCA) perspective, extraction in an aqueous medium was identified as the approach associated with the lowest environmental burden (Bouhzam et al., 2023). The lower chlorogenic acid performance observed for the synthetic solvent strategy was attributed to the localisation of this compound within SCG regions that are more favourably accessed by water (Bouhzam et al., 2023). Overall, these findings underline that solvent choice is a critical determinant in the recovery of polar compounds (Bouhzam et al., 2023).

9.6 Sound-wave-assisted extraction

A protocol based solely on ultrasonic energy achieved a yield of 14.45%, placing it between Soxhlet extraction (15.07%) and the combined approach (20.53%) in the reported comparative dataset (Pandhi and Kumar, 2025). When acoustic radiation (typically above ~20 kHz) is applied to solid matrices, cavitation phenomena generate microcavities and disruptive effects within the structure, which can enhance solvent penetration and mass transfer (Pandhi and Kumar, 2025). Although ultrasound-based extraction can approach the performance of other advanced techniques in terms of product quality indicators, its efficiency remains lower than that of the ultrasound–microwave combination (UMAE) when ultrasound is used without an additional energy mode (Pandhi and Kumar, 2025).

The ultrasonic method was reported to be particularly effective when targeting nonpolar compounds (Pandhi and Kumar, 2025). When energy sources are combined—especially in UMAE—overall performance was reported to be the highest among the evaluated methods (Pandhi and Kumar, 2025; Ribeiro et al., 2024). Noted practical advantages include relatively low equipment cost, operation near room temperature, and a degree of specificity toward selected compound groups (Pandhi and Kumar, 2025).

9.7 Simple heat-based extraction

Heating coffee material in a hot-water medium yielded 13.5 ± 0.41% in the referenced comparison (Pandhi and Kumar, 2025). Historically, hot-water processing has been one of the earliest preparation methods for coffee beverages and is operationally straightforward (Petracco, 2001). The underlying mechanism involves transfer of soluble bioactive constituents from the solid matrix into the aqueous phase during contact at elevated temperature; however, prolonged heating can increase the risk of degradation of thermolabile compounds (Pandhi and Kumar, 2025).

9.8 Combined heating and mechanical stirring

A protocol integrating increased temperature with simultaneous mechanical stirring produced a yield of 10.41 ± 0.14% in the comparative results (Pandhi and Kumar, 2025). This performance was intermediate within the evaluated set. The concurrent use of heat and stirring promotes more homogeneous contact between the solid matrix and solvent, thereby supporting mass transfer (Pandhi and Kumar, 2025). Nevertheless, insufficient control of thermal conditions may promote undesired chemical reactivity and structural alteration of sensitive constituents (Pandhi and Kumar, 2025).

9.9 Homogenisation-assisted extraction

Extraction performed with a high-speed mechanical stirrer (homogenizer) resulted in a yield of 9.74 ± 0.32% (Pandhi & Kumar, 2025). This approach is based on the concept that high shear forces can disrupt cellular or particulate structures, increasing accessibility of extractable components (Pandhi & Kumar, 2025). While the method can be effective for relatively soft biological tissues, the reported yield suggests limitations when applied to rigid matrices such as coffee beans (Pandhi and Kumar, 2025).

9.10 Simple mixing without added energy

When no external energy source was applied, and only mechanical mixing was used, the reported yield was 9.62 ± 0.13% (Pandhi & Kumar, 2025). This baseline approach reflects agitation of the solid matrix in solvent at approximately ambient conditions. Its relatively low performance is consistent with reliance primarily on molecular diffusion rather than enhanced mass-transfer mechanisms (Pandhi and Kumar, 2025). Accordingly, it can serve as a control condition for establishing baseline extraction performance (Pandhi and Kumar, 2025).

9.11 Long-term maceration

Maceration, defined here as maintaining the material in solvent for extended periods (hours to days) without additional intervention, produced a yield of 9.42 ± 0.27% and represented the lowest performance among the nine evaluated methods (Pandhi and Kumar, 2025). The process requires no physical or energetic input beyond time and is driven mainly by diffusion-based transfer (Pandhi and Kumar, 2025). Although energy consumption is minimal and costs can be low, the long processing time has been reported as a key barrier to industrial applicability (Pandhi and Kumar, 2025).

9.12 Pressurised liquid extraction under high pressure

Pressurised liquid extraction (PLE) is conducted by maintaining the solvent in a liquid state under elevated pressure, enabling operation at temperatures above the normal boiling point while limiting solvent loss through evaporation (Ribeiro et al., 2024). Comparative outcomes reported higher values for total polyphenolics and phytosterols under PLE-relevant conditions (Ribeiro et al., 2024). However, this technique is less broadly implemented in routine practice and typically requires specialised equipment (Ribeiro et al., 2024).

9.13 Mechanically assisted soxhlet: Extrusion + soxhlet hybrid approach

Recent studies have demonstrated that applying an extrusion pretreatment before Soxhlet extraction can significantly enhance extraction efficiency (Ribeiro et al., 2024). Under controlled conditions (68 °C, 60 rpm), extrusion followed by Soxhlet extraction yielded 16.42% oil from green Arabica material, whereas Soxhlet extraction performed without extrusion yielded only 9.05% (Ribeiro et al., 2024). These results support the interpretation that mechanical disruption can open the rigid coffee structure, substantially enhancing subsequent solvent penetration and recovery efficiency (Ribeiro et al., 2024). Therefore, this hybrid workflow represents a notable improvement in yield performance (Ribeiro et al., 2024).

9.14 Sensory analysis of coffee brewing methods

Nine brewing protocols were systematically examined with respect to sensory properties and chemical profiles (Gloess et al., 2013). The evaluated methods included semi-automatic and fully automatic espresso/lungo machines, single-use capsules, Moka (Bialetti) pots, French presses, regional German methods, and a standard system without a paper filter (Gloess et al., 2013).

Each sample was analysed using modern instrumental approaches, including headspace-SPME-GC/MS for volatile profiling (38 aroma compounds), HPLC determination of caffeine and chlorogenic acid, as well as measurements of total dissolved solids, Brix index, acidity metrics, and fatty-acid profile characterisation (Gloess et al., 2013; Blank et al., 1992). A trained panel ($n = 7$) evaluated hedonic and sensory attributes—such as appearance (cream/foam), odour, taste, mouthfeel (body/texture), and aftertaste—on a 0–5 scale (Gloess et al., 2013).

The dataset indicated quantitative relationships between chemical composition and perceived properties (Gloess et al., 2013). Specifically, an association was reported between total dissolved solids and mouthfeel ($r^2 = 0.69$), and between aroma intensity and odour perception ($r^2 = 0.73$) (Gloess et al., 2013; Blank et al., 1992). Correlations were also reported between alkaloid-related measurements and perceptions of bitterness and astringency, with r^2 values of 0.69 and 0.75, respectively (Gloess et al., 2013). In contrast, acidity measures (pH and titratable acidity) did not correlate with perceived sourness among participants, suggesting that shifts in organic-acid composition and matrix interactions can limit the explanatory power of bulk acidity metrics for sensory experience (Gloess et al., 2013). Additional observations indicated that the presence of cream—an espresso-related oil–protein foam—substantially influenced perception, while such differences were not always fully captured by instrumental profiling (Gloess et al., 2013).

Overall, the study supports the view that each technological option yields a distinctive profile and that no single "best" method exists across all criteria (Gloess et al., 2013). Espresso-type preparations provide a high concentration, whereas lungo approaches extract a broader set of compounds at a lower intensity. Paper-filter systems tend to yield balanced profiles, and French press preparations may stand out due to their higher oil content, which affects perceived cream-related qualities (Gloess et al., 2013).

9.15 Cross-method performance comparison and general evaluation

Nine techniques were evaluated quantitatively and ranked by yield performance (Pandhi & Kumar, 2025). The highest yield was reported for the combined energy approach (UMAE, 20.53%), followed by Soxhlet (15.07%), ultrasound (14.45%), microwave (14.28%), thermal extraction (13.5 ± 0.41%), thermal + mechanical stirring (10.41 ± 0.14%), homogenisation (9.74 ± 0.32%), simple mixing (9.62 ± 0.13%), and diffusion-driven maceration (9.42 ± 0.27%) (Pandhi & Kumar, 2025). The highest performance was achieved when two distinct energy modes—acoustic and electromagnetic—were applied simultaneously (Pandhi & Kumar, 2025; Osorio-Tobon, 2020).

When selecting an appropriate method, the intended industrial use, environmental impact considerations, economic constraints, and final product quality expectations should be taken into account (Ribeiro et al., 2024). While mechanical pressing remains prevalent in industry, Soxhlet extraction retains importance as a laboratory standard reference. Supercritical CO_2 extraction, despite higher equipment costs, continues to gain relevance in food and pharmaceutical sectors due to its environmental compatibility and ability to support high-purity targets (Ribeiro et al., 2024; Andrade et al., 2012). Encouraging results have been reported for newer hybrid strategies (e.g., extrusion + Soxhlet), which may guide future technological development (Ribeiro et al., 2024). In summary, each technology presents distinct advantages and limitations, and practical success is ultimately determined by context-specific requirements (Gloess et al., 2013; Ribeiro et al., 2024).

10. Analytical chromatographic techniques applied to the chemical characterisation of coffee samples

Chromatographic and hyphenated analytical techniques constitute the cornerstone of chemical characterisation studies in coffee research, enabling the qualitative and quantitative determination of a wide range of bioactive, volatile, and quality-related compounds. Due to the complex and highly heterogeneous chemical matrix of coffee, no single analytical method is sufficient to comprehensively describe its composition, making the combined use of liquid- and gas-phase chromatographic techniques essential for a comprehensive understanding. High-performance liquid chromatography-based methods are primarily employed for the analysis of non-volatile constituents, such as caffeine, chlorogenic acids, and other phenolic compounds. In contrast, gas chromatography coupled with mass spectrometry is indispensable for profiling volatile and semi-volatile aroma compounds. Advanced platforms, including LC–MS/MS and high-resolution mass spectrometry, further enhance analytical selectivity and sensitivity, allowing structural elucidation and trace-level detection of metabolites relevant to quality, authenticity, and bioactivity assessments. A concise overview of the principal chromatographic and analytical approaches applied to coffee and coffee-derived matrices, together with their targeted compound classes and significant analytical advantages, is provided in Table 2.

10.1 High-performance liquid chromatography

Liquid chromatography-based techniques are among the most widely used analytical tools for determining caffeine and other alkaloid compounds in coffee matrices (Yashin et al., 2017; dos Santos and Boffo, 2021). These methods are favoured due to their relatively simple analytical workflows, short analysis times, high reproducibility, and reliable quantitative performance. In addition, compared to more advanced spectrometric platforms, HPLC systems generally present lower operational and maintenance costs, making them suitable for routine analysis and quality control applications (Yashin et al., 2017; dos Santos and Boffo, 2021).

Under reversed-phase chromatographic conditions, caffeine can be readily quantified using ultraviolet detection at wavelengths between 270 and 280 nm. Within the same analytical run, other quality-related coffee constituents such as trigonelline, nicotinic acid, N-methylpyridinium, chlorogenic acids, and 5-hydroxymethylfurfural can also be simultaneously determined (Yashin et al., 2017). The use of diode-array detectors (DAD) enables multi-wavelength acquisition, facilitating the concurrent monitoring of multiple analytes and providing additional spectral confirmation for compound identification (Yashin et al., 2017).

Due to its robustness and analytical reliability, HPLC has been systematically integrated into quality control protocols within the coffee industry and is extensively employed in academic research for chemical profiling purposes (Yashin et al., 2017; dos Santos and Boffo, 2021). Consequently, liquid chromatographic approaches are widely recognised by international regulatory and standardisation bodies for the routine analysis of coffee composition.

10.2 Gas chromatography–mass spectrometry

Gas chromatography coupled with mass spectrometry (GC–MS) is a key analytical technique for characterising volatile and semi-volatile organic compounds in coffee. It is extensively applied to aroma profiling, chemical fingerprinting, and the discrimination of varietal or origin of coffee samples (Yashin et al., 2017). Before GC–MS analysis, several sample preparation strategies may be employed, including static or dynamic headspace analysis, solid-phase microextraction (SPME), supercritical fluid extraction (SFE), and steam distillation. These approaches enable the selective isolation and enrichment of volatile constituents from the complex coffee matrix (Yashin et al., 2017).

Dirinck et al. (2001) demonstrated that GC–MS analysis allows effective differentiation of coffee samples based on their volatile aroma profiles. Similarly, Costa-Freitas et al. (2001) evaluated different SPME fibres in combination with GC–MS and identified optimal extraction conditions for classifying coffee samples based on their volatile composition. Through multivariate analysis, Rocha et al. (2004) further showed that headspace SPME in conjunction with GC-MS offers a quick and accurate method for differentiating coffee brews.

Table 2: Analytical and characterisation techniques applied in coffee chemistry

Analytical technique	Working principle	Target compounds	Strengths	Limitations	References
HPLC-UV/HPLC-DAD	Reversed-phase separation with UV/DAD detection	Caffeine, trigonelline, chlorogenic acids, HMF	Routine quality control, robust and reliable	Possible matrix interferences	Yashin et al., 2017; dos Santos and Boffo, 2021
GC-MS (headspace / SPME)	Separation and MS identification of volatile compounds	Aroma constituents	Excellent aroma profiling and origin discrimination	Not suitable for non-volatile compounds	Dirinck et al., 2001; Costa-Freitas et al., 2001
GC-MS/MS	Tandem MS detection of volatile contaminants	PAHs, acrylamide	Very high sensitivity and selectivity	High instrumentation cost	Guatemala-Morales et al., 2016
LC-MS/MS/ UHPLC-MS/MS	Liquid chromatography coupled with tandem MS	Methylxanthines, phenolics, contaminants	Multicomponent, trace-level analysis	Complex validation procedures	Yashin et al., 2017; dos Santos and Boffo, 2021
UV-Vis spectrophotometry	Beer–Lambert law-based absorbance measurement	Caffeine (and CGA via deconvolution)	Fast, low-cost, field-applicable	Limited specificity	Demissie et al., 2016; Navarra et al., 2017
Ion-pair RP-HPLC	Ion-pairing reagents improve separation	Caffeine, theobromine, theophylline	Effective separation of structurally similar compounds	Mobile-phase optimisation required	Gennano and Abrigo, 1992
High-resolution MS (HRMS)	Accurate mass measurement and fragmentation	Phenolics, diterpenes, metabolites	Structural elucidation, metabolomics	High operational cost	Yashin et al., 2017; dos Santos and Boffo, 2021
SFE coupled to GC-MS/HPLC	Supercritical extraction followed by chromatographic analysis	Aromatic and lipid fractions	Superior aroma preservation, green approach	Specialised equipment	Ribeiro et al., 2024

In addition to aroma characterisation, GC–MS techniques are also applied to contaminant analysis in coffee products. Harmful compounds such as polycyclic aromatic hydrocarbons (PAHs) and acrylamide can be detected at trace levels using GC–MS or GC–MS/MS methodologies (Guatemala-Morales et al., 2016). Furthermore, coffee-specific diterpenes, including cafestol and kahweol, can be characterised using GC-based analytical platforms, contributing to comprehensive assessments of coffee quality and safety (Yashin et al., 2017; dos Santos & Boffo, 2021).

10.3 Liquid chromatography–mass spectrometry

Liquid chromatography coupled with tandem mass spectrometry (LC–MS/MS), particularly in its ultra-high-performance configuration (UHPLC–MS/MS), has become increasingly prevalent in recent years for the comprehensive chemical characterisation of coffee samples (dos Santos and Boffo, 2021). This analytical platform combines the high separation efficiency of liquid chromatography with the sensitivity and selectivity of mass spectrometric detection, enabling reliable identification and quantification of multiple analytes within complex matrices (dos Santos and Boffo, 2021).

UHPLC–MS/MS enables the simultaneous determination of methylxanthines, including caffeine, theobromine, and theophylline, as well as a wide range of other bioactive and quality-related compounds. Method optimisation can be tailored to different coffee matrices, including green, roasted, and soluble coffee products (Yashin et al., 2017; dos Santos and Boffo, 2021). In addition, LC–MS/MS techniques are particularly effective for analysing trace-level contaminants, such as pesticides, mycotoxins, and other undesirable compounds, thereby supporting food safety and regulatory compliance (Yashin et al., 2017; dos Santos and Boffo, 2021).

Owing to its high analytical performance, LC–MS/MS is increasingly regarded as a reference technique for advanced chemical profiling, authenticity assessment, and safety evaluation in coffee research and industrial applications (dos Santos and Boffo, 2021).

10.4 Ultraviolet–visible spectrophotometry

Ultraviolet–visible (UV–Vis) spectrophotometry provides a cost-effective and practical alternative for determining caffeine in coffee samples. The method is based on the Beer–Lambert law, in which the absorbance of a molecule is measured at specific wavelengths corresponding to electronic transitions (Yashin et al., 2017).

Demissie et al. (2016) successfully applied UV–Vis spectrophotometric analysis to Ethiopian green coffee samples for the quantification of caffeine. In another study, Navarra et al. (2017) developed a spectral deconvolution approach enabling the simultaneous determination of caffeine and chlorogenic acid in coffee matrices. In this method, ethanol extraction was employed before spectrophotometric analysis to improve analyte solubility and signal resolution.

Although UV–Vis spectrophotometry is rapid, inexpensive, and suitable for field or screening studies, its analytical specificity is inherently limited due to overlapping absorbance of coexisting coffee constituents. Consequently, chromatographic techniques are preferred in applications requiring high selectivity and accurate quantification.

10.5 Ion-pair reversed-phase chromatography

Ion-pair reversed-phase chromatography is an effective approach for the separation of coffee compounds exhibiting weak ionic characteristics. This technique is particularly useful for the resolution of structurally similar methylxanthines, such as caffeine, theobromine, and theophylline (Yashin et al., 2017).

In the study conducted by Gennaro and Abrigo (1992), methylxanthines were successfully separated and quantified using a reversed-phase ion-pair chromatographic system. The method relies on the addition of an ion-pairing reagent to the mobile phase, which forms electrostatic interactions with analytes displaying partial ionic character. These interactions modify analyte retention behavior, thereby enhancing chromatographic resolution among closely related compounds.

10.6 High-resolution mass spectrometry

High-resolution mass spectrometry (HRMS) is an advanced analytical technique used for accurate molecular mass determination and structural elucidation of complex compounds. Soft ionisation methods, such as electrospray ionisation (ESI) and atmospheric pressure chemical ionisation (APCI), are particularly well suited for metabolomic investigations of coffee matrices (dos Santos and Boffo, 2021).

Beyond caffeine and other methylxanthines, HRMS enables detailed characterisation of phenolic compounds, diterpenes, polyphenols, and a wide range of secondary metabolites present in coffee. Structural elucidation is achieved through tandem mass spectrometry (MS/MS), where fragmentation patterns provide diagnostic information about molecular substructures.

These analytical capabilities render HRMS indispensable for in-depth chemical profiling of coffee, discovery of previously unreported compounds, and studies exploring the relationship between chemical composition and biological activity (Yashin et al., 2017; dos Santos and Boffo, 2021).

10.7 Supercritical fluid extraction

Supercritical fluid extraction (SFE), most commonly employing supercritical carbon dioxide ($scCO_2$), enables the selective extraction of volatile and semi-volatile constituents from coffee matrices. The resulting extracts can subsequently be analysed using GC–MS or HPLC–UV techniques (Yashin et al., 2017).

Compared to conventional solvent-based extraction methods, SFE is regarded as an environmentally friendly and efficient alternative. It allows the simultaneous recovery of diterpenes, such as cafestol and kahweol, along with aromatic compounds from the lipid fraction of coffee.

Extracts obtained via SFE are characterised by superior preservation of aroma compounds, making this technique particularly valuable for quality assessment, flavour research, and novel coffee product development.

10.8 Multicomponent quality analysis

The use of HPLC systems equipped with diode array detectors (HPLC–DAD) enables the simultaneous determination of multiple coffee quality indicators within a single analytical run. Compounds commonly assessed using this approach include caffeine, trigonelline, nicotinic acid, N-methylpyridinium, chlorogenic acids, and 5-hydroxymethylfurfural (Yashin et al., 2017).

The resulting compositional profiles vary according to geographical origin, commercial grade, processing method, and degree of roasting of coffee samples. Multi-wavelength detection, provided by DAD systems, allows for data acquisition at the optimal absorption wavelength for each compound, thereby improving analytical sensitivity and reliability.

Due to these advantages, multicomponent chromatographic profiling is widely applied in coffee industry quality control protocols and product authentication studies (dos Santos and Boffo, 2021).

11. Emerging trends and future perspectives in coffee chromatography

Recent developments in coffee chromatography indicate a decisive shift from single-technique analytical workflows toward integrated, hybrid, and data-driven platforms designed to meet increasing demands for authenticity, safety, and sustainability. While classical chromatographic methods, such as HPLC and GC-MS, remain indispensable for the quantitative determination of caffeine, diterpenes, and contaminants, future perspectives emphasise their combination with spectroscopic fingerprinting and advanced chemometric interpretation to achieve faster and more comprehensive quality assessment (Sim et al., 2023; Aung Moon et al., 2024). In this context, chromatography is no longer viewed as an isolated analytical endpoint but rather as a reference anchor within multi-modal analytical systems.

A prominent emerging trend involves the incorporation of artificial intelligence (AI) and machine-learning-based models to interpret complex chromatographic and spectroscopic datasets. Studies integrating chromatographic outputs with deep learning and computer vision approaches have demonstrated enhanced performance in purity assessment, defect detection, and classification of specialty coffees, including Luwak and origin-specific beans (Hendrawan et al., 2019, 2022; Septiarini et al., 2024). These approaches are expected to play a crucial role in automating quality control pipelines, reducing analyst dependency, and enabling real-time decision-making in industrial environments.

From a regulatory and safety perspective, chromatographic methods will continue to serve as the gold standard for contaminant monitoring, particularly for mycotoxins such as ochratoxin A. Nevertheless, future frameworks are likely to integrate rapid spectroscopic screening tools with confirmatory chromatographic analysis, thereby balancing throughput with analytical robustness (Pernica et al., 2023). Overall, the future of coffee chromatography lies in the convergence of high-resolution separation science, non-destructive spectroscopic sensing, and intelligent data analytics, forming comprehensive platforms that can address both scientific and industrial challenges in coffee quality evaluation (Sagita et al., 2025).

12. AI-driven chromatographic data analysis in coffee science

The application of artificial intelligence to coffee analysis represents a significant paradigm shift in the interpretation of chromatographic data. Instead of relying solely on peak-based quantification, recent studies demonstrate that machine learning and deep learning algorithms can extract latent patterns from chromatographic and spectroscopic datasets, significantly improving classification accuracy and robustness. In particular, convolutional neural networks and supervised learning models have been successfully applied to identify purity levels, detect adulteration, and classify specialty coffee beans using analytical fingerprints as input features (Hendrawan et al., 2019, 2022).

These AI-assisted approaches enable the handling of high-dimensional datasets generated by modern chromatographic systems, overcoming limitations associated with manual data processing and subjective interpretation. Furthermore, the integration of chromatographic data with image-based and spectral features has shown strong potential for automated quality grading and origin authentication (Septiarini et al., 2024). As analytical instrumentation continues to generate increasingly complex datasets, AI-driven analytics are expected to become a central component of next-generation coffee chromatography, particularly in high-throughput and industrial applications.

Recent developments in coffee chromatography have increasingly focused on integrating high-resolution separation techniques with chemometrics and explainable artificial intelligence (XAI) to address the inherent chemical and sensory complexity of coffee matrices. Headspace solid-phase microextraction combined with gas chromatography–mass spectrometry (HS-SPME/GC-MS) remains one of the most powerful and widely applied techniques for profiling volatile organic compounds responsible for coffee aroma. This approach enables the generation of robust chemical fingerprints that can be directly linked to coffee origin, roasting degree, and sensory descriptors, providing an objective complement to traditional sensory analysis (Bressanello et al., 2017; Liberto et al., 2013).

In recent years, multivariate statistical tools and machine learning models have been increasingly applied to chromatographic datasets to enhance discrimination power and predictive accuracy. Support Vector Machines, partial least squares regression, and hierarchical clustering have demonstrated strong performance in classifying single-origin coffees and predicting aroma intensity from GC-MS data, despite the high dimensionality and collinearity typically found in volatile profiles (Bressanello et al., 2018). However, the interpretability of these models has long remained a critical limitation, particularly in quality control and regulatory contexts where transparent decision-making is essential.

The introduction of explainable artificial intelligence represents a major methodological shift in coffee chromatography. By integrating SHapley Additive exPlanations (SHAP) with supervised learning models, it has become possible to identify the specific volatile compounds that contribute most strongly to classification outcomes and sensory predictions. This approach enables chromatographic variables to be directly linked to the chemical drivers of aroma intensity and origin specificity, thereby transforming complex GC-MS datasets into interpretable chemical knowledge (Felizzato et al., 2025). Such interpretability is increasingly recognised as a prerequisite for the practical deployment of artificial intelligence in food quality assessment systems (Angelov et al., 2021; Minh et al., 2022).

From a future perspective, coffee chromatography is expected to evolve toward holistic, data-driven quality assessment frameworks that merge advanced analytical chemistry with transparent computational intelligence. The combination of standardised chromatographic workflows, robust chemometric modelling, and XAI-based interpretation paves the way for reproducible "aroma blueprints" capable of capturing terroir effects, processing conditions, and roasting dynamics within a single analytical platform. These integrated strategies are anticipated to support more reliable origin authentication, enhanced quality certification schemes, and improved consumer trust through scientifically grounded traceability mechanisms (Felizzato et al., 2025; Bressanello et al., 2017).

In this context, chromatography is no longer limited to compound identification and quantification but emerges as a central component of intelligent quality control pipelines. The future of coffee analysis lies in the synergistic use of chromatographic data and explainable computational models, enabling both high analytical performance and meaningful chemical interpretation for industrial, regulatory, and research applications.

13. Conclusion

The findings addressed in this chapter clearly demonstrate that the chemical composition and quality attributes of coffee are too complex to be explained by a single analytical approach. All stages, from production and processing to extraction and analytical characterisation, directly influence the chemical profile and sensory properties of the final product. Therefore, a fundamental conclusion of this work is that coffee research should move beyond reductionist perspectives and adopt multi-layered and holistic methodological frameworks.

As summarised in Table 3, the chemical and sensory complexity of coffee emerges from the cumulative interaction of raw material characteristics, extraction strategies, analytical methodologies, and data interpretation frameworks rather than from any single experimental step. This integrated perspective highlights that extraction techniques, chromatographic platforms, and interpretative models should not be evaluated in isolation, but instead considered as interconnected components of a unified analytical workflow.

The comparative data presented indicate that the selection of an extraction technique should not be based solely on extraction yield. Instead, it should be evaluated by jointly considering target compounds, sustainability, energy consumption, and post-analytical interpretability. In particular, green extraction strategies and hybrid techniques have emerged as robust alternatives, offering a reduced environmental

Table 3: Conceptual summary of analytical workflows and decision stages in coffee chemistry

Stage	Key Focus	Main Insight
Raw material and processing	Origin, roasting, matrix complexity	Coffee chemistry is shaped by cumulative processing steps rather than a single factor
Extraction strategies	Conventional, green, and hybrid techniques	Method selection must be goal-oriented, not yield-driven
Analytical characterisation	GC–MS, LC–MS/MS, HPLC-DAD, HRMS	Multiplatform chromatography is essential for comprehensive profiling
Data interpretation	Chemometrics, multivariate analysis, XAI	Interpretability is as critical as analytical accuracy
Quality control and applications	Authentication, sustainability, and industry use	Integrated analytical workflows improve robustness and comparability

impact while preserving bioactive compounds. In this context, future studies are encouraged to focus on purpose-driven, optimised extraction strategies rather than seeking a single "best" method.

The results obtained from analytical chromatography further highlight the indispensable role of gas and liquid chromatography systems coupled with mass spectrometry in the identification and characterisation of coffee constituents. However, the increasing volume and complexity of analytical data necessitate approaches that extend beyond conventional statistical evaluations. At this stage, chemometric methods and explainable artificial intelligence models not only enhance classification performance but also enable the interpretation of chemical data within a cause–and–effect framework.

In conclusion, the approach presented here advocates that future research in coffee science should progress along an axis that simultaneously emphasises analytical accuracy and interpretability. This holistic perspective, which allows the chemical complexity of coffee samples to be described in a more transparent, comparable, and sustainable manner, is expected to provide valuable guidance for both academic research and industrial quality control applications.

References

Ajila, C., Brar, K., Verma, M., Tyagi, R. D., Godbout, S., and Valero, J. R. (2010). Extraction and analysis of polyphenols: Recent trends. *Critical Reviews in Biotechnology,* 31, 227–249.

Alara, O. R., Abdurahman, N. H., and Ukaegbu, C. I. (2018). Soxhlet extraction of phenolic compounds from Vernonia cinerea leaves and its antioxidant activity. *Journal of Applied Research on Medicinal and Aromatic Plants*, 11, 12–17.

Alliance for Coffee Excellence. (2019). *Cup of Excellence Results: Brazil 2019*. Retrieved July 20, 2020, from https://allianceforcoffeeexcellence.org/brazil-2019/

Altemimi, A., Watson, D. G., Choudhary, R., Dasari, M. R., and Lightfoot, D. A. (2016). Ultrasound assisted extraction of phenolic compounds from peaches and pumpkins. *PloS One*, 11(2), e0148758.

Andrade, K. S., Goncalvez, R. T., Maraschin, M., Ribeiro-do-Valle, R. M., Martínez, J., and Ferreira, S. R. (2012). Supercritical fluid extraction from spent coffee grounds and coffee husks: Antioxidant activity and effect of operational variables on extract composition. *Talanta*, 88, 544–552.

Angelov, P. P., Soares, E. A., Jiang, R., Arnold, N. I., and Atkinson, P. M. (2021). Explainable artificial intelligence: An analytical review. *Wiley Interdisciplinary Reviews: Data Mining and Knowledge Discovery*, 11(5). https://doi.org/10.1002/widm.1424

Aung Moon, S., Wongsakul, S., Kitazawa, H., Kittiwachana, S., and Saengrayap, R. (2024). Application of ATR-FTIR for green Arabica bean shelf-life determination in accelerated storage. *Foods*, 13(15), 2331. https://doi.org/10.3390/foods13152331

Awuchi, C. (2019). The biochemistry, toxicology, and uses of the pharmacologically active phytochemicals: Alkaloids, terpenes, polyphenols and glycosides. *Journal of Medicinal and Aromatic Plants*, 7, 131–150.

Azwanida, N. N. (2015). A review on the extraction methods use in medicinal plants, principle, strength and limitation. *Med Aromat Plants*, 4(196), 2167-0412.

Belo, Y. N., Al-Hamimi, S., Chimuka, L., and Turner, C. (2019). Ultrahigh-pressure supercritical fluid extraction and chromatography of *Moringa oleifera* and *Moringa peregrina* seed lipids. *Analytical and Bioanalytical Chemistry*, 411(16), 3685.

Bilge, T., and Aksoylu, Ö. Z. (2020). Optimisation of microwave effect on bioactives contents and colour attributes of aqueous green tea extracts by central composite design. *Journal of Food Measurement & Characterization*, 14(4), 2240–2252.

Bispo, M. S., Veloso, M. C. C., Pinheiro, H. L. C., De Oliveira, R. F., Reis, J. O. N., and De Andrade, J. B. (2002). Simultaneous determination of caffeine, theobromine, and theophylline by high-performance liquid chromatography. *Journal of Chromatographic Science*, 40(1), 45–48.

Bouhzam, I., Cantero, R., Margallo, M., Aldaco, R., Bala, A., Fullana-i-Palmer, P., and Puig, R. (2023). Extraction of bioactive compounds from spent coffee grounds using ethanol and acetone aqueous solutions. *Foods*, 12(24), 4400.

Bouhzam, I., Cantero, R., Balcells, M., Margallo, M., Aldaco, R., Bala, A., ... & Puig, R. (2023). Environmental and yield comparison of quick extraction methods for caffeine and chlorogenic acid from spent coffee grounds. *Foods*, 12(4), 779.

Brachet, A., Rudaz, S., Mateus, L., Christen, P., and Veuthey, J. L. (2001). Optimisation of accelerated solvent extraction of cocaine and benzoylecgonine from coca leaves. *Journal of Separation Science*, 24(10–11), 865–873.

Brazilian Association of Coffee Industries (ABIC). (2004). *Coffee Quality Program (PQC)*. Retrieved July 20, 2020, from https://www.abic.com.br/certificacao/

Brazilian Association of Coffee Industries (ABIC). (2018). *Norma de qualidade (PQC)*. Retrieved September 23, 2020, from https://www.abic.com.br/wp-content/uploads/2018/04/Norma-PQC-26-06-2018-rev-30.pdf

Brazilian Association of Coffee Industries (ABIC). (2019). *16th National Coffee Quality Contest: Origins of Brazil: 2019 Crop*. Retrieved July 20, 2020, from https://www.abic.com.br/concursonacional/leilao/

Brazilian Association of Coffee Industries (ABIC). (2020). *Coffee Quality Categories*. Retrieved July 20, 2020, from https://www.abic.com.br/recomendacoes-tecnicas/categorias-de-qualidade-do-cafe/

Bressanello, D., Liberto, E., Cordero, C., Rubiolo, P., Pellegrino, G., Ruosi, M. R., and Bicchi, C. (2017). Coffee aroma: Chemometric comparison of the chemical information provided by three different samplings combined with GC–MS to describe the sensory properties in cup. *Food Chemistry*, 214, 218–226. https://doi.org/10.1016/j.foodchem.2016.07.088

Bressanello, D., Liberto, E., Cordero, C., Sgorbini, B., Rubiolo, P., Pellegrino, G. et al. (2018). Chemometric modeling of coffee sensory notes through their chemical signatures: Potential and limits in defining an analytical tool for quality control. *Journal of Agricultural and Food Chemistry*, 66(27), 7096–7109. https://doi.org/10.1021/acs.jafc.8b01340

Bücking, M., and Steinhart, H. (2002). Headspace and sensory analysis characterization of the influence of different milk additives on the flavor release of coffee beverages. *Journal of Agricultural and Food Chemistry*, 50, 1529–1534. https://doi.org/10.1021/jf011117p

Bytof, G., Knopp, S. E., Schieberle, P., Teutsch, I., and Selmar, D. (2005). Influence of processing on the generation of γ-aminobutyric acid in green coffee beans. *European Food Research and Technology*, 220(3), 245–250.

Cavin, C., Holzhaeuser, D., Scharf, G., Constable, A., Huber, W. W., and Schilter, B. (2002). Cafestol and kahweol, two coffee specific diterpenes with anticarcinogenic activity. *Food and Chemical Toxicology*, 40, 1155–1169. https://doi.org/10.1016/S0278-6915(02)00029-7

Chin, F. S., Chong, K. P., Markus, A., and Wong, N. K. (2013). Tea polyphenols and alkaloids content using soxhlet and direct extraction methods. *World J Agric Sci*, 9(3), 266–270.

Chu, R., Wan, L.-S., Peng, X.-R., Yu, M.-Y., Zhang, Z.-R., Zhou, L. et al. (2016). Characterization of new ent-kaurane diterpenoids of Yunnan Arabica coffee beans. *Natural Product Bioprospecting*, 6, 217–223.

Cittan, M., Altuntaş, E., and Çelik, A. (2018). Evaluation of antioxidant capacities and phenolic profiles in Tilia cordata fruit extracts: A comparative study to determine the efficiency of traditional hot water infusion method. *Industrial Crops and Products*, 122, 553–558.

Clarke, R. J., and Macrae, R. (1985). *Coffee: Volume 1—Chemistry*. Elsevier Applied Science.

Coffee Research Institute. (2020). *Classification of Green Coffee Beans: Brazil/New York Method.* Retrieved December 4, 2020, from http://www.coffeeresearch.org/coffee/brazilclass.htm

CONAB (Companhia Nacional de Abastecimento). (2020). *1° Levantamento de Café—Safra 2020.* Retrieved July 19, 2020, from https://www.conab.gov.br/info-agro/safras/cafe/boletim-da-safra-de-cafe

Conde, T., and Mussatto, S. I. (2016). Isolation of polyphenols from spent coffee grounds and silverskin by mild hydrothermal pretreatment. *Preparative Biochemistry and Biotechnology*, 46(4), 406–409.

Costa-Freitas, A. M., Parreira, C., and Vilas-Boas, L. (2001). Comparison of two SPME fibers for differentiation of coffee by analysis of volatile compounds. *Chromatographia*, 54(9), 647–652.

Ćujić, N., Šavikin, K., Janković, T., Pljevljakušić, D., Zdunić, G., and Ibrić, S. (2016). Optimization of polyphenols extraction from dried chokeberry using maceration as traditional technique. *Food Chemistry*, 194, 135–142.

Czerny, M., Mayer, F., and Grosch, W. (1999). Sensory study on the character impact odorants of roasted Arabica coffee. *Journal of Agricultural and Food Chemistry*, 47, 695–699.

de Andrade Lima, M., Andreou, R., Charalampopoulos, D., and Chatzifragkou, A. (2021). Supercritical carbon dioxide extraction of phenolic compounds from potato (*Solanum tuberosum*) peels. *Applied Sciences*, 11(8), 3410.

Debastiani, R., Iochims dos Santos, C. E., Maciel Ramos, M., Sobrosa Souza, V., Amaral, L., Yoneama, M. L. et al. (2019). Elemental analysis of Brazilian coffee with ion beam techniques: From ground coffee to the final beverage. *Food Research International*, 119, 297–304.

Del Campo, G., Berregi, I., Caracena, R., and Zuriarrain, J. (2010). Quantitative determination of caffeine, formic acid, trigonelline and 5-(hydroxymethyl)furfural in soluble coffees by ^{1}H NMR spectrometry. *Talanta*, 81, 367–371.

Demissie, Z. A., Chano, J., and Steenkamp, P. A. (2016). Determination of caffeine in Ethiopian green coffee beans. *Chemistry & Chemical Engineering*, 5(2), 45–52.

Dirinck, I., van Leuven, I., and Dirinck, P. (2001). SDE-GC-MS analysis and titratable acidity for objective flavour characterization of coffee. *Speciality Publications – Royal Society of Chemistry*, 269, 248–252.

Domínguez-Rodríguez, G., Marina, M. L., and Plaza, M. (2021). Enzyme-assisted extraction of bioactive non-extractable polyphenols from sweet cherry (*Prunus avium* L.) pomace. *Food Chemistry*, 339, 128086.

dos Santos, H. D., and Boffo, E. F. (2021). Coffee beyond the cup: Analytical techniques used in chemical composition research—A review. *European Food Research and Technology*, 247(4), 749–775.

Dulsat-Serra, N., Quintanilla-Casas, B., and Vichi, S. (2016). Volatile thiols in coffee: A review on their formation, degradation, assessment and influence on coffee sensory quality. *Food Research International*, 89, 982–988. https://doi.org/10.1016/j.foodres.2016.02.008

Elik, A., Yanık, D. K., and Göğüş, F. (2020). Microwave-assisted extraction of carotenoids from carrot juice processing waste using flaxseed oil as a solvent. *Lwt*, 123, 109100.

Felizzato, G., Bagnulo, E., Botta, G., Tapparo, G., Cordero, C., Navarini, L. et al. (2025). Explainable artificial intelligence for coffee quality control: From coffee origins to aroma intensity. SSRN. https://doi.org/10.2139/ssrn.5942461

Fernández-Marín, R., Fernandes, S. C. M., Andrés, M. A., and Labidi, J. (2021). Microwave-assisted extraction of *Curcuma longa* L. oil: Optimization, chemical structure and composition, antioxidant activity and comparison with conventional soxhlet extraction. *Molecules*, 26, 1516.

Gennaro, M. C., and Abrigo, C. (1992). Caffeine and theobromine in coffee, tea and cola-beverages: Simultaneous determination by reversed-phase ion-interaction HPLC. *Fresenius' Journal of Analytical Chemistry*, 343(6), 523–525.

Giahi, E., Jahadi, M., and Khosravi-Darani, K. (2021). Enzyme-assisted extraction of glycyrrhizic acid from licorice roots using heat reflux and ultrasound methods. *Biocatalysis and Agricultural Biotechnology*, 33, 101953.

Gloess, A. N., Schönbächler, B., Klopprogge, B., D'Ambrosio, L., Chatelain, K., Bongartz, A. et al. (2013). Comparison of nine common coffee extraction methods: Instrumental and sensory analysis. *European Food Research and Technology*, 236, 607–627.

Gomez-Ruiz, J. A., Leake, D. S., and Ames, J. M. (2007). In vitro antioxidant activity of coffee compounds and their metabolites. *Journal of Agricultural and Food Chemistry*, 55, 6962–6969.

González-Centeno, M. R., Knoerzer, K., Sabarez, H., Simal, S., Rosselló, C., and Femenia, A. (2014). Effect of acoustic frequency and power density on the aqueous ultrasonic-assisted extraction of grape pomace (*Vitis vinifera* L.) – A response surface approach. *Ultrasonics Sonochemistry*, 21(6), 2176–2184.

Goyeneche, R., Fanovich, A., Rodrigues, C. R., Nicolao, M. C., and Di Scala, K. (2018). Supercritical CO_2 extraction of bioactive compounds from radish leaves: Yield, antioxidant capacity and cytotoxicity. *The Journal of Supercritical Fluids*, 135, 78–83.

Granvogl, M., and Schieberle, P. (2007). Quantification of 3-aminopropionamide in cocoa, coffee and cereal products. *European Food Research and Technology*, 225, 857–863. https://doi.org/10.1007/s00217-006-0492-9

Grumezescu, A. M., and Holban, A. M. (Eds.). (2019). *Caffeinated and Cocoa-Based Beverages: Volume 8. The Science of Beverages.* Elsevier.

Guatemala-Morales, G. M., Beltrán-Medina, E. A., Murillo-Tovar, M. A., Ruiz-Palomino, P., Corona-González, R. I., and Arriola-Guevara, E. (2016). Validation of analytical conditions for determination of polycyclic aromatic hydrocarbons in roasted coffee by gas chromatography–mass spectrometry. *Food Chemistry*, 197, 747–753.

Hassim, N., Markom, M., Rosli, M. I., and Harun, S. (2021). Scale-up approach for supercritical fluid extraction with ethanol–water modified carbon dioxide on *Phyllanthus niruri* for safe enriched herbal extracts. *Scientific Reports*, 11(1), 15818.

Heemann, A. C. W., Heemann, R., Kalegari, P., Spier, M. R., and Santin, E. (2019). Enzyme-assisted extraction of polyphenols from green yerba mate. *Brazilian Journal of Food Technology*, 22, e2017222.

Heleno, S. A., Prieto, M. A., Barros, L., Rodrigues, A., Barreiro, M. F., and Ferreira, I. C. (2016). Optimization of microwave-assisted extraction of ergosterol from *Agaricus bisporus* L. by-products using response surface methodology. *Food and Bioproducts Processing*, 100, 25–35.

Hendrawan, Y., Widyaningtyas, S., and Sucipto, S. (2019). Computer vision for purity, phenol, and pH detection of Luwak coffee green bean. *Telkomnika*, 17(6), 3073–3082.

Hendrawan, Y., Widyaningtyas, S., Fauzy, M. R., Sucipto, S., Damayanti, R., Al Riza, D. F. et al. (2022). Deep learning to detect and classify the purity level of Luwak coffee green beans. *Pertanika Journal of Science & Technology*, 30(1), 1–15.

Herrera, M. C., and Luque de Castro, M. D. (2004). Ultrasound-assisted extraction for the analysis of phenolic compounds in strawberries. *Analytical and Bioanalytical Chemistry*, 379(7), 1106–1112.

Hofler, F., Jensen, D., Ezzell, J., and Richter, B. (1996). Rapid preparation of environmental samples by accelerated solvent extraction (ASE). *Polycyclic Aromatic Compounds*, 9(1-4), 233–240.

Hu, B., Xi, X., Li, H., Qin, Y., Li, C., Zhang, Z., ... & Luo, Q. (2021). A comparison of extraction yield, quality and thermal properties from Sapindus mukorossi seed oil between microwave assisted extraction and Soxhlet extraction. *Industrial Crops and Products*, 161, 113185.

Hussein, R. A., and El-Anssary, A. A. (2018). Plants secondary metabolites: The key drivers of the pharmacological actions of medicinal plants. *In:* Herbal Medicine. IntechOpen. https://doi.org/10.5772/intechopen.76139

International Coffee Organization (ICO). (2018). *National Quality Standards.* Retrieved November 24, 2020, from http://www.ico.org/documents/cy2017-18/icc-122-12e-national-quality-standards.pdf

International Coffee Organization (ICO). (2019). *Total Production by All Exporting Countries.* Retrieved February 26, 2019, from http://www.ico.org/prices/po-production.pdf

International Coffee Organization (ICO). (2020). *Coffee Market Report (June).* Retrieved September 18, 2020, from http://www.ico.org/

International Organization for Standardization (ISO). (2020). *ISO Homepage.* Retrieved July 20, 2020, from https://www.iso.org/home.html

Iwamoto, H., Izumi, K., Natsagdorj, A., Naito, R., Makino, T., Kadomoto, S. et al. (2019). Coffee diterpenes kahweol acetate and cafestol synergistically inhibit the proliferation and migration of prostate cancer cells. *Prostate*, 79, 468–479.

Jentzer, J. B., Alignan, M., Vaca-Garcia, C., Rigal, L., and Vilarem, G. (2015). Response surface methodology to optimise Accelerated Solvent Extraction of steviol glycosides from Stevia rebaudiana Bertoni leaves. *Food Chemistry*, 166, 561–567.

Jeszka-Skowron, M., Zgoła-Grześkowiak, A., and Grześkowiak, T. (2015). Analytical methods applied for the characterization and the determination of bioactive compounds in coffee. *European Food Research and Technology*, 240, 19–31. https://doi.org/10.1007/s00217-014-2356-z

Jin, S., Yang, M., Kong, Y., Yao, X., Wei, Z., Zu, Y., and Fu, Y. (2011). Microwave-assisted extraction of flavonoids from Cajanus cajan leaves. *Zhongcaoyao,* 42(11), 2235–2239.

K. Habeebullah, S. F., Alagarsamy, S., Sattari, Z., Al-Haddad, S., Fakhraldeen, S., Al-Ghunaim, A., and Al-Yamani, F. (2020). Enzyme-assisted extraction of bioactive compounds from brown seaweeds and characterization. *Journal of Applied Phycology*, 32(1), 615–629.

Kassama, L. S., Shi, J., and Mittal, G. S. (2008). Optimization of supercritical fluid extraction of lycopene from tomato skin with central composite rotatable design model. *Separation and Purification Technology*, 60(3), 278–284.

Kim, H. G., Hwang, Y. P., and Jeong, H. G. (2009). Kahweol blocks STAT3 phosphorylation and induces apoptosis in human lung adenocarcinoma A549 cells. *Toxicology Letters*, 187, 28–34. https://doi.org/10.1016/j.toxlet.2009.01.022

Knopp, S., Bytof, G., and Selmar, D. (2006). Influence of processing on the content of sugars in green Arabica coffee beans. *European Food Research and Technology*, 223(2), 195–201.

Kovačević, D. B., Maras, M., Barba, F. J., Granato, D., Roohinejad, S., Mallikarjunan, K., ... & Putnik, P. (2018). Innovative technologies for the recovery of phytochemicals from Stevia rebaudiana Bertoni leaves: A review. *Food Chemistry*, 268, 513–521.

Ky, C. L., Lonarn, J., Dussert, S., Guyot, B., Hamon, S., and Noirot, M. (2001). Caffeine, trigonelline, chlorogenic acids and sucrose diversity in wild coffee *Coffea arabica* L., *C. canephora* P. accessions. *Food Chemistry*, 75, 223–230. https://doi.org/10.1016/S0308-8146(01)00204-7

Liberto, E., Ruosi, M. R., Cordero, C., Rubiolo, P., Bicchi, C., and Sgorbini, B. (2013). Non-separative headspace solid-phase microextraction–mass spectrometry profile as a marker to monitor coffee roasting degree. *Journal of Agricultural and Food Chemistry*, 61(8), 1652–1660. https://doi.org/10.1021/jf303067q

Lima, C. S., Spindola, D. G., and Bechara, A. (2017). Cafestol, a diterpene molecule found in coffee, induces leukemia cell death. *Biomedicine & Pharmacotherapy*, 92, 1045–1054.

Liu, J. J., Gasmalla, M. A. A., Li, P., and Yang, R. (2016). Enzyme-assisted extraction processing from oilseeds: Principle, processing and application. *Innovative Food Science & Emerging Technologies*, 35, 184–193.

López-Fernández-Sobrino, R., Margalef, M., Torres-Fuentes, C., Avila-Roman, J., Aragones, G., Muguerza, B., and Bravo, F. I. (2021). Enzyme-assisted extraction to obtain phenolic-enriched wine lees with enhanced bioactivity in hypertensive rats. *Antioxidants*, 10(4), 517.

Mabrok, A. F., and Deatheroge, F. E. (1956). *Organic Acid in Brewed Coffee*. New York.

Macedo, G. A., Santana, A. L., Crawford, L. M., Wang, S. C., Dias, F. F., and de Moura Bell, J. M. (2021). Integrated microwave-and enzyme-assisted extraction of phenolic compounds from olive pomace. *Lwt*, 138, 110621.

Machado, A. P. D. F., Sumere, B. R., Mekaru, C., Martinez, J., Bezerra, R. M. N., and Rostagno, M. A. (2019). Extraction of polyphenols and antioxidants from pomegranate peel using ultrasound: Influence of temperature, frequency and operation mode. *International Journal of Food Science and Technology*, 54(9), 2792–2801.

Maeztu, L., Sanz, C., Andueza, S., de Peña, M. P., Bello, J., and Cid, C. (2001). Characterization of espresso coffee aroma by static headspace GC–MS and sensory flavor profile. *Journal of Agricultural and Food Chemistry*, 49(11), 5437–5444. https://doi.org/10.1021/jf0107959

Mahmudati, N., Wahyono, P., and Djunaedi, D. (2020, June). Antioxidant activity and total phenolic content of three varieties of Ginger (*Zingiber officinale*) in decoction and infusion extraction method. *In:* Journal of Physics: Conference Series (Vol. 1567, No. 2, p. 022028). IOP Publishing.

MAPA (Ministry of Agriculture, Livestock, and Supply, Brazil). (2003). *Normative Instruction 08/03 (11 June 2003): Regulamento técnico de identidade e de qualidade para a classificação do café beneficiado grão cru.* Retrieved May 30, 2019, from https://www.jusbrasil.com.br/diarios/580954/pg-4-secao-1-diario-oficial-da-uniao-dou-de-13-06-2003

Martins, S. C. V., Araújo, W. L., Tohge, T., Fernie, A. R., and DaMatta, F. M. (2014). In high-light-acclimated coffee plants the metabolic machinery is adjusted to avoid oxidative stress rather than to benefit from extra light enhancement in photosynthetic yield. *PLoS ONE*, 9, e94862.

Matshediso, P. G., Cukrowska, E., and Chimuka, L. (2015). Development of pressurised hot water extraction (PHWE) for essential compounds from Moringa oleifera leaf extracts. *Food Chemistry*, 172, 423–427.

Meinhart, A. D., Bizzotto, C. S., Ballus, C. A., Prado, M. A., Bruns, R. E., Teixeira Filho, J., and Godoy, H. T. (2010). Optimisation of a CE method for caffeine analysis in decaffeinated coffee. *Food Chemistry*, 120(4), 1155–1161. https://doi.org/10.1016/j.foodchem.2009.11.048

Mellbye, F. B., Jeppesen, P. B., Hermansen, K., and Gregersen, S. (2015). Cafestol, a bioactive substance in coffee, stimulates insulin secretion and increases glucose uptake in muscle cells: Studies in vitro. *Journal of Natural Products*, 78, 2447–2451.

Minh, D., Wang, H. X., Li, Y. F., and Nguyen, T. N. (2022). Explainable artificial intelligence: A comprehensive review. *Artificial Intelligence Review*, 55(5), 3503–3568. https://doi.org/10.1007/s10462-021-10088-y

Ministry of Agriculture, Livestock and Food Supply (MAPA). (2019). *In: Coffees of Brazil: Research, Sustainability and Innovation.* Embrapa.

Morton, I. D., and Macleod, A. J. (1986). *Food Flavours. Part B. The Flavour of Beverages* (p. 239). Amsterdam: Elsevier.

Motilva, M.-J., Serra, A., and Macià, A. (2013). Analysis of food polyphenols by ultra-high-performance liquid chromatography coupled to mass spectrometry: An overview. *Journal of Chromatography A*, 1292, 66–82.

Nakhmedov, F. G. (1984). *Technology of Coffee Products.* Consumer Goods and Food Industry. Moscow.

Namieśnik, J. (2000). Przygotowanie próbek materiału roślinnego do analizy chromatograficznej. *Chemia i Inżynieria Ekologiczna*, 7(8-9), 807–821.

Nascimento, M. M., Santos, H. M., Coutinho, J. P., Lôbo, I. P., da Silva Junior, A. L. S., Santos, A. G., and de Jesus, R. M. (2020). Optimization of chromatographic separation and classification of artisanal and fine chocolate based on its bioactive compound content through multivariate statistical techniques. *Microchemical Journal*, 152, 104342. https://doi.org/10.1016/j.microc.2019.104342

Navarra, G., Iacono, L., and Migliore, M. (2017). Simultaneous determination of caffeine and chlorogenic acid in coffee by spectral deconvolution. *Journal of Chemistry*, 2017, 8634728.

Navarra, G., Moschetti, M., Guarrasi, V., Mangione, M. R., Militello, V., and Leone, M. (2017). Simultaneous determination of caffeine and chlorogenic acids in green coffee by UV/Vis spectroscopy. *Journal of Chemistry*, 2017(1), 6435086.

Nguyen Tram Anh, M., Van Hung, P., & Thi Lan Phi, N. (2021). Optimized conditions for flavonoid extraction from pomelo peel byproducts under enzyme- and ultrasound-assisted extraction using response surface methodology. *Journal of Food Quality*, 2021, 6666381.

Nishad, J., Saha, S., and Kaur, C. (2019). Enzyme- and ultrasound-assisted extractions of polyphenols from Citrus sinensis (cv. Malta) peel: A comparative study. *Journal of Food Processing and Preservation*, 43(8), e14046.

Nuapia, Y., Chimuka, L., and Cukrowska, E. (2018). Assessment of heavy metals in raw food samples from open markets in two African cities. *Chemosphere*, 196, 339–346.

Nuhu, A. A. (2014). Bioactive micronutrients in coffee: Recent analytical approaches for characterization and quantification. *ISRN Nutrition*, 2014, 1–13. https://doi.org/10.1155/2014/384230

Oh, S. H., Hwang, Y. P., Choi, J. H., Jin, S. W., Lee, G. H., Han, E. H. et al. (2018). Kahweol inhibits proliferation and induces apoptosis by suppressing fatty acid synthase in HER2-overexpressing cancer cells. *Food and Chemical Toxicology*, 121, 326–335.

Oniszczuk, A., and Podgórski, R. (2015). Influence of different extraction methods on the quantification of selected flavonoids and phenolic acids from Tilia cordata inflorescence. *Industrial Crops and Products*, 76, 509–514.

Pandhi, S., and Kumar, A. (2025). Assessing the influence of extraction techniques on the phytochemical composition of green coffee (*Coffea arabica*) using Principal Component Analysis (PCA) and Hierarchical Cluster Analysis (HCA). *Journal of the Indian Chemical Society*, 102111.

Pellicanò, T. M., Sicari, V., Loizzo, M. R., Leporini, M., Falco, T., and Poiana, M. (2019). Optimizing the supercritical fluid extraction process of bioactive compounds from processed tomato skin by-products. *Food Science and Technology*, 40(3), 692–697.

Perera, P. R. D., Ekanayake, S., and Ranaweera, K. K. D. S. (2017). Antidiabetic compounds in Syzygium cumini decoction and ready to serve herbal drink. *Evidence-Based Complementary and Alternative Medicine*, 2017(1), 1083589.

Pernica, M., Martiník, J., Boško, R., Svoboda, Z., Cwiková, O., Běláková, S. et al. (2023). Ochratoxin A in roasted coffee marketed in the Czech Republic: Determination by liquid chromatography and fluorescence detection. *Journal of Food and Nutrition Research*, 62, 191–199.

Petracco, M. (2001). Technology IV: Beverage preparation: Brewing trends for the new millennium. *Coffee: Recent Developments*, 140–164.

Puri, M., Sharma, D., and Barrow, C. J. (2012). Enzyme-assisted extraction of bioactives from plants. *Trends in Biotechnology*, 30(1), 37–44.

Rahmalia, W., Fabre, J. F., and Mouloungui, Z. (2015). Effects of cyclohexane/acetone ratio on bixin extraction yield by accelerated solvent extraction method. *Procedia Chemistry*, 14, 455–464.

Rezvankhah, A., Emam-Djomeh, Z., Safari, M., Askari, G., and Salami, M. (2019). Microwave-assisted extraction of hempseed oil: Studying and comparing of fatty acid composition, antioxidant activity, physiochemical and thermal properties with Soxhlet extraction. *Journal of Food Science and Technology*, 56(9), 4198–4210.

Ribeiro, R. C., Mota, M. F. S., Silva, R. M., Silva, D. C., Novaes, F. J., da Veiga Jr, V. F., ... & Rezende, C. M. (2024). Coffee oil extraction methods: A review. *Foods*, 13(16), 2601.

Rocha, S., Maeztu, L., Barros, A., Cid, C., and Coimbra, M. A. (2004). Screening and distinction of coffee brews based on headspace solid-phase microextraction/gas chromatography/principal component analysis. *Journal of the Science of Food and Agriculture*, 84, 4

Romanik, G., Gilgenast, E., Przyjazny, A., and Kamiński, M. (2007). Techniques of preparing plant material for chromatographic separation and analysis. *Journal of Biochemical and Biophysical Methods*, 70(2), 253–261.

Roos, B., van der Weg, G., Urgert, R., Bovenkamp, P., Charrier, A., & Katan, M. (1997). Levels of cafestol, kahweol and related diterpenoids in wild species of the coffee plant. *Journal of Agricultural and Food Chemistry*, 45, 3065–3069. https://doi.org/10.1021/jf9700900

Rosas-Quina, Y. E., and Mejía-Nova, F. C. (2021). Supercritical fluid extraction with cosolvent of alkaloids from Lupinus mutabilis Sweet and comparison with conventional method. *Journal of Food Process Engineering*, 44(4), e13657.

Ryan, D., Shellie, R., Tranchida, P., Casilli, A., Mondello, L., and Marriott, P. (2004). Analysis of roasted coffee bean volatiles using comprehensive two-dimensional gas chromatography–time-of-flight mass spectrometry. *Journal of Chromatography A*, 1054, 57–65. https://doi.org/10.1016/S0021-9673(04)01408-6

Sagita, D., Mardjan, S. S., Widodo, S., Suparlan, S., and Purwandoko, P. B. (2025). Emerging trends and developments in coffee quality assessment and characterization techniques: A systematic review and bibliometric analysis. *Journal of Food Measurement and Characterization*, 1–21.

Seo, H.-Y., Kim, M.-K., Lee, S.-H., Hwang, J. S., Park, K.-G., and Jang, B. K. (2018). Kahweol ameliorates liver inflammation through the inhibition of NF-κB and STAT3 activation in primary Kupffer cells and primary hepatocytes. *Nutrients*, 10, 863.

Septiarini, A., Hamdani, H., Sela, E. I., Hidayat, N., and Afuan, L. (2024). Analysis of shape features by applying gain ratio and machine learning for coffee bean classification. *Coffee Science*, 19, e192206.

Sharif, K. M., Rahman, M. M., Azmir, J., Mohamed, A., Jahurul, M. H. A., Sahena, F., and Zaidul, I. S. M. (2014). Experimental design of supercritical fluid extraction – A review. *Journal of Food Engineering*, 124, 105–116.

Shrivas, K., & Wu, H.-F. (2007). Rapid determination of caffeine in one drop of beverages and foods using drop-to-drop solvent microextraction with gas chromatography/mass spectrometry. *Journal of Chromatography A*, 1170, 9–14.

Sim, J., McGoverin, C., Oey, I., Frew, R., and Kebede, B. (2023). Near-infrared reflectance spectroscopy accurately predicted isotope and elemental compositions for origin traceability of coffee. *Food Chemistry*, 427, 136695. https://doi.org/10.1016/j.foodchem.2023.136695

Sliwinski, A. (1993). Ultradzwieki i ich zastosowania. Warszawa: Wydawnictwa Naukowo-Techniczne.

Szumny, A., Figiel, A., Gutiérrez-Ortíz, A., and Carbonell-Barrachina, Á. A. (2010). Composition of rosemary essential oil (*Rosmarinus officinalis*) as affected by drying method. *Journal of Food Engineering*, 97(2), 253–260.

Taha, S. H., El-Sherbiny, I. M., Salem, A. S., Abdel-Hamid, M., Hamed, A. H., and Ahmed, G. A. (2019). Antiviral activity of curcumin loaded milk proteins nanoparticles on potato virus Y. *Pakistan Journal of Biological Sciences*, 22, 614–622.

United States Department of Agriculture (USDA). (2019, December). *Coffee: World Markets and Trade.* Retrieved November 23, 2020, from https://downloads.usda.library.cornell.edu/usda-esmis/files/m900nt40f/sq87c919h/8w32rm91m/coffee.pdf

Urgert, R., van der Weg, G., Kosmeijer-Schuil, T., van de Bovenkamp, P., Hovenier, R., and Katan, M. (1995). Levels of the cholesterol-elevating diterpenes cafestol and kahweol in various coffee brews. *Journal of Agricultural and Food Chemistry*, 43, 2167–2172. https://doi.org/10.1021/jf00056a039

Uwineza, P. A., and Waśkiewicz, A. (2020). Recent advances in supercritical fluid extraction of natural bioactive compounds from natural plant materials. *Molecules*, 25(17), 3847.

Vági, E., Balázs, M., Komoczi, A., Mihalovits, M., and Szekely, E. (2020). Fractionation of phytocannabinoids from industrial hemp residues with high-pressure technologies. *The Journal of Supercritical Fluids*, 164, 104898.

Vergara-Salinas, J. R., Pérez-Jiménez, J., Torres, J. L., Agosin, E., and Pérez-Correa, J. R. (2012). Effects of temperature and time on polyphenolic content and antioxidant activity in the pressurized hot water extraction of deodorized thyme (*Thymus vulgaris*). *Journal of Agricultural and Food Chemistry*, 60(44), 10920–10929.

Vinatoru, M., Mason, T. J., and Calinescu, I. (2017). Ultrasonically assisted extraction (UAE) and microwave assisted extraction (MAE) of functional compounds from plant materials. *TrAC Trends in Analytical Chemistry*, 97, 159–178.

Wang, X., Peng, X.-R., Lu, J., Hu, G.-L., and Qiu, M.-H. (2018). New dammarane triterpenoids, caffruones A–D, from the cherries of *Coffea arabica*. *Natural Product Bioprospecting*, 8, 413–418.

Wanyika, H. N., Gotebe, E. G., Gitu, L. M., Ngumba, E. K., and Maritim, C. W. (2010). Determination of caffeine content of tea and instant coffee brands found in the Kenyan market. *African Journal of Food Science*, 4, 353–358.

Weldegebreal, B., Redi-Abshiro, M., and Chandravanshi, B. S. (2017). Development of new analytical methods for the determination of caffeine content in aqueous solution of green coffee beans. *Chemical Central Journal*, 11, 126–134.

Yang, N., Lui, C., Liu, X., Degn, T. K., Munchow, M., and Fisk, I. (2016). Determination of volatile marker compounds of common coffee roast defects. *Food Chemistry*, 211, 206–214. https://doi.org/10.1016/j.foodchem.2016.04.124

Yang, Y., Wang, Z., Hu, D., Xiao, K., and Wu, J. Y. (2018). Efficient extraction of pectin from sisal waste by combined enzymatic and ultrasonic process. *Food Hydrocolloids*, 79, 189–196.

Yashin, A., Yashin, Y., Xia, X., and Nemzer, B. (2017). Chromatographic methods for coffee analysis: A review. *Journal of Food Research*, 6(4), 60–82.

Yenduri, S., Venkatesh, K. G., and Prashant, K. N. (2025). Recent advances in extraction of polyphenols by advanced extraction methods. *Phytochemical Analysis*, 36, 1875–1892.

Yi, C., Shi, J., Xue, S. J., Jiang, Y., and Li, D. (2009). Effects of supercritical fluid extraction parameters on lycopene yield and antioxidant activity. *Food Chemistry*, 113(4), 1088–1094.

Zhang, F., Chen, B., Xiao, S., and Yao, S. Z. (2005). Optimization and comparison of different extraction techniques for sanguinarine and chelerythrine in fruits of Macleaya cordata (Willd) R. Br. *Separation and Purification Technology*, 42(3), 283–290.

Zhang, L., Song, J., Kong, L., Yuan, T., Li, W., Zhang, W. et al. (2020). The strategies and techniques of drug discovery from natural products. *Pharmacology & Therapeutics*, 216, 107686. https://doi.org/10.1016/j.pharmthera.2020.107686

Zhang, Y. H., Fu, X. P., and Liang, W. J. (2016). Antioxidant activity and composition of anthocyanins of crude extracts from Yunnan Arabica coffee husk. *Food Science and Technology*, 41, 219–223.

Ziyatdinova, G., Aytuganova, I., Nizamova, A., and Budnikov, H. (2013). Differential pulse voltammetric assay of coffee antioxidant capacity with MWNT-modified electrode. *Food Analytical Methods*, 6, 132–139.

Zuorro, A., Fidaleo, M., and Lavecchia, R. (2011). Enzyme-assisted extraction of lycopene from tomato processing waste. *Enzyme and Microbial Technology*, 49(6–7), 567–573.

CHAPTER

7

Utilising Coffee and Coffee Waste Products in Baked Goods Production: Innovation, Sustainability, and New Flavours

Zhivka Goranova

Department of Food Technology, Institute of Food Preservation and Quality, Plovdiv, Agricultural Academy, Bulgaria

The coffee industry generates substantial quantities of by-products throughout cultivation, processing, roasting, and brewing, including coffee pulp, husk, silverskin, and spent coffee grounds. Traditionally regarded as waste, these materials are increasingly recognised as valuable resources rich in dietary fibre, polyphenols, melanoidins, residual lipids, and bioactive compounds. Their valorisation aligns with global sustainability goals, circular economy principles, and the growing demand for functional and environmentally responsible food products. This chapter critically examines the utilisation of coffee and coffee-derived waste products in baked goods production, with a focus on technological functionality, nutritional enhancement, sensory innovation, and sustainability outcomes.

The chapter first outlines the chemical composition and functional properties of major coffee by-products relevant to bakery applications. Quantitative data are presented on fibre content, polyphenol levels, water-binding capacity, and residual caffeine, highlighting their influence on dough rheology, gas retention, texture, and shelf life. Formulation guidelines are systematically discussed for yeast-leavened products, biscuits, cakes, and gluten-free systems, supported by experimental evidence and industrial case studies. Particular attention is given to the optimisation of inclusion levels, hydration adjustments, and mitigation strategies to maintain product quality.

Beyond technological considerations, the nutritional and health-related implications of coffee-enriched baked goods are reviewed, including fibre contribution, antioxidant potential, and emerging evidence on bioaccessibility after baking. The chapter further explores environmental and economic benefits,

E-mail: jivka_goranova@abv.bg

DOI: 10.1201/9781003689393-7

demonstrating how food-grade valorisation of coffee by-products can significantly reduce greenhouse gas emissions, landfill disposal, and resource consumption. Regulatory frameworks in major markets are also addressed, guiding safe and compliant industrial implementation.

By integrating food chemistry, processing technology, sensory science, and sustainability assessment, this chapter positions coffee by-products as multifunctional ingredients capable of delivering novel flavours, enhanced nutritional profiles, and measurable environmental benefits. The findings underscore their strong potential for innovation in modern bakery systems and highlight key research directions for future development.

1. Introduction

Global food systems are under increasing pressure to balance productivity with sustainability, resource efficiency, and waste reduction, particularly in the context of climate change, population growth, and tightening environmental regulations. According to the Food and Agriculture Organisation of the United Nations (FAO), approximately one-third of all food produced globally is lost or wasted, representing not only a substantial economic loss but also a significant environmental burden in terms of land use, water consumption, and greenhouse gas emissions (FAO, 2023). Within this framework, agro-industrial by-products have emerged as a critical focus of research and innovation, as their valorisation offers tangible opportunities to reduce waste while generating added value.

Coffee-related residues constitute one of the most abundant and underutilised agro-industrial by-product streams. The International Coffee Organisation (ICO) reports that global coffee production exceeded 10.5 million tonnes of green coffee beans in 2023, supporting the livelihoods of more than 125 million people worldwide (ICO, 2024). However, coffee processing and consumption generate substantial quantities of by-products, including coffee pulp, husk, silverskin, and spent coffee grounds. It is estimated that for every kilogram of green coffee produced, up to 2 kg of by-products are generated across the value chain (Nichifor et al., 2025). When inadequately managed, these residues contribute to methane emissions, soil and water contamination, and increased disposal costs, particularly in producing countries and urban consumption centres.

In parallel, the global bakery sector is undergoing rapid transformation driven by shifts in consumer behaviour and regulatory frameworks. Demand is increasing for clean-label products, fibre-enriched formulations, plant-based ingredients, and foods with demonstrable sustainability credentials. Baked goods represent one of the most widely consumed food categories worldwide, with high per capita intake across diverse socioeconomic and cultural contexts (OECD, 2023). Their formulation versatility and scalable manufacturing processes make them an ideal vehicle for incorporating functional and bioactive ingredients derived from agro-industrial residues.

The integration of coffee and coffee-derived waste products into baked goods, therefore, represents a convergence of two strategic imperatives: sustainable waste valorisation and innovation in bakery product development. Coffee by-products are rich in dietary fibre, polyphenols, melanoidins, and residual lipids, all of which exhibit technological functionality and potential health relevance. Recent food science research demonstrates that these materials can enhance nutritional profiles, contribute to flavour complexity, and improve oxidative stability when processed and formulated appropriately (Navajas-Porras et al., 2025; Arias et al., 2023).

This chapter examines the utilisation of coffee and coffee waste products in baked goods production from a multidisciplinary perspective, integrating insights from food chemistry, processing technology, sensory science, and sustainability assessment. By critically evaluating composition, processing strategies, formulation guidelines, and industrial feasibility, the chapter positions coffee-derived ingredients as promising contributors to next-generation baked goods that align with circular economy principles, consumer expectations, and environmental sustainability goals.

2. Coffee production and waste streams

Coffee production is a globally distributed agro-industrial system characterised by complex processing pathways and significant material flows. From cultivation and primary processing at origin to roasting, distribution, and final beverage preparation, coffee undergoes multiple transformations that generate substantial quantities of organic residues. These waste streams differ markedly in their physicochemical properties, spatial distribution, and environmental impact; yet, collectively, they represent a significant challenge and opportunity within the context of sustainable food systems.

Unlike many agricultural commodities, coffee generates by-products at both the production and consumption stages. Primary processing residues, such as pulp, mucilage, and husk, are concentrated in producing regions, often in proximity to sensitive ecosystems and rural communities. In contrast, secondary residues—including coffee silverskin and spent coffee grounds—are predominantly generated in urban and industrial settings, where large-scale disposal and waste management infrastructures are required. This spatial and logistical separation complicates conventional waste management strategies and contributes to the underutilisation of coffee-derived residues.

Recent sustainability assessments have highlighted that coffee by-products are frequently treated as waste, despite their high organic matter content and rich composition in dietary fibre, polyphenols, melanoidins, and residual lipids (Navajas-Porras et al., 2025; Arias et al., 2023). Inadequate handling of these materials can result in environmental externalities such as greenhouse gas emissions, soil and water contamination, and increased landfill burdens. Conversely, their valorisation within food systems offers a pathway to reduce environmental impact while generating functional ingredients with nutritional and technological value.

Understanding the origin, scale, and characteristics of coffee waste streams is therefore a prerequisite for their effective utilisation in food applications. This section provides an overview of the major by-products generated along the coffee value chain, quantifies their production volumes, and highlights the structural challenges associated with their management. By establishing this production and waste context, the section lays the foundation for subsequent discussion on the technological, nutritional, and sustainability-driven incorporation of coffee-derived materials into baked goods.

2.1 Origin and scale of coffee by-products

Coffee production is a multi-stage process that generates substantial quantities of by-products at each step of the value chain, from primary processing at origin to final beverage preparation. The nature, composition, and volume of these residues vary considerably depending on the processing method, geographic region, and scale of operation; yet, collectively, they represent one of the most significant agro-industrial waste streams associated with a single commodity.

At the production origin, coffee cherries are processed using either wet (washed) or dry (natural) methods. In wet processing, which is widely used for Arabica coffee, depulping and fermentation generate coffee pulp and mucilage, while dry processing produces coffee husk as the main residue. These primary by-products account for approximately 40–50% of the fresh coffee cherry mass, with coffee pulp alone representing up to 29–43% of the total weight of harvested cherries (Mussatto et al., 2011; Navajas-Porras et al., 2025). Given the current global coffee cherry production levels, this translates into tens of millions of tonnes of organic residues annually, particularly concentrated in producing countries across Latin America, Africa, and Southeast Asia.

Following the export of green coffee, roasting constitutes the next major transformation stage. During roasting, the coffee silverskin—a thin tegument layer surrounding the green bean—is detached and collected via cyclone systems in industrial roasters. Although silverskin represents a relatively small fraction by mass, typically 1–2% of roasted coffee bean weight, its large-scale generation makes it a significant residue at the global level. With worldwide roasted coffee production exceeding 10 million tonnes annually, silverskin generation is estimated to be around 100,000–200,000 tonnes per year (Arias et al., 2023). Importantly, silverskin is produced under food-grade conditions, which facilitates its potential direct valorisation in food applications.

The final and quantitatively most significant waste stream emerges during coffee brewing. Spent coffee grounds (SCG) are generated after extraction in households, cafés, offices, and industrial food-service operations. SCG typically accounts for 45–50% of the original roasted coffee mass, retaining significant moisture (60–70%) and residual organic matter. In urban environments, SCG constitute a significant fraction of organic municipal waste, with large cities generating thousands of tonnes annually (Makiso et al., 2023). Their decentralised generation and high moisture content present logistical and environmental challenges, including rapid microbial spoilage and methane emissions if landfilled.

Overall, it is estimated that for every kilogram of roasted coffee consumed, 1.8–2.2 kg of by-products and residues are generated across the value chain (Nichifor et al., 2025). Historically, these materials have been underutilised, commonly disposed of as waste or applied as low-value compost and animal feed. This practice persists despite extensive evidence that coffee by-products are rich in dietary fibre, polyphenols, melanoidins, and residual lipids, conferring both technological functionality and potential health relevance (Navajas-Porras et al., 2025; Arias et al., 2023).

The scale and composition of coffee-related waste streams, therefore, highlight a critical opportunity for sustainable valorisation. Redirecting even a fraction of these residues into food-grade applications—such as baked goods—could significantly reduce environmental burdens while creating added economic and nutritional value.

2.2 Classification of coffee-derived materials for food use

From a food science and technological perspective, coffee-derived materials suitable for incorporation into bakery products can be systematically classified according to their origin within the coffee value chain, chemical composition, technological functionality and regulatory status. Such classification is essential, as each category exhibits distinct physicochemical properties that directly influence processing behaviour, formulation requirements and final product quality.

- Roasted coffee powders and coffee extracts – Roasted coffee powders and concentrated coffee extracts represent the most established coffee-derived ingredients in food applications. They are primarily used for flavouring, colouring, and aroma enhancement in bakery products, confectionery, and desserts. Their functional role is largely sensory rather than structural. Roasted coffee powders contribute volatile aroma compounds, melanoidins, and dark colour pigments formed during roasting. Coffee extracts—either aqueous or hydroalcoholic—allow more precise dosing of flavour compounds and are widely used in industrial bakery systems due to their consistency and ease of incorporation (Clarke and Macrae, 2012). However, their nutritional contribution is limited, and their use does not address waste valorisation objectives, as they are derived from primary coffee products rather than by-products.
- Coffee silverskin – Coffee silverskin is a thin layer of skin released during the roasting of green coffee beans. Although it represents only 1–2% of the roasted bean mass, its large-scale generation and food-grade origin make it a beautiful ingredient for bakery applications.
- Chemically, silverskin is characterised by: a high total dietary fibre content (60–70%), predominantly insoluble fibre; moderate polyphenol levels (12–25 mg GAE/g); melanoidins formed during roasting, which contribute to antioxidant activity; and low fat and sugar content, enhancing formulation stability.

  From a technological standpoint, silverskin functions primarily as a fibre-enriching and texture-modifying ingredient, increasing water absorption and influencing crumb structure. Its roasted flavour profile also contributes subtle coffee-like notes, which can enhance sensory complexity when appropriately dosed (Navajas-Porras et al., 2025; Arias et al., 2023).
- Spent coffee grounds (SCG) – Spent coffee grounds are generated after the brewing process and represent the most significant coffee-related waste stream in urban and industrial environments. Despite having undergone aqueous extraction, spent coffee grounds retain a substantial proportion of the original chemical constituents of coffee beans. Their composition is characterised by high levels of dietary fibre, typically ranging between 45 and 55%, alongside

residual lipids in the order of 10–15%, including biologically active diterpenes. In addition, spent coffee grounds contain measurable amounts of proteins, generally reported between 12 and 15%, as well as residual polyphenols and melanoidins formed during roasting.

- Within bakery systems, spent coffee grounds function as multifunctional ingredients. Their high fibre content contributes to nutritional enrichment and modifies dough hydration and texture, while their dark colour and roasted flavour compounds enhance sensory complexity. The residual lipid fraction can partially substitute for shortening in specific formulations, particularly in low-moisture baked goods. Their polyphenolic compounds have been shown to delay lipid oxidation, thereby improving oxidative stability and shelf life in fat-rich products such as cookies and crackers (Makiso et al., 2023).

  Nevertheless, the application of spent coffee grounds in food products requires appropriate technological pre-treatment. Due to their high moisture content after brewing, drying and stabilisation steps are essential to prevent microbial growth and ensure product safety. Subsequent milling is necessary to achieve uniform particle size and consistent functional performance in bakery formulations. Although residual caffeine levels in spent coffee grounds are relatively low, typically between 0.2 and 0.6%, this factor may still require consideration in products intended for caffeine-sensitive consumer groups or specific population segments.
- Coffee pulp and coffee husk – Coffee pulp and coffee husk are generated during primary processing at origin and constitute the largest coffee-derived by-products by mass, accounting for approximately 40–50% of the fresh coffee cherry weight. These materials are characterised by a high content of carbohydrates, including pectins and hemicelluloses, alongside appreciable levels of phenolic compounds and organic acids. Owing to this composition, coffee pulp and husk exhibit significant nutritional and functional potential, particularly as sources of dietary fibre and bioactive compounds.

Despite these advantages, incorporating coffee pulp and husk into food systems presents greater challenges compared to other coffee-derived materials. Their composition may include anti-nutritional factors, and they are more susceptible to microbial contamination and mycotoxin formation, especially when drying and storage conditions are inadequate. In addition, substantial variability in chemical composition arises from differences in processing methods, cultivars, and geographic origins. These factors complicate standardisation and raise safety concerns that must be carefully addressed before food application.

As a consequence, the utilisation of coffee pulp and husk in bakery products typically necessitates targeted pre-treatment strategies, such as controlled fermentation, washing or thermal processing, to reduce anti-nutritional components and improve microbiological safety. Comprehensive risk assessment and regulatory approval are also required to ensure compliance with food safety standards. While these materials hold considerable promise, their adoption in food systems remains less widespread than that of coffee silverskin and spent coffee grounds, which benefit from more favourable safety profiles and established processing practices (Mussatto et al., 2011; Navajas-Porras et al., 2025).

From a functional standpoint, the classification of coffee-derived materials highlights the distinct technological roles they play within bakery formulations. Roasted coffee powders and extracts primarily contribute sensory attributes such as flavour and colour, whereas coffee silverskin functions mainly as a fibre-enriching ingredient with water-binding and antioxidant properties. Spent coffee grounds exhibit multifunctionality, providing dietary fibre, residual lipids, flavour-active compounds and enhanced oxidative stability. In contrast, coffee pulp and husk represent emerging ingredients with substantial functional potential but increased complexity related to safety and processing requirements.

This functional differentiation underscores the necessity for tailored processing and formulation strategies to maximise technological performance, sensory quality, and consumer safety. A clear understanding of these distinctions provides a critical foundation for the rational integration of coffee-derived materials into baked goods, supporting targeted innovation aligned with sustainability objectives and circular economy principles.

3. Chemical composition and functional potential

The chemical composition is the primary determinant of the technological, sensory, and nutritional functionality of coffee-derived materials, as the relative proportions of dietary fibre, polyphenols, lipids, and other bioactive compounds govern their behaviour in bakery systems and their contribution to product quality.

3.1 Dietary fibre

Dietary fibre represents the dominant macronutrient in most coffee-derived by-products and is a primary driver of their technological and nutritional functionality in baked goods. Among these materials, coffee silverskin exhibits the highest total dietary fibre content, typically ranging between 60 and 70% on a dry matter basis. This fibre fraction is predominantly insoluble and composed mainly of cellulose, hemicelluloses, and lignin, reflecting the structural role of silverskin in the coffee bean (Navajas-Porras et al., 2025). Spent coffee grounds contain slightly lower but still substantial fibre levels, generally reported to be between 45 and 55%. In contrast, coffee pulp and husk display more variable fibre contents, depending on cultivar, processing method, and degree of fermentation during primary processing.

From a functional perspective, the fibre fraction of coffee by-products exhibits a high water-binding capacity, often exceeding that of traditional cereal brans. Water-holding capacities of 3–5 g water per gram of dry fibre have been reported for coffee silverskin and spent coffee grounds, values that significantly influence dough hydration and mixing behaviour (Suanno et al., 2025). In bakery formulations, this property contributes to increased dough viscosity, modified gas cell stability, and altered crumb structure in the final product.

In yeast-leavened wheat-based systems, the high concentration of insoluble fibre can interfere with gluten network development through both physical dilution and competitive water absorption. This interaction often results in reduced dough extensibility, lower gas retention, and decreased loaf volume if formulation adjustments are not implemented. Consequently, optimisation strategies such as increased hydration, extended mixing times, or supplementation with vital gluten are commonly required to maintain technological performance when coffee-derived fibres are incorporated.

In chemically leavened products such as cakes, muffins, and biscuits, the effects of coffee-derived fibre differ markedly. In these systems, where gluten development is less critical, fibre-induced increases in batter viscosity can enhance air incorporation and stabilisation, contributing to improved crumb uniformity and textural stability. In low-moisture baked goods, such as crackers and cookies, fibre enrichment frequently leads to increased crispness and structural strength, attributes that are often perceived positively by consumers.

Beyond technological functionality, dietary fibre from coffee by-products contributes to the nutritional profile of baked goods by increasing total fibre intake and promoting satiety. Emerging in vitro digestion studies suggest that a substantial fraction of coffee-derived fibre resists upper gastrointestinal digestion and may exert prebiotic effects in the colon, particularly through interactions with melanoidins and associated polyphenols (LIczbiński and Bukowska, 2022). These characteristics position coffee by-products as valuable ingredients for the development of fibre-enriched and functionally enhanced bakery products.

3.2 Polyphenols and antioxidant activity

Polyphenolic compounds represent one of the most biologically and technologically significant fractions of coffee-derived materials. Coffee beans are naturally rich in phenolic acids, with chlorogenic acids (CGAs)—including caffeoylquinic, feruloylquinic and dicaffeoylquinic acids—constituting the dominant class. Although the roasting and subsequent brewing processes induce partial degradation and transformation of these compounds, coffee by-products, such as coffee silverskin and spent coffee grounds, retain substantial levels of polyphenols and their thermally derived products.

Recent analytical studies employing LC–MS/MS and spectrophotometric methods report total polyphenol contents ranging from approximately 12 to 35 mg gallic acid equivalents (GAE) per gram of dry matter in spent coffee grounds and silverskin, depending on roast degree, extraction conditions and post-processing treatments (Arias et al., 2023; Andrade et al., 2022; Fernandez-Rosillo et al., 2025). In addition to native CGAs, these materials contain phenolic degradation products, melanoidin-bound polyphenols, and low-molecular-weight antioxidant compounds formed during roasting, all of which contribute to overall antioxidant capacity.

From a technological perspective, polyphenols in coffee-derived materials play a key role in improving the oxidative stability of baked goods. Their radical-scavenging activity can delay lipid oxidation, particularly in fat-rich products such as cookies, crackers, and cakes, where oxidative rancidity is a significant factor limiting shelf life. Experimental baking trials have demonstrated that incorporating spent coffee grounds at levels of 5–10% can reduce peroxide formation and secondary oxidation products by 20–30% during storage compared to control formulations (Silva et al., 2021).

However, the behaviour of polyphenols during baking is influenced by complex interactions with other food components. During thermal processing, polyphenols may interact with proteins, starches, and dietary fibres through hydrogen bonding and covalent associations, which can reduce their extractability but do not necessarily eliminate their biological activity. Indeed, emerging evidence suggests that polyphenols bound to the food matrix or incorporated into melanoidin structures may exhibit delayed release during digestion, contributing to sustained antioxidant effects in the gastrointestinal tract (LIczbiński and Bukowska, 2022).

From a nutritional standpoint, the presence of polyphenols enhances the functional profile of coffee-enriched baked goods. In vitro digestion models indicate that a significant proportion of these compounds remains bioaccessible after baking and gastrointestinal simulation, supporting their potential contribution to antioxidant intake. Nevertheless, the extent of polyphenol retention and bioactivity is highly formulation-dependent, underscoring the importance of optimised processing conditions and ingredient interactions.

Overall, the polyphenolic fraction of coffee-derived materials represents a multifunctional component that simultaneously contributes to oxidative stability, sensory depth, and potential health benefits in baked goods, reinforcing the value of these ingredients within sustainable and functional bakery product development.

3.3 Lipids and flavour-active compounds

In addition to dietary fibre and polyphenols, coffee-derived materials—particularly spent coffee grounds—contain a significant lipid fraction that contributes both to technological functionality and sensory quality in baked goods. Spent coffee grounds typically retain between 10% and 15% lipids on a dry matter basis, depending on the coffee species, roast degree, and brewing method. This lipid fraction is composed primarily of triacylglycerols, along with minor components such as sterols and diterpenes, including cafestol and kahweol (Makiso et al., 2023).

From a technological perspective, these residual lipids influence several quality attributes in bakery products. They contribute to crumb softness and mouthfeel by interfering with starch retrogradation and gluten aggregation, thereby slowing the firming of the crumb during storage. In low-moisture baked goods, such as biscuits and crackers, coffee-derived lipids can partially replace conventional shortening, contributing to crispness and structural integrity while reducing the need for added fats. Their presence also enhances aroma retention by acting as a reservoir for volatile compounds, moderating their release during baking and consumption.

Coffee by-products also retain a diverse array of flavour-active compounds formed during the roasting process. Volatile classes such as pyrazines, furans, pyrroles, and phenolic derivatives are particularly relevant to bakery applications, as they impart characteristic roasted, cocoa-like, nutty, and caramel notes. These compounds persist through baking due to their thermal stability and interaction with the food matrix, enabling flavour differentiation without reliance on synthetic flavour additives (Genovese, 2025).

The interaction between lipids and flavour compounds is especially important in determining sensory outcomes. Lipophilic volatiles preferentially partition into the lipid phase, enhancing flavour persistence and reducing aroma loss during high-temperature baking. This mechanism helps explain why coffee-enriched bakery products often exhibit more profound and more complex flavour profiles compared with formulations relying solely on added coffee powders or extracts (Franca and Oliveira, 2022; Meerasri and Sothornvit, 2022).

Nevertheless, lipid oxidation poses a potential technological challenge, particularly in formulations with extended shelf lives. While polyphenols and melanoidins present in coffee-derived materials provide intrinsic antioxidant protection, careful control of processing, storage conditions, and formulation balance remains essential to prevent the development of rancid or off-flavours.

Overall, the lipid and flavour-active fractions of coffee-derived materials play a crucial role in enhancing mouthfeel, aroma complexity, and sensory differentiation in baked goods. Their multifunctional contribution supports the development of clean-label bakery products with distinctive flavour profiles and reduced dependence on synthetic additives.

3.4 Quantitative composition of coffee by-products relevant to bakery applications

Beyond qualitative descriptions, quantitative characterisation of coffee by-products is essential for their rational and reproducible use in baked goods formulations. Precise knowledge of macronutrient composition, fibre fractions, and bioactive compound levels allows formulators to predict technological behaviour, optimise hydration, and balance sensory outcomes. Table 1 summarises representative compositional ranges reported in recent studies published between 2023 and 2025.

Coffee silverskin exhibits an exceptionally high total dietary fibre content, typically ranging from 60 to 70% on a dry matter basis. This fibre fraction is dominated by insoluble polysaccharides, including cellulose, hemicelluloses, and lignin, while the soluble fibre fraction generally represents only 3–7%. Such composition confers high water-binding capacity and structural rigidity, explaining the pronounced effects of silverskin on dough hydration, viscosity, and crumb firmness in bakery products (Andrade et al., 2022; Lee et al., 2023; Zhao et al., 2023; Navajas-Porras et al., 2025).

Spent coffee grounds display a more balanced macronutrient profile. Dietary fibre content typically ranges between 45 and 55%, accompanied by a substantial lipid fraction of approximately 10–15% and residual protein levels between 12 and 15%. This combination of fibre, lipids, and proteins underpins the multifunctional behaviour of spent coffee grounds in bakery systems, where they simultaneously contribute to texture modification, flavour development, and nutritional enrichment (Arias et al., 2023). Compared with conventional cereal brans, spent coffee grounds often provide higher lipid-associated flavour retention and enhanced antioxidant protection.

The polyphenol content varies considerably depending on the coffee species, roasting intensity, and brewing conditions. Light-to-medium roasted spent coffee grounds retain the highest levels of extractable polyphenols, commonly reported in the range of 18–35 mg gallic acid equivalents (GAE) per gram of dry matter. Coffee silverskin typically exhibits slightly lower but still significant polyphenol

Table 1: Chemical composition of major coffee by-products relevant to bakery applications

Coffee-derived material	Total dietary fibre (%)	Lipids (%)	Proteins (%)	Polyphenols (mg GAE/g d.m.)	Residual caffeine (%)	Key functional implications in baked goods
Coffee silverskin	60–70	2–4	15–18	12–25	<0.1	Fibre enrichment, water binding, antioxidant activity
Spent coffee grounds	45–55	10–15	12–15	18–35	0.2–0.6	Fibre, partial fat replacement, flavour, shelf-life
Coffee pulp	30–45	2–6	8–12	20–40	0.8–1.2	Carbohydrate source, antioxidant potential
Coffee husk	35–50	1–3	7–10	15–30	0.6–1.0	Fibre enrichment requires safety pre-treatment

Sources: Mussatto et al. (2011); Navajas-Porras et al. (2025); Silva et al. (2021); Arias et al. (2023).

concentrations, generally ranging from 12 to 25 mg GAE/g. These values are comparable to, or exceed, those reported for widely used fibre-rich ingredients such as wheat bran or oat bran, positioning coffee by-products as competitive functional ingredients from both technological and nutritional perspectives (Silva et al., 2021).

Residual caffeine content represents an additional compositional parameter of relevance for food formulation. In spent coffee grounds, caffeine levels are generally low, typically ranging from 0.2 to 0.6% on a dry matter basis, as a substantial proportion of the caffeine is extracted during brewing. While these concentrations are unlikely to pose a safety concern for the general population, they may influence formulation decisions for products intended for children, pregnant women, or caffeine-sensitive consumers, necessitating appropriate labelling and dosage control.

The quantitative composition of coffee by-products clearly demonstrates their suitability for use as multifunctional ingredients in bakery formulations (Table 1). High fibre content supports nutritional claims, residual lipids and proteins contribute to textural and sensory quality, and polyphenols enhance oxidative stability. Such compositional profiles provide a robust scientific basis for the targeted incorporation of coffee-derived materials into baked goods within sustainable and clean-label product development strategies.

4. Processing and stabilisation for bakery applications

The successful utilisation of coffee and coffee-derived by-products in baked goods depends not only on their chemical composition but also on the use of appropriate processing and stabilisation strategies. Fresh coffee by-products are inherently unstable due to their high moisture content, residual enzymatic activity, and susceptibility to microbial growth. Without adequate pre-treatment, these factors can compromise food safety, technological performance, and sensory quality. Consequently, processing and stabilisation represent critical steps in transforming coffee residues into safe, standardised, and functional ingredients suitable for bakery applications.

From a technological standpoint, stabilisation aims to achieve three primary objectives: microbiological safety, compositional consistency, and functional reliability. These objectives must be met while preserving desirable bioactive compounds, minimising nutrient losses, and maintaining compatibility with industrial bakery processes. The selection of processing methods is therefore guided by both safety considerations and targeted functional outcomes.

4.1 Drying and microbial safety

Fresh coffee by-products are inherently highly perishable due to their elevated moisture content, which typically ranges from 60 to 75% immediately after generation, particularly in the case of spent coffee grounds and coffee pulp. Such conditions create a favourable environment for microbial growth, enzymatic activity, and rapid quality deterioration. Drying is therefore a critical stabilisation step to ensure microbial safety, extend shelf-life, and enable the subsequent use of coffee-derived materials as food-grade ingredients in bakery applications.

Conventional hot-air drying remains the most widely applied method owing to its operational simplicity, scalability, and relatively low capital cost. Drying temperatures between 60 and 90°C are commonly employed to reduce moisture content to below 10–12%, a level generally considered safe for long-term storage. However, prolonged exposure to high temperatures may promote lipid oxidation and partial degradation of thermolabile polyphenols, particularly in spent coffee grounds with higher residual fat content (Klingel et al., 2020; Viegas et al., 2022).

To address these limitations, alternative drying technologies have been investigated. Infrared drying has attracted increasing attention due to its ability to achieve rapid moisture removal with improved energy efficiency and reduced thermal gradients within the material. Studies report that infrared-assisted drying can shorten drying times by up to 30–40% while preserving higher levels of antioxidant activity compared with conventional hot-air drying (Arias et al., 2023). Freeze-drying offers superior retention of bioactive compounds and volatile flavour constituents, but its high energy demand and processing cost currently restrict its use to high-value or niche applications.

In addition to moisture reduction, thermal treatments play a crucial role in mitigating microbiological hazards. Drying processes significantly reduce bacterial and fungal loads and can limit the formation of mycotoxins associated with mould growth, particularly in coffee pulp and husk. When appropriately validated, thermal stabilisation enables compliance with food safety requirements and supports regulatory approval for food use. Nevertheless, drying must be carefully controlled to avoid post-process contamination, highlighting the importance of hygienic handling, packaging, and storage conditions.

Overall, drying serves as a foundational step in the safe valorisation of coffee by-products for bakery applications. The choice of drying method involves a balance between microbial safety, preservation of functional and bioactive compounds, energy efficiency, and economic feasibility, all of which must be considered when designing industrial-scale processing strategies.

4.2 Milling and particle size control

Following drying and stabilisation, milling represents a critical processing step that determines the functional performance of coffee-derived ingredients in bakery systems. Particle size distribution directly influences water absorption capacity, dough rheology, gas retention, crumb structure, and sensory perception, as consistently reported for fibre-rich plant by-products incorporated into wheat-based matrices (Suanno et al., 2025; Navajas-Porras et al., 2025). Inadequate control of particle size may therefore result in technological challenges such as poor dough handling, uneven baking, reduced loaf volume, and undesirable textural attributes.

Coffee by-products such as silverskin and spent coffee grounds are characterised by rigid, lignocellulosic structures that require sufficient mechanical energy for effective size reduction. Conventional hammer mills, pin mills, and roller mills are commonly used at an industrial scale; however, the selection of milling technology must balance size reduction efficiency with heat generation. Excessive frictional heating during milling can accelerate lipid oxidation, increase peroxide values, and degrade thermolabile bioactive compounds, particularly polyphenols and volatile aroma constituents (Arias et al., 2023).

Research indicates that finely milled powders with particle sizes below approximately 250 μm are generally most suitable for incorporation into wheat-based baked goods, including bread, cakes and biscuits. Within this size range, coffee-derived particles are more uniformly dispersed throughout the dough matrix, promoting consistent hydration behaviour and reducing the risk of gritty mouthfeel (Suanno et al., 2025). Conversely, coarse particles exceeding 500 μm may disrupt gluten network formation by acting as physical discontinuities, creating localised stress points that impair gas retention during fermentation and baking, ultimately reducing loaf volume and crumb uniformity (Cavanagh et al., 2023; Lee et al., 2023; Navajas-Porras et al., 2025).

Particle size also strongly affects water-binding behaviour and dough rheology. Finer particles exhibit increased specific surface area and enhanced hydration capacity, leading to greater competition for water between fibre fractions and gluten proteins. While this effect may be advantageous for improving moisture retention, crumb softness, and shelf-life extension, excessive fineness can increase dough viscosity, reduce extensibility, and require formulation adjustments, such as increased water addition, modified mixing times, or the use of dough conditioners (Suanno et al., 2025).

Advanced milling approaches, including cryogenic milling, have been explored to mitigate thermal damage during the grinding process. By maintaining sub-ambient processing temperatures, cryogenic milling limits lipid oxidation and preserves volatile aroma compounds and polyphenols more effectively than conventional milling. Nevertheless, the high operational cost, increased energy demand, and technical complexity of cryogenic systems currently restrict their application to high-value or speciality bakery products rather than large-scale commodity production (Arias et al., 2023).

In practical formulation scenarios, optimal particle size selection depends on the targeted bakery application and desired sensory profile. Fine powders are preferred for cakes, cookies, and biscuits, where smooth texture and uniform crumb are essential quality attributes. Slightly coarser fractions may be acceptable or even desirable in fibre-enriched breads, crackers, and artisanal products, where a more rustic texture is tolerated by consumers (Navajas-Porras et al., 2025). Regardless of application, standardisation of particle size distribution is essential to ensure reproducible functional performance, predictable water absorption, and consistent sensory quality across production batches.

Overall, controlled milling and particle size optimisation are indispensable for the successful integration of coffee-derived materials into bakery formulations. By aligning particle characteristics with product-specific requirements and processing conditions, manufacturers can maximise technological functionality, maintain sensory acceptability, and ensure consistent product quality while exploiting the sustainability potential of coffee by-products.

4.3 Rheological effects of coffee-derived ingredients on dough systems

The incorporation of coffee-derived ingredients into bakery formulations induces significant modifications in dough rheology, primarily through interactions with water, gluten proteins, and starch components. These effects are primarily governed by the high dietary fibre content, particle morphology, and surface chemistry of coffee by-products, which alter hydration dynamics, mixing behaviour, and viscoelastic properties. Understanding these rheological changes is essential for optimising dough handling, fermentation performance, and final product quality.

From a rheological perspective, coffee-derived materials act as non-gluten, water-active fillers within the dough matrix. Their presence modifies the balance between elastic and viscous components, influencing dough development time, stability, and resistance to deformation. Such effects have been widely reported for fibre-rich by-products and are particularly pronounced when coffee ingredients are incorporated at levels exceeding 5–10% flour substitution (Suanno et al., 2025; Navajas-Porras et al., 2025).

4.3.1 Water absorption and dough development

Water absorption is one of the most immediate and influential rheological consequences of incorporating coffee-derived ingredients into dough systems. Coffee silverskin and spent coffee grounds exhibit high water-binding capacity due to their elevated content of insoluble polysaccharides, porous microstructure, and large specific surface area. Studies consistently report increases in farinograph water absorption of 3–8% for each 5% substitution of wheat flour with coffee by-products, depending on particle size and fibre composition (Navajas-Porras et al., 2025).

This increased water demand directly affects dough development and mixing behaviour. Higher hydration levels are required to achieve optimal dough consistency, which prolongs dough development time and alters energy input during mixing. In farinograph and mixograph analyses, the addition of coffee-derived fibres has been associated with delayed dough peak development and reduced dough stability, reflecting competition for water between gluten proteins and fibre fractions (Suanno et al., 2025).

Spent coffee grounds, which contain residual lipids in addition to fibre, may partially offset excessive water absorption by contributing a plasticising effect. This lipid fraction can improve dough extensibility and reduce mixing resistance at moderate inclusion levels (≤5%), particularly in cookie and cracker formulations where gluten development is less critical (Arias et al., 2023). In contrast, coffee silverskin, characterised by a highly lignified fibre matrix and minimal lipid content, exerts a more substantial water-binding effect and more pronounced dilution of the gluten network.

At higher inclusion levels, excessive water absorption may result in overly viscous doughs with reduced extensibility, negatively impacting gas retention during fermentation and limiting loaf volume in yeast-leavened products. Consequently, formulation adjustments are often required, including increased water addition, modified mixing times, or the use of dough conditioners and emulsifiers to restore dough balance (Martinez-Saez et al., 2017; Trà et al., 2021; Ahanchi et al., 2024; Suanno et al., 2025).

Significantly, the impact of coffee-derived ingredients on water absorption and dough development is strongly influenced by particle size. Finely milled fractions (<250 μm) exhibit greater hydration capacity and more uniform dispersion, while coarser particles tend to create localised hydration gradients that impair dough homogeneity (Navajas-Porras et al., 2025). These observations highlight the necessity of integrating milling optimisation with rheological assessment during product development.

Overall, water absorption behaviour represents a critical control point in the formulation of bakery products enriched with coffee-derived ingredients. Precise adjustment of hydration and mixing parameters is essential to ensure acceptable dough handling properties and consistent product quality.

4.3.2 Gluten dilution and gas retention

One of the most critical rheological consequences of incorporating coffee-derived ingredients into wheat-based dough systems is the dilution and disruption of the gluten network. Coffee by-products are inherently gluten-free and function as inert or semi-active fillers within the dough matrix, replacing a fraction of gluten-forming proteins when used as partial flour substitutes. This substitution effect directly influences gas retention, dough elasticity, and final product volume.

Gluten dilution occurs when wheat flour is partially replaced by fibre-rich coffee materials, reducing the overall concentration of glutenin and gliadin proteins available for network formation. Studies on fibre-enriched dough systems demonstrate that substitution levels above 5–7% typically lead to measurable reductions in gluten network continuity, as evidenced by decreased dough extensibility and elasticity in alveograph and dynamic rheological measurements (Suanno et al., 2025). In the case of coffee silverskin, the effect is particularly pronounced due to its high proportion of rigid, insoluble fibre and negligible lipid content.

Beyond simple dilution, physical interference further compromises gluten development. Coarse coffee-derived particles may act as mechanical discontinuities within the dough, interrupting gluten strand alignment and weakening the viscoelastic matrix. This disruption reduces the dough's ability to trap carbon dioxide produced during yeast fermentation, resulting in diminished gas retention and a reduced loaf volume (Navajas-Porras et al., 2025).

Spent coffee grounds may exert a slightly less detrimental effect on gluten functionality compared to silverskin, owing to their residual lipid fraction. These lipids can partially plasticise the dough matrix, improving extensibility and compensating for some loss of elasticity at low to moderate inclusion levels (≤5%). Nevertheless, at higher substitution rates, gluten dilution remains the dominant factor governing dough performance (Arias et al., 2023).

Reduced gas retention manifests during proofing and baking as a lower specific volume, increased crumb density, and an irregular cell structure. Rheological studies report a shift toward more viscous-dominated behaviour (lower storage modulus-to-loss modulus ratios), indicating reduced elastic recovery and weaker gas-holding capacity in fibre-enriched doughs (Suanno et al., 2025).

To mitigate the effects of gluten dilution, several formulation strategies can be employed. These include limiting inclusion levels, optimising particle size distribution, increasing dough hydration, and incorporating functional additives such as emulsifiers (e.g., DATEM), hydrocolloids, or vital wheat gluten. Such approaches can partially restore gas retention and improve loaf volume without negating the nutritional and sustainability benefits of coffee-derived ingredients.

In summary, gluten dilution and impaired gas retention represent key technological challenges in the formulation of bakery products enriched with coffee by-products. A clear understanding of these mechanisms is essential for balancing nutritional enhancement with acceptable dough performance and product quality.

4.3.3 Dough viscoelasticity and texture optimisation

Dough viscoelasticity represents a critical determinant of processability, gas retention, and final product texture in bakery systems. The incorporation of coffee-derived ingredients modifies the balance between elastic (solid-like) and viscous (liquid-like) behaviour through combined effects of gluten dilution, fibre–water interactions, and particle-induced structural heterogeneity. Understanding and optimising these changes is essential for achieving acceptable texture in coffee-enriched baked goods.

Dynamic rheological measurements consistently show that the addition of fibre-rich coffee by-products increases dough stiffness, reflected by elevated storage modulus (G'), while simultaneously reducing extensibility and elastic recovery at higher substitution levels (Suanno et al., 2025). This behaviour is primarily attributed to the high content of insoluble dietary fibre, which acts as a rigid filler within the dough matrix, restricting gluten network mobility and limiting strain tolerance during fermentation and baking.

At low inclusion levels (≤3–5%), coffee-derived ingredients may contribute positively to dough structure by increasing viscosity and improving shape retention, particularly in cookies, crackers, and laminated products where excessive elasticity is undesirable. In these systems, increased viscous behaviour (higher loss modulus, G'')

enhances dough handling and limits excessive spread during baking, contributing to improved dimensional stability (Navajas-Porras et al., 2025).

As substitution levels increase beyond 5–7%, however, viscoelastic imbalance becomes more pronounced. Excessive fibre content leads to a disproportionate increase in dough rigidity, reduced extensibility, and impaired gas cell expansion. This results in dense crumb structures, reduced loaf volume, and firmer texture in yeast-leavened products (Suanno et al., 2025). Spent coffee grounds generally exert a milder stiffening effect than coffee silverskin due to their residual lipid fraction, which contributes a plasticising influence and partially offsets fibre-induced rigidity (Arias et al., 2023).

Texture optimisation strategies aim to restore a favourable viscoelastic balance while maintaining nutritional and sustainability benefits. Key approaches include:

- **Optimising inclusion level**, typically maintaining coffee-derived ingredients at 3–5% flour substitution for yeast-leavened products and up to 10% for cookies and crackers;
- **Particle size reduction**, as finer particles (<250 μm) distribute more evenly within the dough matrix and reduce stress concentration points;
- **Hydration adjustment**, increasing water addition to compensate for fibre water-binding and restore dough extensibility;
- **Use of functional additives**, such as emulsifiers, hydrocolloids, or enzymes, to enhance gluten network functionality and gas retention.

Rheological optimisation has been shown to improve textural outcomes significantly. Studies report that targeted adjustments can restore up to 80–90% of the specific volume and softness of control formulations, even at moderate inclusion levels of coffee by-products (Navajas-Porras et al., 2025). These findings underscore the feasibility of producing high-quality bakery products enriched with coffee-derived ingredients when rheological principles are carefully applied (Koay et al., 2023; Oliveira Batista et al., 2023).

In summary, coffee-derived ingredients exert predictable and manageable effects on dough viscoelasticity. Through informed formulation and processing adjustments, their impact on texture can be optimised, enabling the development of bakery products that combine enhanced nutritional value, sustainability, and consumer-acceptable texture.

5. Formulation strategies and bakery product development

The successful integration of coffee and coffee-derived by-products into baked goods requires carefully designed formulation strategies that balance technological functionality, sensory quality, nutritional enhancement, and processing feasibility. Unlike conventional bakery ingredients, coffee-derived materials exhibit high dietary fibre content, complex flavour profiles, and pronounced water-binding capacity, necessitating targeted adjustments in formulation and process parameters. This section outlines practical strategies for incorporating coffee-based ingredients into a range of bakery products, from bread to biscuits, while maintaining product quality and consumer acceptance.

5.1 Selection of coffee-derived ingredients according to product category

Different bakery products exhibit distinct structural and sensory requirements, which strongly influence ingredient selection. Coffee-derived ingredients should therefore be chosen based on their functional compatibility with the target product.

Yeast-leavened breads are particularly sensitive to gluten dilution and gas retention loss. In these systems, finely milled spent coffee grounds or silverskin powders are best incorporated at low inclusion levels, typically not exceeding 3–5% flour substitution. At these levels, nutritional enrichment can be achieved without compromising loaf volume or crumb structure excessively (Suanno et al., 2025; Navajas-Porras et al., 2025).

In contrast, cookies, crackers, and biscuits tolerate higher levels of fibre and non-gluten materials. These products rely less on gluten network development and benefit from the flavour, colour, and textural contributions of coffee-derived ingredients. Inclusion levels of 5–10%, and in some cases up to 15%, have been reported without detrimental effects on product quality, particularly when particle size and hydration are appropriately controlled (Arias et al., 2023).

Cakes and muffins represent an intermediate category, where aeration and batter viscosity are critical. Here, coffee extracts or finely milled powders are preferred to minimise structural disruption, while providing flavour differentiation and antioxidant enrichment.

5.2 Optimisation of hydration and mixing parameters

One of the most critical formulation adjustments involves water management. Due to their high fibre content and porous structure, coffee-derived ingredients significantly increase water absorption. Failure to compensate for this effect results in stiff doughs, incomplete hydration, and reduced product volume.

Empirical studies indicate that water addition typically needs to be increased by 2–8% for every 5% replacement of wheat flour with coffee by-products, depending on particle size and fibre composition (Navajas-Porras et al., 2025). Mixing time may also need to be extended to ensure adequate hydration and dispersion of coffee-derived particles.

In batter-based systems, such as cakes, increased liquid content and modified emulsification are often necessary to maintain proper viscosity and air incorporation.

5.3 Management of gluten dilution and structure

Gluten dilution remains one of the principal technological challenges when incorporating coffee-derived ingredients into wheat-based products. Several formulation strategies can be employed to mitigate this effect:

- Limiting inclusion levels, particularly in yeast-leavened products;
- Using finer particle sizes to reduce physical disruption of the gluten network;

- Supplementing formulations with vital wheat gluten when higher fibre enrichment is desired;
- Incorporating emulsifiers or hydrocolloids to enhance gas retention and dough elasticity.

Such approaches have been shown to restore up to 80–90% of the specific volume of control bread formulations at moderate inclusion levels (Suanno et al., 2025).

5.4 Quantified formulation guidelines for bakery products

To facilitate practical implementation, Table 2 presents recommended inclusion ranges, hydration adjustments, and expected technological effects for major bakery product categories. These guidelines are derived from recent experimental studies and provide a framework for rational product development.

In yeast-leavened breads, coffee by-products are most effective at 3–5% substitution, where fibre enrichment and flavour enhancement are achieved with minimal volume loss. In cookies and crackers, inclusion levels of up to 10% allow substantial fibre and antioxidant enrichment while contributing desirable roasted and cocoa-like flavour notes. Cakes and muffins typically tolerate 2–6% inclusion, provided batter viscosity and aeration are carefully controlled (Arias et al., 2023).

Table 2: Recommended inclusion levels of coffee-derived ingredients in selected bakery products

Bakery product category	Coffee-derived ingredient	Typical inclusion level (% flour basis)	Technological impact	Sensory impact
Wheat bread	Coffee silverskin	2–5	Increased water absorption, reduced loaf volume at high levels	Mild roasted notes, darker crumb
Cookies	Spent coffee grounds	3–8	Improved dough stability, fat-mimicking effect	Chocolate–coffee flavour, increased bitterness at >8%
Cakes	Coffee extract/fine SCG	2–6	Moisture retention, softer crumb	Enhanced aroma, colour intensification
Crackers	Silverskin or SCG	5–10	Increased crispness, reduced spread	Toasted, nutty flavour
Gluten-free bakery	SCG (fine)	3–7	Improved structure, water binding	Balanced bitterness, cocoa-like notes

5.5 Sensory optimisation and flavour integration

Coffee-derived ingredients contribute distinctive roasted, bitter, and cocoa-like notes that can enhance product differentiation when properly balanced. Sensory optimisation strategies include pairing coffee ingredients with complementary flavours such as cocoa, nuts, spices, or caramel, and adjusting sugar levels to counterbalance bitterness.

The roasting degree of the source coffee and the processing of the by-products significantly influence flavour intensity. Light-to-medium roasted materials provide milder, more aromatic profiles, whereas darker roasts contribute stronger bitterness and colour. Selection should therefore align with the desired sensory profile of the final product.

5.6 Nutritional positioning and clean-label considerations

From a nutritional standpoint, coffee-derived ingredients enable fibre enrichment, antioxidant enhancement, and reduced reliance on artificial colourants or flavourings. Their use supports clean-label and sustainability-driven product positioning, which consumers increasingly value.

However, residual caffeine content, particularly in spent coffee grounds, must be considered in product labelling and formulation, especially for products intended for children or sensitive populations. Regulatory compliance and transparent communication are essential for successful market adoption.

5.7 Industrial scalability and product development pathways

From an industrial perspective, the adoption of coffee-derived ingredients requires consistent quality, reliable supply, and compatibility with existing production lines. Standardisation of particle size, moisture content, and compositional variability is essential to ensure reproducible performance.

Pilot-scale trials indicate that coffee by-products can be integrated into conventional bakery processes with minimal equipment modification, provided formulation adjustments are made. Their utilisation aligns strongly with circular economy principles, offering both environmental and economic benefits through waste valorisation (Navajas-Porras et al., 2025; Arias et al., 2023).

In conclusion, formulation strategies for coffee-enriched bakery products must be product-specific, data-driven, and grounded in an understanding of rheological and sensory interactions. When appropriately designed, such strategies enable the development of innovative baked goods that combine sustainability, nutritional value and novel flavour profiles.

6. Sensory evaluation and consumer acceptance

Sensory evaluation represents a critical interface between technological feasibility and market success in the development of bakery products enriched with coffee and coffee-derived by-products. While formulation strategies may achieve desirable nutritional and sustainability outcomes, consumer acceptance ultimately determines the commercial viability of such innovations. The integration of coffee-derived ingredients introduces complex sensory changes that affect appearance, aroma, flavour, texture, and aftertaste, necessitating systematic sensory assessment throughout product development.

6.1 Sensory attributes influenced by coffee-derived ingredients

Coffee and coffee by-products impart distinctive sensory characteristics to baked goods. The most immediately perceptible change is colour, as coffee-derived materials introduce brown to dark-brown hues that intensify with increasing inclusion level and roasting degree. In many bakery products, a darker crumb and crust colour is perceived positively and associated with whole-grain, cocoa-rich, or indulgent profiles, particularly in cookies, brownies, and cakes.

Aroma and flavour are strongly influenced by volatile compounds, such as pyrazines, furans, and phenolic derivatives, which contribute roasted, cocoa-like, nutty, and caramel notes. At moderate inclusion levels, these attributes enhance product differentiation and complexity. However, excessive addition may introduce bitterness, astringency, or burnt notes, particularly when darker-roasted materials or coarsely milled fractions are used (Arias et al., 2023).

Texture perception is also affected by coffee-derived ingredients. A high fibre content can increase firmness, reduce perceived softness, and modify mouthfeel. In cookies and crackers, this effect may be desirable, contributing to crispness, whereas in bread and cakes, it may negatively affect consumer liking if not correctly managed through formulation optimisation (Silva et al., 2021; Koay et al., 2023; Ahmed et al., 2023; Suanno et al., 2025).

6.2 Descriptive sensory analysis and trained panels

Descriptive sensory analysis using trained panels is essential for characterising the specific sensory changes induced by coffee-derived ingredients. Standardised protocols, including quantitative descriptive analysis (QDA), enable the identification and quantification of key attributes such as roasted aroma intensity, bitterness, sweetness balance, crumb firmness and aftertaste persistence.

Studies using trained sensory panels report strong correlations between the inclusion level of coffee by-products and increases in roasted aroma intensity and bitterness, alongside decreases in perceived sweetness at higher substitution rates (greater than 7–10%). Such findings support the need for precise formulation control to maintain sensory balance (Navajas-Porras et al., 2025).

Instrumental–sensory correlations further enhance understanding of these effects. For example, polyphenol concentration and melanoidin content have been linked to bitterness and colour intensity, while fibre content correlates with firmness and reduced elasticity. These relationships allow predictive modelling of sensory outcomes based on compositional data.

6.3 Consumer acceptance and hedonic evaluation

Consumer studies provide insight into real-world acceptance of coffee-enriched bakery products. Hedonic testing consistently demonstrates that moderate inclusion levels of coffee-derived ingredients (typically 3–5% in bread and 5–10% in cookies and cakes) achieve the highest overall liking scores. At these levels, consumers perceive enhanced flavour complexity and novelty without excessive bitterness or textural penalties (Arias et al., 2023).

Importantly, consumer acceptance is strongly context-dependent. Products explicitly positioned as "coffee-flavoured", "cocoa-like" or "sustainably enriched" tend to receive higher acceptance, as consumer expectations align with the sensory profile. Conversely, unexpected coffee notes in traditionally neutral products may reduce liking unless appropriately communicated through product labelling and marketing.

Texture remains a critical driver of acceptance. Consumer rejection is more likely when coffee-derived fibres significantly increase hardness or dryness, highlighting the importance of rheological optimisation discussed in Section 5.

6.4 Cross-cultural and demographic considerations

Consumer perception of coffee flavour varies across cultures and demographic groups. In coffee-consuming cultures, roasted and bitter notes are often associated with sophistication and quality, whereas in other markets, bitterness may be less tolerated. Age, habitual coffee consumption, and familiarity with wholegrain or functional foods further influence acceptance.

Studies indicate that younger consumers and frequent coffee drinkers show higher willingness to accept coffee-enriched bakery products, particularly when sustainability and waste valorisation are emphasised (Nichifor et al., 2025). Such insights are valuable for targeted product positioning and market segmentation.

6.5 Sensory optimisation strategies

To maximise consumer acceptance, several sensory optimisation strategies may be employed:

- Limiting inclusion levels to ranges supported by sensory data;
- Selecting the appropriate roasting degree of coffee-derived materials to control bitterness.
- Pairing coffee ingredients with complementary flavours (e.g., cocoa, vanilla, nuts, spices);
- Adjusting sweetness and fat content to balance bitterness and mouthfeel;
- Using fine particle sizes to minimise gritty texture.

Iterative sensory testing, combining trained panel analysis with consumer feedback, is essential for refining formulations and ensuring market readiness.

6.6 Implications for product development

Sensory evaluation provides critical feedback that guides formulation adjustments and supports evidence-based product development. When sensory optimisation is integrated early in the innovation process, coffee-derived ingredients can be successfully incorporated into bakery products that meet both consumer expectations and sustainability goals.

In conclusion, sensory evaluation and consumer acceptance are decisive factors in the successful utilisation of coffee and coffee waste products in baked goods. Systematic sensory methodologies, combined with informed formulation strategies and clear consumer communication, enable the development of innovative bakery products that are both sensorially appealing and aligned with circular economy principles.

7. Nutritional, health, and sustainability implications

The incorporation of coffee and coffee-derived by-products into baked goods has implications that extend beyond technological performance and sensory quality. These ingredients significantly contribute to the nutritional profile of bakery products, offer potential health-related benefits, and align strongly with sustainability and circular economy objectives. As such, their utilisation reflects broader trends in food system transformation toward functional nutrition and resource efficiency.

7.1 Nutritional enhancement of bakery products

One of the most significant nutritional impacts of coffee-derived ingredients is the enrichment of dietary fibre. Conventional bakery products, particularly those made from refined wheat, are often characterised by a low fibre content. Coffee silverskin and spent coffee grounds, which contain between 45 and 70% total dietary fibre, provide an effective means of increasing fibre intake without significantly altering product format or consumer habits (Navajas-Porras et al., 2025).

Increased fibre content contributes to improved gastrointestinal function, enhanced satiety and modulation of postprandial glycaemic response. From a nutritional labelling perspective, even modest inclusion levels (3–5%) may allow fibre-related claims, depending on regulatory frameworks. In addition, coffee-derived ingredients contribute minerals and residual proteins, further improving the nutritional density of baked goods (Salazar-López et al., 2020; Bevilacqua et al., 2023; Ahmed et al., 2024).

Polyphenols and melanoidins present in coffee byproducts also contribute to the antioxidant capacity. Although thermal processing during baking reduces some phenolic content, studies demonstrate that a substantial proportion remains bioaccessible, supporting the nutritional relevance of these compounds in finished products (Arias et al., 2023).

7.2 Potential health-related benefits

Beyond basic nutrition, coffee-derived bioactive compounds have been linked to a range of health-related benefits. Chlorogenic acids and related polyphenols exhibit antioxidant and anti-inflammatory properties, which may contribute to reduced oxidative stress when consumed as part of a balanced diet. Melanoidins formed during roasting possess antioxidant activity and may act as fermentable substrates for beneficial gut microbiota, functioning similarly to dietary fibre (Navajas-Porras et al., 2025).

Residual caffeine levels in spent coffee grounds are generally low and unlikely to exert significant physiological effects at typical bakery inclusion rates. However, cumulative intake and consumer sensitivity should be considered, particularly for products intended for children, pregnant women, or individuals sensitive to caffeine. Transparent labelling and appropriate formulation control are therefore essential.

While bakery products should not be positioned as therapeutic foods, the incorporation of coffee-derived ingredients supports the development of products with enhanced functional value relative to conventional formulations.

7.3 Glycaemic response and metabolic considerations

The addition of high-fibre coffee by-products may favourably influence the glycaemic response of baked goods by slowing starch digestion and glucose absorption. Insoluble fibre fractions increase matrix density and reduce enzymatic accessibility of starch granules, potentially lowering postprandial blood glucose peaks. Such effects have been reported for fibre-enriched bakery products and are likely applicable to coffee-derived fibres, although further targeted clinical studies are required to substantiate specific claims (Suanno et al., 2025).

7.4 Sustainability and circular economy benefits

From a sustainability perspective, the utilisation of coffee by-products represents a clear example of waste valorisation within a circular bioeconomy framework. Coffee production generates millions of tonnes of by-products annually, much of which is underutilised or disposed of as waste, contributing to greenhouse gas emissions and environmental burden (Nichifor et al., 2025).

Redirecting these materials into food applications reduces waste streams, lowers demand for virgin raw materials, and enhances overall resource efficiency. Life cycle assessment studies indicate that upcycling food industry by-products into ingredients for human consumption offers greater environmental benefits than composting or energy recovery alone (Arias et al., 2023).

In addition, local utilisation of spent coffee grounds from urban brewing operations can reduce transportation-related emissions and support decentralised, sustainable ingredient supply chains.

7.5 Economic and social implications

The valorisation of coffee by-products also carries economic and social benefits. Creating higher-value food ingredients from waste streams can generate additional revenue for coffee processors, roasters, and food manufacturers. At the same time, it supports innovation in small and medium-sized enterprises and aligns with consumer demand for sustainable, ethically produced foods.

For consumers, products incorporating upcycled ingredients may enhance perceived value and brand trust, particularly when sustainability credentials are clearly communicated (Nichifor et al., 2025).

7.6 Regulatory and communication considerations

Despite their potential benefits, coffee-derived ingredients must comply with food safety and regulatory requirements. Clear definitions, compositional consistency, and documented safety assessments are essential for market approval. In some regions, novel food regulations may apply, particularly for coffee pulp and husk, necessitating thorough risk assessment and regulatory engagement.

Effective communication of nutritional and sustainability benefits must be evidence-based and transparent to avoid misleading claims. When properly managed, such communication can enhance consumer acceptance and drive market success.

7.7 Overall implications for sustainable food systems

In summary, incorporating coffee and coffee waste products into baked goods offers multifaceted benefits that span nutrition, health, and sustainability. By enhancing fibre intake, introducing bioactive compounds, and reducing food system waste, coffee-derived ingredients contribute to more resilient and sustainable food systems. Their successful implementation, however, depends on careful formulation, regulatory compliance, and responsible communication.

As global food systems continue to seek solutions that reconcile environmental responsibility with consumer satisfaction, the utilisation of coffee by-products in bakery products represents a scientifically grounded and practically feasible pathway toward sustainable innovation.

8. Sustainability and circular economy implications

The utilisation of coffee and coffee-derived waste products in baked goods production exemplifies a practical and scalable application of circular economy principles within the agri-food sector. By transforming underutilised by-products into value-added food ingredients, this approach addresses multiple sustainability challenges simultaneously, including waste reduction, resource efficiency, climate impact mitigation, and economic resilience across the food value chain.

8.1 Coffee by-products as secondary raw materials

Coffee production and consumption generate substantial volumes of organic residues at different stages of the value chain, including pulp, husk, silverskin, and spent coffee grounds. Traditionally, these materials have been treated as waste or low-value inputs for composting or energy recovery, despite their rich biochemical composition. From a circular economy perspective, such by-products represent secondary raw materials with significant untapped potential for integration into the food system (Navajas-Porras et al., 2025).

Valorising coffee by-products for bakery applications shifts their role from waste to functional ingredients, thereby extending material life cycles and reducing reliance on primary agricultural resources. This approach aligns with the waste hierarchy principle, prioritising prevention and reuse over disposal (Lee et al., 2023; Zhao et al., 2023; Ahmed et al., 2024).

8.2 Environmental impact reduction

The diversion of coffee byproducts from waste streams to food applications yields measurable environmental benefits. Disposal of coffee residues in landfills or uncontrolled environments is associated with the release of methane and the generation of leachate. Incorporating food products reduces the volume of organic waste and associated greenhouse gas emissions (Nichifor et al., 2025).

Life cycle assessment (LCA) studies indicate that upcycling food processing by-products into human food ingredients typically results in lower environmental impact compared with composting or bioenergy production, particularly when the functional substitution of conventional ingredients (e.g., cereal fibres) is considered (Arias et al., 2023). In bakery systems, coffee-derived ingredients can partially replace refined flour, cocoa, or synthetic additives, further enhancing environmental efficiency.

8.3 Resource efficiency and water footprint considerations

Coffee cultivation and processing are resource-intensive, particularly in terms of water use during wet processing. While by-product utilisation does not directly reduce water consumption at origin, it improves overall resource efficiency by maximising the value extracted from existing inputs. Each additional food application derived from coffee residues increases the return on investment in environmental sustainability embedded in coffee production.

Moreover, the use of locally sourced spent coffee grounds from urban brewing operations can reduce transportation distances and associated emissions, supporting decentralised and low-impact ingredient supply chains.

8.4 Economic sustainability and value creation

From an economic perspective, integrating coffee by-products into bakery products creates new value streams for actors across the supply chain, including coffee processors, roasters, food manufacturers, and small enterprises. Upcycling initiatives can reduce waste management costs while generating marketable ingredients aligned with consumer demand for sustainable products.

For bakeries and food brands, the use of coffee-derived ingredients presents opportunities for differentiation through sustainability-driven innovation. Products incorporating upcycled materials often command higher perceived value, particularly when supported by transparent communication and credible sustainability claims (Nichifor et al., 2025).

8.5 Social and ethical dimensions

Circular utilisation of coffee by-products also has social implications. By enhancing the economic value of coffee residues, such practices may contribute indirectly to improved income stability in coffee-producing regions, particularly when by-products such as pulp and husk are processed locally. While this potential is still emerging, it aligns with broader goals of inclusive and sustainable agricultural development.

At the consumer level, products featuring upcycled ingredients support increased awareness of food waste issues and encourage more responsible consumption patterns.

8.6 Regulatory and policy alignment

The valorisation of food by-products aligns with international sustainability frameworks and policy initiatives, including the United Nations Sustainable Development Goals (notably SDG 12: Responsible Consumption and Production). In the European context, it is consistent with the Farm to Fork Strategy and circular bioeconomy action plans.

However, regulatory clarity remains essential. Food safety assessments, compositional standardisation, and traceability are prerequisites for market adoption. Clear regulatory pathways facilitate innovation while safeguarding consumer health and trust.

9. Regulatory and safety aspects

The integration of coffee and coffee-derived by-products into baked goods requires strict compliance with food safety and regulatory frameworks to ensure consumer protection and market legitimacy. Although the valorisation of agro-industrial by-products aligns with sustainability and circular economy strategies, coffee-derived materials must meet the same safety and quality standards as conventional food ingredients.

Food safety concerns vary depending on the type and origin of the coffee by-product. Spent coffee grounds and coffee silverskin generally present lower microbiological risk, as they originate from thermally processed materials. Nevertheless, these materials may contain residual moisture levels exceeding 60% immediately after generation, creating favourable conditions for microbial growth if drying is delayed. Studies report that reducing moisture content below 10–12% effectively inhibits microbial proliferation and ensures microbiological stability during storage (Arias et al., 2023). Regular monitoring of total aerobic plate counts, yeast and mould levels, and water activity is therefore essential.

Chemical safety considerations include the presence of heavy metals, pesticide residues, and mycotoxins. While roasted coffee and its by-products typically comply with established regulatory limits, variability in agricultural practices and post-harvest handling can influence contaminant levels. Coffee pulp and husk, in particular, may be more susceptible to mycotoxin contamination if drying and storage conditions are inadequate. As a result, these materials require validated pre-treatment strategies and comprehensive compositional testing before food use (Navajas-Porras et al., 2025).

Residual caffeine content represents an additional safety and labelling consideration. Spent coffee grounds typically contain between 0.2% and 0.6% caffeine on a dry matter basis, reflecting partial extraction during the brewing process. At typical bakery inclusion levels (3–10%), the resulting caffeine intake from baked goods remains low and is unlikely to pose health risks to the general population. However, cumulative exposure and consumer sensitivity necessitate transparent labelling and cautious formulation for products intended for children, pregnant women or caffeine-sensitive individuals (Klingel et al., 2020).

From a regulatory perspective, certain coffee by-products may be classified as novel foods in jurisdictions where evidence of significant historical consumption is limited. In the European Union, coffee silverskin and spent coffee grounds have increasingly been evaluated for food applications. In contrast, coffee pulp and husk are more likely to require novel food authorisation due to their limited history of use. Novel food approval processes typically require detailed dossiers that cover production methods, compositional analysis, toxicological assessment, and intended use levels, ensuring a high level of consumer protection.

Labelling and consumer communication are integral to regulatory compliance and acceptance. Coffee-derived ingredients must be clearly declared in ingredient lists, and any claims related to fibre enrichment, antioxidant content, or sustainability must be substantiated and aligned with authorised claim frameworks. Sustainability-related descriptors, such as "upcycled" or "circular", should be communicated carefully to avoid consumer misunderstanding, while maintaining transparency regarding ingredient origins.

Quality assurance and traceability systems are essential for managing batch-to-batch variability in composition, moisture content, and particle size of coffee-derived ingredients. As highlighted in Table 3, standardised sourcing, validated processing protocols, and documented quality control measures support regulatory compliance and reduce safety risks across the supply chain. Such systems are particularly critical for industrial-scale implementation, where consistency, reproducibility, and process reliability directly influence product performance and market acceptance. Overall, regulatory and safety considerations should not be viewed as barriers but rather as enabling frameworks for responsible innovation. When appropriately addressed through validated processing strategies, rigorous analytical testing, and transparent documentation, coffee-derived by-products can be safely and legally incorporated into baked goods, thereby supporting sustainability objectives while maintaining consumer confidence and regulatory alignment.

Table 3: Sustainability and safety considerations for coffee-derived ingredients used in baked goods

Aspect	Key considerations	Mitigation strategies
Microbial safety	High initial moisture, mould risk	Controlled drying (<10% moisture), thermal treatment
Chemical safety	Mycotoxins (pulp/husk), caffeine	Raw material screening, HACCP, regulatory compliance
Oxidative stability	Lipid oxidation in SCG	Low-temperature milling, antioxidant retention
Regulatory status	Novel food classification (EU)	Safety dossiers, EFSA evaluation
Environmental impact	Waste reduction, resource efficiency	Integration into circular economy models

10. Conclusions

The utilisation of coffee and coffee-derived waste products in baked goods production represents a scientifically robust and practically feasible approach to ingredient innovation within contemporary food systems. This chapter has demonstrated that coffee by-products—such as spent coffee grounds, coffee silverskin, and, to a lesser extent, coffee pulp and husk—possess compositional and functional properties that enable their effective incorporation into a wide range of bakery products.

From a technological perspective, the high dietary fibre content, residual lipids, and bioactive compounds of coffee-derived ingredients significantly influence dough rheology, water absorption, gluten network development, and texture formation. When these effects are properly understood and managed through optimised particle size, hydration control, and formulation strategies, coffee-derived materials can be integrated into bakery systems without compromising product quality. Quantified inclusion thresholds and rheological optimisation approaches provide a practical framework for product developers and manufacturers.

Sensory evaluation and consumer studies further indicate that moderate incorporation levels of coffee-derived ingredients enhance flavour complexity, colour, and perceived product differentiation, particularly in cookies, crackers, cakes, and speciality breads. Consumer acceptance is highest when sensory profiles align with product positioning and when sustainability attributes are clearly communicated. These findings underscore the importance of integrating sensory science into the formulation development process.

Nutritionally, coffee-derived ingredients contribute meaningful fibre enrichment and antioxidant capacity, supporting the development of bakery products with improved nutritional profiles relative to conventional formulations. While bakery products should not be positioned as functional or therapeutic foods, the inclusion of coffee by-products aligns with broader public health goals related to fibre intake and dietary diversity.

From an environmental and socio-economic standpoint, the valorisation of coffee by-products exemplifies the principles of a circular economy in action. Redirecting these materials from waste streams to food applications reduces environmental burden, improves resource efficiency, and creates new value streams across the coffee and bakery sectors. Such practices contribute to climate mitigation efforts and support the transition toward more sustainable food systems.

Regulatory and safety considerations remain central to successful implementation. Compliance with food safety standards, appropriate processing, and transparent labelling are essential to ensure consumer trust and market acceptance. While certain coffee by-products may require regulatory approval as novel foods, clear pathways exist to support responsible innovation.

In conclusion, integrating coffee and coffee waste products into baked goods presents a compelling opportunity to combine sustainability, nutritional enhancement, and sensory innovation. Continued interdisciplinary research, coupled with industrial collaboration and regulatory engagement, will be crucial for scaling these solutions and fully realising their potential within global food systems.

References

Ahanchi, M., Sugianto, E. C., Chau, A., and Khoddami, A. (2024). Quality properties of bakery products and pasta containing spent coffee grounds (SCGs): A review. *Foods*, 13(22), 3576. https://doi.org/10.3390/foods13223576

Ahmed, A. R., Alqahtani, N. K., Ramadan, K. M., Mohamed, H. I., Mahmoud, M. A., and Elkatry, H. O. (2023). The bioactive substances in Spent Black Tea and Arabic Coffee could improve the nutritional value and extend the shelf life of sponge cake after fortification. *ACS Omega*, 8, 33593–33609.

Ahmed, H., Abolore, R. S., Jaiswal, S., and Jaiswal, A. K. (2024). Toward circular economy: Potentials of spent coffee grounds in bioproducts and chemical production. *Biomass*, 4(2), 286–312. https://doi.org/10.3390/biomass4020014

Andrade, C., Perestrelo, R., and Câmara, J. S. (2022). Bioactive compounds and antioxidant activity from spent coffee grounds as a powerful approach for its valorisation. *Molecules*, 27(21), 7504. https://doi.org/10.3390/molecules27217504

Arias, A., Ioannidou, S. M., Giannakis, N., Feijoo, G., Moreira, M. T., and Koutinas, A. (2023). Review of potential and prospective strategies for the valorization of coffee grounds within the framework of a sustainable and circular bioeconomy. *Ind. Crops Prod.*, 205, 117504.

Bevilacqua, E., Cruzat, V., Singh, I., Rose'Meyer, R. B., Panchal, S. K., and Brown, L. (2023). The potential of spent coffee grounds in functional food development. *Nutrients*, 15, 994.

Cavanagh, Q., Brooks, M. S.-L., and Upasinghe, H. V. (2023). Innovative technologies used to convert spent coffee grounds into new food ingredients: Opportunities, challenges, and prospects. *Future Foods*, 8, 100255.

FAO. (2023). The State of Food and Agriculture 2023 – Revealing the true cost of food to transform agrifood systems. Rome. Food and Agriculture Organization of the United Nations. https://doi.org/10.4060/cc7724en

Fernandez-Rosillo, F., Cabrejos-Barrios, E. M., Chávez-Quintana, S. G., and Quiñones-Huatangari, L. (2025). Polyphenol degradation kinetics of specialty coffee in different presentations. *Foods*, 14(21), 3600. https://doi.org/10.3390/foods14213600

Franca, A. S., and Oliveira, L. S. (2022). Potential uses of spent coffee grounds in the food industry. *Foods*, 11, 2064.

International Coffee Organization (ICO). (2024). Available online: https://www.ico.org/documents/cy2020-21/cmr-0221-e.pdf

Klingel, T., Kremer, J. I., Gottstein, V., Rajcic de Rezende, T., Schwarz, S., and Lachenmeier, D. W. (2020). A review of coffee by-products including leaf, flower, cherry, husk, silver skin, and spent grounds as novel foods within the European Union. *Foods*, 9, 665.

Koay, H. Y., Azman, A. T., Zin, Z. M., Portman, K. L., Hasmadi, M., Rusli, N. D., and Zainol, M. K. (2023). Assessing the impact of Spent Coffee Ground (SCG) concentrations on shortbread: A study of physicochemical attributes and sensory acceptance. *Future Foods*, 8, 100245.

Lee, Y.-G., Cho, E.-J., Maskey, S., Nguyen, D.-T., and Bae, H.-J. (2023). Value-added products from coffee waste: A review. *Molecules*, 28(8), 3562. https://doi.org/10.3390/molecules28083562

LIczbiński, P., and Bukowska, B. (2022). Tea and coffee polyphenols and their biological properties based on the latest in vitro investigations. *Industrial Crops and Products*, 175, 114265.

Makiso, M. U., Tola, Y. B., Ogah, O., and Endale, F. L. (2023). Bioactive compounds in coffee and their role in lowering the risk of major public health consequences: A review. *Food Sci Nutr.*, 12(2), 734–764. doi: 10.1002/fsn3.3848. PMID: 38370073; PMCID: PMC10867520.

Martinez-Saez, N., García, A. T., Pérez, I. D., Rebollo-Hernanz, M., Mesías, M., Morales, F. J., and Del Castillo, M. D. (2017). Use of spent coffee grounds as food ingredient in bakery products. *Food Chem.*, 216, 114–122.

Meerasri, J., and Sothornvit, R. (2022). Novel development of coffee oil extracted from spent coffee grounds as a butter substitute in bakery products. *J. Food Process. Preserv.*, 46, e16687.

Mussatto, S. I., Machado, E. M. S., Martins, S., and Teixeira, J. A. (2011). Production, composition, and application of coffee and its industrial residues. *Food and Bioprocess Technology*, 4(5), 661–672.

Navajas-Porras, B., Castillo-Correa, M., Navarro-Hortal, M. D., Montalbán-Hernández, C., Peña-Guzmán, D., Hinojosa-Nogueira, D., and Romero-Márquez, J. M. (2025). The valorization of coffee by-products and waste through the use of green extraction techniques: A bibliometric analysis. *Applied Sciences*, 15(3), 1505. https://doi.org/10.3390/app15031505

Nichifor, B., Zait, L., and Timiras, L. (2025). Drivers, barriers, and innovations in sustainable food consumption: A systematic literature review. *Sustainability*, 17(5), 2233. https://doi.org/10.3390/su17052233

O'Sullivan, M. G. (2017). CH14. Sensory properties of bakery and confectionary products. *In:* A Handbook for Sensory and Consumer Driven New Product Development: Innovative Technologies for the Food and Beverage Industry (pp. 305–324). Woodhead Publishing Ltd., UK. https://doi.org/10.1016/B978-0-08-100352-7.00014-2

OECD. (2023). *Global Food Consumption Patterns and Sustainability Challenges.* Organisation for Economic Co-operation and Development.

Oliveira Batista, J., Car Cordeiro, C., Klososki, S. J., Mongruel Eleutério Dos Santos, C., Leão, G. M. C., Pimentel, T. C., and Rosset, M. (2023). Spent coffee grounds improve the nutritional value and technological properties of gluten-free cookies. *J. Culin. Sci. Technol.*, 21, 994–1004.

Salazar-López, N. J., López-Rodríguez, C. V., Hernández-Montoya, D. A., and Campos-Vega, R. (2020). Health benefits of spent coffee grounds. *Food Wastes and By-Products: Nutraceutical Health Potential*, 327, 51.

Sharma, H. (2020). A Detail Chemistry of Coffee and Its Analysis. IntechOpen. doi: 10.5772/intechopen.91725

Silva, C. T., Corrêa Souza, M., da Fonseca Machado, A. P., de Paula Nascimento, R., Cunha, D. T., Bezerra, R. M. N., and Rostagno, M. A. (2021). Thermal stability and sensory evaluation of a bioactive extract from roasted coffee (Coffea arabica) beans added at increasing concentrations to conventional bread. *Journal of Food Processing and Preservation*, 45(11), e15955. https://doi.org/10.1111/jfpp.15955

Suanno, C., Marincich, L., Corneti, S., Aloisi, I., Pincigher, L., Papi, E. et al. (2025). Biochemical analysis of wheat milling by-products for their valorization as potential food ingredients. *International Journal of Molecular Sciences*, 26(12), 5830. https://doi.org/10.3390/ijms26125830

Sugianto, E. C. (2023). Incorporating spent coffee grounds into Semolina Pasta: Effects on its cooking quality. Honours Thesis, School of Life and Environmental Sciences, Faculty of Science, The University of Sydney, Camperdown, Australia.

Trà, T. T. T., Yến, V. T. N., Thu, N. T. A., and Nguyệt, T. N. M. (2021). Use of wheat flour and spent coffee grounds in the production of cookies with high fiber and antioxidant content: Effects of spent coffee grounds ratio on the product quality. *IOP Conf. Ser. Earth Environ. Sci.*, 947, 012044.

Viegas, C., Gomes, B., Oliveira, F., Dias, M., Cervantes, R., Pena, P. et al. (2022). Microbial contamination in the coffee industry: An occupational menace besides a food safety concern. *Int. J. Environ. Res. Public. Health*, 19, 13488.

Zhao, N., Liu, Z., and Yu, T. (2023). Spent coffee grounds: Present and future of an environmentally friendly applications on industries – A review. *Trends Food Sci. Technol.*, 143, 104312.

CHAPTER

8

Nanotechnology in Coffee Science: Applications and Innovations

Gabriela Geraldo Sangoi, Thuany Lura Teixeira Bandeira, Aline Krüger Batista, Taíse Schuster Montagner, Alencar Kolinski Machado and Aline Ferreira Ourique*

Nanoscience Graduate Program, Franciscan University (UFN), Santa Maria, RS, Brazil

Coffee (*Coffea arabica* L.) holds great cultural and economic importance worldwide, with its chemical matrix comprising a wide range of molecules exhibiting high antioxidant content and other bioactive properties. However, the applicability of these molecules is limited by their sensitivity to pH, light, and temperature, which constrains their use in various applications. Furthermore, coffee by-products, such as spent coffee grounds and husks, also possess untapped potential. As an alternative to enhance molecular stability, improve cultivation, and enable the reuse of by-products, nanotechnology has emerged as a promising approach. Nanocarriers can protect the chemical matrix of coffee from degradation and allow targeted delivery, while also reducing the effective dosage of the active compounds due to increased permeability, thus finding applications in pharmaceutical, cosmetic, and functional nutrition fields. Metallic nanoparticles can be synthesised through green synthesis methods using coffee by-products, as these residues retain high antioxidant potential, making such nanoparticles promising candidates for environmental remediation. Additionally, nanotechnology can be applied to improve coffee cultivation and enhance its antifungal properties for other crops. Thus, integrating chemical knowledge of coffee with the innovations offered by nanotechnology represents a promising pathway toward sustainable development and the full utilisation of this valuable natural resource.

1. Introduction

Nanotechnology is a modern field of research that has been gaining market applications at an exponential rate, impacting several areas, including academia,

*Corresponding author: alineourique@gmail.com

DOI: 10.1201/9781003689393-8

industry, and entrepreneurship, among others. According to ISO/TC 229, nanotechnology is the application and comprehensive study of nanomaterial properties changes, from material control and nanoscale process, with a scale of less than 100 nm in one, two, or three dimensions, with the scale changes, providing new properties, such as biological, electronic, magnetic, and/or optical. More than 65 years after nanotechnology first emerged as a theory, we can now observe many products already being commercialised that use this knowledge as a fundamental principle in their production processes (Rambaran and Schirhagl, 2022). Working with materials at the nanoscale is neither easy nor simple; however, the results obtained, regardless of the objective, can be highly promising, and this approach can also be applied to natural health products. Nowadays, it is increasingly common to find studies involving nanotechnology based on natural products, including applications using coffee, for example.

Coffee has been heavily marketed worldwide for many years, boosting the economies of several countries, including Brazil, Vietnam, and Colombia (Correia et al., 2024). Moderate coffee consumption is associated with various health benefits, including antioxidant, anti-inflammatory, and neuroprotective effects. These biological properties can be explained by coffee's complex chemical composition, which includes caffeine, chlorogenic acids, caffeic acid, trigonelline, and diterpenes (Makiso et al., 2023). On the other hand, coffee's chemical matrix is susceptible to degradation due to environmental conditions, including exposure to oxygen (oxidation), light (photosensitivity), temperature (thermal instability), and variable pH levels (changes that occur during preparation), among other factors. Additionally, poor solubility in its more natural forms and low bioavailability pose challenges to researchers and producers (Fernandez-Rosillo et al., 2025). Furthermore, industrial coffee processing generates substantial amounts of waste, which can have a negative impact on the environment (Ishimwe and Mupenzi, 2025). Nanotechnology has been explored as a strategy to address these issues (Figure 1).

There is substantial evidence indicating that using nanocarriers to encapsulate coffee derivatives is a promising approach, as it can enhance intestinal absorption, gastrointestinal stability, and preservation of the antioxidant activity of phenolic compounds (Do Carmo et al., 2022; Rojas-Orduña et al., 2023; Tavares et al., 2024). Nanotechnology has also been used to enhance coffee production. For example, zinc oxide nanoparticles (Meléndez-Mori, 2025), extract-based nanoformulations combined with silver nanoparticles (Soundhararajan and Srinivasan, 2025), and nanofertilizers (Kah et al., 2018) have shown potential in mitigating salinity stress, fungal infections, and poor soil nutrition, thereby improving coffee agriculture.

Interestingly, nanotechnology has also served as an important tool for addressing waste generated by coffee cultivation and processing through biosynthesis or green synthesis techniques, using natural product derivatives to produce metallic nanoparticles. Kusumahastuti and Gintu (2025) and Drummer et al. (2024) provide excellent examples of using organic coffee waste in green synthesis processes. However, this remains an area with much room for improvement, especially considering that coffee is one of the most consumed beverages in the world.

In this context, the objective of this chapter is to present a narrative review of how nanotechnology is being applied worldwide in coffee science, with a focus on aspects of application and innovation.

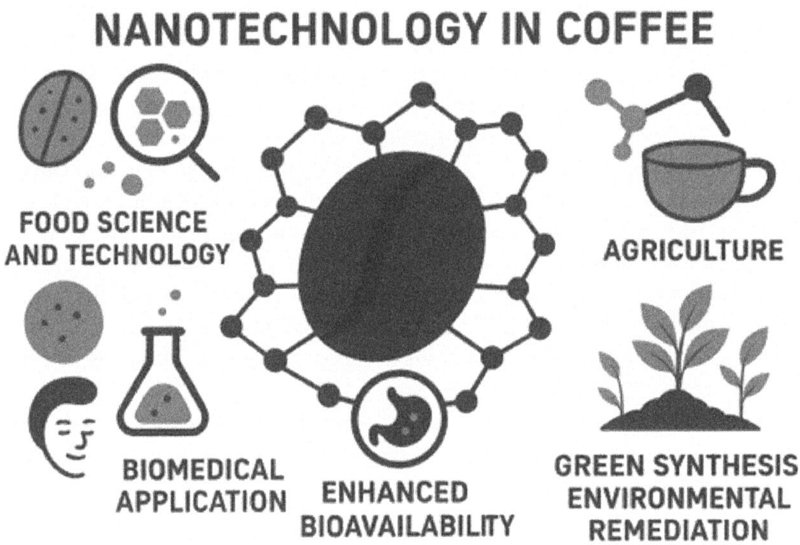

Figure 1: Coffee and its byproducts: Perspectives and innovation

2. Coffee bioactive compounds and nanotechnology applications

Coffee beans concentrate several bioactive compounds of scientific and technological relevance and interest, such as caffeine, chlorogenic acids, caffeic acid, trigonelline, and diterpenes, among them cafestol. These metabolites exhibit important functional properties, including antioxidant, anti-inflammatory, neuroprotective, antimicrobial, and stimulant effects (Makiso et al., 2023). However, these compounds face intrinsic limitations, such as low solubility, rapid metabolism, chemical instability, and reduced permeability (Fernandez-Rosillo et al., 2025), which restrict their full use in pharmaceutical, cosmetic, and nutraceutical formulations. In this context, nanotechnology emerges as an effective alternative to overcome these challenges, enabling greater bioavailability, stability, and controlled release.

Caffeine, the primary alkaloid component of coffee beans (Saud and Salamatullah, 2021), is recognised for its lipolytic action, stimulation of blood circulation, and ability to inhibit 5α-reductase, thereby favouring hair growth. Additionally, it acts on the central nervous system, enhancing attention, physical and mental performance, and exerting a thermogenic effect. However, its low stability and rapid metabolism can limit its clinical application. To overcome these barriers, several nanotechnological strategies—such as deformable liposomes, mesoporous silica nanoparticles, and lecithin-based nano-phytosomes—have been employed to enhance skin permeation, improve therapeutic and cosmetic effects, and expand investigations into antitumor applications (Fathi et al., 2022; Siddique et al., 2023; Park et al., 2024). Deformable liposomes, for instance, are elastic lipid vesicles containing an edge activator, typically a single-chain surfactant, that destabilises the lipid bilayer and increases its flexibility, allowing the vesicle to traverse the narrow pores of the stratum corneum.

Propylene glycol not only functions as an effective edge activator but also enhances skin penetration, promoting deeper delivery of active compounds (Park et al., 2024).

Similarly, nanophytosomes are nanovesicles formed by the molecular association of phytochemicals with phospholipids, which improves the permeation, stability, and bioavailability of plant-derived compounds. Lecithin provides structural similarity to the cellular membrane, facilitating absorption, while the incorporation of cholesterol can further increase vesicle stability (Fathi et al., 2022).

Fathi et al. (2022) enriched extracts of coffee silverskin, a brownish-coloured film regarded as a by-product (Costa et al., 2014) of coffee, with caffeine and chlorogenic acid, resulting in an increase of approximately 2–3% in these compounds compared to the original extract. The enriched fraction was subsequently loaded into nanophytosomes at a 1% (w/v) extract concentration, prepared by the thin-film hydration method followed by ultrasonication to reduce particle size. These nanocarriers were then coated with either cholesterol or natural polymers, such as pectin, gum arabic, maltodextrin, or starch, forming protective layers designed to enhance stability against gastrointestinal barriers and improve the absorption and, consequently, the bioavailability of caffeine and chlorogenic acid in the body. This approach shows promising potential for enhancing both mental and physical performance.

A study conducted by Siddique et al. (2023) reported the formulation of MCM-41-type mesoporous silica nanoparticles (MSNPs, ~120 nm), coated with caffeine, with a loading efficiency of 28%, developed for antitumor applications. *In silico* analyses were performed to identify potential apoptotic targets, with a focus on modulating the NF-κB and TRAIL pathways for further investigation. The results demonstrated a reduction in cell survival factors associated with the NF-κB pathway and an increase in pro-apoptotic markers related to the TRAIL pathway, indicating that the nanoformulation significantly enhances the anticancer action of caffeine.

Chlorogenic acids (CGA), especially 5-caffeoylquinic acid, exhibit proven antioxidant and anti-inflammatory properties, protecting against the development and progression of metabolic, cardiovascular, and neurodegenerative diseases. This is achieved through the modulation of lipid and glucose metabolism, in addition to its potential hepatoprotective activity (Naveed et al., 2018). However, they undergo rapid degradation and present low bioavailability. Formulations such as lecithin-based nano-phytosomes and polymeric nanoparticles have shown positive results, enhancing stability and biological utilisation of these compounds (Naveed et al., 2018; Park et al., 2024), for example, the study conducted by Park et al. (2024), in which liposomal vesicles containing propylene glycol were developed, improving the stability of coffee-derived bioactives and increasing skin permeability by more than 15-fold; in this work, transdermal permeation was assessed *ex vivo* using Franz diffusion cells with human skin, enabling quantification of caffeine permeated over 24 hours by HPLC.

Caffeic acid is another phenolic compound present in coffee, with recognised antioxidant, anti-inflammatory, and photoprotective activities. Its topical use is promising in preventing skin ageing, alterations induced by environmental pollutants, and damage caused by UV radiation. Nanoencapsulation in solid lipid nanoparticles (SLN/NLS), biocompatible and non-toxic lipid carriers that can incorporate compounds with diverse physicochemical properties and control their release in aqueous physiological environments, has proven effective in enhancing the stability and antioxidant activity of the compound (Sguizzato et al., 2020).

Sguizzato et al. (2020) developed SLN incorporated into a thermosensitive poloxamer-407 (p407) gel for the cutaneous delivery of caffeic acid. The nanoparticles exhibited an average size of approximately 200 nm, with an encapsulation efficiency of about 88% of the active compound within the lipid phase and around 12% dispersed in the aqueous phase. Moreover, the use of nanotechnology enhanced the water solubility of caffeic acid by approximately 1.7-fold. The nanoencapsulated form demonstrated enhanced permeability and prolonged retention of the active compound in the skin, ensuring sustained release for at least 24 hours due to the combination of the SLN lipid matrix with the thermosensitive p407 gel, which reduced the diffusion rate by up to eight times. In addition, the antioxidant activity of caffeic acid–loaded nanoparticles was confirmed through an ex vivo human skin model exposed to cigarette smoke, reinforcing the relevance of SLN-based semi-solid systems, whose viscosity—achieved through polymeric gelation—is crucial for maintaining the formulation on the skin surface and optimising topical performance.

Trigonelline, a pyridine alkaloid found in green beans, as well as in coffee and its by-products (Konstantinidis et al., 2023), exhibits antioxidant properties and efficacy in the treatment of alopecia. Nanocrystals, which are drug particles reduced to sizes below 1 μm, markedly enhance solubility and bioavailability. By decreasing particle size to the nanoscale, their surface area and dissolution rate increase substantially, thereby improving the absorption of poorly soluble compounds. In this context, studies have demonstrated that formulations containing nanocrystals significantly enhance hair growth (Shetty et al., 2024), in addition to showing potential effects against glioblastomas, with greater selectivity and safety compared to the free forms (Sathiyaseelan et al., 2020).

A study conducted by Shetty et al. (2024) proposed an alternative to the reference treatments for alopecia, minoxidil and finasteride, through the use of trigonelline nanocrystals. The nanocrystals were produced using the sonoprecipitation technique and incorporated into a topical cream formulation containing 79% of the active compound. *In vivo* studies demonstrated that the cream was non-toxic in Wistar rats and showed a significant effect in promoting hair length and density compared to the control group and a commercial hair oil. However, its efficacy did not surpass that of the reference treatment with minoxidil after 30 days.

Further exploring the use of trigonelline, Sathiyaseelan et al. (2020) employed chitosan nanoparticles loaded with trigonelline to evaluate their antitumor activity against neuroblastoma cells (C6 cell line) and to assess biocompatibility in adrenal pheochromocytoma cells (PC12 cell line). The nanoparticles were found to be stable, sustained by a zeta potential of +30.9 mV; however, the encapsulation efficiency decreased as the trigonelline concentration increased, with an encapsulation efficiency of approximately 83% at a trigonelline concentration of 0.5 $mg.mL^{-1}$ in the produced nanoparticles. The efficacy against the C6 cell line was considered significant by the authors, showing an IC_{50} value of 34 $\mu g.mL^{-1}$. Moreover, the biocompatibility tests on the adrenal cell line indicated positive results, with no observed cytotoxic damage to the cells.

Table 1: Coffee bioactive compounds: Properties, technological challenges, and applied nanotechnological systems

Coffee bioactive compound	Properties/Functions	Technological challenges	Examples of applied nanotechnological systems	References
Caffeine	Immunomodulator; lipolytic activity; improves blood circulation; stimulates hair growth through 5α-reductase inhibition; central nervous system stimulant, improving attention, physical and mental performance; thermogenic effect.	Low stability and poor skin permeation.	Deformable liposomes; mesoporous silica nanoparticles.	Fathi et al., 2022; Siddique et al., 2023; Park et al., 2024
Chlorogenic acids (CGA)	Anti-inflammatory and antioxidant properties; promising for the treatment of neurodegenerative disorders; contributes to protection against metabolic and cardiovascular diseases.	Low stability, low bioavailability, and inadequate release.	Polymeric nanoparticles; lecithin-based nano-phytosomes.	Naveed et al., 2018; Park et al., 2024
Caffeic acid	Antioxidant; anti-inflammatory; immunomodulatory.	Rapid oxidation; photosensitivity; low stability; need for controlled release; poor skin permeation.	Solid lipid nanoparticles.	Sguizzato et al., 2020
Trigonelline	Antioxidant; anti-inflammatory; neuroprotective; potential for alopecia treatment.	Rapid oxidation, low solubility, need for high concentrations, and slow bioactivity.	Water-soluble chitosan nanoparticles; nanocrystals.	Shetty et al., 2024; Sathiyaseelan et al., 2020
Diterpenes (cafestol)	Antibacterial activity.	Low solubility in aqueous medium.	Zinc oxide nanoparticles.	Ballıca et al., 2020

Another bioactive compound present in the chemical matrix of coffee is cafestol, a diterpene-class compound that has shown limited antibacterial activity when used alone. However, when combined with chitosan-coated zinc oxide nanoparticles, they exhibited strong antimicrobial action against Gram-positive and Gram-negative bacteria, including resistant strains, revealing potential for use in medical devices and surfaces requiring bacterial control, as demonstrated in the study by Ballıca et al. (2020), in which the authors developed such nanoparticles using the chemical precipitation method and demonstrated antibacterial activity of the compounds within the concentration range of 20–1000 $\mu g.mL^{-1}$, showing no bacterial selectivity, that is, the compounds were effective against both Gram-positive and Gram-negative bacteria.

Overall, the application of nanotechnology to coffee bioactive compounds appears promising in broadening their action spectrum and enhancing their therapeutic, cosmetic, and pharmaceutical effects. Furthermore, recent advances highlight the importance not only of coffee beans but also of their by-products, such as coffee silverskin, in the development of innovative and sustainable high-value-added products.

Table 1 provides a synthesis of the main properties, technological challenges, and nanotechnological systems applied to these coffee bioactive compounds, as reported in the literature.

The combined analysis of available results shows that coffee bioactive compounds possess high functional potential but also present limitations that hinder their direct application. From Table 1, it is evident that challenges such as low stability, low solubility, and reduced bioavailability are common to most of these compounds. In this context, nanotechnology serves as an overcoming tool, offering innovative solutions such as liposomes, polymeric nanoparticles, solid lipid systems, and nanocrystals.

These advances not only expand the use of bioactives in cosmetic and pharmaceutical products but also pave the way for more complex biomedical applications, such as antitumor therapy, neuroprotection, and treatment of metabolic diseases. It is also worth highlighting the role of coffee by-products, such as coffee silverskin, which, when incorporated into nano-structured systems, demonstrate high added value, aligning technological innovation with sustainability.

3. Coffee and its by-products in the pharmaceutical and cosmetic industries

Nanotechnology has gained increasing prominence in the pharmaceutical and cosmetic industries by enabling the development of efficient and innovative nanostructured systems. These systems favour the design of formulations with enhanced stability, controlled release, and targeted delivery, thereby amplifying the therapeutic effects of drugs/actives (Islam et al., 2025). In addition, they contribute to improving the performance of cosmetics, driving the production of modern, functional, and high-performance products (Gupta et al., 2022).

In the pharmaceutical field, nanostructured systems used as drug carriers allow more efficient cellular absorption and ensure greater selectivity in delivering the drug to its site of action, thus enhancing therapeutic efficacy while reducing potential side effects (Islam et al., 2025). In cosmetic applications, nanotechnology improves properties such as durability, water resistance, and skin penetration capacity, while overcoming common limitations related to the bioavailability of active compounds (Oliveira et al., 2022).

The use of nanotechnology has sparked interest in harnessing natural sources rich in bioactive compounds derived from plants and foods, which, although non-essential nutrients, exert beneficial effects on health by exhibiting antioxidant, anti-inflammatory, and regenerative properties. However, the use of these compounds is limited due to challenges such as the low stability of their chemical matrix, poor solubility, rapid degradation, and difficulty in permeating physical and biological barriers, such as the skin. Thus, nanotechnology emerges as an effective strategy to overcome these limitations and enhance the utilisation of natural compounds, expanding the therapeutic and cosmetic potential of these bioactives (Mehta et al., 2025).

In this context, caffeine, found in coffee, for instance, has its therapeutic effects enhanced when incorporated into nanostructured systems, such as nanoemulsions (NE). The NEs are colloidal systems composed of nanometric oil droplets dispersed in water and stabilised by surfactants, which are known for improving the solubility, stability, and cellular uptake of bioactive compounds, thereby enhancing the biological activity of encapsulated molecules. This potential was demonstrated in the study by Amirizadeh et al. (2025), in which the K562 cancer cell line (cells from chronic myelogenous leukaemia) was exposed to nanoemulsions containing caffeine. The nanoformulation was produced using the spontaneous emulsification technique, exhibiting a polydispersity index of approximately 0.2 and an average diameter of around 168 nm. The cellular viability of the cancer cell line decreased by 50% within 48 hours of evaluation; however, in its nanoemulsified form, caffeine was more effective in reducing cellular viability, with a concentration difference of approximately 270 μg/mL. In contrast, the biocompatibility of both free caffeine and its nanoemulsion was promising, showing no cytotoxicity in PBMC (peripheral blood mononuclear cells) and RBC (red blood cells) lines.

Furthermore, Siddique et al. (2023) developed mesoporous silica nanoparticles coated with caffeine and observed significant antitumor activity in an acute myeloid leukaemia (AML) model. In silico analyses first demonstrated the ability of nanoparticles to bind to potential cellular targets related to apoptosis, mainly with the TRAIL-DR5 protein, a pro-apoptotic complex, and with the binding site of the NF-kB p52/RelB/DNA complex, directly related to antiproliferative action. These interactions were subsequently confirmed with *in vitro* and *in vivo* assays. The *in vitro* tests were specifically designed to determine the antioxidant potential of the nanoparticles, which demonstrated significant reducing potential, especially when combined with the antitumor drug doxorubicin, showing an increase of approximately 16% compared to the drug alone. In addition, there was an increase in antioxidant capacity compared to caffeine alone, demonstrating improved delivery of the active ingredient when it is in a nanomedicine format. For *in vivo* trials, female Sprague-Dawley rats were induced to develop leukaemia through the administration of benzene. Haematological and enzymatic analyses were performed, along with analyses of STMN1 and S1009A gene expression markers, directly related to leukaemia, to evaluate the antitumor potential of this nanomedicine, where it was demonstrated that there was normalisation of the gene expression of the aforementioned genes, as well as the GAPDH and mTOR genes, which were elevated after the administration of benzene. In addition, the compound could increase the expression of pro-apoptotic TRAIL and DR5 genes and decrease the expression of the anti-apoptotic cFLIP protein, demonstrating significant antitumor potential, especially against AML.

Complementarily, Ratliff et al. (2019) reported neuroprotective effects and improvements in behavioural and neuronal parameters following the use of coffee-derived nanoparticles in an experimental model of traumatic brain injury. The nanoparticles produced in this study were obtained using the sonication method after diluting and boiling coffee in water. In general, brain damage was induced in rats, which were then treated with intraperitoneal injections of 200 μL of the nanocoffee solution administered at 30 minutes, 3 days, and 7 days after injury. The results obtained were promising in terms of improving cognitive deficits, as observed through behavioural tests, in addition to improved neuronal and general health of the animals, as evidenced by protein analyses, such as the inhibition of PARP (poly (ADP-ribose) polymerase), a type of enzyme crucial for DNA repair and genomic stability, with inhibition related to decreased microglial activation and neurological deficits. In contrast, another protein that was also reduced after treatment with coffee nanoparticles was β-catenin, a protein related to cellular adhesion and signalling pathways, which, when decreased, can be detrimental to the repair process, as postulated by the study authors.

Moreover, residues generated during the roasting process, such as coffee grounds and coffee silverskin, have also been incorporated into lipidic and polymeric nano/microcarriers, demonstrating high encapsulation efficiency, controlled release, and improved physicochemical stability, as in the study conducted by Fathi et al. (2022), cited above. These systems enhance the therapeutic potential of coffee while promoting the sustainable utilisation of industrial by-products.

In this regard, Kanai et al. (2024) reused spent coffee grounds to produce holocellulose nanofibers as potential additives in cosmetic formulations, owing to their antioxidant, biocompatible properties, and high water retention capacity. The authors used the chemical composition of coffee grounds, which are rich in polysaccharides such as mannose, galactose, and glucose, to produce nanocellulose fibres. These fibres were produced using an ultra-pressure wet jet mill, resulting in fibres 2-3 nm wide and between 0.7 and 1 nm long. Thus, the sustainable reuse of coffee residues demonstrates significant potential for the development of innovative pharmaceutical, cosmetic, and food additive products.

In the cosmetic field, caffeine has gained prominence as a highly versatile active ingredient. The study by Fathi et al. (2022) focuses specifically on the sustainable valorisation of coffee by-products, utilising the enriched extract of coffee silverskin (the fine skin released during roasting) as a source of caffeine and polyphenols. To optimise the use of this extract, the researchers employed an advanced methodological process, formulating nano/micro lipid carriers (such as liposomal nanocarriers, or "nano-phytosomes") modified with cholesterol and coated with polymeric nanoparticles. This complex nanostructure was designed to fulfil multiple objectives: enhance skin penetration, protect the active compound from degradation, and enable a sustained, controlled release of the active ingredient (confirmed by *in vitro* assays). The results confirmed the high stability of the particles, with the core size mainly ranging between 226 and 808 nm. Such nanotechnological advances enable the development of more effective cosmetics, such as anti-dark circle serums, firming creams, and hair care products, that combine high performance and functionality, while simultaneously promoting the circular economy through the innovative utilisation of coffee agro-industrial waste.

Recent studies have also demonstrated caffeine's potential in the treatment of androgenetic alopecia. Thepphankulngarm et al. (2024) developed hollow mesoporous silica nanoparticles loaded with caffeine at a concentration of 0.01 g and coated with ultradeformable liposomes to improve delivery to hair follicles. This strategy promoted enhanced follicular penetration, stimulating hair growth and reducing hair loss. The nanoparticles produced were between 140 and 160 nm in size and had good thermal stability and a slow release rate. Approximately 65% of the encapsulated caffeine was released over 8 hours, with the highest release occurring around the seventh hour, a behaviour related to the double encapsulation of these nanoparticles. Furthermore, the materials produced were considered non-toxic by the authors, as they did not indicate a reduction in cell viability and showed better cell proliferation conditions in human hair follicle dermal papilla cells than pure caffeine, highlighting the improved biocompatibility of the nanocomposite.

Beyond facial and hair care, caffeine has also been explored in body treatments, such as combating cellulite. Lipid-based nanocarriers can increase the lipolytic activity of fat by promoting its breakdown in adipocytes. This improves the penetration of active substances and prolongs their effects. These systems have shown positive results in improving skin appearance and reducing measurements in specific body areas (Kassem et al., 2023; Fouad et al., 2024).

Kassen et al. (2023) produced nanostructured lipid carriers (NLCs) to carry caffeine molecules for the topical treatment of cellulite. The NLCs are composed of a combination of solid and liquid lipids that form nanostructures with increased drug-loading capacity, enhanced stability, and reduced premature drug release during storage (Kassen et al., 2023). These nanostructures were produced through high-speed homogenisation and ultrasound, presenting a caffeine content of 97.57% in the optimized formulation. *In vivo* trials with Wistar rats demonstrated that a single topical application of the product containing the nanoformulation had approximately twice the skin retention of caffeine compared to the simple gel without nanotechnology. In addition, the anti-cellulite capacity was also greater, as demonstrated by increased production of collagen, elastin, and hyaluronic acid, as well as increased thermogenesis, due to increased expression of UCP1 (thermogenin), PPAR-γ (related to adipocyte differentiation), and lipolytic capacity, through increased HSL (hormone-sensitive lipase) and decreased pro-inflammatory cytokines.

Table 2: Applications of nanotechnology in pharmaceutical and cosmetic systems derived from coffee and its by-products

Application area	Nanosystem type/technology	Principal function	Observed benefits	References
Pharmaceutical				
	Nanoparticles, liposomes, polymeric systems	Drug and bioactive carrier	Increases cellular absorption, promotes controlled and targeted release, and reduces side effects	Islam et al., 2025
	Silica nanoparticles, ultra-deformable liposomes	Therapeutic potentiation	Anticancer action without cytotoxicity; neuroprotective effect	Amirizadeh et al., 2025; Siddique et al., 2023; Ratliff et al., 2019
Cosmetics				
	Lipid nanoparticles, liposomes, nanoemulsions	Improved stability and skin penetration	Increases durability, water resistance, and bioavailability of active ingredients	Oliveira et al., 2022
	Nanocarriers of bioactive compounds	Overcoming low solubility and stability	Improved antioxidant, anti-inflammatory, and regenerative action	Gupta et al., 2022
	Holocellulose nanofibers	Natural cosmetic additive	Antioxidant properties, biocompatible, and high water retention	Kanai et al., 2024
	Mesoporous silica nanoparticles and ultra-deformable liposomes	Stimulation of hair growth	Improved follicular penetration and reduced hair loss	Thepphankulngarm et al., 2024
	Lipid nanocarriers with caffeine	Lipolytic stimulation	Improved skin appearance and reduction in measurements	Kassem et al., 2023; Fouad et al., 2024

Furthermore, a study conducted by Fouad et al. (2024) produced solid lipid nanoparticles, obtained through homogenisation followed by shear ultrasonication, with active ingredient encapsulation efficiency greater than 75% and successful permeation through the transdermal route. The work combined *in vitro* characterisation and permeation experiments with *in vivo* histological evaluation in rats. In addition, the gel containing the nanoparticles promoted a greater reduction in adipose tissue than free caffeine and untreated tissue.

In this context, studies associating nanotechnology with coffee and its by-products have increasingly stood out as innovative and sustainable approaches, expanding potential applications in various aesthetic treatments. This combination ranges from facial and hair care to more advanced body protocols, offering enhanced results and emphasising the use of natural products. Examples of such applications already available on the market include the nCaffeine Complex from NanoScoping (2024), whose label describes the encapsulation of caffeine, combined with other active ingredients, in lipid nanoparticles; the Nano Caffeine active from *Tulípia Cosméticos* (2022) and the body massage cream with nano caffeine from *D'Água Natural* (2025), all illustrating the practical application of nanotechnology in cosmetic formulations. Table 2 summarises the advancements in the application of coffee and its by-products in conjunction with nanotechnology in the cosmetic and pharmaceutical fields.

In summary, the application of nanotechnology to coffee and its by-products represents a promising strategy to enhance the stability, bioavailability, and therapeutic efficacy of caffeine and other bioactive compounds. Nanostructured systems demonstrate superior delivery performance, leading to more effective outcomes across different experimental models.

Furthermore, the use of coffee residues, such as spent coffee grounds, silver skin, and holocellulose-rich fibres, strengthens the integration between technological innovation and sustainability. This approach adds value to agro-industrial by-products, enabling the development of more efficient, versatile, and environmentally responsible formulations.

4. Nanotechnology in food applications of coffee and its by-products

Coffee, one of the most widely consumed beverages worldwide, generates several by-products rich in bioactive compounds, such as chlorogenic acids. However, the direct application of these extracts in foods presents several important limitations, including instability during processing and potential undesirable changes in flavour and aroma (Husni et al., 2021; Garcia-Solis et al., 2022). In this context, nanotechnology has emerged as a promising strategy for preserving the biological activity of these compounds and expanding their applicability in the food sector.

A relevant example is the study by Tavares et al. (2024), which investigated the co-encapsulation of roasted coffee and coffee husk extracts with the probiotic Kluyveromyces lactis, extracted from canastra cheese, in microcapsules. This approach resulted in greater retention of phenolic compounds, leading to increased antioxidant activity in the coffee extract, as determined by tests such as DPPH and ABTS, which are widely used to measure the antioxidant capacity of molecules and compounds. The authors highlight that the nano-structured system effectively integrated the antioxidant properties of coffee compounds with the functional benefits of the yeast, reinforcing the potential of nanoencapsulated coffee in functional and symbiotic foods. In addition, co-encapsulation promoted higher probiotic viability during refrigerated storage, resulting from the optimal encapsulation efficiency of the yeast within the microcapsule, which was around 97%. In addition, the yeasts also showed high viability, around 8.5 log CFU/mL, which is recommended for probiotic foods, thus providing good gastrointestinal digestion conditions.

Similar advances have been observed in the use of encapsulating agents for protecting and delivering bioactive compounds from green coffee. In the study by Do Carmo et al. (2022), polydextrose and inulin were used as wall materials at a concentration of 30%, previously dissolved in water and mixed with the coffee extract for microencapsulation. Green coffee extract was processed to obtain soluble solids, and spray drying and freeze-drying techniques were used to produce the microparticles, being average diameter varies according to the method used to produce the microparticles, whether through lyophilisation, ranging from 11 to 183 μm when using polydextrose or 5 to 103 μm when using inulin, or through pulverisation, ranging from 1 to 11 μm regardless of the material used as a base. This method is the most accurate for producing the most homogeneous microparticles. Furthermore, when incorporated into dairy beverages, these microcapsules led to higher total phenolic content, which was not degraded regardless of storage time, as proven by the antioxidant capacity test, ABS, which also demonstrated increased antioxidant capacity. Furthermore, the different techniques and agents used for the synthesis of microparticles do not significantly affect the appearance and aroma of the product generated, in this case, the milk drink, with all of them receiving good scores in terms of overall product acceptance. *In vitro* digestion assays confirmed the protection and controlled release of phenolics based on the tests performed, it can be observed that the main phenolics found in green coffee extract, caffeine, chlorogenic acid, and trigonelline, had similar concentrations in all samples, around 90%, which is considered a good amount of recovery, corroborating the fact that *in vitro* digestion releases large amounts of antioxidant compounds. These findings reinforce the role of microencapsulation in improving the stability and bioaccessibility of coffee bioactive compounds.

Beyond grain extracts, the valorisation of residues generated along the production chain, such as coffee pulp (*Coffea arabica* var. *Caturra*), has gained prominence as a functional and sustainable alternative. In the study by Rojas-Orduña et al. (2023), red and yellow pulps were analysed, with the yellow variety exhibiting the highest fibre and carotenoid content. Ultrasonication was employed to extract carotenoids, and the carotenoid-rich extract was then microencapsulated by spray drying using maltodextrin as the wall material. The emulsion presented an encapsulation rate of approximately 45.5% for carotenoids, a low encapsulation factor; however, according to the authors, this is justifiable given the technique used. Furthermore, the emulsion was incorporated into dairy products, which, upon mixing, resulted in a decrease in the viscosity of the products, the only property that changed. The microencapsulated extracts demonstrated suitable applicability in fermented dairy products, enhancing their nutritional value without significantly compromising their sensory properties.

In food applications, challenges such as the instability and bitterness of bioactive extracts, particularly those derived from coffee, have been overcome through the use of nanotechnology. In the study by Guillén-Enríquez et al. (2023), ultrasound was applied to finely ground coffee powder, reducing particle size to the nanoscale, with an average of approximately 101 nm. This size reduction generated a significant increase in surface area, resulting in greater extraction efficiency of bioactive compounds. Specifically, it increased caffeine from 0.54 to 3.21 $mg.g^{-1}$, total phenolics from 426.7 to 1825.3 μg $GAE.g^{-1}$, and improved the solubility of the powder from 40.7% to 75.9%. The sensory improvements observed in the study, including enhancements in aroma, flavour, and texture, were attributed to more efficient extraction and preservation of functional compounds. Together, these changes contribute to a more balanced sensory profile and a less bitter beverage. The literature suggests that bitterness modulation can occur through more efficient extraction, which adjusts the balance between volatile and phenolic compounds, as well as through the encapsulation of molecules responsible for bitter taste, such as caffeine and chlorogenic acids, thereby promoting their controlled release and delaying gustatory perception (Aree, 2018). Thus, nanotechnology applied to coffee not only improves the quality of the beverage but also serves as an effective strategy to modulate and soften bitterness, making the final product more pleasant to the palate.

Advancing the application of nanotechnology in food packaging materials, the study by Saidi et al. (2025) innovatively demonstrates how coffee processing residues can be utilised, contributing to the development of functional biofilms. In this work, cellulose nanofibrils were extracted directly from spent coffee grounds and employed as nano-reinforcement in carboxymethylcellulose-based films, combined with grape seed oil. The role of the nanofibrils was crucial: they acted in a dual capacity, both as nano-fillers and as stabilisers for the oil droplets, facilitating efficient dispersion within the polymeric matrix. This nano-reinforcement strategy resulted in a significant improvement in the material's water resistance. Consequently, the final film structure was optimised, resulting in enhanced characteristics for food preservation. The resulting material proved to be more hydrophobic, less soluble, and had a lower moisture content. Furthermore, the nanostructure conferred to the film a notably greater mechanical resistance and an important antimicrobial potential, essential elements for ensuring the safety and extending the product's shelf life. Such advances reinforce the applicability of nanotechnology in the sustainable valuation of waste, positioning coffee byproducts as high-value ingredients for the creation of more efficient and responsible food packaging.

Nanotechnology has driven the use of byproducts, including husk, pulp, spent coffee grounds, and silver skin, which pose environmental problems, but also have potential as functional materials. Wardana et al. (2025) demonstrated that these byproducts can be transformed into high-value-added ingredients through nanostructuring. The process of valuing coffee husks, for example, is typically performed by techniques such as nanoprecipitation or high-energy milling, which are used to obtain nanomaterials like nanocrystalline cellulose or nanofibrils.

Once obtained, these nanomaterials are incorporated into sustainable biofilm and edible film systems. The strategic addition of these cellulose nanostructures led to a significant improvement in the films' mechanical resistance and, crucially, in their barrier properties. Such barrier properties are essential, as they confer efficient protection against gas permeation and UV radiation, thereby extending the useful life and maintaining the quality of the final product.

Collectively, these studies highlight how micro- and nanotechnologies applied to the food sector significantly expand the possibilities for the use of coffee and its by-products. In addition to enhancing health benefits and improving sensory acceptance, these approaches strengthen sustainable practices, contributing to a more efficient and environmentally responsible production chain.

5. Nanotechnological strategies for coffee cultivation and postharvest management

Coffee (*Coffea arabica* L.), one of the most important tropical crops for global consumption and of great socioeconomic impact (Alamri et al., 2022), faces several challenges, such as the demand for higher-quality beans, the inefficient use of agricultural inputs, and the high susceptibility to fungal diseases, which compromise productivity and may result in contamination by toxins (Lu et al., 2022; Rege et al., 2022). In response, recent studies have highlighted the great potential of nanotechnology as a tool to overcome these limitations and improve coffee management in agriculture.

Recently, a nanotechnological approach was explored to enhance coffee production. Meléndez-Mori et al. (2025) demonstrated that zinc oxide nanoparticles synthesised through a green method using banana peel extract can help mitigate the adverse effects of salt stress in *Coffea arabica*. Physiological and biochemical assays performed on treated seedlings under stress indicated that the nanoparticles promoted beneficial responses. This was evidenced by increased biomass and photosynthetic efficiency, as well as reduced oxidative stress. This action suggests that the nanoparticles improve membrane stability and assist in osmotic regulation, providing greater resilience to the plant.

Another promising study (Soundhararajan and Srinivasan, 2025) focused on phytosanitary protection. It demonstrated that an antifungal nanoformulation combining Triphala extract with silver nanoparticles is highly effective against foliar pathogens in *Coffea arabica.* The nanoparticles were synthesised through a green process using Triphala extract, creating a synergistic formulation. *In vitro* and *in vivo* assays on coffee leaves against fungi such as *Cladorrhinum flexuosum* and *Rigidoporus* vinctus revealed high inhibition rates of mycelial growth and severe morphological damage to fungal hyphae, leading to their death. These results highlight the nanoformulation as an ecologically potent alternative for disease control, capable of reducing the severity of plant infections.

Regarding nutrition, nanofertilizers offer advantages by increasing fertilisation efficiency and reducing environmental impacts, as they allow greater nutrient absorption and availability at lower doses compared to conventional fertilisers

(Kah et al., 2018). The nanoscale size (<100 nm) of the particles is the primary differential, as it provides a larger surface area and facilitates nutrient absorption through both foliar and root uptake. This mechanism results in greater nutrient availability and translocation within the plant, allowing the use of much lower doses compared to conventional fertilisers and consequently minimising the risk of runoff and environmental pollution.

In the scope of application, the study by Rossi et al. (2019) demonstrated the impact of foliar application of zinc nanoparticles on *Coffea arabica*. The implementation process was carefully designed under controlled conditions, using coffee seedlings, and compared the application of zinc nanoparticles (at a dose of 10 mg/L of elemental zinc) with that of conventional fertiliser (zinc sulfate). The foliar spraying method, applied in two sessions over 45 days, is crucial for nanotechnology, as it allows the zinc nanoparticles (with a size smaller than 100 nm) to more efficiently overcome the leaf's protective barriers, such as the cuticle and stomata. The results showed significant physiological improvements, notably an increase in photosynthetic efficiency and stomatal conductance. The implementation was validated by X-ray microanalysis maps, which confirmed the superior penetration and distribution of the nutrient in the leaf tissue, resulting in a significantly higher zinc content in the leaves. This efficient delivery of zinc strengthens the plant's physiological state, ensuring the availability of crucial resources for successful flowering and, consequently, impacting fruit quality, which presented better classification, higher weight, and improved beverage characteristics. From a foundational perspective, the synthesis and agricultural potential of these biogenic nanoparticles are systematically reviewed by authors, such as Rani et al. (2022). This work confirms the essential role of zinc in plant reproductive processes, encompassing flowering and pollen viability. It establishes the theoretical basis for utilising zinc nanoparticles as an efficient and sustainable strategy for optimising the flowering stage of the coffee plant.

In disease management, nanotechnology has also shown promising results, especially against coffee leaf rust (*Hemileia vastatrix*), the most relevant disease affecting the crop. Research indicates that copper and silver nanoparticles reduce uredospore germination, offering alternatives for integrated control programs (Ding et al., 2023; Bhaskar et al., 2024). In addition to direct effects on pathogens, some formulations act as defence inducers: Gomes et al. (2023) reported that chitosan nanoparticles associated with copper activated resistance mechanisms in *Coffea arabica*, showing greater efficiency and lower toxicity compared with conventional copper-based fungicides, such as copper oxychloride and copper hydroxide.

In terms of residue valorisation and product preservation, Lozada-Ramírez et al. (2023) demonstrated that the conjugation of anthocyanins extracted from coffee husks, by-products generated during the processing of *Coffea arabica* fruits, with zinc oxide nanoparticles increased antioxidant stability and preserved bioactive compounds, highlighting the potential of nanotechnology to add value and extend the shelf life of these residues. Complementing these strategies, Andrea et al. (2025) found that nano-structured chitosan coatings incorporating spent coffee ground extracts were effective in controlling *Rhizopus stolonifer* in tomatoes and prolonging the shelf life of treated fruits, underscoring the sustainable applicability of nanotechnology in postharvest management.

To complement these strategies and ensure final quality, postharvest monitoring is also being revolutionised by nanotechnological tools. The development of chemical nanosensors and electronic noses (*e-noses*) based on metal oxide sensors has enabled rapid and non-destructive detection of volatile organic compounds. The preparation of these tools involves synthesising metal oxide nanoparticles to maximise reactive surface area. The detection process is based on changes in the electrical conductivity of the metal oxide sensor surface when interacting with volatile organic compounds released by coffee. This response, analysed through chemometric modelling, can be used to discriminate beans based on their flavour and aroma profiles, classify the final product, and validate postharvest processing claims in real time (Santos-Rivera et al., 2025).

Beyond sensory classification, nanosensors offer a crucial mechanism for ensuring food safety by detecting chemical contaminants. Heavy metal pollution in soil, which is absorbed by plants, is a growing concern in coffee production. In this context, the development of nano-structured electrochemical biosensors has shown promise for ultrasensitive and rapid detection of heavy metal ions (such as cadmium and lead) directly in coffee and soil samples (Mahajan et al., 2024). This application allows producers and regulatory agencies to monitor raw material quality with precision and in real time, providing superior safety control and reducing the risk of contamination along the supply chain.

In summary, nanotechnology has established itself as a strategic ally in revolutionising the coffee value chain. Its applications range from enhancing plant resilience and optimising nutritional efficiency with nanofertilizers to supporting the circular economy through residue valorisation. Additionally, the development of nanosensors represents a strong technological trend for ultrasensitive contaminant detection and sensory quality control. Despite the advances, the need for large-scale studies and continuous environmental assessments remains crucial. Thus, the strategic integration of nanotechnology may drive a more efficient, innovative, and responsible coffee production.

6. The use of coffee by-products in green synthesis and environmental remediation

Biosynthesis, or green synthesis, refers to the process of obtaining metallic nanoparticles without the use of solvents in the reaction, instead employing microorganisms, natural products in their raw form or their derivatives (Xu et al., 2023). This process is considered environmentally friendly due to the absence of waste disposal into the environment and the potential for reusing organic matter in the synthesis (Alsaiari et al., 2023).

Summarising the reaction process, a metallic precursor is used in conjunction with a compound or microorganism that acts as a reducing agent in the reaction. In the case of natural products, their antioxidant potential, attributed to their chemical matrix, enables this function, following the typical stages of nucleation and stabilisation of the reaction (Kong et al., 2021). The chemical matrix of plants and their byproducts contributes to the reduction of metal ions and their stabilisation during nanoparticle formation (Gowda et al., 2022).

The use of plants and their derivatives in the biosynthesis process not only contributes to the reduction of solvent emissions into the environment but also promotes the reutilisation of organic matter, an ongoing challenge in many countries that are major producers of organic waste (Velden et al., 2021; Nitisuk et al., 2025). In this context, the use of agricultural by-products such as coffee proves to be highly relevant, since coffee and its residues are rich in tannins, caffeine, and polyphenols, which enable their function as reducing and stabilising agents in green synthesis processes (Manasa et al., 2021; Kusumahastuti and Gintu, 2025).

Spent coffee grounds are one of the most notable coffee by-products. Kusumahastuti and Gintu (2025) utilised them for the synthesis of silver nanoparticles, while Drummer et al. (2024) employed them in the production of cobalt oxide nanoparticles. The strategy adopted by these authors focuses on the substantial quantity of coffee waste discarded globally on a daily basis, considering that each kilogram of coffee generates at least twice its weight in wet organic matter, or spent coffee grounds (Miranda et al., 2022).

Kusumahastuti and Gintu (2025) emphasise the significance of tannins and caffeic acid present in spent coffee grounds, which facilitate the reduction of metal ions to their zero-valent state, such as silver. However, this characteristic also applies to other metals, as the stabilisation reaction results from the presence of hydroxyl groups in phenolic compounds (El-Desouky et al., 2023).

The study conducted by Drummer et al. (2024) reports the use of phytochemicals extracted from spent coffee grounds in green synthesis for wastewater treatment, with the synthesised nanoparticles demonstrating efficiency in the degradation and removal of dyes. In this regard, environmental remediation through metallic nanoparticles derived from coffee or its byproducts has also been addressed by other research groups in recent years (Yen et al., 2022; Yang et al., 2022; Xiao et al., 2024; Lan et al., 2025).

Yen et al. (2022) also employed coffee waste to adsorb fluorescent polystyrene particles, known as nanoplastics, achieving a maximum adsorption rate of 74%, using a concentration of 25 $g.L^{-1}$ of spent coffee grounds. The authors further postulate that the adsorptive capacity of coffee grounds is directly associated with hydrogen bonding between the functional groups present in coffee, particularly carboxyl and hydroxyl groups (Anastopoulos et al., 2017; Azouaou et al., 2010; Dai et al., 2019). In addition, electrostatic interactions also contribute to this property (Yen et al., 2022), while the highly porous surface structure of the material further enhances adsorption efficiency (Dai et al., 2018).

Regarding nanoparticles, Yang et al. (2023) developed a magnetic biochar nano-hybridised with magnesium oxide, using coffee grounds as the biomass source for the purpose of remediating lead-contaminated water. Biochar, a carbon-rich compound derived from organic matter, exhibits high stability and significant adsorptive potential due to its abundant functional groups (Li et al., 2019). The adsorption capacity of this nanomaterial was reported to be 321.9 $mg.g^{-1}$ of Pb^{2+}, the metal adsorbed behind biochar use, with the primary mechanism of action attributed to ion exchange. This is consistent with the well-established suitability of biochar for adsorbing contaminants, owing to its large surface area and high porosity (Li et al., 2019), as well as surface precipitation, which has been identified as a key factor in the adsorption process for this nanomaterial (Yang et al., 2023). In addition to coffee grounds, coffee husks can also be employed in the synthesis of biochar, as demonstrated in the study conducted by Lan et al. (2025), in which the resulting biochar was impregnated with copper nanoparticles for the decontamination of copper cyanide. The maximum adsorption capacity of the material reached 132.21 $mg.g^{-1}$.

Three-dimensional nanocarbon materials have also been synthesised from coffee grounds (Xiao et al., 2024). The authors reported the application of the obtained material for the removal of pollutants from wastewater, including antibiotics, radioactive compounds, and microplastics. The nanocarbon exhibited a maximum adsorption capacity of 373.1 $mg.g^{-1}$, a property attributed to various structural features. The role of polyphenols in the synthesis was given special attention because their inclusion of functional groups like hydroxyl, carboxyl, and aromatic moieties greatly increased the adsorption efficiency (Anastopoulos et al., 2017; Azouaou et al., 2010; Dai et al., 2019).

In summary, the use of coffee waste for environmental remediation is both cost-effective, owing to its wide availability, and efficient, given its high capacity for chemical reduction and stabilisation in green synthesis processes. This efficiency stems from the presence of functional groups such as hydroxyl and carboxyl, found, for instance, in phenolic compounds within the coffee's chemical matrix. These characteristics also endow its derivatives, such as biochar, with a high adsorption capacity.

Figure 2 shows the possibilities of applying coffee and its byproducts for environmental remediation and green synthesis.

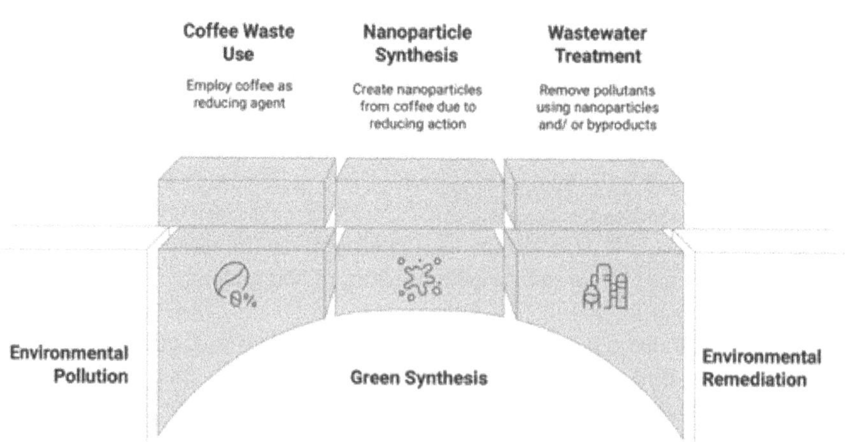

Figure 2: Properties of coffee and its byproducts for environmental remediation and green synthesis

7. Regulatory landscape of nanomaterials in consumer products

The characteristics of nanomaterials can lead to technological benefits, as described in this chapter, but they also raise concerns regarding their potential toxicity, bioaccumulation, and environmental impacts. Faced with this duality, countries have advanced in regulating nanomaterials in consumer products. However, disparities among regions are evident, with the European Union adopting more restrictive approaches, while the United States and Brazil continue to make gradual progress. The absence of specific regulations in certain areas reinforces the need for regulatory harmonisation and greater consumer transparency.

The European Union (EU) is considered a global reference in the regulation of nanomaterials. For cosmetic products, Regulation (EC) N°. 1223/2009 applies, defining nanomaterials as insoluble or biopersistent substances intentionally manufactured on a scale of 1 to 100 nm. This regulation undergoes periodic updates, particularly in the annexes that address prohibited substances, restricted substances, or labelling requirements. The main requirements include: mandatory notification, in which every product containing nanomaterials must be reported to the Cosmetic Products Notification Portal (CPNP) six months in advance; labelling, in which nano-form ingredients must be identified with the word "nano" in parentheses after the ingredient name; and safety assessment, which requires specific toxicological evaluation dossiers.

In 2024, with Regulation (EU) 2024/858, several nano-scale substances were restricted or banned, such as colloidal silver, colloidal gold, and certain copolymers. In the food sector, Regulation (EU) 2015/2283 established that nano-structured ingredients must be treated as "novel foods," requiring prior approval by the European Food Safety Authority (EFSA). Packaging containing nanoparticles, such as antimicrobial silver, must also be evaluated regarding nanoparticle migration into food. Nanopesticides and nanofertilizers are assessed under Regulation (EC) N°. 1107/2009 and REACH (Regulation 1907/2006). EFSA requires detailed studies on environmental behaviour, persistence, effects on non-target organisms, and soil mobility.

The United States does not have specific and mandatory legislation exclusively focused on nanomaterials in cosmetics, foods, or agriculture. The regulatory approach is guidance-based and sectoral. The Food and Drug Administration (FDA) recommends specific safety testing for nanomaterials, considering alterations in physicochemical properties at the nanoscale, such as enhanced skin penetration and reactivity. There is no requirement for distinct nano-labeling. The FDA evaluates nanomaterials on a case-by-case basis under the regulatory framework for food additives and food-contact materials, and the term "nano" is not mandatory on labels. The Environmental Protection Agency (EPA) regulates nanopesticides under the Federal Insecticide, Fungicide, and Rodenticide Act (FIFRA), with evaluations including toxicity, bioaccumulation, and environmental impact.

Brazil has made advances in regulating cosmetics, but it still lacks specific standards for nanomaterials in food and agriculture. The National Health Surveillance

Agency (ANVISA) regulates cosmetic products through notification or registration, depending on the risk associated with each product. In 2024, RDC N° 894 strengthened cosmetovigilance and post-market monitoring, although it does not establish specific guidelines for nanomaterials. ANVISA regulates food packaging and ingredients, but it does not yet require specific evaluation for nanomaterials or mandatory labelling. Nanopesticides, nanofertilizers, and nanocarriers are assessed according to traditional pesticide and fertiliser regulations, without dedicated nanotechnology guidelines. Research highlights concerns regarding their effects on soil organisms, beneficial microbiota, plant absorption, and bioaccumulation throughout the food chain.

8. Conclusion

The latest advances demonstrate the trend toward using natural products and their derivatives not only to take advantage of their properties, but also to provide environmental remediation, for example, through the reuse of inputs. In this sense, coffee and its by-products are no exception, with their use being improved and even intensified through the use of nanotechnology. The properties of nanotechnological materials and structures protect the molecules in the chemical matrix of coffee, thereby increasing the bioavailability and bioactivity of the active ingredients. This enables, for example, the use of coffee in the production of nanocosmetics, medicines, and functional foods. Similarly, the antioxidant action of coffee molecules and their byproducts can be utilised as chemical reducers for the green synthesis of nanoparticles, which in turn can facilitate environmental remediation through the decontamination of effluents. Also in the environmental sphere, nanotechnology favours and improves coffee cultivation, and its by-products can be used as antifungals in other crops. Thus, the potential for using nanotechnology in both the reuse of organic coffee waste and the preservation of its chemical matrix stands out, favouring and improving its application in various scientific fields.

References

Alamri, E., Rozan, M., and Bayomy, H. (2022). A study on chemical composition, antioxidants, and volatile compounds in roasted Coffea arabica. *Saudi Journal of Biological Sciences*, 29(5), 3133–3139. https://doi.org/10.1016/j.sjbs.2022.03.025

Alsaiari, N. S., Alzahrani, F. M., Amari, A., Osman, H., Harharah, H. N., Elboughdiri, N. et al. (2023). Plant and microbial approaches as green methods for the synthesis of nanomaterials: Synthesis, applications, and future perspectives. *Molecules*, 28(1), 463. https://doi.org/10.3390/molecules28010463

Amirizadeh, A., Sheikhzadeh, S., and Delirezh, N. (2025). Synthesis, characterization, and evaluation of the cytotoxic effects of caffeine nanoparticles on K562 cancer cell line. *Medical Oncology*, 42(7), 31. https://doi.org/10.1007/s12032-025-02784-4

Anastopoulos, I., Karamesouti, M., Mitropoulos, A. C., and Kyzas, G. Z. (2017). A review for coffee adsorbents. *Journal of Molecular Liquids*, 229, 555–565. https://doi.org/10.1016/j.molliq.2016.12.096

Andrea, M. J., Mariana, P. G., Mónica, H. L., Nacary, C. P. Z., Silvia, B. B., and Laura, B. N. (2025). Nanostructured chitosan coating with a coffee residue extract for the preservation of tomato and controlling pre- and postharvest disease caused by Rhizopus stolonifer. *Processes*, 13(1), 220. https://doi.org/10.3390/pr13010220

Aree, T. (2019). Understanding structures and thermodynamics of β-cyclodextrin encapsulation of chlorogenic, caffeic and quinic acids: Implications for enriching antioxidant capacity and masking bitterness in coffee. *Food Chemistry*, 293, 550–560. https://doi.org/10.1016/j.foodchem.2019.04.084

Azouaou, N., Sadaoui, Z., Djaafri, A., and Mokaddem, H. (2010). Adsorption of cadmium from aqueous solution onto untreated coffee grounds: Equilibrium, kinetics and thermodynamics. *Journal of Hazardous Materials*, 184, 126–134. https://doi.org/10.1016/j.jhazmat.2010.08.014

Ballıca, G., Çevikbaş, H., Ulusoy, S. and Yıldırım, Y. (2020). The synthesis of novel cafestol loaded zinc oxide nanoparticles and their characterization. *Applied Nanoscience*, 10, 4263-4272. https://doi.org/10.1007/s13204-020-01525-5

Bhaskar, P., Pandey, S. P., Kumar, U., Kim, H., Jayakodi, S. K., Gupta, M. K. et al. (2024). Nanobionics for sustainable crop production: Recent advances and prospects. *Frontiers in Plant Science*, 15, 1399586. https://doi.org/10.1016/j.onano.2023.100198

Brazil. National Health Surveillance Agency (ANVISA). (2024). Collegiate Board Resolution – RDC No. 894 of 27 August 2024: Good Practices in Cosmetovigilance. https://www.gov.br/anvisa

Correia, P. F. C., dos Reis, J. G. M., Amorim, P. S., da Costa, J. S. and da Silva, M. T. (2024). Impacts of Brazilian green coffee production and its logistical corridors on the international coffee market. *Logistics*, 8 (2), 39-51. https://doi.org/10.3390/logistics8020039

Costa, A. S., Alves, R. C., Vinha, A. F., Barreira, S. V. P., Nunes, M. A., Cunha, L. M. et al. (2014). Optimization of antioxidants extraction from coffee silverskin, a roasting by-product, having in view a sustainable process. *Industrial Crops and Products*, 53, 350–357. https://doi.org/10.1016/j.indcrop.2014.01.006

D'Água Natural. (2025). *Caffeine 7-actives Massage Cream.* Retrieved October 15, 2025, from https://www.daguanatural.com.br/cafeinaativos

Dai, Y., Sun, Q., Wang, W., Lu, L., Liu, M., Li, J. et al. (2018). Utilizations of agricultural waste as adsorbent for the removal of contaminants: A review. *Chemosphere*, 211, 235–253. https://doi.org/10.1016/j.chemosphere.2018.06.179

Dai, Y., Zhang, K., Meng, X., Li, J., Guan, X., Sun, Q. et al. (2019). New use for spent coffee ground as an adsorbent for tetracycline removal in water. *Chemosphere*, 215, 163–172. https://doi.org/10.1016/j.chemosphere.2018.09.150

Ding, Y., Li, Y., Zhang, Y., Liu, Y., Li, H., Wang, Q. et al. (2023). Recent trends in foliar nanofertilizers: A review. *Plants*, 12(13), 2482. https://doi.org/10.3390/nano13212906

do Carmo, L. B., Benincá, D. B., Grancieri, M., Pereira, L. V., Lima Filho, T., Saraiva, S. H. et al. (2022). Green coffee extract microencapsulated: Physicochemical characteristics, stability, bioaccessibility, and sensory acceptability through dairy beverage consumption. *International Journal of Environmental Research and Public Health*, 19(20), 13221. https://doi.org/10.3390/ijerph192013221

Drummer, S., Mkhari, O., and Chowdhury, M. (2024). Green synthesis of Co_3O_4 nanoparticles using spent coffee: Application in catalytic and photocatalytic dye degradation. *Next Nanotechnology*, 6, 100069. https://doi.org/10.1016/j.nxnano.2024.100069

El-Desouky, N., Shoueir, K., El-Mehasseb, I., and El-Kemary, M. (2023). Synthesis of silver nanoparticles using bio valorization coffee waste extract: Photocatalytic flow-rate performance, antibacterial activity, and electrochemical investigation. *Biomass Conversion and Biorefinery*, 13(17), 15871–15885. https://doi.org/10.1007/s13399-021-02256-5

European Commission. (2024). Cosmetic products – Specific topics: Nanomaterials. https://single-market-economy.ec.europa.eu/sectors/cosmetics/cosmetic-products-specific-topics/nanomaterials

European Commission. (2024). Regulation (EU) 2024/858 of 14 March 2024 amending Regulation (EC) No 1223/2009 regarding nanomaterials. https://nanotechia.org/news/european-commission-published-amendment-cosmetic-regulation

European Food Safety Authority (EFSA). (2023). Guidance on risk assessment of nanomaterials to be applied in the food and feed chain. https://health.ec.europa.eu/publications/sccs-guidance-safety-assessment-nanomaterials-cosmetics-2nd-revision_en

European Union. (2015). Regulation (EU) 2015/2283 of the European Parliament and of the Council of 25 November 2015 on novel foods. *Official Journal of the European Union*. https://eur-lex.europa.eu

Environmental Protection Agency (EPA). (n.d.). Nanomaterial pesticides overview. https://www.epa.gov

Fathi, F., Ebrahimi, S. N., Prior, J. A. V., Machado, S. M. L., Kouchaksaraee, R. M., Oliveira, M. B. P. P. et al. (2022). Formulation of nano/micro-carriers loaded with an enriched extract of coffee silverskin: Physicochemical properties, in vitro release mechanism and in silico molecular modeling. *Pharmaceutics*, 14, 112–142. https://doi.org/10.3390/pharmaceutics14010112

Fernandez-Rosillo, F., Cabrejos-Barrios, E. M., Chávez-Quintana, S. G. and Quiñones-Huatangari, L. (2025). Polyphenol degradation kinetics of specialty coffee in different presentations. *Foods*, 14(21), 3600–3621. https://doi.org/10.3390/foods14213600

Food and Drug Administration (FDA). (n.d.). Guidance for industry: Safety of nanomaterials in cosmetic products. https://www.fda.gov

Fouad, S. A., Badr, T. A., Abdelbary, A., Fadel, M., Abdelmonem, R., Jasti, B. R. et al. (2024). New insight for enhanced topical targeting of caffeine for effective cellulite treatment: In vitro characterization, permeation studies, and histological evaluation in rats. *AAPS PharmSciTech*, 25(7), 237. https://doi.org/10.1208/s12249-024-02943-2

Garcia-Solis, S. E., Perez-Perez, V., Tapia-Maruri, D., Villalobos-Castillejos, F., Arenas-Ocampo, M. L., Camacho-Diaz, B. H. et al. (2022). Microencapsulation of the green coffee waste extract with high antioxidant activity by spray-drying. *Journal of Food Processing and Preservation*, 46(10), e16864. https://doi.org/10.1111/jfpp.16864

Gomes, D. G., Sanada, K., Pieretti, J. C., Shigueoka, L. H., Sera, G. H., Seabra, A. B. et al. (2023). Nanoencapsulation boosts the copper-induced defense responses of a susceptible Coffea arabica cultivar against Hemileia vastatrix. *Antibiotics*, 12(2), 249–271. https://doi.org/10.3390/antibiotics12020249

Gowda, B. H. J., Ahmed, M. G., Chinnam, S., Paul, K., Ashrafuzzaman, M., Chavali, M. et al. (2022). Current trends in bio-waste mediated metal/metal oxide nanoparticles for drug delivery. *Journal of Drug Delivery Science and Technology*, 71, 103305. https://doi.org/10.1016/j.jddst.2022.103305

Guillén-Enríquez, J. C., Rivera-Alvarez, M. A., Ramirez-Alvarado, M., García-Cerna, J. A., and Rivera-Alvarez, R. M. (2023). Effects of foliar application of zinc sulfate and zinc nanoparticles in coffee (*Coffea arabica* L.). *Notulae Botanicae Horti Agrobotanici Cluj-Napoca*, 51(2), 12835. https://doi.org/10.15835/nbha51212835

Gupta, V., Mohapatra, S., Mishra, H., Farooq, U., Kumar, K., Ansari, M. J. et al. (2022). Nanotechnology in cosmetics and cosmeceuticals—A review of latest advancements. *Gels*, 8(3), 173–204. https://doi.org/10.3390/gels8030173

Husni, M. A., Nugroho, A. K., Sulaiman, T. N. S., and Fakhrudin, N. (2021). Microencapsulation of green coffee beans (*Coffea canephora*) extract using whey protein concentrate. *Baghdad Journal of Pharmaceutical Sciences*, 19(1), 88–96. https://bijps.uobaghdad.edu.iq/index.php/bijps/article/view/1394

International Organization for Standardization. (2005). ISO/TC 229 – Nanotechnologies. Available on: https://www.iso.org/committee/381983.html

Ishimwe, M. and Mupenzi, C. (2025). Assessing the environmental impact of post-harvest coffee processing: A case study of Kanzu Coffee Washing Station in Karambi Sector (2022–2024). *East African Journal of Environmental and Natural Research*, 8(2), 244–252. eajenr.8.2.3294

Islam, S., Ahmed, M. S., Islam, M. A., Hassain, N., & Chowdhury, M. A. (2025). Advances in nanoparticles in targeted drug delivery—A review. *Results in Surfaces and Interfaces*, 19, 100529. https://doi.org/10.1016/j.rsurfi.2025.100529

Kah, M., Kookana, R. S., Gogos, A., and Bucheli, T. D. (2018). A critical evaluation of nanopesticides and nanofertilizers against their conventional analogues. *Nature Nanotechnology*, 5(1), 20–35. https://doi.org/10.1038/s41565-018-0131-1

Kanai, N., Yamada, K., Sumida, C., Tanzawa, M., Ito, Y., Saito, T. et al. (2024). Mannan-rich holocellulose nanofibers mechanically isolated from spent coffee grounds: Structure and properties. *Carbohydrate Polymer Technologies and Applications*, 8, 100539. https://doi.org/10.1016/j.carpta.2024.100539

Kassem, A. A., Asfour, M. H., Abd El-Alim, S. H., Khattab, M., and Salama, A. (2023). Topical caffeine-loaded nanostructured lipid carriers for enhanced treatment of cellulite: A 32 full factorial design optimization and in vivo evaluation in rats. *International Journal of Pharmaceutics*, 643, 123271. https://doi.org/10.1016/j.ijpharm.2023.123271

Kong, Y., Paray, B. A., Al-Sadoon, M. K., and Albeshr, M. F. (2021). Novel green synthesis, chemical characterization, toxicity, colorectal carcinoma, antioxidant, anti-diabetic, and anticholinergic properties of silver nanoparticles: A chemopharmacological study. *Arabian Journal of Chemistry*, 14, 202106. https://doi.org/10.1016/j.arabjc.2021.103193

Konstantinidis, N., Franke, H., Schwarz, S., and Lachenmeier, D. W. (2023). Risk assessment of trigonelline in coffee and coffee by-products. *Molecules*, 28(8), 3460–3482. 10.3390/molecules28083460

Kusumahastuti, D. K. A., and Gintu, A. R. (2025). Green nanotechnology: Silver nanoparticle synthesis via brewed coffee grounds extract. *Journal of Inclusion Phenomena and Macrocyclic Chemistry*, 105(1), 241–249. https://doi.org/10.1007/s10847-025-01294-0

Lan, Y., Lu, Y., Li, B., Li, Y., Guo, J.,Yang, W. et al. (2025). Response surface method design of nano-copper modified coffee shell biochar to treat copper cyanide wastewater: Characterization and catalytic oxidation adsorption mechanism. *Materials Chemistry and Physics*, 343, 130915. https://doi.org/10.1016/j.matchemphys.2025.130915

Li, R., Deng, H., Zhang, X., Wang, J. J., Awasthi, M. K., Wang, Q. et al. (2019). High-efficiency removal of Pb(II) and humate by a CeO_2–MoS_2 hybrid magnetic biochar. *Bioresource Technology*, 273, 335–340. https://doi.org/10.1016/j.biortech.2018.10.053

Lozada-Ramírez, J. D., Guerrero-Moras, M. C., González-Peña, M. A., Silva-Pereira, T. S., Anaya de Parrodi, C., and Ortega-Regules, A. E. (2023). Stabilization of anthocyanins from coffee (*Coffea arabica* L.) husks and in vivo evaluation of their antioxidant activity. *Molecules*, 28(3), 1353–1373.https://doi.org/10.3390/molecules28031353

Lu, L., Tibpromma, S., Karunarathna, S. C., Jayawardena, R. S., Lumyong, S., Xu, J. et al. (2022). Comprehensive review of fungi in coffee. *Pathogens*, 11(4), 411–428. https://doi.org/10.3390/pathogens11040411

Mahajan, P., Khanna, V., Singh, A., and Singh, K. (2024). Advances in nanomaterial-based biosensors for heavy metal detection and remediation in soil. *Journal of the Electrochemical Society*, 171(11), 117527. https://iopscience.iop.org/article/10.1149/1945-7111/ad9413

Makiso, M. U., Tola, W. B., Ogah, O., and Endale, F. L. (2023). Bioactive compounds in coffee and their role in lowering the risk of major public health consequences: A review. *Food Science & Nutrition*, 12, 734–764. https://doi.org/10.1002/fsn3.3848

Manasa, V., Padmanabahn, A., and Appaiah, K. A. A. (2021). Utilization of coffee pulp waste for rapid recovery of pectin and polyphenols for sustainable material recycle. *Waste Management*, 120, 762–771. https://doi.org/10.1016/j.wasman.2020.10.045

Mehta, J., Pathania, K., and Pawar, S. V. (2025). Recent overview of nanotechnology based approaches for targeted delivery of nutraceuticals. *Sustainable Food Technology*, 3, 947–978. https://doi.org/10.1039/D5FB00122F

Meléndez-Mori, J. B., Lapiz-Culqui, Y. K., Huaman-Huaman, E., Zuta-Puscan, M., and Oliva-Cruz, M. (2025). Can zinc oxide nanoparticles alleviate the adverse effects of salinity stress in Coffea arabica? *Agronomy*, *15*, 1239–1257. https://doi.org/10.3390/agronomy15051239

Miranda, A., Akpobolokemi. T., Chung, E., Ren, G., & Raimi-Abraham, B. T. (2022). pH alteration in plant-mediated green synthesis and its resultant impact on antimicrobial properties of silver nanoparticles (AgNPs). *Antibiotics*, 11(11), 1592–1607. https://doi.org/10.3390/antibiotics11111592

NanoScoping. (2024). *nCaffeine Complex.* Retrieved October 15, 2025, from https://www.nanoscoping.com.br/produtos/ncaffeine-complex/

Naveed, M., Hejazi, V., Abbas, M., Kamboh, A. A., Khan, G. J., Shumzaid, M. et al. Chlorogenic acid (CGA): A pharmacological review and call for further research. *Biomedicine & Pharmacotherapy*, 97, 67–74. 10.1016/j.biopha.2017.10.064

Nitisuk, P., Wanyo, P., Chamsai, T., & Charoenjit, K. (2025). Sustainable valorization of tropical fruit peels for sustainable production of natural antioxidants and functional food ingredientes. *Sustainable Food Technology*, 3, 1189–1202. https://doi.org/10.1039/D4FB00371C

Oliveira, C., Coelho, C., Teixeira, J. A., Ferreira-Santos, P., and Botelho, C. M. (2022). Nanocarriers as active ingredients enhancers in the cosmetic industry—The European and North America regulation challenges. *Molecules*, 27(5), 1669–1704. https://doi.org/10.3390/molecules27051669

Park, S., Park, S. H., and Park, S. I. (2024). Analysis of Coffea canephora bean supercritical extract and raffinate and development of active material. *International Journal of Pharmacology*, 20(3), 373–382. https://doi.org/10.3923/ijp.2024.373.382

Rambaran, T., and Schirhagl, R. (2022). Nanotechnology from lab to industry – A look at current trends. *Nanoscale Advances*, 4(18), 3664–3675. 10.1039/d2na00439a

Ratliff, W. A., Saykally, J. N., Mervis, R. F., Lin, X., Cao, C., and Citron, B. A. (2019). Behavior, protein and dendritic changes after model traumatic brain injury and treatment with nanocoffee particles. *BMC Neuroscience*, 20(44), 46. https://doi.org/10.1186/s12868-019-0525-5

Rege, A., and Lee, J. S. H. (2022). State agricultural subsidies drive the expansion of cashew monoculture in northern Western Ghats, India. *PLoS ONE*, 17(6), e0269092. https://doi.org/10.1371/journal.pone.0269092

Rojas-Orduña, E.,Hernández-Carrión, M., Gómez-Franco, J. D., Narváez-Cuenca, C. E. and Sánchez-Camargo, A. P. (2023). Utilization of red and yellow Coffea arabica var. Caturra pulp: Macronutrient analysis, carotenoid extraction, and encapsulation for dairy product enrichment. *Frontiers in Nutrition*, 10, 1–14. https://doi.org/10.3389/fnut.2023.1231049

Rossi, L., Fedenia L. N., Sharifan, H., Ma, X., and Lombardini, L. (2019). Effects of foliar application of zinc sulfate and zinc nanoparticles in coffee (*Coffea arabica* L.) plants. *Plant Physiology and Biochemistry*, 135, 160–166. https://doi.org/10.1016/j.plaphy.2018.12.005

Saidi, L., Zia, M. B., Gullins, A., Wang, Y., Wich, P. R., and Selomulya, C. (2025). Enhancing the water resistance of carboxymethyl cellulose films with cellulose nanofibers from spent coffee grounds and grapeseed oil. *Food and Bioprocess Technology*, 18, 10739–10756. https://doi.org/10.1007/s11947-025-04068-1

Santos-Rivera, M., Viswanathan, L., and Sheibani, F. (2025). Enhancing coffee quality and traceability: Chemometric modeling for post-harvest processing classification using near-infrared spectroscopy. *Spectroscopy Journal*, 3(2), 20–37. https://doi.org/10.3390/spectroscj3020020

Sathiyaseelan, A., Saravanakumar, K., Jayalakshmi, J., Gopi, M., Shajahan, A., Barathikannan, K. et al. (2020). Trigonelline-loaded chitosan nanoparticles prompted antitumor activity on glioma cells and biocompatibility with pheochromocytoma cells. *International Journal of Biological Macromolecules*, 163, 36–43. https://doi.org/10.1016/j.ijbiomac.2020.06.165

Saud, S., and Salamatullah, A. M. (2021). Relationship between the chemical composition and the biological functions of coffee. *Molecules*, 26(24), 7634–7650. https://doi.org/10.3390/molecules26247634

Sguizzato, M., Mariani, P., Ferrara, F., Drechsler, M., Hallan, S. S., Huang, N. et al. (2020). Nanoparticulate gels for cutaneous administration of caffeic acid. *Nanomaterials*, 10(5), 961–980. https://doi.org/10.3390/nano10050961

Shetty, K. S., Koland, M., and Ashwini, M. (2024). Trigonelline nanocrystal-based topical creams for the treatment of hair loss. *Journal of Advanced Pharmaceutical Technology & Research*, 15(4), 297–303. 10.4103/JAPTR.JAPTR_26_24

Siddique, M. H., Bukhari, S., Khan, U. I., Essa, A., Ali, Z., Sabir, U. et al. (2023). In silico, in vitro, and in vivo evaluation of caffeine-coated nanoparticles as a promising therapeutic avenue for AML through NF-kappa B and TRAIL pathways modulation. *Pharmaceuticals*, 16(12), 1742–1761. https://doi.org/10.3390/ph16121742

Soundhararajan, R., & Srinivasan, H. (2023). Influence of polyherbal nanoformulation on plant growth and biochemical constituents in legume seedlings. *Journal of Plant Growth Regulation*, 42, 7142–7150. https://doi.org/10.1007/s00344-023-11003-1

Tavares, D. G., De Souza, M. A. M., Santos, T. L. d., Silva, A. d. A. D., Abreu, D. J. M. d., and Duarte, W. F. (2024). Co-encapsulation of coffee and coffee by-product extracts with probiotic *Kluyveromyces lactis*. *Foods*, 16(19), 3056–3074. https://doi.org/10.3390/foods13193056

Thepphankulngarm, N., Rattanawaleedirojn, P., and Kraisit, P. (2024). Nanotechnology-driven delivery of caffeine using ultradeformable liposomes-coated hollow mesoporous silica nanoparticles for enhanced follicular delivery and treatment of androgenetic alopecia. *International Journal of Molecular Sciences*, 25(22), 12170–12200. https://doi.org/10.3390/ijms252212170

Tulípia Cosméticos. (2022). *Nano Caffeine.* Retrieved October 15, 2025. https://tulipia.com.br/principios-ativos-cosmeticos/nano-caffeine?

Van der Velden, R., da Fonseca-Zang, W., Zang, J. W., Clyde-Smith, D., Leandro, W. M., Parikh, P. et al. (2021). Closed-loop organic waste management systems for family farmers in Brazil. *Environmental Technology*, 43(15), 2252–2269. https://doi.org/10.1080/09593330.2021.1871660

Wardana, A. A, Febrianto, N. A., Marcellino, V., Wigati, L. P., Nkede, F. N., Tanaka, F. et al. (2025) Potency of coffee by-products as sustainable added material in bio/edible film development. *Frontiers in Sustainable Food Systems*, 9, 1667255. https://doi.org/10.3389/fsufs.2025.1667255

Xiao, G., Guo, J., Zheng, M., Zhang, J., Chen, L., Dai, M. et al. (2024). Tea/coffee sustainable nanoarchitectures purify wastewater. *Nano Letters*, 24(49), 15509–15516. https://doi.org/10.1021/acs.nanolett.4c03070

Xu, J., Yıldıztekin, M., Han, D., Keskin, C., Baran, A., Baran, M. F. et al. (2023). Biosynthesis, characterization, and investigation of antimicrobial and cytotoxic activities of silver nanoparticles using Solanum tuberosum peel aqueous extract. *Heliyon*, 9(8), e19061. https://doi.org/10.1016/j.heliyon.2023.e19061

Yang, S., Wang, C., Li, B., Chen, H., and Wang, J. (2023) Removal of Pb^{2+} from aqueous solution using an MgO nano-hybridized magnetic biochar from spent coffee grounds. *Chemical Physics Letters*, 833, 140894. https://doi.org/10.1016/j.cplett.2023.140894

Yen, P., Hsu, C., Huang, M., and Liau, V. H-C. (2022). Removal of nano-sized polystyrene plastic from aqueous solutions using untreated coffee grounds. *Chemosphere*, 286, 131863. https://doi.org/10.1016/j.chemosphere.2021.131863

CHAPTER

9

Nanoparticles Derived from *Coffea arabica*: Emerging Strategies for Antimicrobial and Cancer Therapy

Gabriel Toneto Druzian, Mariele Samuel Nascimento, Altevir Rossato Viana, Fabio Andrei Duarte, and Erico Marlon Moraes Flores*

Department of Chemistry, Federal University of Santa Maria, 97105-900, Santa Maria, Brazil

This chapter discusses the potential of *Coffea arabica* as a sustainable and multifunctional resource for the green synthesis of nanoparticles with biomedical applications. Emphasis is given to the global challenges posed by antimicrobial resistance and the limitations of conventional cancer therapies, highlighting the urgent need for innovative and selective therapeutic strategies. The chapter reviews the fundamental principles of nanotechnology, the biological mechanisms underlying the antimicrobial and anticancer activity of nanoparticles, and the advantages of plant-mediated synthesis routes over conventional physical and chemical methods. The chemical composition of *Coffea arabica* extracts, rich in phenolic compounds, carbohydrates, and organic acids, is presented as a key factor enabling nanoparticle reduction, stabilisation, and biological functionality. Experimental studies demonstrating the antimicrobial and anticancer performance of nanoparticles synthesised from coffee extracts are critically analysed, with attention to particle size, morphology, synthesis conditions, and biological outcomes. Finally, the chapter discusses current limitations related to reproducibility and standardisation, emphasising the importance of systematic evaluation of experimental parameters to advance the practical implementation of coffee-derived nanoparticles in green nanomedicine.

*Corresponding author: ericommf@gmail.com

DOI: 10.1201/9781003689393-9

1. Introduction

In recent years, nanotechnology has emerged as one of the most important scientific fields for the development of alternative therapeutic approaches (dos Santos et al., 2024; Silva et al., 2023; Szczyglewska et al., 2023). Nanoparticles (NPs) have gained increasing attention due to their unique physicochemical properties compared to bulk materials (Khan et al., 2019). These properties include high surface area and functionalisation capability, as well as specific nanoscale behaviour (Joudeh and Linke, 2022; Khan et al., 2019). Due to these characteristics, NPs have been studied for antibacterial (Hancharova et al., 2024; Luo et al., 2024), antiviral (Yasamineh et al., 2022), antitumor (Alves et al., 2024; da Silva et al., 2019), antifungal (Slavin and Bach, 2022; Umamaheswari and Abirami, 2023), and antiparasitic (Alrefaee et al., 2025; Attia et al., 2022) activities.

Conventional methods for synthesising NPs generally employ physical and chemical processes, which are well established (Khan et al., 2019; Kumari et al., 2023). However, these methods often require high investments, which increases the production of NPs, and, in some cases, the use of toxic chemical reagents is necessary (Khan et al., 2019). Depending on the production route, undesirable byproducts can be obtained, and combined with intense energy consumption, contribute to increased environmental impacts (Kumari et al., 2023; Saxena et al., 2025). For this reason, approaches based on the use of plant extracts for the synthesis and stabilisation of NPs have emerged as a scientific and technological advance, offering important strategies for new, more cost-effective, environmentally friendly, and sustainable synthetic routes (Osman et al., 2024; Ying et al., 2022).

Among the plant extracts that can be used for the synthesis of NPs, the *Coffea arabica* has emerged. This plant is one of the most cultivated coffee species in the world. The potential of coffee for the synthesis of NPs may be related to the presence of organic compounds, such as phenolic compounds, polysaccharides, and organic acids (Abdelmigid et al., 2022; Ashraf et al., 2022; Bekmukhametova et al., 2022; Makiso et al., 2024; Sunoqrot et al., 2021; Wang et al., 2022). The presence of these constituents can facilitate the reduction of metal precursors and confer biological functionalities to the synthesised NPs (Chien et al., 2019; Makiso et al., 2024). Thus, recent research has explored *C. arabica* extracts as multifunctional agents in NP synthesis, highlighting their ability to act simultaneously as a reducer, stabiliser, and source of bioactive compounds, which expands their potential for therapeutic applications (Ashraf et al., 2022; Bekmukhametova et al., 2022; Chien et al., 2019; El-Desouky et al., 2023; Panzella et al., 2020; Trejo-Teniente et al., 2024).

The biomedical importance of NPs synthesised from *C. arabica* is justified by two major global challenges: antimicrobial resistance (AMR) and the limitations of conventional cancer therapies. Antimicrobial resistance is a global health threat that compromises the effectiveness of traditional treatments, leading to numerous deaths (Ruckert et al., 2024; Sartorius et al., 2024; Tacconelli et al., 2018). Regarding cancer treatment, traditional therapies often lack selectivity, resulting in many adverse effects and impacting quality of life. In this scenario, the development of NPs derived from *C. arabica* presents an innovative and sustainable alternative to fill these gaps (Sulastri et al., 2025; Trejo-Teniente et al., 2024).

Considering the aforementioned aspects, this chapter discusses the potential of *C. arabica* extracts for the synthesis of NPs. Its role in the development of NPs with antimicrobial and anticancer activity is particularly notable, reinforcing the importance of *C. arabica* as a sustainable and functional resource in the production of these materials.

2. Antimicrobial resistance: Current challenges

Antimicrobial resistance represents a global health challenge. It is characterised by the ability of microorganisms to resist the action of antimicrobial agents that were previously effective drugs, allowing them to survive and multiply (Ferrara et al., 2024; Ho et al., 2025). MacNair et al. (2024) have reported that within a short period, more than 70% of pathogenic bacteria exhibit resistance to at least one antimicrobial drug (MacNair et al., 2024). Recent data attribute approximately five million deaths annually to bacterial AMR, a number that is expected to increase considerably by 2050 (Murray et al., 2022).

The COVID-19 pandemic has further intensified the problem of AMR, as widespread and often inappropriate antibiotic use accelerated the emergence and dissemination of resistant pathogens (Ho et al., 2025). In addition, the misuse of antibiotics extends beyond clinical settings, encompassing livestock and agricultural practices (Ferrara et al., 2024). This scenario poses a threat to public health, food security, and the healthcare system. In addition to these behavioural and epidemiological factors, the pharmaceutical landscape also reflects the growing difficulty in combating AMR (Panzella et al., 2020).

Since 2010, twenty-nine antibiotics have been approved for clinical use, most of them being modifications of existing classes, including carbapenems, aminoglycosides, and macrolides (Ho et al., 2025). Despite efforts to overcome AMR mechanisms through structural changes, cross-resistance between older and newer antibiotics within the same class frequently occurs. This is due to the abundance of resistance genes within resistomes, which limits the efficacy of new compounds against specific multidrug-resistant pathogens (Ho et al., 2025).

A global assessment estimated that AMR contributes to approximately 25.4 million hospital admissions annually, corresponding to 3.5% of all hospitalisations worldwide. The aggregate cost of these hospitalisations is estimated at US$66.4 billion per year (Laurence et al., 2025). However, these economic burdens are not evenly distributed: high-income countries have higher expenditures, reflecting the adoption of more intensive and resource-demanding treatment protocols. This disparity highlights the complex interplay between AMR, health infrastructure, and economic capacity, highlighting the need for targeted interventions to reduce the clinical and financial impacts on global health systems (Laurence et al., 2025). In addition to the clinical and economic burdens, AMR underscores the urgent need for innovative antibacterial strategies that circumvent existing resistance mechanisms and restore therapeutic efficacy (Ho et al., 2025).

3. Cancer disease: Current challenges

According to data from the World Health Organization (WHO) for 2022, approximately 20 million new cases of cancer and 9.7 million deaths associated with the disease were recorded worldwide (Organization, 2024). The WHO estimates that one in five individuals will develop cancer at some point in their lifetime. This situation is projected to worsen, with estimates suggesting that by 2050 the annual number of new cases could reach 35 million (an increase of 77% compared to 2022) (Organization, 2024). In Sweden, for example, the number of diagnoses in individuals under 50 years of age has nearly tripled since 1990 (Organization, 2024). Among the different types, breast cancer was the most diagnosed in women in more than 157 countries, while lung cancer ranked as the most common cancer among men (Organization, 2024). In addition to the high global incidence, several systemic and socioeconomic factors further exacerbate cancer problems worldwide.

The high incidence of cancer is exacerbated by organisational challenges, a lack of molecular diagnostics, limited access to advanced therapies (particularly in developing countries), the unavailability of essential medicines, bureaucratic hurdles in approving new treatments, low patient adherence, and political barriers (Barrios et al., 2023). Collectively, these issues contribute to making cancer a global crisis of elevated magnitude (Barrios et al., 2023). Another critical factor limiting cancer treatment is the relatively high cost of oncology drugs. According to a study by Leighl et al. (2021) since 1995, the prices of cancer drugs have increased significantly, resulting in financial hardship for patients. In many cases, this financial burden can lead patients to interrupt or discontinue treatment, compromising the therapeutic efficacy of the drugs (Leighl et al., 2021).

In addition to the socioeconomic challenges associated with cancer diagnosis and treatment, conventional therapeutic treatments also face limitations of a biological and clinical nature. The main approaches currently employed include chemotherapy, radiotherapy, immunotherapy, and targeted therapies. Despite the recent advances, these strategies still have shortcomings, particularly in terms of low selectivity between tumour cells and healthy tissue (Anand et al., 2023; Leighl et al., 2021; Organisation, 2024). This lack of specificity can cause a range of systemic effects, including nausea, vomiting, intense fatigue, anaemia, generalised inflammation, and, in some cases, severe autoimmune reactions (Anand et al., 2023; Leighl et al., 2021; Organisation, 2024).

Another challenge concerns the ability of tumours to develop resistance mechanisms to both chemotherapy and immunotherapy. This resistance can result from genetic mutations, changes in the tumour microenvironment, or overexpression of cellular efflux pumps, which reduce intracellular drug concentrations. As a result, therapeutic efficacy tends to decrease progressively, leading to the need for higher doses, which intensify side effects, or replacement with alternative therapies, which are often more toxic and expensive (Anand et al., 2023; Leighl et al., 2021; Organization, 2024).

Considering the limitations of current cancer treatments, the development of alternative and more precise therapeutic strategies is essential. These approaches should aim not only to minimise effects and overcome drug resistance but also to enhance therapeutic selectivity and patient quality of life. Recent advances in nanotechnology, gene therapy, and personalised medicine offer new approaches to addressing persistent challenges (Anand et al., 2023; Peer et al., 2007; Tian et al., 2022).

4. Nanotechnology and applications

Some researchers even suggest that the level of nanotechnology development within a country may serve as a predictor of its position in the global economy (Bayda et al., 2019; Szczyglewska et al., 2023). Numerous types of nanoscale materials (with dimensions smaller than 100 nm) have been developed, with NPs being among the most notable outcomes of this progress (Guerra et al., 2024; Joudeh and Linke, 2022; Khan et al., 2019).

The NPs can be classified according to their morphology, size, and chemical characteristics, encompassing categories such as carbon-based, ceramic, semiconductor, polymeric, lipid-based, and metallic NPs (Guerra et al., 2024; Joudeh and Linke, 2022; Khan et al., 2019). In general, due to their extremely small size and high surface-to-volume ratio compared to the corresponding bulk materials, the NPs show unique optical, mechanical, electronic, magnetic, and thermal properties (da Silva et al., 2019; Hancharova et al., 2024; Mohamed et al., 2023). For this reason, numerous studies have reported their applications in agriculture, food science, cosmetics, and chemical industries, materials engineering, military and security technologies, as well as in medicine (da Silva et al., 2019; Hancharova et al., 2024; Szczyglewska et al., 2023). These diverse applications highlight the versatility of nanotechnology and its increasing importance in both industrial and biomedical fields (Guerra et al., 2024; Joudeh and Linke, 2022; Khan et al., 2019).

4.1 Nanotechnology and the search for emerging strategies for antimicrobial treatment

In the medical field, NPs have also been explored for the biological control of several microorganisms, standing out for their antimicrobial efficiency (Hancharova et al., 2024; Luo et al., 2024), as well as their antiviral (Yasamineh et al., 2022), antifungal (Slavin and Bach, 2022; Umamaheswari and Abirami, 2023), and antiparasitic (Alrefaee et al., 2025; Attia et al., 2022) potentials. Nanoparticles of silver (AgNPs), zinc oxide (ZnONPs), gold (AuNPs), and copper (CuNPs), e.g., can exhibit antimicrobial activity and have proven to be effective against a wide range of microorganisms, including *Escherichia coli*, *Proteus vulgaris*, *Staphylococcus aureus*, *Pseudomonas aeruginosa*, *Micrococcus luteus*, *Shigella sonnei*, *Vibrio cholera*, and *Enterococcus faecalis*, among others (Dhand et al., 2016; More et al., 2023; Raghupathi et al., 2011). Such findings reinforce the potential of nanotechnology as a promising alternative to conventional antimicrobials, offering new possibilities to address the global challenge of drug resistance (Dhand et al., 2016; More et al., 2023; Raghupathi et al., 2011).

In relation to the mechanisms of action of NPs, AgNPs can adhere to the bacterial cell membrane through electrostatic interactions between the positively charged NPs and the negatively charged phosphorus and sulfur groups present on the membrane surface. This interaction leads to increased membrane permeability, the formation of "pits" (pores), leakage of essential biomolecules, and ultimately, cell rupture (More et al., 2023). After crossing the cell membrane, AgNPs can penetrate the cytoplasm and cause damage to intracellular organelles, such as ribosomes, as well as to genetic material, including DNA (More et al., 2023). After crossing the cell membrane, AgNPs can penetrate the cytoplasm and damage organelles such as ribosomes and DNA. This disruption compromises vital cellular functions, including protein synthesis and the replication of genetic material (More et al., 2023).

The AgNPs can also induce the generation of reactive oxygen species (ROS, as hydroxyl radicals, superoxide ions, and hydrogen peroxide), which further damage cellular components such as membranes, proteins, lipids, and DNA. This oxidative stress overload ultimately leads to cell death (More et al., 2023). Moreover, the Ag^+ ions released from NPs can bind to thiol groups (–SH) in respiratory enzymes, leading to their inactivation. This process disrupts ATP production and compromises cellular metabolism. AgNPs can also bind to or intercalate with DNA, resulting in hydrogen bond breakage, condensation, and inhibition of replication and transcription (More et al., 2023). The controlled release of Ag^+ ions through a process known as oxidative dissolution plays a central role in these mechanisms, as these ions are highly toxic to microorganisms and account for many of the mentioned antimicrobial effects (More et al., 2023).

Other types of NPs have also demonstrated important antimicrobial activity. Among them, ZnONPs stand out, showing effectiveness against *S. aureus*, *E. coli*, *Bacillus subtilis*, and *Klebsiella pneumoniae* (Raghupathi et al., 2011). The antimicrobial effects of ZnONPs are mainly attributed to the generation of ROS, particularly under UV light exposure, the release of Zn^{2+} ions, and damage to the bacterial cell membrane (Raghupathi et al., 2011). The CuNPs have also demonstrated notable antimicrobial efficacy, being effective against *E. coli*, *S. aureus*, and Candida species (Ruparelia et al., 2008). Their antimicrobial activity is likely due to oxidative damage to membranes and proteins, as well as binding to DNA, leading to the disruption of essential biochemical functions (Ruparelia et al., 2008). Similarly, a critical discussion on the efficacy of metallic nanoparticle-based treatments for cutaneous leishmaniasis was conducted (Guerra et al., 2024). Their activity is primarily driven by the generation of ROS and nitrogen species, which induce DNA damage (genotoxicity) and protein denaturation, inhibit the antioxidant defence system, disrupt mitochondrial function, alter calcium homeostasis, and promote lipid peroxidation (Guerra et al., 2024).

Besides antibacterial and antiparasitic efficacy, the NPs also exhibit antiviral activity (Yasamineh et al., 2022). Certain NPs, such as AgNPs, possess direct antiviral properties by interfering with viral entry, replication, and structural integrity (Yasamineh et al., 2022). Studies have shown that AgNPs can inhibit the replication of viruses, including SARS-CoV-2, representing a promising therapeutic approach (Yasamineh et al., 2022). Beyond direct antiviral effects, NPs can serve as platforms for the development of viral vaccines, offering advantages such as enhanced antigenic stability and the ability to elicit robust immune responses (Yasamineh et al., 2022). Additionally, NPs can be utilised to capture and neutralise viruses, thereby facilitating their clearance by the immune system (Yasamineh et al., 2022). These NPs-based structures can mimic host cells, attracting viruses through specific receptors and promoting their phagocytosis (Yasamineh et al., 2022). Controlled-release systems based on NPs can also enhance the bioavailability and efficacy of antiviral drugs, enabling localised delivery while reducing systemic side effects (Yasamineh et al., 2022).

In general, the mechanisms of action of NPs against viruses are diverse. Nanoparticles can bind to viral surface proteins, such as the SARS-CoV-2 spike protein, preventing their interaction with host cell receptors. They can also interact with the viral membrane or capsid, causing structural destabilisation, inhibiting fusion with the host cell, and blocking viral replication. Additionally, NPs can also induce the generation of ROS, further contributing to viral inactivation (Yasamineh et al., 2022). Several NPs-based strategies have been evaluated against viral infections, including H1N1, H3N2, influenza, HIV-1, HBV, EBOV, and IVA (Yasamineh et al., 2022).

4.2 Nanotechnology and the search for emerging strategies for cancer treatment

The NPs have also emerged as a promising alternative in cancer therapy. They can offer greater specificity, minimising adverse effects, and can be used in combination with existing treatment modalities (Alves et al., 2024; da Silva et al., 2019; Kavin et al., 2025; Trejo-Teniente et al., 2024; Wang et al., 2024). In addition, NPs hold potential as contrast agents, enabling high-resolution tumour detection through advanced imaging techniques (Alves et al., 2024; da Silva et al., 2019; Kavin et al., 2025; Trejo-Teniente et al., 2024; Wang et al., 2024). In the therapeutic field, NPs have demonstrated innovative results and are considered strong candidates for medical applications, and these NPs (such as liposomes) can be applied in cancer treatment (Wang et al., 2024):

(i) NPs can be loaded with chemotherapeutic agents and release them in a controlled manner at the tumour site, thereby increasing efficacy and reducing side effects;
(ii) certain NPs, such as AgNPs and AuNPs, can absorb light and generate heat, leading to localised hyperthermia and tumour cell destruction;
(iii) NPs can enhance the radiosensitivity of tumour cells, improving the effectiveness of radiotherapy;
(iv) When combined with photosensitising agents, NPs can mediate photodynamic therapy, in which light exposure generates ROS that damage tumour cells;
(v) Under the influence of external magnetic fields, NPs can be directed toward specific tumours, enabling localised treatment;
(vi) NPs can serve as carriers of therapeutic genes or interfering RNA molecules (siRNA, miRNA), thereby silencing oncogenes or inducing apoptosis;
(vii) NPs can overcome tumour resistance mechanisms, such as efflux pumps, thus increasing the intracellular concentration of drugs (Allen and Cullis, 2013; Kanasty et al., 2013; Peer et al., 2007; Wang et al., 2024).

In recent years, several nanopharmaceuticals have already been approved by the U.S. Food and Drug Administration (FDA), including Doxil® (for ovarian cancer), DaunoXome® (for Kaposi sarcoma), Onivyde® (for metastatic pancreatic cancer), and Arikayce® (for pulmonary infections caused by *Mycobacterium avium* complex) (Administration). Given the remarkable physicochemical properties of NPs and the growing need for novel therapeutic approaches in cancer treatment, numerous studies have highlighted the application of nanotechnology in this scientific field (Kavin et al., 2025; More et al., 2023; Wang et al., 2022; Yasamineh et al., 2022). These advances underscore nanotechnology as a crucial tool in modern oncology, enabling the integration of conventional treatments and targeted therapeutic solutions.

5. Nanoparticle synthesis for antimicrobial and anticancer applications

Considering the remarkable applications of NPs, it is crucial to develop synthetic routes with high quality standards, particularly for biological applications (Silva et al., 2023; Szczyglewska et al., 2023). However, conventional methods commonly employed in NPs synthesis present several limitations, including high costs, challenges in large-scale implementation, the use of toxic reagents, and, in some cases, the requirement of cleanroom facilities (Gandlevskiy et al., 2025; Khan et al., 2019). Moreover, relatively large volumes of waste may also be generated. For these reasons, alternative synthesis approaches have been investigated (Saxena et al., 2025; Ying et al., 2022). Among them, aqueous plant extracts have gained special attention. Due to the presence of organic reducing compounds (e.g., flavonoids, alkaloids, polyphenols, sugars, terpenoids), plant extracts can act simultaneously as reducing agents of metal ions and stabilisers of NPs, making them a valuable alternative to conventional synthesis methods (Gandlevskiy et al., 2025; Saxena et al., 2025).

In addition to the mentioned advantages, the use of aqueous plant extracts is aligned with the principles of green chemistry and circular chemistry, offering more sustainable and environmentally friendly synthetic routes with relatively lower costs (Ashraf et al., 2022; Chien et al., 2019). This approach minimises the use of hazardous chemicals and reduces energy consumption. Furthermore, it promotes the valorisation of renewable natural resources and, in some cases, enables the utilisation of agricultural byproducts or plant residues, thereby contributing to waste reduction and resource circularity. Collectively, these aspects highlight the potential of plant-mediated synthesis not only as a safer and more economical alternative but also as a pathway consistent with global efforts toward sustainable development (Abdelmigid et al., 2022; Ashraf et al., 2022; Chien et al., 2019; Dhand et al., 2016).

6. *Coffea arabica* as a green source for NP synthesis against AMR and limitations of conventional cancer therapies

The aqueous extracts of coffee (*C. arabica*) contain various biomolecules in their composition, such as carbohydrates, amino acids, proteins, and phenolic compounds, making them a promising and sustainable route for the green synthesis and stabilisation of NPs. Furthermore, the use of aqueous coffee extracts offers a low-cost, widely available, and easy-to-prepare alternative, which reinforces their potential for biotechnological and industrial applications. Nanoparticles obtained from coffee have demonstrated excellent physicochemical properties and biological activities, including antimicrobial, antioxidant, and catalytic actions, which expand their potential applications in areas such as medicine, pharmacology, and agriculture (Makiso et al., 2024; Sulastri et al., 2025; Sunoqrot et al., 2021).

The biomedical importance of NPs obtained from *C. arabica* is notable in light of two major current challenges: AMR and the limitations of conventional cancer therapies. In this context, NPs derived from *C. arabica* emerge as an innovative, sustainable, and multifunctional alternative, with the potential to overcome current therapeutic limitations and contribute to the advancement of green nanomedicine. Figure 1 illustrates a schematic representation of *C. arabica* extracts for the synthesis of NPs with potential applications against AMR and cancer therapies.

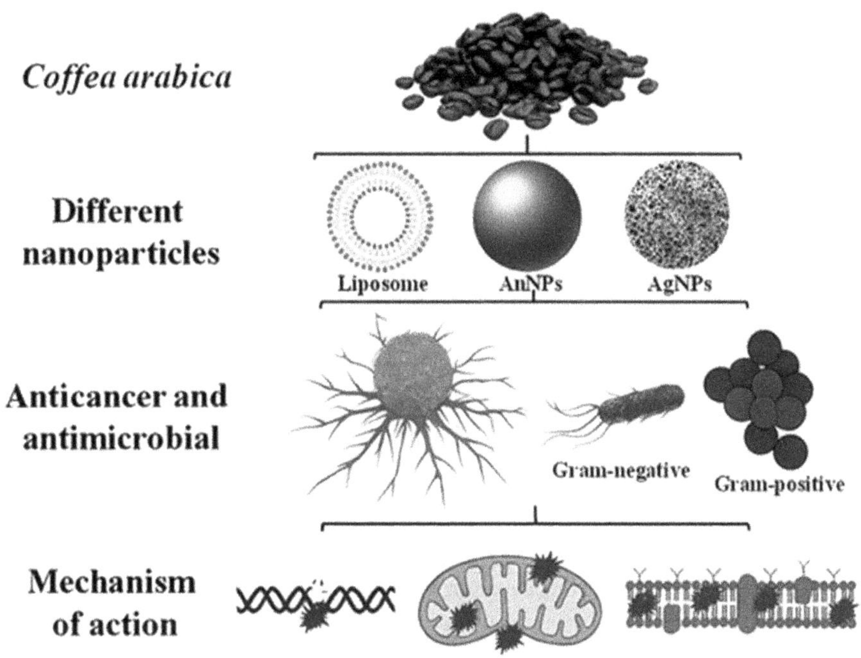

Figure 1: Schematic representation of *C. arabica* extract-mediated synthesis of NPs and their antimicrobial and cancer treatment applications

As illustrated in Figure 1, *C. arabica* extracts demonstrate the ability to mediate the synthesis of various types of NPs, including AgNPs and AuNPs (Khan et al., 2019; More et al., 2023). The diversity of nanomaterials derived from coffee extracts is attributed to the complex composition of organic compounds that participate in the reduction and stabilisation processes of NPs. The biological potential of *C. arabica* is also evidenced by *in vitro* cytotoxic evaluation of liposomal formulations against pancreatic cancer cell lines, showing promising outcomes for biomedical applications, particularly in liposome production (Zaremba-Czogalla et al., 2020). Furthermore, NPs synthesised from *C. arabica* extracts exhibit remarkable antibacterial activity against both Gram-positive and Gram-negative bacteria (Dhand et al., 2016; Kavin et al., 2025; Raghupathi et al., 2011). This antimicrobial effect is primarily attributed to their ability to generate ROS, which interact with cellular components such as proteins and lipids, leading to membrane damage, leakage of intracellular contents, and ultimately bacterial cell death (Guerra et al., 2024; More et al., 2023; Raghupathi et al., 2011; Slavin and Bach, 2022). Collectively, these findings underscore the versatile and sustainable nature of *C. arabica* as a valuable source for the biosynthesis of functional nanomaterials with broad biomedical potential. These applications will be discussed in more detail in section 6.1.

6.1 Antimicrobial applications of NPs synthesised from *C. arabica*

Several studies have investigated the use of coffee in NPs synthesis for antimicrobial applications. These studies are summarised in Table 1.

According to Table 1, a variety of NPs (Ag, ZnO, Fe_3O_4, Au, polymeric, and chitosan-based systems) have been successfully synthesised using coffee extracts as a green and sustainable reducing agent (Dhand et al., 2016; Kavin et al., 2025; Panzella et al., 2020). The resulting NPs exhibited mean particle sizes ranging from 20 to 250 nm and demonstrated important antimicrobial activity against bacterial, fungal, and parasitic microorganisms (Ashraf et al., 2022; Kavin et al., 2025). Most studies on NPs synthesised using coffee aqueous extract have focused on antibacterial assays (75%), followed by investigations against fungi (20%) and parasites (5%). This emphasis on antibacterial applications may be attributed to the global concern over bacterial resistance and the availability of standardised in vitro evaluation models.

Among the ten studies summarised in Table 1, *E. coli* and *S. aureus* were used as model microorganisms in eleven and twelve studies, respectively, followed by *P. aeruginosa* (six studies), *B. subtilis* (three studies), *E. aerogenes*, *B. cereus*, *P. amygdali*, *P. savastanoi*, *P. corrugate*, *P. syringae*, and *K. pneumoniae* (one study each). In contrast, relatively few studies have investigated antifungal and antiparasitic effects, and no reports were found on the antiviral potential of these NPs, highlighting a research gap. Expanding studies to include parasitic, fungal, and viral pathogens is essential, as these microorganisms pose major challenges in both agricultural and clinical contexts. Such efforts could not only extend the antimicrobial spectrum attributed to *C. arabica*-mediated NPs but also provide valuable insights into their underlying mechanisms and broaden their potential applications in medicine and crop protection.

Despite the promising antimicrobial activity reported for NPs synthesised using *C. arabica* extracts, it is essential to acknowledge the limitations and factors that influence their efficacy. Not all synthesised NPs exhibit uniform bioactivity; for instance, ZnO NPs synthesised from coffee extracts showed negligible antimicrobial effects against *P. aeruginosa*, *S. aureus*, *E. aerogenes*, and *K. pneumonia* (Abdelmigid et al., 2022). Such variations can be attributed to multiple parameters, including

Table 1: Overview of selected studies on the green synthesis of MNPs using *C. arabica* for antimicrobial applications

NPs	Synthesis conditions	Mean size	Application	Effective concentration	References
Ag	Extract concentration: 1% (w v^{-1}) $AgNO_3$ concentration: 1 mmol L^{-1} Proportion between extract and $AgNO_3$ solution: 1:9 v v^{-1}	147 nm	Antibacterial	IC_{50}: 16.7 µg mL^{-1} Inhibition zone of 8.2, 5.3, 4.0, and 3.8 mm for *P. aeruginosa, B. subtilis, E. coli* and *S. aureus,* respectively	Kavin et al., 202)
Fe_3O_4	$Fe(NO_3)_3.9H_2O)$ concentration: 0.3 mol L^{-1} Extract volume: 15 mL	BET: 23.90 m^2 g^{-1} Pore size: 6.3171 nm	Antibacterial activity	Bactericidal activity against *S. aureus*: 125 µg mL^{-1} Bactericidal activity against *E. coli:* 150 µg mL^{-1} Inhibited *S. aureus* at 20 mm Inhibited *E. coli* at 16 mm	Mohamed et al., 2023
Ag	Extract concentration: 100 µg mL^{-1} (10 mL) $AgNO_3$ concentration: 0.1%	20 to 70 nm	Antileishmanial and antibacterial activity	IC_{50} values of the AgNPs were 65.4 and 47.70 µg mL^{-1} against promastigotes and amastigotes of *Leishmania major* Minimum inhibitory concentration (MIC) values of the AgNPs were obtained for *Bacillus subtilis* and *Staphylococcus aureus* (204 µg mL^{-1})	Sharifi et al., 2023
Polyphenol-based polymeric nanoparticles	Coffee beans mass: 2 g Water volume: 25 mL Oxidant extracts: $NaIO_4$ or $CuSO_4$ mass: 2 g (25 mL boiling water)	44 to 250 nm	Antibacterial and antifungal	The NPs were inactive against bacterial (*S. aureus, E. coli,* and *P. aeruginosa*) and fungal (*C. albicans*)	Sunoqrot et al., 2021
Ag	Extract volume: 100 mL $AgNO_3$ concentration: 5 mmol L^{-1} (100 mL)	20 nm	Antibacterial activity	Inhibition zone of 24.3 ± 0.2, 20.2 ± 0.3, 11.4 ± 0.3, and 10.2 ± 0.3 for *P. aeruginosa, S. aureus, E. coli,* and *B. subtilis,* respectively	El-Desouky et al., 2023

(*Contd.*)

(*Contd.*)

NPs	Synthesis conditions	Mean size	Application	Effective concentration	References
ZnO	Extract volume: 5 mL $Zn(C_2H_3O_2)_2.2H_2O$ concentration: 0.01 mol L^{-1} (95 mL)	115.7 nm	Antibacterial activity	IC_{50}: 120, 100, and 90 µg mL^{-1} after 24, 48, and 72 h. The ZnONPs did not show any antibacterial activity against *E. aerogenes, P. aeruginosa,* and *K. pneumoniae* at the applied concentrations (2 mg mL^{-1}, 4 mg mL^{-1}, and 8 mg mL^{-1})	Abdelmigid et al., 2022
Fe_3O_4	Mass of nanopowder: 4 g (dissolved in 400 mL water) Extract volume: 10 mL	20 nm	Antifungal potential	Inhibiting the growth *of F. oxysporum* (*in vitro* and *in vivo* assays); The induced resistance by Fe_3O_4NPs in tomato plants is interrelated with improved enzyme activities and the gene expression of PAL, POD, and polyphenol oxidase; Fe_3O_4 NPs cannot only control tomato wilt disease but also help to enhance the growth parameters of the tomato plant by strengthening its defense system	Ashraf et al., 2022
Ag	$AgNO_3$ concentration: 0.1 mol L^{-1} Extract volume: 100 mL	20 to 30 nm	Antibacterial activity	Inhibition zone: 3.1 and 2.7 cm for *E. coli* and *S. aureus*, respectively, using 0.1 mol L^{-1} AgNPs MIC: 0.05 mol L^{-1} (0.2675 mg L^{-1})	Dhand et al., 2016
Ag	Mass of spent coffee grounds: 1 mg $AgNO_3$ concentration: 50 mmol L^{-1} (10 mL)	24.6 ± 3.3 nm	Antibacterial activity	Inhibition zone in *S. aureus*: 9.8 ± 0.9 mm; Inhibition zone in *E. coli*: 8.3 ± 0.3 mm; Reduction for *S. aureus:* 95% (5 to 100 mg mL^{-1}); Reduction for *E. coli*: no statistically significant difference vs. buffer	Chien et al., 2019

Ag-HSCG	Hydrolysed spent coffee grounds: 1 g $AgNO_3$ concentration: 10 mmol L^{-1} (100 mL)	20 nm	Preparation of chitosan films functionalised with AgNP-HSCG for green antibacterial devices	MIC of Ag-HSCG: < 0.1 mg mL^{-1} against *E. coli* and *P. aeruginosa.* Antimicrobial activity of the composite films: 8 mm for *S. aureus* and around 9 mm for *E. coli* and *P. aeruginosa*	Panzella et al., 2020
ZnO	Volume of spent coffee grounds: 12.5 mL $ZnCl_2$ concentration: 0.37 mol L^{-1}	33.5 to 34.4 nm	Antibacterial activity	Inhibition zone of 22 ± 0.25 and 28 ± 0.38 for *E. coli* and *S. aureus*, respectively, using 5 mg mL^{-1} of ZnONPs	Ostovar et al., 2023
Ag	*C. Arabica* extract: 15 mL Volume of $AgNO_3$: 5 mL	73.08 nm	Antibacterial activity	Inhibition zone of 3.33 mm and 4.05 mm for *E. coli* and *S. aureus*, respectively, using AgNPs and photodynamic laser. The bacterial death for *E. coli* and *S. aureus* was 95.98 ± 4.82 and 78.96 ± 7.25%	Astuti et al., 2025
Fe_2O_3	Volume of clove or green coffee extract: 15 mL $Fe(NO_3)_3.9H_2O$ concentration: 0.3 mol L^{-1} (5 mL)	BET: 18.18 and 23.90 M^2 g^{-1} using clove and green coffee, respectively	Antibacterial and antifungal activity	Fe_2O_3NPs from clove extract exhibited effective bactericidal activity against the studied pathogens, with a minimum inhibitory concentration (MIC) of 20 µg mL^{-1} against *S. aureus* and 40 µg mL^{-1} against *Bacillus cereus*. The inhibition zones for *Aspergillus niger*, *Penicillium reqfortii*, and *C. albicans* fungi were 4.7, 7.0, and 6.4 mm, respectively, using 75 µg 100 mL^{-1} of Fe_2O_3 from clove extract.	Atta et al., 2025

(Contd.)

(Contd.)

NPs	Synthesis conditions	Mean size	Application	Effective concentration	Ref.
Ag	Volume of grams of coffee extract: 2 mL $AgNO_3$ concentration: 10 mL of 1 mmol L^{-1}	50 nm	Antibacterial activity	The synthesised AgNPs showed strong antimicrobial activity against *P. aeruginosa* and *E. coli* bacteria	Baghaienezhad et al., 2020
FeO	Volume of coffee extract: 100 mL $FeSO_4 \cdot 7H_2O$ concentration: 150 mmol L^{-1}	30 to 50 nm	Antibacterial and antifungal activity	The synthesised NPs demonstrated significant antimicrobial activity against *E. coli* and *S. aureus*, as well as antifungal activity against *A. niger*	Alangari et al., 2024
Ag	Coffee extract concentration: 0.1% (v v^{-1}) Ag+ concentration: 100 mg L^{-1}	16.92 ± 3.78 nm	Antibacterial activity	The AgNPs exhibited potent bactericidal activity against *Pseudomonas amygdali, Pseudomonas savastanoi, Pseudomonas corrugata,* and *Pseudomonas syringae*. The *P. amygdali* showed the highest susceptibility, with MICs of 0.5 to 1.5 μg mL-1, using a siphon and a slow dripper as the apparatus for coffee brew extract	Khan et al., 2025

EC_{50}: Concentration equivalent to 50% antioxidant activity; MIC: Minimum inhibitory concentration; DOX: Doxorubicin drug; MUA: Mercaptoundecanoic acid; BET: Brunauer-Emmett-Teller method; MIC: Minimum inhibitory concentration particle size, which affects surface area and interaction with microbial cells, pH and composition of the extract, which influence the reduction and stabilisation processes, concentration of precursors, which determines the yield and uniformity of NPs, and stability of the final colloidal suspension, impacting their biological activity over time. Acknowledging these factors is crucial for a thorough evaluation of the green synthesis approach and for informing future studies that aim to optimise conditions to maximise antimicrobial potential. Moreover, understanding these limitations helps identify research gaps, such as the underexplored antiviral activity and the need for standardised evaluation protocols across different microbial models.

6.2 Cancer treatment applications of NPs synthesised from *C. arabica*

NPs have emerged as promising tools in cancer therapy due to their ability to increase treatment efficacy while minimising the adverse effects of conventional treatments (Tian et al., 2022). Their unique physicochemical properties enable targeted drug delivery, improved imaging contrast, and modulation of the tumour microenvironment (Peer et al., 2007; Trejo-Teniente et al., 2024). Among the various green synthesis approaches, *C. arabica* aqueous extracts have been successfully employed to produce ZnO, Au, Ag, and polyphenol-based polymeric NPs for anticancer purposes. These studies are summarised in Table 2.

As summarised in Table 2, coffee-derived NPs exhibit particle sizes ranging from 8.29 to 250 nm. Particle size is a critical parameter influencing their biological performance, as it can directly affect penetration, distribution, and accumulation within tumour tissues (Gandlevskiy et al., 2025; Peer et al., 2007; Xu et al., 2020). Therefore, synthesising NPs with controlled dimensions is crucial to optimise their therapeutic efficacy (Peer et al., 2007).

It is noteworthy that the studies listed in Table 2 employed diverse cell lines, such as MCF-7 (breast cancer cells), HEK 293 (embryonic kidney cells), A172 (glioblastoma cells), and H69 (small cell lung carcinoma cells). Most applications were developed based on evaluation against the MCF-7 cell line (66.6%). The MCF-7 cell line serves as a well-established model in cancer research due to its complete characterisation and ER expression. The majority of breast cancer patients (approximately 73%) with a known hormone receptor (HR) and HER2 status are $HR^+/HER2^-$. Because MCF-7 cells share this same profile, they are considered an essential model for investigating the biology and treatment of this type of cancer (Booms et al., 2019).

Trejo-Teniente et al. (2024) demonstrated that the combination of AuNPs with doxorubicin and mercaptoundecanoic acid (used as a ligand) showed promise for cancer treatment. These NPs exhibited increased specificity for cancer cells, increased cytotoxicity against tumour cells, and reduced damage to normal cells (Trejo-Teniente et al., 2024). Moreover, this system enabled a pH-responsive drug release profile, favouring drug delivery within the acidic tumour microenvironment (Trejo-Teniente et al., 2024).

In another study, the synthesis of ZnONPs using an aqueous extract of coffee leaves and ultrasound was performed, and the resulting nanoparticles were subsequently

Table 2: Overview of selected studies on the green synthesis of MNPs using *Coffea arabica* for anticancer applications

NPs	Synthesis conditions	Mean size	Application	Effective concentration	References
Ag	*C. arabica* extract: 1% (w v^{-1}) $AgNO_3$ concentration: 1 mmol L^{-1} Proportion between extract and $AgNO_3$ solution: 1:9 v v^{-1}	147 nm	Anti-cancer	Cancer cell lines: MCF-7 breast cancer cells IC_{50}: 16.7 µg mL^{-1}	Kavin et al., 2025
Ag	Extract concentration: 100 µg mL^{-1} (10 mL) $AgNO_3$ concentration: 0.1%	20 to 70 nm	Anticancer	Cancer cell lines: Hek293, MCF-7, and A172. CC_{50}: 437.2, 116.8, and 72.9 µg mL^{-1}, respectively	Sharifi et al., 2023
Polyphenol-based polymeric nanoparticles	Coffee beans mass: 2 g Oxidant extracts: $NaIO_4$ or $CuSO_4$ mass: 2 g (25 mL boiling water)	44 to 250 nm	Anticancer	Cancer cell lines: MCF-7 breast cancer cells IC50: Values of 42, 40, and 57 µg mL^{-1} for green, medium-roasted, and dark-roasted coffee NPs, respectively Polyphenol content: 44.2 GAE g^{-1} of medium-roasted coffee extract, 38.3 GAE g^{-1} of green coffee extract, and 35.2 GAE g^{-1} of dark-roasted coffee extracts	Sunoqrot et al., 2021
Au-MUA-DOX	Extract volume: 1 mL Commercial Au solution concentration: 0.13 mmol L^{-1} (19 mL) Proportion between AuNPs:doxorubicin: 10:1	20 to 40 nm	Anticancer	Cancer cell lines: H69 cells reduced up to 40% at 50 ppm	Trejo-Teniente et al., 2024
ZnO	$C_4H_6O_4Zn{\cdot}2H_2O$ concentration: 0.1 mol L^{-1} (100 mL) Mass ratio between extract and $C_4H_6O_4Zn{\cdot}2H_2O$: 1.71	8.29 ± 1.38 nm	Nanocarriers for mangiferin	Release percentage of mangiferin after 132 h: 96.24 ± 1.90%	Wang et al., 2022

EC_{50}: Concentration equivalent to 50% antioxidant activity; MIC: Minimum inhibitory concentration; DOX: Doxorubicin drug; MUA: Mercaptoundecanoic acid.

employed to encapsulate mangiferin (MGF) (Wang et al., 2022). Mangiferin is a bioactive compound recognised for its antioxidant, anti-inflammatory, antiviral, and anticancer properties (Wang et al., 2022). The authors reported cumulative release percentages of MGF at 12, 84, 120, and 132 h as 55.46 ± 2.06%, 88.41 ± 3.25%, 96.05 ± 1.92%, and 96.24 ± 1.90%, respectively. These findings demonstrated that the MGF-ZnONPs exhibited excellent sustained-release behaviour. Therefore, it can be inferred that the greenly synthesised ZnONPs show potential as drug delivery systems or bioactive carriers for future biomedical applications.

Despite the positive results obtained to date, further research is needed to elucidate how the bioactive compounds present in coffee and its residues influence the formation of NPs. These studies should also focus on improving the physicochemical properties of these materials and enhancing their therapeutic performance. The adoption of sustainable and scalable methods for nanoparticle synthesis could further contribute to this progress. Such approaches open up new opportunities for biomedical applications, such as oncology.

7. Importance of evaluating experimental parameters during NPs synthesis for biomedical applications

As shown in Tables 1 and 2, NPs synthesised from aqueous extracts of *C. arabica* exhibited biological activity against both microorganisms and tumour cells. However, for their effective application in the biomedical field, it is important to ensure the standardisation of their morphological characteristics and size. Parameters such as pH, metal precursor concentration, reaction time, extract type, and specific plant part used in the preparation can influence the size, morphology, and bioactivity of the obtained NPs. Therefore, the development of methodologies that allow systematic investigation of these parameters must be planned and critically discussed to establish more reproducible, efficient, and suitable synthetic routes for biomedical use.

Among the studies presented, one study deserves particular attention. In this study, the authors proposed a novel synthesis route for ZnONPs using an aqueous extract of *C. arabica* leaves assisted by ultrasound, while systematically investigating key parameters that can affect the physicochemical properties of the nanomaterial. This investigation included pH, temperature, the ratio of the metal precursor to the plant extract, as well as the frequency, power, and exposure time of the ultrasound. According to the authors, the highest yield (43.59 ± 0.13%) of ZnONPs was achieved under the following optimised conditions: pH 8, mass ratio of coffee leaves to $C_4H_6O_4Zn \cdot 2H_2O$ of 1.71, ultrasound application time of 10 min, ultrasound frequency of 28/40 kHz, ultrasound power of 180 W, and synthesis temperature of 30 °C. According to the authors, spherical NPs with an average particle size of 8.29 ± 1.38 nm were obtained (Wang et al., 2022).

Other studies cited in Tables 1 and 2 did not evaluate the experimental parameters for NP synthesis together, making it difficult to standardise the characteristics of the synthesised materials and ensure the reproducibility of these methodologies in research laboratories for subsequent studies. This limitation highlights a promising research opportunity with *C. arabica*, a plant species with recognised therapeutic potential for antibacterial and anticancer applications.

8. Conclusions

The use of *C. arabica* in the green synthesis of NPs for biomedical applications represents a promising approach in this field of research. In addition to reducing, stabilising, and conferring biological functionality to the synthesised NPs, *C. arabica* offers a more sustainable synthetic route. This chapter demonstrates that NPs synthesised from their extracts exhibit remarkable antimicrobial activity against various bacterial strains and in oncological therapies.

In the biomedical field, the antimicrobial potential of coffee-derived NPs may help address the growing challenge of antimicrobial resistance. At the same time, their anticancer activity emerges as an alternative for the development of more selective and biocompatible therapeutic strategies. However, further studies are needed to elucidate their mechanisms of action, improve particle stability and targeting efficiency, and evaluate their in vivo safety.

Considering the importance of biological applications of NPs synthesised from plant extracts, the synthesis routes must ensure the reproducibility of the physicochemical characteristics of the synthesised materials, thereby facilitating their efficient application. However, variability between studies hinders direct comparisons and hinders the standardisation of results. Factors such as pH, metal precursor concentration, reaction time, type of plant extract, and the specific part of the plant used to prepare the extract can influence the size, morphology, and bioactivity of the resulting NPs. In this way, future multidisciplinary research will be important for the implementation of these materials in practical biomedical applications worldwide.

References

Abdelmigid, H. M., Hussien, N. A., Alyamani, A. A., Morsi, M. M., AlSufyani, N. M., and Kadi, H. A. (2022). Green synthesis of zinc oxide nanoparticles using pomegranate fruit peel and solid coffee grounds vs. chemical method of synthesis, with their biocompatibility and antibacterial properties investigation. *Molecules*, 27, 1236. https://doi.org/10.3390/molecules27041236

Administration, U. S. Food and Drug. FDA-Approved Drugs. https://www.accessdata.fda.gov/scripts/cder/daf/ Accessed 2025 Nov 04.

Alangari, A., Alqahtani, M. S., Shahid, M., Syed, R., Goel, M., Lakshmipathy, R., and Kaur, K. (2024). Green synthesis of FeO nanoparticles from coffee and its application for antibacterial, antifungal, and anti-oxidation activity. *Green Processing and Synthesis*, 13, 20230268. https://doi.org/10.1515/gps-2023-0268

Allen, T. M., and Cullis, P. R. (2013). Liposomal drug delivery systems: From concept to clinical applications. *Advanced Drug Delivery Reviews*, 65, 36. https://doi.org/10.1016/j.addr.2012.09.037

Alrefaee, S. H., Aljohani, F. S., El-Khatib, M., Shahin, Y. H., Elwakil, B. H., Shahin, S. H., and Darwish, A. A.-E. (2025). Evaluation of potential antiparasitic effect of ZnO nanoparticles on experimental cryptosporidiosis in immunosuppressed mice. *BioMetals*, 38, 647. https://doi.org/10.1007/s10534-025-00669-7

Alves, K. C., Costa, C., Remédios, C. M. R., Calcagno, D. Q., Lima, M. O., Silva, J. R. A., and Alves, C. N. (2024). LDH-indomethacin nanoparticles antitumoral action: A possible coadjuvant drug for cancer therapy. *Molecules*, 29, 3353. https://doi.org/10.3390/molecules29143353

Anand, U., Dey, A., Chandel, A. K. S., Sanyal, R., Mishra, A., Pandey, D. K., and Pérez de la Lastra, J. M. (2023). Cancer chemotherapy and beyond: Current status, drug candidates, associated risks and progress in targeted therapeutics. *Genes & Diseases*, 10, 1367. https://doi.org/10.1016/j.gendis.2022.02.007

Ashraf, H., Batool, T., Anjum, T., Ilyas, A., Li, G., Naseem, S., and Riaz, S. (2022). Antifungal potential of green synthesized magnetite nanoparticles black coffee–magnetite nanoparticles against wilt infection by ameliorating enzymatic activity and gene expression in *Solanum lycopersicum* L. *Frontiers in Microbiology*, 13, 1. https://doi.org/10.3389/fmicb.2022.754292

Astuti, S., Yessica, A., Nurdin, D., Yaqubi, A. K., Arifianto, D., Susilo, Y., and Abd Razak, N. A. (2025). The effect of coffee as a bio-reductant in the synthesis of silver nanoparticles combined with laser photodynamics for bacteria inactivation. *Brazilian Journal of Biology*, 85, 1. https://doi.org/10.1590/1519-6984.288042

Atta, R. R., Eweis, A. A., Mohammed, D. A., El-Halim, A. A., El-Tawab, S. A., Attallah, M. I., and Mahmoud, R. (2025). Iron oxide nanoparticles from clove and green coffee for antioxidant and antimicrobial activities and analytical application in caffeine removal. *Scientific Reports*, 15, 35265. https://doi.org/10.1038/s41598-025-16744-5

Attia, M. M., Yehia, N., Mohamed Soliman, M., Shukry, M., El-Saadony, M. T., and Salem, H. M. (2022). Evaluation of the antiparasitic activity of the chitosan-silver nanocomposites in the treatment of experimentally infested pigeons with *Pseudolynchia canariensis*. *Saudi Journal of Biological Sciences*, 29, 1644. https://doi.org/10.1016/j.sjbs.2021.10.067

Baghaienezhad, M., Boroghani, M., and Anabestani, R. (2020). Silver nanoparticles synthesis by coffee residues extract and their antibacterial activity. *Nanomedicine Research Journal*, 5, 29. https://doi.org/10.22034/nmrj.2020.01.004

Barrios, C., de Lima Lopes, G., Yusof, M. M., Rubagumya, F., Rutkowski, P., and Sengar, M. (2023). Barriers in access to oncology drugs – A global crisis. *Nature Reviews Clinical Oncology*, 20, 7. https://doi.org/10.1038/s41571-022-00700-7

Bayda, S., Adeel, M., Tuccinardi, T., Cordani, M., and Rizzolio, F. (2019). The history of nanoscience and nanotechnology: From chemical-physical applications to nanomedicine. *Molecules*, 25, 112. https://doi.org/10.3390/molecules25010112

Bekmukhametova, A., Uddin, M. M. N., Houang, J., Malladi, C., George, L., Wuhrer, R., and Lauto, A. (2022). Fabrication and characterization of chitosan nanoparticles using the coffee-ring effect for photodynamic therapy. *Lasers in Surgery and Medicine*, 54, 758. https://doi.org/10.1002/lsm.23530

Booms, A., Coetzee, G. A., and Pierce, S. E. (2019). MCF-7 as a model for functional analysis of breast cancer risk variants. *Cancer Epidemiology, Biomarkers & Prevention*, 28, 1735. https://doi.org/10.1158/1055-9965.EPI-19-0066

Chien, H. W., Kuo, C. J., Kao, L. H., Lin, G. Y., and Chen, P. Y. (2019). Polysaccharidic spent coffee grounds for silver nanoparticle immobilization as a green and highly efficient biocide. *International Journal of Biological Macromolecules*, 140, 168. https://doi.org/https://doi.org/10.1016/j.ijbiomac.2019.08.131

da Silva, P. B., Machado, R. T. A., Pironi, A. M., Alves, R. C., de Araújo, P. R., Dragalzew, A. C., and Chorilli, M. (2019). Recent advances in the use of metallic nanoparticles with antitumoral action – Review. *Current Medicinal Chemistry*, 26, 2108. https://doi.org/10.2174/0929867325666180214102918

Dhand, V., Soumya, L., Bharadwaj, S., Chakra, S., Bhatt, D., and Sreedhar, B. (2016). Green synthesis of silver nanoparticles using *Coffea arabica* seed extract and its antibacterial activity. *Materials Science and Engineering: C*, 58, 36. https://doi.org/https://doi.org/10.1016/j.msec.2015.08.018

dos Santos, M. S., Silva, J. M., Barbieri, M. B., Filho, S. A., & Backx, B. P. (2024). Bionanotechnology and its applications: The plurality of science is fundamental for the search for solutions. *Plant Nano Biology*, 7, 100060. https://doi.org/https://doi.org/10.1016/j.plana.2024.100060

El-Desouky, N., Shoueir, K., El-Mehasseb, I., and El-Kemary, M. (2023). Synthesis of silver nanoparticles using bio valorization coffee waste extract: Photocatalytic flow-rate performance, antibacterial activity, and electrochemical investigation. *Biomass Conversion and Biorefinery*, 13, 15871. https://doi.org/10.1007/s13399-021-02256-5

Ferrara, F., Castagna, T., Pantolini, B., Campanardi, M. C., Roperti, M., Grotto, A., and Langella, R. (2024). The challenge of antimicrobial resistance (AMR): Current status and future prospects. *Naunyn-Schmiedeberg's Archives of Pharmacology*, 397, 9603. https://doi.org/10.1007/s00210-024-03318-x

Gandlevskiy, N., Viana, A. R., Druzian, G. T., Oliveira, D. K., Schuch, A. P., Barge, A., & Flores, E. M. M. (2025). Ultrasound-assisted green synthesis of silver nanoparticles using *Ruta graveolens* L. extract and antitumor evaluation. *Ultrasonics Sonochemistry*, 117, 107340. https://doi.org/https://doi.org/10.1016/j.ultsonch.2025.107340

Guerra, R. O., do Carmo Neto, J. R., da Silva, P. E. F., Franco, P. I. R., Barbosa, R. M., de Albuquerque Martins, T., and Silva, M. V. (2024). Metallic nanoparticles and treatment of cutaneous leishmaniasis: A systematic review. *Journal of Trace Elements in Medicine and Biology*, 83, 127404. https://doi.org/https://doi.org/10.1016/j.jtemb.2024.127404

Hancharova, M., Halicka-Stępień, K., Dupla, A., Lesiak, A., Sołoducho, J., and Cabaj, J. (2024). Antimicrobial activity of metal-based nanoparticles: A mini-review. *BioMetals*, 37, 773. https://doi.org/10.1007/s10534-023-00573-y

Ho, C. S., Wong, C. T. H., Aung, T. T., Lakshminarayanan, R., Mehta, J. S., Rauz, S., and Ting, D. S. J. (2025). Antimicrobial resistance: A concise update. *The Lancet Microbe*, 6, 100947. https://doi.org/https://doi.org/10.1016/j.lanmic.2024.07.010

Joudeh, N., and Linke, D. (2022). Nanoparticle classification, physicochemical properties, characterization, and applications: A comprehensive review for biologists. *Journal of Nanobiotechnology*, 20, 262. https://doi.org/10.1186/s12951-022-01477-8

Kanasty, R., Dorkin, J. R., Vegas, A., and Anderson, D. (2013). Delivery materials for siRNA therapeutics. *Nature Materials*, 12, 967. https://doi.org/10.1038/nmat3765

Kavin, T., Murugaiyah, V., Tan, J. K., Kassim, M. N. I., Ramakrishna, S., & Vigneswari, S. (2025). Eco-friendly synthesis of silver nanoparticles using *Coffea arabica* husk for enhanced antibacterial and anti-cancer applications. *Biomass and Bioenergy*, 194, 107625. https://doi.org/https://doi.org/10.1016/j.biombioe.2025.107625

Khan, I., Saeed, K., and Khan, I. (2019). Nanoparticles: Properties, applications and toxicities. *Arabian Journal of Chemistry*, 12, 908. https://doi.org/https://doi.org/10.1016/j.arabjc.2017.05.011

Khan, M. A., Dzimitrowicz, A., Prusinski, M., Motyka-Pomagruk, A., Jamroz, P., Cyganowski, P., and Pohl, P. (2025). Coffee-mediated synthesis of silver nanoparticles showing antibacterial properties against economically important phytopathogens from the genus *Pseudomonas*. *Colloids and Surfaces A: Physicochemical and Engineering Aspects*, 718, 136802. https://doi.org/https://doi.org/10.1016/j.colsurfa.2025.136802

Kumari, S., Raturi, S., Kulshrestha, S., Chauhan, K., Dhingra, S., András, K., and Singh, T. (2023). A comprehensive review on various techniques used for synthesizing nanoparticles. *Journal of Materials Research and Technology*, 27, 1739. https://doi.org/https://doi.org/10.1016/j.jmrt.2023.09.291

Laurence, T., Lamberti, O., Smith, R., Drake, T., and McDonnell, A. (2025). The Global Direct Inpatient Cost of Antimicrobial Resistance: A Modelling Study. https://www.cgdev.

org/publication/global-direct-inpatient-cost-antimicrobial-resistance-modelling-study. Accessed 2025 Nov 04.

Leighl, N. B., Nirmalakumar, S., Ezeife, D. A., & Gyawali, B. (2021). An arm and a leg: The rising cost of cancer drugs and impact on access. *ASCO Educational Book*, 41, e1-e12. https://doi.org/10.1200/edbk_100028

Luo, L., Huang, W., Zhang, J., Yu, Y., and Sun, T. (2024). Metal-based nanoparticles as antimicrobial agents: A review. *ACS Applied Nano Materials*, 7, 2529. https://doi.org/10.1021/acsanm.3c05615

MacNair, C. R., Rutherford, S. T., and Tan, M.-W. (2024). Alternative therapeutic strategies to treat antibiotic-resistant pathogens. *Nature Reviews Microbiology*, 22, 262. https://doi.org/10.1038/s41579-023-00993-0

Makiso, M. U., Tola, Y. B., Ogah, O., and Endale, F. L. (2024). Bioactive compounds in coffee and their role in lowering the risk of major public health consequences: A review. *Food Science & Nutrition*, 12(2), 734. https://doi.org/https://doi.org/10.1002/fsn3.3848

Mohamed, A., Atta, R. R., Kotp, A. A., Abo El-Ela, F. I., Abd El-Raheem, H., Farghali, A., and Mahmoud, R. (2023). Green synthesis and characterization of iron oxide nanoparticles for the removal of heavy metals (Cd^{2+} and Ni^{2+}) from aqueous solutions with antimicrobial Investigation. *Scientific Reports*, 13, 7227. https://doi.org/10.1038/s41598-023-31704-7

More, P. R., Pandit, S., Filippis, A. D., Franci, G., Mijakovic, I., and Galdiero, M. (2023). Silver nanoparticles: Bactericidal and mechanistic approach against drug resistant pathogens. *Microorganisms*, 11, 369. https://doi.org/10.3390/microorganisms11020369

Murray, C. J. L., Ikuta, K. S., Sharara, F., Swetschinski, L., Robles Aguilar, G., Gray, A., and Naghavi, M. (2022). Global burden of bacterial antimicrobial resistance in 2019: A systematic analysis. *The Lancet*, 399, 629. https://doi.org/10.1016/S0140-6736(21)02724-0

Organization, W. H. (2024). Global Cancer Burden Growing, Amidst Mounting Need for Services. https://www.iarc.who.int/news-events Accessed 2025 Nov 04.

Osman, A. I., Zhang, Y., Farghali, M., Rashwan, A. K., Eltaweil, A. S., Abd El-Monaem, E. M., and Yap, P.-S. (2024). Synthesis of green nanoparticles for energy, biomedical, environmental, agricultural, and food applications: A review. *Environmental Chemistry Letters*, 22, 841. https://doi.org/10.1007/s10311-023-01682-3

Ostovar, N., Mohammadi, N., and Khodadadeh, F. (2023). Photocatalytic, antioxidant and antibacterial potential of bio-synthesized ZnO nanoparticles derived from espresso spent coffee grounds: Optimization by central composite design. *Inorganic and Nano-Metal Chemistry*, 53, 938. https://doi.org/10.1080/24701556.2023.2187419

Panzella, L., Cerruti, P., Aprea, P., Paolillo, R., Pellegrino, G., Moccia, F., and Napolitano, A. (2020). Silver nanoparticles on hydrolyzed spent coffee grounds (HSCG) for green antibacterial devices. *Journal of Cleaner Production*, 268, 122352. https://doi.org/https://doi.org/10.1016/j.jclepro.2020.122352

Peer, D., Karp, J. M., Hong, S., Farokhzad, O. C., Margalit, R., and Langer, R. (2007). Nanocarriers as an emerging platform for cancer therapy. *Nature Nanotechnology*, 2, 751. https://doi.org/10.1038/nnano.2007.387

Raghupathi, K. R., Koodali, R. T., and` Manna, A. C. (2011). Size-dependent bacterial growth inhibition and mechanism of antibacterial activity of zinc oxide nanoparticles. *Langmuir*, 27, 4020. https://doi.org/10.1021/la104825u

Ruckert, A., Lake, S., and Van Katwyk, S. R. (2024). Developing a protocol on antimicrobial resistance through WHO's pandemic treaty will protect lives in future pandemics. *Globalization and Health*, 20, 10. https://doi.org/10.1186/s12992-024-01015-1

Ruparelia, J. P., Chatterjee, A. K., Duttagupta, S. P., and Mukherji, S. (2008). Strain specificity in antimicrobial activity of silver and copper nanoparticles. *Acta Biomaterialia*, 4, 707. https://doi.org/https://doi.org/10.1016/j.actbio.2007.11.006

Sartorius, B., Gray, A. P., Davis Weaver, N., Robles Aguilar, G., Swetschinski, L. R., Ikuta, K. S., and Naghavi, M. (2024). The burden of bacterial antimicrobial resistance in the WHO African region in 2019: A cross-country systematic analysis. *The Lancet Global Health*, 12, e201. https://doi.org/https://doi.org/10.1016/S2214-109X(23)00539-9

Saxena, R., Kotnala, S., Bhatt, S. C., Uniyal, M., Rawat, B. S., Negi, P., and Riyal, M. K. (2025). A review on green synthesis of nanoparticles toward sustainable environment. *Sustainable Chemistry for Climate Action*, 6, 100071. https://doi.org/https://doi.org/10.1016/j.scca.2025.100071

Sharifi, F., Mohamadi, N., Tavakoli Oliaee, R., Sharifi, I., Doostmohammadi, M., Soltanian, S., and Sharififar, F. (2023). The potential effect of silver nanoparticles synthesized with *Coffea arabica* green seeds on Leishmania major proliferation, cytotoxicity activity, and cytokines expression level. *Journal of Parasitic Diseases*, 47, 131. https://doi.org/10.1007/s12639-022-01549-4

Silva, D. F., Melo, A. L. P., Uchôa, A. F. C., Pereira, G. M. A., Alves, A. E. F., Vasconcellos, M. C., and Passos, M. F. (2023). Biomedical approach of nanotechnology and biological risks: A mini-review. *International Journal of Molecular Sciences*, 24, 16719. https://doi.org/10.3390/ijms242316719.

Slavin, Y. N., and Bach, H. (2022). Mechanisms of antifungal properties of metal nanoparticles. *Nanomaterials*, 12, 4470. https://doi.org/10.3390/nano12244470

Sulastri, Ahmad, A., Karim, A., Wahid, I., Rauf, W., Karim, H., and Farid, A. M. (2025). Green synthesized zinc oxide nanoparticles from *Coffea arabica*: Bioprospecting and functional potential as an antioxidant and larvicidal agent against *Aedes aegypti*. *Tropical Journal of Natural Product Research*, 9, 90. https://doi.org/10.26538/tjnpr/v9i1.13

Sunoqrot, S., Al-Shalabi, E., Al-Bakri, A. G., Zalloum, H., Abu-Irmaileh, B., Ibrahim, L. H., and Zeno, H. (2021). Coffee bean polyphenols can form biocompatible template-free antioxidant nanoparticles with various sizes and distinct colors. *ACS Omega*, 6, 2767. https://doi.org/10.1021/acsomega.0c05061

Szczyglewska, P., Feliczak-Guzik, A., and Nowak, I. (2023). Nanotechnology – General aspects: A chemical reduction approach to the synthesis of nanoparticles. *Molecules*, 28, 4932. https://doi.org/10.3390/molecules28134932

Tacconelli, E., Carrara, E., Savoldi, A., Harbarth, S., Mendelson, M., Monnet, D. L., and Zorzet, A. (2018). Discovery, research, and development of new antibiotics: The WHO priority list of antibiotic-resistant bacteria and tuberculosis. *The Lancet Infectious Diseases*, 18, 318. https://doi.org/https://doi.org/10.1016/S1473-3099(17)30753-3

Tian, H., Zhang, T., Qin, S., Huang, Z., Zhou, L., Shi, J., and Shen, Z. (2022). Enhancing the therapeutic efficacy of nanoparticles for cancer treatment using versatile targeted strategies. *Journal of Hematology & Oncology*, 15, 132. https://doi.org/10.1186/s13045-022-01320-5

Trejo-Teniente, I., Jaramillo-Loranca, B. E., Vargas-Hernández, G., Villanueva-Ibáñez, M., Tovar-Jiménez, X., Olvera-Venegas, P. N., and Tapia-Ramírez, J. (2024). Synthesis and toxicity assessment of *Coffea arabica* extract-derived gold nanoparticles loaded with doxorubicin in lung cancer cell cultures. *Frontiers in Bioengineering and Biotechnology*, 12, 1. https://doi.org/10.3389/fbioe.2024.1378601

Umamaheswari, K., and Abirami, M. (2023). Assessment of antifungal action mechanism of green synthesized gold nanoparticles (AuNPs) using *Allium sativum* on *Candida* species. *Materials Letters*, 333, 133616. https://doi.org/https://doi.org/10.1016/j.matlet.2022.133616

Wang, B., Hu, S., Teng, Y., Chen, J., Wang, H., Xu, Y., and Gao, X. (2024). Current advance of nanotechnology in diagnosis and treatment for malignant tumors. *Signal Transduction and Targeted Therapy*, 9, 200. https://doi.org/10.1038/s41392-024-01889-y

Wang, Q., Mei, S., Manivel, P., Ma, H., & Chen, X. (2022). Zinc oxide nanoparticles synthesized using coffee leaf extract assisted with ultrasound as nanocarriers for mangiferin. *Current Research in Food Science*, 5, 868. https://doi.org/10.1016/j.crfs.2022.05.002

Xu, L., Wang, Y. Y., Huang, J., Chen, C. Y., Wang, Z. X., and Xie, H. (2020). Silver nanoparticles: Synthesis, medical applications and biosafety. *Theranostics*, 10, 8996. https://doi.org/10.7150/thno.45413

Yasamineh, S., Kalajahi, H. G., Yasamineh, P., Yazdani, Y., Gholizadeh, O., Tabatabaie, R., & Dadashpour, M. (2022). An overview on nanoparticle-based strategies to fight viral infections with a focus on COVID-19. *Journal of Nanobiotechnology*, 20, 440. https://doi.org/10.1186/s12951-022-01625-0

Ying, S., Guan, Z., Ofoegbu, P. C., Clubb, P., Rico, C., He, F., and Hong, J. (2022). Green synthesis of nanoparticles: Current developments and limitations. *Environmental Technology & Innovation*, 26, 102336. https://doi.org/https://doi.org/10.1016/j.eti.2022.102336

Zaremba-Czogalla, M., Jaromin, A., Sidoryk, K., Zagórska, A., Cybulski, M., and Gubernator, J. (2020). Evaluation of the in vitro cytotoxic activity of caffeic acid derivatives and liposomal formulation against pancreatic cancer cell lines. *Materials*, 13, 5813. https://doi.org/10.3390/ma13245813

CHAPTER

10

Safety of Coffee-Based Nanomaterials and Their Derivatives

Altevir Rossato Viana[1], Mariele Samuel Nascimento[1], Gabriel Toneto Druzian[1], Marcia Foster Mesko[2] and Erico Marlon Moraes Flores[1*]

[1] Department of Biochemistry and Molecular Biology, Federal University of Santa Maria, Santa Maria, RS 97105-900, Brazil

[2] Center for Chemical, Pharmaceutical and Food Sciences, Federal University of Pelotas, RS 96160-000, Brazil

Coffee-based nanomaterials have emerged as a promising class of green nanotechnologies, combining the chemical richness of *Coffea arabica* with the functional advantages of nanoscale systems. Driven by increasing concerns regarding the safety, biocompatibility, and long-term impacts of engineered nanomaterials, this chapter critically reviews current knowledge on the safety profiles of coffee-derived nanoparticles and their derivatives, with a particular focus on biomedical and biotechnological applications. Coffee and its by-products, including spent coffee grounds, pulp, and silver skin, are rich in bioactive compounds such as chlorogenic acids, caffeine, polyphenols, and melanoidins, which can act as natural reducing, stabilising, or functional agents during nanoparticle synthesis. These properties make coffee an attractive and sustainable alternative to conventional chemical synthesis routes.

This chapter provides an overview of the historical development of nanotechnology in medicine, followed by a discussion of early-stage safety assessment strategies, including in silico, in vitro, and in vivo approaches. Special attention is given to regulatory perspectives and international guidelines that govern nanomaterial safety evaluation. The available experimental evidence on coffee-based nanoparticles published within the last five years is systematically examined, encompassing metal nanoparticles synthesised using coffee extracts, nanoemulsions containing green coffee, and caffeine-loaded nanocarriers. These studies indicate that

*Corresponding author: ericommf@gmail.com

DOI: 10.1201/9781003689393-10

coffee-derived nanomaterials generally exhibit favourable biocompatibility profiles at low to moderate doses, although dose-dependent toxicity and organ-specific effects have been reported in vivo at higher exposures.

Key physicochemical factors influencing nanoparticle toxicity, such as size, shape, surface charge, composition, and surface functionalisation, are also discussed in the context of coffee-based systems. Despite encouraging preliminary findings, the chapter highlights significant knowledge gaps, particularly regarding long-term exposure, biodistribution, and chronic toxicity. Overall, this work underscores the need for standardised, multidisciplinary safety assessments to support the responsible development and translation of coffee-derived nanomaterials into clinical, cosmetic, and environmental applications.

1. Introduction

Nanotechnology has revolutionised the healthcare sector, enabling innovative approaches to diagnosis, therapy, and disease monitoring (Gujjar et al., 2025). These advances have enabled the development of nanomaterials with distinct physicochemical properties that can selectively interact with biological targets, thereby optimising drug delivery to specific tissues (Lu et al., 2023). The application of nanomaterials can make treatments more precise, effective, and personalised, transforming the way healthcare is delivered (Haleem et al., 2023; Kazi et al., 2025). However, as the use of these materials increases, concerns about their biocompatibility, degradation, and potential adverse effects associated with prolonged exposure also arise, highlighting the fundamental importance of comprehensive safety studies (Fortune et al., 2025).

Considering the broad range of applications of nanomaterials in the biomedical field, it is essential to explore alternative synthesis routes that reduce the use of toxic reagents, minimise toxicity, and enhance the therapeutic efficacy of the resulting products. In this context, nanotechnology based on the use of plant extracts has emerged as a promising and sustainable approach (Kazemi et al., 2023; Karnwal et al., 2024). Bioactive compounds present in plants can act as natural reducing and stabilising agents during the formation of nanomaterials, replacing hazardous solvents and catalysts, or they can result in nanoparticles with therapeutic properties (Kazemi et al., 2023). Plant-derived nanoparticles are distinguished by their biocompatibility, multifunctionality, and adherence to environmentally responsible practices (Chanthapong et al., 2025). Unlike conventional synthesis methods, this approach aligns with the principles of green chemistry (Lithi et al., 2025). The term "green chemistry" refers to a set of principles and practices designed to make chemical processes more sustainable, thereby preventing or minimising the generation and use of substances that are harmful to human health, animal life, and the environment (Kadam and Bhangale, 2024). Nevertheless, even nanomaterials obtained through green routes require rigorous safety evaluation, as factors such as size, surface charge, and composition can influence their biological interactions and potential toxicity (Pathak et al., 2025).

Among the plants investigated for the synthesis of green nanomaterials, *Coffea arabica* stands out as a promising source due to its global availability and chemical richness (Ali et al., 2022). In addition to its economic and nutritional relevance as the most widely consumed raw material for beverages globally, *C. arabica* is recognised as an abundant source of bioactive compounds, particularly polyphenols, chlorogenic acids, and melanoidins, which exhibit strong antioxidant and anti-inflammatory activities (Abdelmigid et al., 2022). Waste from the coffee industry, such as spent coffee grounds, is also rich in (hemi)cellulose, proteins, melanoidins, and polyphenols. Aqueous extracts obtained from *C. arabica* beans, and even from their residues, have shown potential for the formation of nanostructures (Campos-Veja et al., 2015; Caprioli et al., 2015). Furthermore, the bioactive compounds present in *C. arabica* and its residues may act in reducing metallic nanoparticles, as well as be encapsulated in various formulations (Castro et al., 2018). Despite these advantages, studies focused on the toxicological evaluation and safety of coffee-based nanomaterials remain limited.

Nanoparticles (NPs) possess unique properties due to their nanometric size, which allows them to cross cellular membranes in living organisms (Sandvig, Iversen, and Skotland 2024). While this characteristic offers advantages for therapeutic applications, it can also pose potential risks, including organ accumulation, oxidative stress, inflammation, and cytotoxicity (Morais et al., 2020; Kher and Kumar, 2022). Well-designed preclinical studies are thus essential, as they provide valuable pharmacodynamic and toxicological data, helping to establish safe initial human doses at which adverse effects are not expected (Polson and Fuji, 2012). In vitro cell-based models, combined with in silico simulations, have emerged as faster and more cost-effective approaches for preliminary safety assessments (Nair and Weiskirchen, 2024). However, animal models remain indispensable tools in biomedical research, as they enable the study of physiology, pathophysiology, and toxicity in complex biological systems before clinical trials can be conducted (Mukherjee et al., 2022; Nair and Weiskirchen, 2025).

In this sense, conducting rigorous safety studies is fundamental for ensuring that the benefits of nanotechnology applied to medicine outweigh its potential risks. The use of natural matrices, such as coffee, represents a sustainable and promising alternative, combining biologically active compounds of recognised therapeutic relevance with the synthesis of nanomaterials that exhibit enhanced functional properties. However, for such advances to be consolidated safely and effectively, it is important to thoroughly evaluate the toxicity and long-term impacts of these nanomaterials, ensuring that the pursuit of innovation remains aligned with human health protection and environmental sustainability. Based on these aspects, this chapter reviews and discusses the available literature addressing the safety of coffee-derived nanomaterials, focusing on studies that evaluate their biological interactions and toxicological profiles in in vitro and/or in vivo models.

2. History of nanotechnology and application in the medical field

Following the initial statements about modern nanotechnology in 1959 and later in 1974 (Feynman, 1960; Taniguchi, 1974), the possibility of miniaturising information was proposed, as well as the use of machines capable of building others on an increasingly smaller scale, ultimately reaching the molecular level. Nanomaterials are materials that have at least one dimension, such as pore size, film thickness, or surface characteristics, in the range of 1 to 100 nanometers (Hughes et al., 2025).

Currently, nanotechnology stands among the most promising innovations of the 21st century. It is the ability to transform the fundamentals of nanoscience into practical applications, utilising techniques such as observation, measurement, manipulation, organisation, control, and production of materials on the nanometer scale (Bayda et al., 2020). Nanoparticles represent an advancement in drug delivery systems. They can be synthesised from various materials, including polymers, lipids, and metals, with a particular emphasis on natural biodegradable polymers, which exhibit excellent biocompatibility (Chenxi et al., 2025). Various types of nanocarriers can be employed in a wide range of biomedical applications, including diagnostic and therapeutic imaging, advanced laboratory analyses, targeted drug and gene delivery, genetic editing tools such as Clustered Regularly Interspaced Short Palindromic Repeats-associated protein 9 (CRISPR-Cas9), anticancer therapies and cancer diagnostics, immune system modulation, as well as incorporation into implants and prostheses with enhanced functionalities (Ferreira et al., 2020; Anik et al., 2021; Senjab et al., 2024). However, most nanotechnology-based drugs approved by regulatory agencies are in oncology (Jia et al., 2023).

The first oncology drug developed using nanotechnology to receive Food and Drug Administration (FDA) approval was the polyethene glycol (PEG) liposomal formulation of doxorubicin (Doxil/Caelyx), which was commercially launched in 1995. Since then, the field of nanotherapies has advanced significantly, leading to the FDA and other regulatory agencies approving several nanomedicines for cancer treatment. Among these formulations, liposomes, both PEGylated and non-PEGylated, as well as other lipid-based nanoparticles, still represent a substantial portion of the nanotherapeutics available on the market. Furthermore, other nanoplatforms, such as polymeric, metallic, and protein-based nanoparticles, have also gained approval for use in cancer therapies (Giri et al., 2023).

Economic factors are one of the main challenges to the success of nanopharmaceuticals. The high costs of raw materials and manufacturing processes result in products that are more expensive than conventional drugs. For instance, the production of medicines such as Abraxane™ and Doxil™ is considerably more costly than that of their non-nano counterparts, paclitaxel and doxorubicin, respectively (Resnik and Tinkle 2007). It is estimated that the entire process of developing and commercialising a new nanomedicine takes between 10 and 15 years and requires an investment of approximately US$1 billion (Bobo et al., 2016). Therefore, the clinical benefits of these formulations must be clearly demonstrated to justify their higher prices compared to traditional drugs (Allen and Cullis, 2013). Furthermore, large-scale manufacturing under good manufacturing practice conditions poses specific challenges, as even small variations in the process can affect critical properties such as size, shape, composition, surface charge, drug release profile, biocompatibility, toxicity, and in vivo performance (Aillon et al., 2009).

3. Importance of early studies for screening the safety of new therapeutic candidates

The field of nanotoxicology was first proposed by Oberdörster et al. (2005) and has since undergone continuous evolution, now also referred to as "nanosafety" (Clift et al., 2020). In this context, nanotechnology has gained prominence, driving research in regulatory science, particularly in identifying potential adverse effects. Particle size is the only characteristic shared by all nanomaterials, although it is not sufficient on its own to determine associated risks (Rasmussen et al., 2025). However, factors such as composition, shape, surface charge, and protein interactions affect their ability to cross biological barriers and form aggregates, making precise analyses more challenging (Gatoo et al., 2014).

Through the Organization for Economic Cooperation and Development (OECD), the European Union established a Working Group on Manufactured Nanomaterials (WPMN) in 2006. One of the objectives of the WPMN is to enhance the understanding of the applicability of OECD test guidelines developed for nanomaterials in general (Hund-Rinke et al., 2016). A technical report was created highlighting the importance of several International Organization for Standardization (ISO) standards for assessing the toxicity of nanomaterials. For example, ISO/TR 16196 and ISO/TR 16197 discuss essential protocols for sample preparation and nanomaterial dosimetry, ensuring reliable toxicological screening. The Technical Specification ISO/TS 19337:2016(E) describes the characterisation of working nanomaterial suspensions when conducting in vitro toxicity tests and applicable measurement methods. In particular, ISO Technical Committee (TC) 229 addresses standards for toxicity assessment (Sharifi et al., 2022).

Since nanotechnology is intrinsically linked to the health sector, toxicity assessment is essential to understand how nanomaterials interact with biological tissues (Ma et al., 2024). Within the body, nanoparticles can act as carriers of therapeutic agents, facilitating their distribution to organs such as the kidneys, liver, and lungs, where they may also undergo metabolism and excretion (Yusuf et al., 2023). However, predicting their long-term effects remains a challenge, as these particles can accumulate in various organs, interact with non-target cells, and undergo degradation and elimination processes influenced by their small size and high surface reactivity (Poon et al., 2019).

When developing new drugs, chemicals, or personal care products, the industry must assess the potential health risks associated with their use. For this purpose, it is necessary to provide regulatory authorities with information on the properties of these substances, results of safety tests, risk assessments, and appropriate risk management measures (Silbergeld et al., 2015). Currently, a variety of models and approaches are available, each with specific advantages and limitations. Early studies can be categorised into three levels of complexity: in silico (utilising computational simulations or experiments), in vitro (outside a living organism, such as in Petri dishes), and in vivo (within living organisms, including both invertebrates and vertebrates) (Paiva et al., 2020).

In vitro and animal studies enable the identification of acute and severe toxicities, as well as, to a lesser extent, chronic toxicities. However, approximately one-third of drug candidates fail in clinical trials due to unexpected toxicities, mainly of a cardiovascular and hepatic nature (Arrowsmith and Miller 2013). This limitation stems from biological differences, as well as the oversimplification of in vitro models, which are generally based on two-dimensional monolayers of cells that do not accurately reproduce human physiological conditions (Pridgeon et al., 2018).

When selecting experimental models, it is essential to acknowledge the advantages and limitations of the chosen method, as well as the associated ethical considerations. Animal testing, for example, involves a large number of individuals and often causes suffering (Burden et al., 2021). In vivo approaches to assess nanotoxicity primarily rely on experimental animal models, including zebrafish (*Danio rerio*), *Caenorhabditis elegans*, *Drosophila melanogaster*, mice, rats, and rabbits (Chen et al., 2020). The basic mechanism of this method operates as follows: when nanomaterials are administered orally, they are absorbed through the gastrointestinal tract and subsequently transported via the bloodstream to various organs, including the liver, spleen, kidneys, heart, lungs, and brain (Taka et al., 2021).

Among the employed methods, acute toxicity tests are widely used for validation. These tests analyse the toxicological response of animals after the administration of a substance over a short period, usually 24 hours, requiring a minimum observation period of 14 days to record effects such as toxicity, recovery time, weight variation, death, or signs of intoxication. The experimental design must consider both the biological characteristics of the test organism and the structural and physicochemical properties of the evaluated nanomaterials (Wang et al., 2025).

It is important to emphasise that nanoparticles can induce toxicity in both in vitro and in vivo systems through different mechanisms. The most common include oxidative stress, cell death pathways (apoptosis, necrosis, and autophagy), genotoxicity, and immune responses, each leading to distinct patterns of nanoparticle-induced toxicity (Domb et al., 2021). Therefore, conducting safety screening before the clinical trial phase is essential, progressively increasing the levels of complexity until reaching human studies, as illustrated in Figure 1.

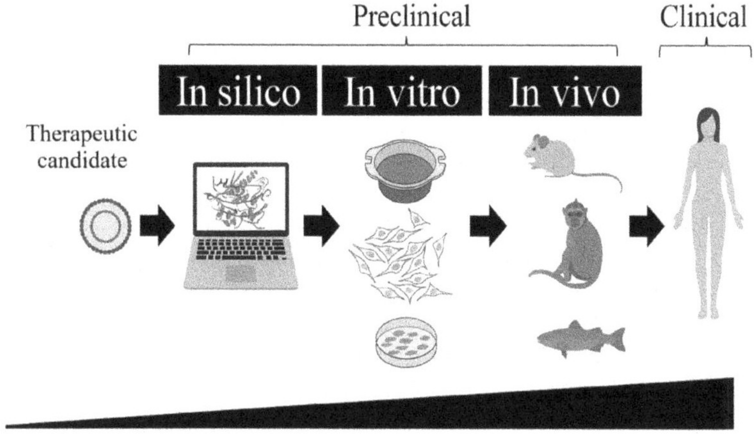

Figure 1: Representation of the synthesis of a coffee-based nanomaterial and the phases necessary to reach experiments involving humans

The drug development process begins with preclinical studies, in which potential drugs are extensively tested in vitro and in vivo to evaluate toxicity, pharmacokinetics, and initial efficacy, determining whether they are viable for human testing (Zurita-Cruz et al., 2019). Clinical trials then proceed through several phases. Phase 0 involves microdosing studies in a small number of volunteers. Phase 1 tests safety, tolerability, and dosage in a small group of healthy volunteers or patients. Phase 2 expands to larger groups to assess biological activity and preliminary efficacy while continuing safety evaluation. Phase 3 involves large-scale trials that compare the new treatment with standard therapies to confirm its effectiveness and monitor any adverse effects, often serving as the basis for regulatory approval. Finally, Phase 4 (post-marketing) studies continue to monitor safety in broader populations to identify rare or long-term side effects and optimise real-world use (Mishra, 2021).

4. Coffee and its benefits

Since around 1400, coffee cultivation has expanded from a small region in East Africa to various tropical regions around the world. Although coffee is a globally significant crop, most of its production is concentrated in the Americas, with Brazil being the primary producer (McCook, 2024). The global relevance of coffee is also associated with the diversity of cultivated species, among which *C. arabica* (Arabica) and *C. canephora* Pierre (Robusta) are the most notable (Galanakis, 2017). Coffee is one of the most widely consumed beverages in the world, appreciated for its physiological effects, flavour, and aroma. It is the second most popular beverage in the world, second only to water in terms of consumption (Kerna et al., 2021). In addition to being an important agricultural product for tropical and subtropical countries, it also plays an essential role for manufacturers in Europe and North America (Jeong et al., 2013).

The chemical composition of coffee ready for consumption is primarily influenced by the processing techniques used on green beans, including the pre-roasting and roasting stages. Furthermore, the methods applied during harvesting, the industrial processing of green coffee, and the ways consumers prepare the beverage also contribute to variations in the concentration of certain compounds in the final product (Pereira et al., 2020; Neto et al., 2021).

The leaves and green seeds of *C. arabica* contain bioactive compounds, including chlorogenic acids (CGAs), flavonoids, and trigonelline. The leaves exhibit higher levels of total phenolics, flavonoids, and CGAs, as well as stronger antioxidant activities compared to the green seeds (Jeon et al., 2024). Coffee pulp is rich in phenolic acids, such as caffeic acid, ferulic acid, and vanillic acid (Clifford, 1985), with 5-O-caffeoylquinic acid being the main phenolic compound (Clifford, 1999). In general, studies indicate that the pulp has beneficial effects for degenerative diseases, such as cancer (Sun et al., 2006), cardiovascular disorders (Kuriyama et al., 2006), and conditions associated with obesity and diabetes (Iso et al., 2006; Odegaard et al., 2008).

According to a 2022 study, coffee's antioxidant capability can be preserved even though high roasting diminishes its phenolic contents (Wu et al., 2023). Another study, in 2023, indicated that during the roasting process, phenolic compounds undergo degradation and the formation of other phenolic compounds, resulting in a slight reduction in total phenolic content (Anh-Dao et al., 2023). Additionally, variations in metabolites, such as caffeine, have been observed among different varieties of *C. arabica* (Happyana, Pratiwi, and Hakim, 2021). Factors such as geographic location and cultivation conditions, including climate and soil properties, influence these differences, resulting in unique characteristics in the beans produced in different regions (Mazzafera and Silvarolla, 2010).

Coffee processing, whether through wet or dry methods, generates various by-products that can be reused in different industrial sectors. Among them are the husk, mucilage, parchment, silver skin, and mainly coffee pulp, which represents the largest volume by-product in wet processing (about 1 ton generated for every 2 tons of coffee produced) (Blinová et al., 2017; Santos et al., 2021). Coffee grounds are rich in bioactive compounds, including polysaccharides, proteins, phenolics, melanoidins, and dietary fibres, the latter accounting for approximately 43% of the dry weight, with potential applications in the formulation of functional foods (Campos-Vega et al., 2015; Vázquez-Sánchez et al., 2018). Another highly relevant by-product is spent coffee grounds, which are continuously generated during beverage preparation through filtration and percolation. Modern technologies enable the full utilisation of this residue, adding value to an abundant and low-cost material while reducing environmental impact, in accordance with the concept of a "zero-waste economy" (Campos-Vega, Loarca-Piña, Vergara-Castañeda, and Oomah, 2015).

Due to the easy availability of raw materials and existing knowledge of their bioactive compounds, *C. arabica* has attracted the attention of researchers for use in the synthesis of metallic nanoparticles or the encapsulation of certain therapeutic constituents (Zuorro et al., 2022; Shaddel et al., 2024).

5. Nanoparticles derived/loaded from *C. arabica*

There has been a growing interest in the development of nanoparticles derived from or functionalised with coffee extracts, particularly those from *C. arabica*, due to their rich content of bioactive compounds, such as chlorogenic acids, caffeine, and flavonoids, which exhibit antioxidant, anti-inflammatory, and antimicrobial properties. Despite the promising potential of these nanostructures for biomedical applications, studies specifically addressing the safety and toxicity profiles of these nanoparticles remain scarce in the literature. A careful evaluation of their biocompatibility, stability, and potential adverse effects is essential to validate their use in therapeutic systems and ensure the translation of these technologies into clinical practice. Table 1 includes three articles that assessed the in vitro safety profile and one in vivo study of coffee-derived nanoparticles, all published within the last five years. Among these studies, two used coffee extract as a reducing agent for the synthesis of metal nanoparticles, one investigated the use of coffee in a nanoemulsion, and another used a caffeine-loaded nanoparticle.

Size, shape, surface charge, and surface functionalisation are factors that contribute to the toxicity of NPs. In addition, their composition, concentration, and

Table 1: In vitro and in vivo studies on the safety profile of coffee-based nanoparticles

Particle type	Size and Zeta potential	Biological model	Main results	Reference
ZnONPs reduced with coffee bean extract	272.8 nm –11.7 to –17.9 mV	VERO E6	IC_{50}: 103 µg mL^{-1}	Abdelmigid et al., 2022
AuNPs derived from coffee extract	20 to 40 nm –24 mV	MRC5	It only states that it did not present cytotoxicity	Trejo-Teniente et al., 2024
Nanoemulsions containing green coffee	200 nm –42.2 to –48.2 mV	BALB/3T3 clone A31	100 µg mL^{-1} is considered non-toxic	Buzanello et al., 2023
Caffeine-loaded chitosan nanoparticles	4 to 12 nm n.i.	Wistar albino rats	20 mg kg^{-1} did not alter different biochemical parameters	Sawie et al., 2023

IC_{50}: Half maximal inhibitory concentration
n.i.: Not informed

the duration and route of exposure, can also influence their toxic effects (Khadanga and Mishra, 2024). In general, the four studies presented different NP sizes (4 to 272.8 nm). Studies report that the toxicity of nanomaterials is directly linked to their size. Smaller particles tend to induce greater toxic effects, especially in aquatic organisms and mammalian cells (Ivask et al., 2014; Kong et al., 2020). This is justified by the fact that smaller particles tend to release more ions (when applicable), interact more easily with cell membranes, and penetrate tissues more easily (Ajdary et al., 2018). Cellular uptake and the production of reactive oxygen species in various organelles are also important deleterious effects (Voigt et al., 2014; Rizzo et al., 2013). Another important issue to note is that while one nanostructure may not be harmful, many others may be, depending on the model used to evaluate biocompatibility. According to certain research, while NPs are harmless in vitro, they have been shown to cause the death of embryos and even adult animals when they are exposed to in vivo models (Voigt et al., 2014; Rizzo et al., 2013).

The surface charge reported in the articles was negative, ranging from –11.7 to –48.2 mV, except for the study by Sawie et al. (2023), which did not provide this data. Positive surface charges and values well above 0 are more likely to cause toxicity in biological media (Valdiglesias et al., 2023; Bhattacharjee et al., 2010). The positive charge facilitates internalisation, increasing cellular uptake and, consequently, causing more cellular damage (Barbalinardo et al., 2025). Cell membranes generally have negative charges (phospholipids, carbohydrate groups, etc.). Positively charged NPs tend to interact more strongly with the membrane, facilitating adsorption or internalisation, which can damage the membrane (Huang et al., 2017).

Abdelmigid et al. (2022) reported that zinc oxide NPs synthesised from coffee waste exhibit a more favourable safety profile compared to those obtained via traditional chemical methods. Cytotoxicity assays in Vero E6 cells showed that the green NPs (ZnONPs_CE) display dose- and time-dependent toxic effects, but with higher IC_{50} values (≈103 μg mL^{-1}) than chemically synthesised ones (≈93 μg mL^{-1}), indicating lower toxicity. This increased biocompatibility is attributed to both the slightly larger particle size and the presence of bioactive coffee compounds that act as reducing and stabilising agents, thereby providing cellular protection. In contrast, the chemically synthesised NPs exhibited a nanorod shape and a greater capacity to induce oxidative stress and membrane damage, enhancing their cytotoxic potential. Thus, the results suggest that the coffee-based green route offers safer NPs for biotechnological and agricultural applications. However, further studies are still needed to assess risks at higher doses and in different biological models.

In another study, the safety of gold NPs synthesised via a green method from *C. arabica* extract and loaded with doxorubicin was evaluated. The results showed that these NPs have controlled sizes (20 to 40 nm), stability under physiological conditions, and the ability to release the drug at tumour pH. In vitro tests demonstrated selectivity, reducing the viability of lung cancer cells (H69) without causing significant harm to healthy cells (MRC5), especially for formulations containing the MUA ligand. Therefore, the study indicates that this nanomaterial is promising and relatively safe as a drug delivery system, with potential to increase therapeutic efficacy and reduce the side effects associated with doxorubicin (Trejo-Teniente et al., 2024).

Buzanello et al. (2023) demonstrated that green coffee nanoemulsions, with an average size of approximately 200 nm and good homogeneity, exhibit an excellent in vitro safety profile for topical cosmetic applications. In cell viability assays (MTT), the formulations maintained over 50% viability in 3T3 fibroblasts, indicating no cytotoxicity according to Organisation for Economic Co-operation and Development criteria. Additionally, they stimulate cell proliferation, suggesting potential beneficial bioactivity. In the alternative ocular irritation test (HET-CAM), no signs of irritation or vascular damage were observed, reinforcing their non-irritating profile. These results indicate a natural, stable, and environmentally friendly formulation (utilising coffee industry waste) that meets safety criteria without the need for animal testing and shows promise for dermatological and cosmetic applications.

Sawie et al. (2023) conducted tests in Wistar rats. They found that zinc oxide NPs synthesised from coffee waste were relatively safe at moderate doses but showed adverse effects at higher concentrations. In acute and subchronic exposures, low doses did not cause changes in haematological, biochemical, or histopathological parameters, indicating good tolerability. However, at higher levels, signs of oxidative stress, liver inflammation, and kidney alterations appeared, suggesting dose-dependent toxicity. Therefore, the study suggests that these NPs have potential for both biomedical and agricultural applications; however, their use requires careful safety evaluation across exposure ranges.

6. Factors influencing the toxicity of NPs

NPs possess unique properties that make them valuable in medicine, industry, and the environment, but their small size and high reactivity can lead to toxicity in biological systems. Factors such as size, shape, surface charge, composition, functionalisation, dose, and exposure duration influence their toxic effects. Smaller particles, elongated shapes or sharp edges, surface charge, and the release of toxic ions or reactive oxygen species can increase cellular stress, inflammation, and apoptosis. Therefore, strict control of the dose and characteristics of NPs is essential to balance efficacy and safety in applications such as drug delivery.

7. Conclusions

Nanotechnology has consolidated itself as one of the most promising and transformative areas of modern science, with wide-ranging applications spanning healthcare, the food industry, and environmental sustainability. However, despite remarkable scientific and technological advances, key knowledge gaps persist, particularly concerning the safety, toxicity, and long-term biocompatibility of nanomaterials. In this regard, the development of NPs derived from plant extracts represents not only a sustainable and eco-friendly alternative but also an opportunity to integrate natural biodiversity into advanced material science.

Coffee, due to its global availability and chemical richness, especially in bioactive compounds such as chlorogenic acids, caffeine, and polyphenols, emerges as a particularly attractive natural source for nanoparticle synthesis. Nevertheless, a critical analysis of the literature indicates that most research in this area remains limited to functional or application-oriented studies. At the same time, the toxicological and safety aspects of coffee-derived NPs are still poorly explored.

Future studies on coffee-related nanomaterials should prioritise comprehensive in vitro and in vivo evaluations to establish standardised safety profiles, as well as a mechanistic understanding of their biological interactions, metabolism, and potential tissue accumulation. Furthermore, integrating predictive in silico modelling, omics-based approaches, and regulatory frameworks (such as OECD and FDA guidelines) could accelerate the translation of these nanomaterials into clinically and industrially viable technologies.

In summary, the scientific production in this field is still in its infancy, but it holds remarkable potential. Strengthening interdisciplinary collaborations between material scientists, toxicologists, and biotechnologists will be crucial to ensure that coffee-derived NPs and other plant-based nanostructures can be safely and effectively applied in therapeutic, cosmetic, and environmental contexts, contributing to the sustainable advancement of nanotechnology.

References

Abdelmigid, H. M., Hussien, N. A., Alyamani, A. A., Morsi, M. M., AlSufyani, N. M., and Kadi, H. A. (2022). Green synthesis of zinc oxide nanoparticles using pomegranate fruit peel and solid coffee grounds vs. chemical method of synthesis, with their biocompatibility and antibacterial properties investigation. *Molecules*, 27(4), 1236. https://doi.org/10.3390/molecules27041236

Aillon, K. L., Xie, Y., El-Gendy, N., Berkland, C. J., and Forrest, M. L. (2009). Effects of nanomaterial physicochemical properties on in vivo toxicity. *Advanced Drug Delivery Reviews*, 61(6), 457–466. https://doi.org/10.1016/j.addr.2009.03.010

Ajdary, M., Moosavi, M., Rahmati, M., Falahati, M., Mahboubi, M., Mandegary, A. et al. (2018). Health concerns of various nanoparticles: A review of their in vitro and in vivo toxicity. *Nanomaterials*, 8(9), 634. https://doi.org/10.3390/nano8090634

Ali, A., Zahid, H. F., Cottrell, J. J., and Dunshea, F. R. (2022). A comparative study for nutritional and phytochemical profiling of *Coffea arabica* (*C. arabica*) from different origins and their antioxidant potential and molecular docking. *Molecules*, 27(16), 5126. https://doi.org/10.3390/molecules27165126

Allen, T. M., and Cullis, P. R. (2013). Liposomal drug delivery systems: From concept to clinical applications. *Advanced Drug Delivery Reviews*, 65(1), 36–48. https://doi.org/10.1016/j.addr.2012.09.037

Anh-Dao, L.-T., Chi-Thien, T., Nhut-Truong, N., Tu-Chi, T., Minh-Huy, D., Thanh-Nho, N. et al. (2024). Changes in the total phenolic contents, chlorogenic acid, and caffeine of coffee cups regarding different brewing methods. *Food Research*, 8(6), 71–79. https://doi.org/10.26656/fr.2017.8(6).279

Anik, M. I., Mahmud, N., Al Masud, A., and Hasan, M. (2021). Gold nanoparticles (GNPs) in biomedical and clinical applications: A review. *Nano Select*, 3(4). https://doi.org/10.1002/nano.202100255

Arrowsmith, J., and Miller, P. (2013). Phase II and Phase III attrition rates 2011–2012. *Nature Reviews Drug Discovery*, 12(8), 569–569. https://doi.org/10.1038/nrd4090

Barbalinardo, M., Chiarini, F., Teti, G., Paganelli, F., Mercadelli, E., Bartoletti, A. et al. (2025). Surface charge overrides protein corona formation in determining the cytotoxicity, cellular uptake, and biodistribution of silver nanoparticles. *ACS Applied Bio Materials*. https://doi.org/10.1021/acsabm.5c00392

Bayda, S., Adeel, M., Tuccinardi, T., Cordani, M., and Rizzolio, F. (2019). The history of nanoscience and nanotechnology: From chemical–physical applications to nanomedicine. *Molecules*, 25(1), 112. https://doi.org/10.3390/molecules25010112

Bhattacharjee, S., Haan, L. H. J. D., Evers, N. M., Jiang, X., Marcelis, A. T. M., Zuilhof, H. et al. (2010). Role of surface charge and oxidative stress in cytotoxicity of organic monolayer-coated silicon nanoparticles towards macrophage NR8383 cells. *Particle and Fibre Toxicology*, 7(1), 25. https://doi.org/10.1186/1743-8977-7-25

Blinová, L., Sirotiak, M., Bartošová, A., and Soldán, M. (2017). Review: Utilization of waste from coffee production. *Research Papers Faculty of Materials Science and Technology*, Slovak University of Technology, 25(40), 91–101. https://doi.org/10.1515/rput-2017-0011

Bobo, D., Robinson, K. J., Islam, J., Thurecht, K. J., and Corrie, S. R. (2016). Nanoparticle-based medicines: A review of FDA-approved materials and clinical trials to date. *Pharmaceutical Research*, 33(10), 2373–2387. https://doi.org/10.1007/s11095-016-1958-5

Burden, N., Clift, M. J. D., Jenkins, G. J. S., Labram, B., and Sewell, F. (2021). Opportunities and challenges for integrating new in vitro methodologies in hazard testing and risk assessment. *Small*, 17(15), 2006298. https://doi.org/10.1002/smll.202006298

Buzanello, E. B., Lopes, S., Coelho, D. S., Voyten, A. P., Fanan, S., Mazzarino, L. et al. (2023). Biological activities of green coffee nanoemulsions evaluated through alternative methods: MTT, cellular proliferation, and HET-CAM assays. *Biotechnology Research & Innovation*, 7(1), e2023011–e2023011. https://doi.org/10.4322/biori.00112023

Campos-Vega, R., Loarca-Piña, G., Vergara-Castañeda, H. A., and Oomah, B. D. (2015). Spent coffee grounds: A review on current research and future prospects. *Trends in Food Science & Technology*, 45(1), 24–36. https://doi.org/10.1016/j.tifs.2015.04.012

Campos-Vega, R., Vázquez-Sánchez, K., López-Barrera, D., Loarca-Piña, G., Mendoza-Díaz, S., and Oomah, B. D. (2015). Simulated gastrointestinal digestion and in vitro colonic fermentation of spent coffee (*Coffea arabica* L.): Bioaccessibility and intestinal permeability. *Food Research International*, 77, 156–161. https://doi.org/10.1016/j.foodres.2015.07.024

Caprioli, G., Cortese, M., Sagratini, G., and Vittori, S. (2015). The influence of different types of preparation (espresso and brew) on coffee aroma and main bioactive constituents. *International Journal of Food Sciences and Nutrition*, 66(5), 505–513. https://doi.org/10.3109/09637486.2015.1064871

Castro, A. C. C. M., Oda, F. B., Almeida-Cincotto, M. G. J., Davanço, M. G., Chiari-Andréo, B. G., Cicarelli, R. M. B. et al. (2018). Green coffee seed residue: A sustainable source of antioxidant compounds. *Food Chemistry*, 246, 48–57. https://doi.org/10.1016/j.foodchem.2017.10.153

Chanthapong, P., Maensiri, D., Rangsrisak, P., Jaiyan, T., Rahaeng, K., Oraintara, A. et al. (2025). Plant-based ZnO nanoparticles for green nanobiocontrol of a highly virulent bacterial leaf blight pathogen: Mechanistic insights and biocompatibility evaluation. *Nanomaterials*, 15(13), 1011–1011. https://doi.org/10.3390/nano15131011

Chen, R. J., Chen, Y. Y., Liao, M. Y., Lee, Y. H., Chen, Z. Y., Yan, S. J. et al. (2020). The current understanding of autophagy in nanomaterial toxicity and its implementation in safety assessment-related alternative testing strategies. *International Journal of Molecular Sciences*, 21, 2387. https://doi.org/10.3390/ijms21072387

Chenxi, Z., Hemmat, A., Thi, N., & Afrand, M. (2025). Nanoparticle-enhanced drug delivery systems: An up-to-date review. *Journal of Molecular Liquids*, 126999. https://doi.org/10.1016/j.molliq.2025.126999

Clifford, M. N. (1985). The composition of green and roasted coffee beans. *In:* Clifford, M. N., Willson, K.C. (Eds.). Coffee Botany, Biochemistry and Production of Beans and Beverage (pp. 457). Chapman and Hall, London, UK.

Clifford, M. N. (1999). Chlorogenic acids and other cinnamates: Nature, occurrence and dietary burden. *Journal of Science of Food Agriculture*, 79, 362–372. https://doi.org/10.1002/(SICI)1097-0010(19990301)79:3<362::AID-JSFA256>3.0.CO;2-D

Clift, M. J. D., Jenkins, G. J. S., and Doak, S. H. (2020). An alternative perspective towards reducing the risk of engineered nanomaterials to human health. *Small*, 16(36), 2002002. https://doi.org/10.1002/smll.202002002

de Melo Pereira, G. V., de Carvalho Neto, D. P., Magalhães Júnior, A. I., do Prado, F. G., Pagnoncelli, M. G. B., Karp, S. G. et al. (2020). Chemical composition and health properties of coffee and coffee by-products. *Advances in Food and Nutrition Research*, 91, 65–96. https://doi.org/10.1016/bs.afnr.2019.10.002

Domb, A. J., Sharifzadeh, G., Nahum, V., and Hosseinkhani, H. (2021). Safety evaluation of nanotechnology products. *Pharmaceutics*, 13(10), 1615. https://doi.org/10.3390/pharmaceutics13101615

Ferreira, D., Fontinha, D., Martins, C., Pires, D., Fernandes, A. R., and Baptista, P. V. (2020). Gold nanoparticles for vectorization of nucleic acids for cancer therapeutics. *Molecules*, 25(15), 3489. https://doi.org/10.3390/molecules25153489

Feynman, R. P. (1960). There's plenty of room at the bottom. *Engineering Science*, 23, 22–36.

Fortune, A., Aime, A., Raymond, D., and Kumar, S. (2025). Nanotechnology in medicine: A double-edged sword for health outcomes. *Health Nanotechnology*, 1(1). https://doi.org/10.1186/s44301-025-00008-2

Galanakis, Charis M. (Ed.). (2017). *Handbook of Coffee Processing By-Products: Sustainable Applications*. 1. ed. London: Academic Press, Elsevier, 426 p. ISBN: 978-0-12-811290-8

Gatoo, M. A., Naseem, S., Arfat, M. Y., Dar, A. M., Qasim, K., and Zubair, S. (2014). Physicochemical properties of nanomaterials: Implication in associated toxic manifestations. *BioMed Research International*, 2014, 498420. https://doi.org/10.1155/2014/498420

Giri, P. M., Banerjee, A., and Layek, B. (2023). A recent review on cancer nanomedicine. *Cancers*, 15, 2256. https://doi.org/10.3390/cancers15082256

Gujjar, S., Kukal, S., Jayabal, P., Balaji, N., Sainger, S., Roy, S. et al. (2025). Nanomaterials for biomedical applications: Addressing regulatory hurdles and strategic solutions. *Nano Trends*, 11, 100127–100127. https://doi.org/10.1016/j.nwnano.2025.100127

Haleem, A., Javaid, M., Singh, R. P., Rab, S., & Suman, R. (2023). Applications of nanotechnology in medical field. *Global Health Journal*, 7(2), 70–77. https://doi.org/10.1016/j.glohj.2023.02.008

Han, P., Li, X., Yang, J., Zhang, Y., & Chen, J. (2024). Advancing toxicity predictions: A review on in vitro to in vivo extrapolation in next-generation risk assessment. *Environment & Health*, 2(7), 499–513. https://doi.org/10.1021/envhealth.4c00043

Happyana, N., Pratiwi, A., and Hakim, E. H. (2021). Metabolite profiles of the green beans of Indonesian arabica coffee varieties. *International Journal of Food Science*, 2021, 1–9. https://doi.org/10.1155/2021/5782578

Huang, Y.-W., Cambre, M., and Lee, H.-J. (2017). The toxicity of nanoparticles depends on multiple molecular and physicochemical mechanisms. *International Journal of Molecular Sciences*, 18(12), 2702. https://doi.org/10.3390/ijms18122702

Hughes, K. J., Ganesan, M., Rumiana Tenchov, Iyer, K. A., Krittika Ralhan, Diaz, L. L. et al. (2025). Nanoscience in action: Unveiling emerging trends in materials and applications. *ACS Omega*, 10(8), 7530–7548. https://doi.org/10.1021/acsomega.4c10929

Hund-Rinke, K., Baun, A., Cupi, D., Fernandes, T. F., Handy, R., Kinross, J. H. et al. (2016). Regulatory ecotoxicity testing of nanomaterials – Proposed modifications of OECD test guidelines based on laboratory experience with silver and titanium dioxide nanoparticles. *Nanotoxicology*, 10(10), 1442–1447. https://doi.org/10.1080/17435390.2016.1229517

Iso, H., Date, C., Wakai, K., Fukui, M., Tamakoshi, A., and JACC Study Group. (2006). The relationship between green tea and total caffeine intake and risk for self-reported type 2 diabetes among Japanese adults. *Annals of Internal Medicine*, 144(8), 554–562. https://doi.org/10.7326/0003-4819-144-8-200604180-00005

Ivask, A., Kurvet, I., Kasemets, K., Blinova, I., Aruoja, V., Suppi, S. et al. (2014). Size-dependent toxicity of silver nanoparticles to bacteria, yeast, algae, crustaceans and mammalian cells in vitro. *PLoS ONE*, 9(7), e102108. https://doi.org/10.1371/journal.pone.0102108

Jeon, Y. A., Natraj, P., Kim, S. C., Moon, J.-K., and Lee, Y. J. (2024). Comparative analysis of phytochemical and functional profiles of arabica coffee leaves and green beans across different cultivars. *Foods*, 13(23), 3744. https://doi.org/10.3390/foods13233744

Jeong, J. H., Jeong, H. R., Jo, Y. N., Kim, H. J., Lee, U., and Heo, H. J. (2013). Antioxidant and neuronal cell protective effects of Columbia arabica coffee with different roasting conditions. *Preventive Nutrition and Food Science*, 18(1), 30–37. https://doi.org/10.3746/pnf.2013.18.1.030

Jia, Y., Jiang, Y., He, Y., Zhang, W., Zou, J., Magar, K. T. et al. (2023). Approved nanomedicine against diseases. *Pharmaceutics*, 15(3), 774. https://doi.org/10.3390/pharmaceutics15030774

Kadam, S. S., and Bhangale, C. J. (2024). A review on challenges and impact of green chemistry on environment. *Journal of Pharmaceutical Research International*, 36, 39–50. https://doi.org/10.9734/jpri/2024/v36i7753

Karnwal, A., Jassim, A. Y., Mohammed, A. A., Sharma, V., Said, M., and Sivanesan, I. (2024). Nanotechnology for healthcare: Plant-derived nanoparticles in disease treatment and regenerative medicine. *Pharmaceuticals*, 17(12), 1711–1711. https://doi.org/10.3390/ph17121711

Kazemi, S., Hosseingholian, A., Gohari, S. D., Feirahi, F., Moammeri, F., Mesbahian, G. et al. (2023). Recent advances in green synthesized nanoparticles: From production to application. *Materials Today Sustainability*, 24, 100500. https://doi.org/10.1016/j.mtsust.2023.100500

Kazi, R. N. A., Hasani, I. W., Khafaga, D. S. R., Kabba, S., Farhan, M., Aatif, M. et al. (2025). Nanomedicine: The effective role of nanomaterials in healthcare from diagnosis to therapy. *Pharmaceutics*, 17(8), 987. https://doi.org/10.3390/pharmaceutics17080987

Kerna, N. A., Rashid, S., Chawla, S., Carsrud, N. D. V., Anderson, I. I. J., Pruitt, K. D. et al. (2021). Coffee drinking and the human heart: Cardio-supportive effects of accustomed coffee-consumption. *EC Cardiology*, 8, 30–38. https://doi.org/10.31080/eccy.2021.08.00858

Khadanga, V., and Mishra, P. C. (2024). A review on toxicity mechanism and risk factors of nanoparticles in respiratory tract. *Toxicology*, 504, 153781. https://doi.org/10.1016/j.tox.2024.153781

Kher, C., and Kumar, S. (2022). The application of nanotechnology and nanomaterials in cancer diagnosis and treatment: A review. *Cureus*. https://doi.org/10.7759/cureus.29059

Kong, I. C., Ko, K.-S., Koh, D.-C., and Chon, C.-M. (2020). Comparative effects of particle sizes of cobalt nanoparticles to nine biological activities. *International Journal of Molecular Sciences*, 21(18), 6767. https://doi.org/10.3390/ijms21186767

Kuriyama, S., Shimazu, T., Ohmori, K., Kikuchi, N., Nakaya, N., Nishino, Y. et al. (2006). Green tea consumption and mortality due to cardiovascular disease, cancer, and all causes in Japan: The Ohsaki study. *JAMA*, 296(10), 1255–1265. https://doi.org/10.1001/jama.296.10.1255

Lithi, I. J., Ahmed Nakib, K. I., Chowdhury, A. M. S., and Sahadat Hossain, Md. (2025). A review on the green synthesis of metal (Ag, Cu, and Au) and metal oxide (ZnO, MgO, Co_3O_4, and TiO_2) nanoparticles using plant extracts for developing antimicrobial properties. *Nanoscale Advances*, 7(9), 2446–2473. https://doi.org/10.1039/d5na00037h

Lu, Q., Yu, H., Zhao, T., Zhu, G., and Li, X. (2023). Nanoparticles with transformable physicochemical properties for overcoming biological barriers. *Nanoscale*, 15(32), 13202–13223. https://doi.org/10.1039/D3NR01332D

Ma, X., Tian, Y., Yang, R., Wang, H., Allahou, L. W., Chang, J. et al. (2024). Nanotechnology in healthcare, and its safety and environmental risks. *Journal of Nanobiotechnology*, 22(1). https://doi.org/10.1186/s12951-024-02901-x

Mazzafera, P., and Silvarolla, M. B. (2010). Caffeine content variation in single green arabica coffee seeds. *Seed Science Research*, 20(3), 163–167. https://doi.org/10.1017/s0960258510000140

McCook, S. (2024). Coffee. *The Oxford Handbook of Agricultural History*, 343–360. https://doi.org/10.1093/oxfordhb/9780190924164.013.18

Mishra, A. (2021). A brief review on pre-clinical studies and phases of clinical trials. *Journal of Pharmacovigilance*, 9, 330. https://doi.org/10.35248/23296887.21.9.330.

Morais, M., Teixeira, A. L., Dias, F., Machado, V., Medeiros, R., and Prior, J. A. V. (2020). Cytotoxic effect of silver nanoparticles synthesized by green methods in cancer. *Journal of Medicinal Chemistry*, 63(23), 14308–14335. https://doi.org/10.1021/acs.jmedchem.0c01055

Mukherjee, P., Roy, S., Ghosh, D., and Nandi, S. K. (2022). Role of animal models in biomedical research: A review. *Laboratory Animal Research*, 38(1). https://doi.org/10.1186/s42826-022-00128-1

Nair, D. G., and Ralf Weiskirchen. (2024). Advanced in vitro models for preclinical drug safety: Recent progress and prospects. *Current Issues in Molecular Biology*, 47(1), 7–7. https://doi.org/10.3390/cimb47010007

Oberdörster, G., Oberdörster, E., and Oberdörster, J. (2005). Nanotoxicology: An emerging discipline evolving from studies of ultrafine particles. *Environmental Health Perspectives*, 113(7), 823–839. https://doi.org/10.1289/ehp.7339

Odegaard, A. O., Pereira, M. A., Koh, W.-P., Arakawa, K., Lee, H.-P., and Yu, M. C. (2008). Coffee, tea, and incident type 2 diabetes: The Singapore Chinese health study. *The American Journal of Clinical Nutrition*, 88(4), 979–985. https://doi.org/10.1093/ajcn/88.4.979

Paiva, P. P., Nunes, J. H. B., Nonato, F. R., Ruiz, A. L. T. G., Zafred, R. R. T., Sousa, I. M. O. et al. (2020). In silico, in vitro, and in vivo antitumor and anti-inflammatory evaluation of a standardized alkaloid-enriched fraction obtained from *Boehmeria caudata* Sw. aerial parts. *Molecules*, 25(17), 4018–4018. https://doi.org/10.3390/molecules25174018

Pathak, G., Swati Mangla, Gupta, G. K., Bhan, V., and Kapoor, R. K. (2025). Toxicological assessment and risk management of nanoparticles mediated composite materials – Critical review: State of the art. *Discover Polymers*, 2(1). https://doi.org/10.1007/s44347-025-00023-7

Polson, A. G., and Fuji, R. N. (2012). The successes and limitations of preclinical studies in predicting the pharmacodynamics and safety of cell-surface-targeted biological agents in patients. *British Journal of Pharmacology*, 166(5), 1600–1602. https://doi.org/10.1111/j.1476-5381.2012.01916.x

Poon, W., Zhang, Y.-N., Ouyang, B., Kingston, B. R., Wu, J. L. Y., Wilhelm, S. et al. (2019). Elimination pathways of nanoparticles. *ACS Nano*, 13(5), 5785–5798. https://doi.org/10.1021/acsnano.9b01383

Pridgeon, C. S., Schlott, C., Wong, M. W., Heringa, M. B., Heckel, T., Leedale, J. et al. (2018). Innovative organotypic in vitro models for safety assessment: Aligning with regulatory requirements and understanding models of the heart, skin, and liver as paradigms. *Archives of Toxicology*, 92(2), 557–569. https://doi.org/10.1007/s00204-018-2152-9

Rasmussen, K., Sayre, P., Kobe, A., Gonzalez, M., and Rauscher, H. (2025). 25 years of research and regulation: Is nanotechnology safe to commercialize? *Frontiers in Toxicology*, 7. https://doi.org/10.3389/ftox.2025.1629813

Resnik, D. B., and Tinkle, S. S. (2007). Ethics in nanomedicine. *Nanomedicine*, 2(3), 345–350. https://doi.org/10.2217/17435889.2.3.345

Rizzo, L. Y., Golombek, S. K., Mertens, M. E., Pan, Y., Laaf, D., Broda, J. et al. (2013). In vivo nanotoxicity testing using the zebrafish embryo assay. *Journal of Materials Chemistry B*, 1, 3918–3925. https://doi.org/10.1039/C3TB20528B

Sandvig, K., Iversen, T. G., and Skotland, T. (2024). Entry of nanoparticles into cells and tissues: Status and challenges. *Beilstein Journal of Nanotechnology*, 15, 1017–1029. https://doi.org/10.3762/bjnano.15.83

Santos, É. M. dos, Macedo, L. M. de, Tundisi, L. L., Ataide, J. A., Camargo, G. A., Alves, R. C. et al. (2021). Coffee by-products in topical formulations: A review. *Trends in Food Science & Technology*, 111, 280–291. https://doi.org/10.1016/j.tifs.2021.02.064

Sawie, H. G., Khadrawy, Y. A., El-Gizawy, M. M., Mourad, H. H., Omara, E. A., and Hosny, E. N. (2023). Effect of alpha-lipoic acid and caffeine-loaded chitosan nanoparticles on obesity and its complications in liver and kidney in rats. *Naunyn-Schmiedeberg S Archives of Pharmacology*, 396(11), 3017–3031. https://doi.org/10.1007/s00210-023-02507-4

Senjab, R. M., Nour AlSawaftah, Abuwatfa, W. H., and Husseini, G. A. (2024). Advances in liposomal nanotechnology: From concept to clinics. *RSC Pharmaceutics*, 1(5), 928–948. https://doi.org/10.1039/d4pm00176a

Shaddel, R., Akbari-Alavijeh, S., Cacciotti, I., Yousefi, S., Tomas, M., Capanoglu, E. et al. (2024). Caffeine-loaded nano/micro-carriers: Techniques, bioavailability, and applications. *Critical Reviews in Food Science and Nutrition*, 64(15), 4940–4965. https://doi.org/10.1080/10408398.2022.2147143.

Sharifi, S., Mahmoud, N. N., Voke, E., Landry, M. P., & Mahmoudi, M. (2022). Importance of standardizing analytical characterization methodology for improved reliability of the nanomedicine literature. *Nano-Micro Letters*, 14(1). https://doi.org/10.1007/s40820-022-00922-5

Silbergeld, E. K., Mandrioli, D., and Cranor, C. F. (2015). Regulating chemicals: Law, science, and the unbearable burdens of regulation. *Annual Review of Public Health*, 36(1), 175–191. https://doi.org/10.1146/annurev-publhealth-031914-122654

Sun, C.-L., Yuan, J.-M., Koh, W.-P., and Yu, M. C. (2005). Green tea, black tea and breast cancer risk: A meta-analysis of epidemiological studies. *Carcinogenesis*, 27(7), 1310–1315. https://doi.org/10.1093/carcin/bgi276

Taka, L. A., Tata, C. M., Klink, M. J., Mbianda, X. Y., Mtunzi, F. M., and Naidoo, E. B. (2021). A review on conventional and advanced methods for nanotoxicology evaluation of engineered nanomaterials. *Molecules*, 26(21), 6536. https://doi.org/10.3390/molecules26216536

Taniguchi., N. (1974). On the basic concept of "nano-technology" in the basic concept of "nano-technology". *Proceedings of the International Conference on Production Engineering Tokyo, Part II*. Japan Society of Precision Engineering, Tokyo.

Trejo-Teniente, I., Jaramillo-Loranca, B. E., Vargas-Hernández, G., Villanueva-Ibáñez, M., Tovar-Jiménez, X., Olvera-Venegas, P. N. et al. (2024). Synthesis and toxicity assessment of *Coffea arabica* extract-derived gold nanoparticles loaded with doxorubicin in lung cancer cell cultures. *Frontiers in Bioengineering and Biotechnology*, 12. https://doi.org/10.3389/fbioe.2024.1378601

Valdiglesias, V., Paz, M., Touzani, A., Baúlde, S., Mosquera, J., Criado, A. et al. (2023). Influence of surface charge on biological behaviour of gold nanoparticles in human SH-SY5Y neuronal cells. *Materials Proceedings*, 14(1), 52. https://doi.org/10.3390/IOCN2023-14516

Vázquez-Sánchez, K., Martinez-Saez, N., Rebollo-Hernanz, M., del Castillo, M. D., Gaytán-Martínez, M., and Campos-Vega, R. (2018). In vitro health promoting properties of antioxidant dietary fiber extracted from spent coffee (*Coffee arabica* L.) grounds. *Food Chemistry*, 261, 253–259. https://doi.org/10.1016/j.foodchem.2018.04.064

Voigt, N., Henrich-Noack, P., Kockentiedt, S., Hintz, W., Tomas, J., and Sabel, B. A. (2014). Toxicity of polymeric nanoparticles in vivo and in vitro. *Journal of Nanoparticle Research*, 16(6), 2379. https://doi.org/10.1007/s11051-014-2379-1

Wang, S., Li, J., Zhang, Z., Cao, S., Zhang, Z., Bian, Y. et al. (2025). Advances in nanomedicine and delivery systems for gastric cancer research. *Frontiers in Bioengineering and Biotechnology*. 13, 1565999. https://doi.org/10.3389/fbioe.2025.1565999

Wu, H., Lu, P., Liu, Z., Sharifi-Rad, J., and Suleria, H. A. R. (2022). Impact of roasting on the phenolic and volatile compounds in coffee beans. *Food Science & Nutrition*, 10(7), 2408–2425. https://doi.org/10.1002/fsn3.2849

Yusuf, A., Almotairy, A. R. Z., Henidi, H., Alshehri, O. Y., and Aldughaim, M. S. (2023). Nanoparticles as drug delivery systems: A review of the implication of nanoparticles' physicochemical properties on responses in biological systems. *Polymers*, 15(7), 1596. https://doi.org/10.3390/polym15071596

Zuorro, A., Iannone, A., Miglietta, S., and Lavecchia, R. (2022). Green synthesis of silver nanoparticles using spent coffee ground extracts: Process modelling and optimization. *Nanomaterials*, 12(15), 2597. https://doi.org/10.3390/nano12152597

Zurita-Cruz, J. N., Barbosa-Cortés, L., and Villasís-Keever, M. Á. (2019). From research to practice: Clinical phases for drug development. *Revista Alergia Mexico*, 66(2), 246–253. https://doi.org/10.29262/ram.v66i2.625

CHAPTER
11

Toxicological Aspects of Nanotechnology in Coffee Science

Ana Carolina Penna dos Santos, Joseana Antunes Porciuncula, Altevir Rossato Viana and Cristiano Rodrigo Bohn Rhoden*
Programa de Pós-graduação em Nanociências, Universidade Franciscana – UFN, Santa Maria, RS, Brazil

Although the coffee production chain plays a significant role in the global economy, it also generates substantial amounts of organic waste that can harm the environment if not properly managed. While these by-products possess valuable physicochemical properties that enable their conversion into functional materials, a clear gap persists between laboratory-scale research and the implementation of reused technologies in real-world, small-scale applications. This chapter aims to demonstrate how advanced processing routes can transform coffee-derived residues into high-value nanomaterials suitable for civil construction, wastewater treatment, packaging, and energy-related applications. Recent studies have highlighted the potential of techniques such as pyrolysis, chemical activation, and hydrothermal synthesis to produce nanostructured materials with high porosity, structural stability, and excellent functional performance. These characteristics make them promising candidates for a range of sustainable technological uses. In summary, the valorisation of coffee waste represents an innovative and environmentally responsible approach that not only mitigates ecological impacts but also supports the circular bioeconomy by converting an abundant residue into economically and socially valuable resources.

1. Introduction

Coffee has played a significant role in the economy and culture, particularly in Brazil, where, since the 19th century, it has been one of the country's most important agricultural products (Figure 1). Globally recognised as one of Brazil's most important agricultural products, its cultivation generates jobs, boosts exports, and

*Corresponding author: cristianorbr@gmail.com

DOI: 10.1201/9781003689393-11

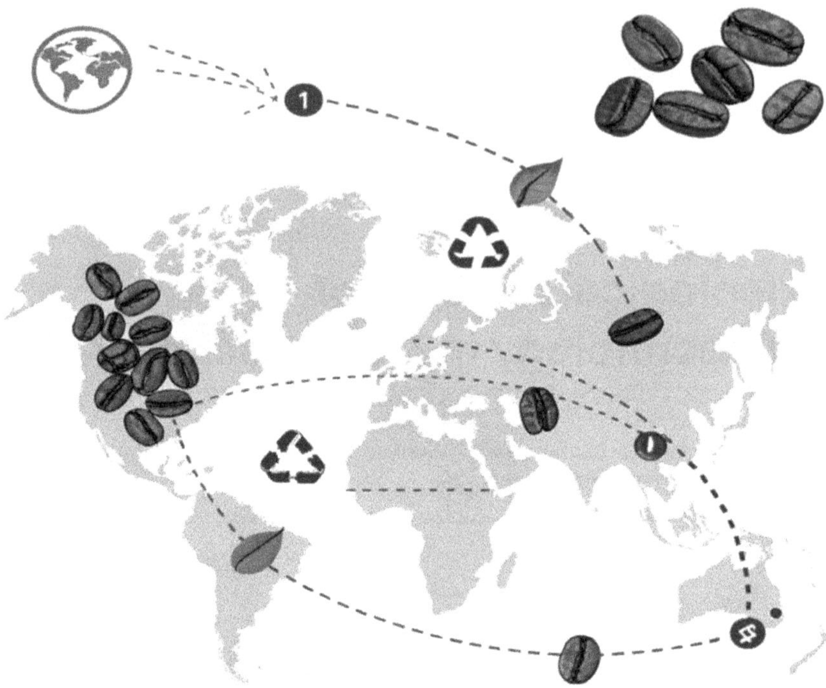

Figure 1: Coffee route

becomes part of the daily life of Brazilians, representing hospitality and tradition. This shows how important it is to Brazilian society in many ways (Barreto Peixoto et al., 2023; Santos et al, 2023).

At the same time, the coffee manufacturing cycle generates a lot of waste that, without proper disposal, can have serious effects on the environment, such as water pollution and the release of greenhouse gases like methane and carbon dioxide (Freitas and Renó, 2022). Pulp, husk, and liquid effluents are byproducts that contain a lot of organic matter and can pollute water sources not only near production sites (Cangussu et al., 2021; Tamilselvan et al., 2024). If these wastes are disposed of untreated, they can cause significant environmental problems and affect local ecosystems (Machado et al., 2023). Liquid waste contains phenolic compounds and caffeine, which may alter aquatic species' biodiversity (Franca and Oliveira, 2022). This reveals a critical gap in the sustainable management of coffee by-products and highlights the need for scalable and low-cost technologies capable of transforming these residues into high-value-added products (Canton, 2021).

It is important to develop long-term plans that encourage the reuse of these residual resources. A new idea is to transform coffee waste into high-value-added nanomaterials. This will help solve environmental problems and promote technological progress in areas such as the environment, energy, health, agriculture, and construction (Rhoden et al., 2021; Pongsiriyakul et al., 2024; Yusufoğlu et al., 2024).

In this way, nanotechnology enables the manipulation of materials at the nanoscale (10^{-9} m), improving their physical, chemical, and mechanical properties compared to full-scale materials. Furthermore, the development of nanoscale materials offers a range of benefits, such as increased surface area, porosity, and the potential for surface functionalization (El-Naggar et al., 2024).

It has been demonstrated that it is possible to make metal and carbon nanoparticles from coffee grounds. Metallic nanoparticles have proven to be extremely effective in water remediation (Vargas et al., 2023), can be easily synthesised (Barbieri et al., 2024), are responsive to magnetic fields (Harres et al., 2023), and exhibit low toxicity (De Oliveira et al., 2023). Thus, encouraging their use in sensors and supercapacitors as photocatalytic and antibacterial agents (El-Desouky et al., 2023; Gabisa and Ratanatamskul, 2024; Reyes-Pérez et al., 2023; De Oliveira et al., 2022).

A possible counter-argument would be that these technologies are difficult to apply and very expensive, especially for small producers who often have limited resources. However, the circular economy is advancing, and new research, cooperative networks, and incentive-based public policies will gradually break down these barriers, facilitating access to sustainable practices and new technologies for everyone.

The valorisation of coffee residues aligns with the principles of the circular bio economy and several United Nations Sustainable Development Goals (SDGs), particularly SDG 12 (Responsible Consumption and Production) and SDG 13 (Climate Action). By promoting waste-to-resource strategies, such initiatives strengthen local economies, generate green jobs, and foster technological sovereignty in developing countries such as Brazil (Tsigkou et al., 2025).

In recent years, studies have focused on finding innovative and sustainable applications for coffee waste, such as the grounds, husk, and silver skin, which contain valuable organic and inorganic compounds (Johnson et al., 2022; Bijla et al., 2022). These by-products have been used as alternative raw materials for the production of biopolymers, adsorbents, and nanocomposites, promoting the concept of a circular economy and reducing dependence on synthetic materials derived from petroleum (Bomfim et al., 2022; Bairwan et al., 2025).

Recent research has demonstrated that incorporating coffee waste into polymer matrices, such as HDPE and ABS, enhances both mechanical and thermal properties, while also reducing the environmental impact of disposal (Amena et al., 2022). This type of innovation is fundamental for the advancement of sustainable materials with potential applications in packaging, construction, and technological devices.

Moreover, other approaches explore the conversion of coffee waste into energy sources and high-value-added chemicals, such as biochars, metal nanoparticles, and biodegradable films (Ko et al., 2025; Zheng et al., 2025; Limand Foo, 2024). Such advances have shown promising results in areas such as water purification, photocatalysis, and biofuel production (Changotra et al., 2024; Mohammadpour et al., 2024).

The transformation of agro-industrial waste into functional and technologically relevant materials reflects a paradigm shift, where waste is reinterpreted as a resource. This sustainable transition not only reduces environmental damage but also stimulates scientific innovation and the creation of more circular and resilient production chains (Onsongo et al., 2025; Maraveas, 2024).

Therefore, this chapter aims to explore the potential of coffee-derived nanomaterials for environmental and technological applications, discussing synthesis routes, material properties, and the sustainable perspectives that connect nanoscience with circular economy strategies.

2. Main wastes and their potential

2.1 Coffee grounds

Compared to other agricultural residues, such as rice husks and sugarcane bagasse, coffee residues demonstrate a higher content of lipids and bioactive compounds, which expands their applicability in biorefinery schemes. For example, while rice husks are often used for silica extraction and sugarcane bagasse for bioethanol production, spent coffee grounds offer the potential for both biofuel production and the synthesis of advanced nanomaterials (Bijla et al., 2022; Machado et al., 2023).

The used and discarded coffee grounds, known as spent coffee grounds (SCG), have a complex chemical composition that makes them highly promising for applications in different sectors. In addition to the already highlighted lignocellulose, recent studies have identified the presence of proteins, minerals (such as potassium, calcium, and magnesium), and high-value-added bioactive compounds, including caffeine and chlorogenic acids. The diversity of molecules present in these residues allows for the exploration of various technological routes: from the production of liquid biofuels (such as biodiesel and bioethanol) to the synthesis of functional nanomaterials aimed at the adsorption of emerging contaminants.

Typically, SCG contain around 45–50% carbohydrates (mainly cellulose and hemicellulose), 10–15% proteins, and 15–20% lipids, with variable levels of phenolic compounds and caffeine depending on the coffee variety and brewing conditions (Kovalcik et al., 2018; Choe, 2025).

Compared with other agricultural biomasses, such as sugarcane bagasse and rice husk, SCG has a higher lipid content, favouring its use in transesterification processes for biodiesel production (Kamil et al., 2019). From an environmental perspective, the valorisation of this waste directly mitigates impacts such as accumulation in landfills and the generation of leachate rich in potentially contaminating organic substances.

Beyond biofuel generation, SCG has emerged as a promising precursor for carbon-based nanomaterials, such as activated carbon, carbon dots, and graphene-like nanosheets. These materials exhibited remarkable adsorption and electrochemical properties, enabling their use in environmental remediation and energy storage devices (Ko et al., 2025; Ngene and Nzihou, 2025).

Pomace is the solid waste left behind after making the drink. For a long time, it was discarded without being put to any meaningful use. Recent studies, on the other hand, have shown that this material has a highly varied composition and can be utilised in creative and environmentally friendly ways. The pomace contains a significant amount of lignocellulose, a group of plant fibres that can be converted into biofuels such as bioethanol. This encourages the use of renewable energy sources (Mukherjee et al., 2022). It also contains lipids, which can be used to produce biofuels and natural cosmetics (Johnson et al., 2022).

Phenolic chemicals are another important component, as they are known for their ability to combat free radicals. Researchers are investigating these molecules for use in the food industry, where they may enhance and improve products with functional elements that support the body's defence against oxidative stress (Pyrzynska, 2025). Coffee grounds contain antibacterial properties, which are believed to originate from the polyphenols that comprise them (Chien and Chen, 2022).

Therefore, the comprehensive valorisation of SCG integrates waste minimisation with resource recovery, embodying the core principles of the circular bioeconomy. Such strategies promote environmental protection while generating economic and technological value for coffee-producing regions (Tsigkou et al., 2025; Ali et al., 2025).

In this sense, coffee grounds are a valuable source of bioactive substances and fibres that can be utilised in various ways, such as generating renewable energy, producing functional meals, and creating environmentally degradable items. Using it not only has a less detrimental effect on the environment, but it also contributes value to the economy, which helps the chain be more productive.

The following sections explore other coffee-derived residues, highlighting their specific characteristics and technological potential within integrated biorefinery frameworks.

2.2 Bark and mucilage

The coffee husk and mucilage represent significant fractions of the solid waste generated during coffee processing. From a chemical standpoint, the husk has a high content of insoluble fibres, lignin, and cellulose, making it a suitable material for the production of biocomposites and biochar. Recent research has explored the carbonisation of the husk to obtain high surface area adsorbents, applicable in the removal of dyes and heavy metals. Mucilage, rich in soluble sugars and pectins, has high potential for use in the formulation of biopolymers and hydrophilic gels with functional properties.

On average, coffee husk contains about 30–35% cellulose, 20–25% hemicellulose, and 20–25% lignin, while mucilage presents up to 85% soluble sugars and 10% pectins (Gil-Gómez et al., 2024; Cruz et al., 2024).

Moreover, studies indicate that the controlled fermentation of mucilage can release organic acids of interest to the food and pharmaceutical industries. From the perspective of nanotechnology, these residues can be used as precursors for obtaining nanocellulose and polysaccharide-loaded nanoparticles, with applications in smart packaging, antimicrobial coatings, and controlled drug release systems.

Although often discarded, both husk and mucilage possess unique structural and chemical properties that make them valuable for sustainable product development. Due to its high content of insoluble fibres, coffee husk can be transformed into functional components for food formulations and eco-friendly biocomposites. These fibres may be used to make functional components that can be used in food and in the making of biocomposites (Amena et al., 2022).

The mucilage, which is rich in soluble components, can be used as a natural thickener and possesses prebiotic properties that promote the growth of beneficial bacteria in the colon (Tamilselvan et al., 2024). They can work together to make functional supplements and bio-packaging (Oliveira et al., 2021; Liu et al., 2023). In addition to these residues from wet processing, coffee roasting also generates by-products with high functional potential, such as the silver skin.

The silver skin is the finest residue generated during the roasting of coffee beans. Although in small quantities, its composition concentrates natural antioxidants such as polyphenols and flavonoids, as well as dietary fibres and lipids. These constituents give the film a high biotechnological value, especially for the formulation of nutritional supplements and cosmetics. From a nanotechnology perspective, silver skin has been investigated as a precursor for fluorescent carbon dots, exhibiting interesting optical properties for bioimaging and chemical sensing applications.

Other routes include pyrolysis to obtain bio-oils and acid hydrolysis to extract nanocellulose, useful in biodegradable films. Recent studies also highlight its potential as a source of bioactive compounds for the formulation of active packaging that slows down food oxidation. Despite the small volume of this waste, its valorisation proves to be strategic as it adds high value in niche applications, connecting technological innovation with sustainability.

Altogether, these by-products exemplify the multidimensional potential of coffee residues, bridging agriculture, biotechnology, and nanoscience toward circular and sustainable innovation.

2.3 The silver screen

During the roasting of coffee beans, a thin outer tegument known as coffee silverskin (CS) is detached and typically discarded as waste. This layer is usually thrown away. On average, coffee silverskin contains about 18–25% cellulose, 10–15% lignin, 12-17% proteins, and 2–5% lipids, along with phenolic compounds and caffeine that account for its antioxidant activity (Oliveira et al., 2025; Koskinakis et al., 2025). This film contains phenolic chemicals, lipids, and fibres that act as antioxidants and anti-inflammatory agents. It may also be utilised as biomass to provide renewable energy (Nolasco et al., 2022; Nzekoue et al., 2022). Crista et al. (2020) state that it can also be used to make bio-oil by pyrolysis, while Reyes-Pérez et al. (2023) state that it can be used to produce cosmetic extracts.

Recent studies have demonstrated that coffee silverskin (CS) is an abundant source of valuable bioactive compounds and structural biopolymers that can be repurposed for industrial applications. Liu et al. (2023) demonstrated that cellulose extracted from silverskin can be utilised to develop composite films incorporating natural pigments for food packaging, thereby enhancing not only mechanical strength but also sensory appeal. These biodegradable films present a sustainable alternative to conventional plastics, aligning with the principles of green chemistry and circular economy. Similarly, Oliveira et al. (2021) highlighted that coffee by-products, including silverskin, can enhance the development of *active food packaging materials* with antimicrobial and oxygen-scavenging properties, extending food shelf life and reducing waste.

In addition to its industrial potential, the silverskin's chemical composition contributes to its health-promoting properties. Nzekoue et al. (2022) found that silverskin is rich in B vitamins, minerals such as potassium and magnesium, and phytosterols, which are beneficial for cardiovascular health. Moreover, Nolasco et al. (2022) emphasised that when properly processed, silverskin does not pose significant biological risks, making it suitable for functional food formulations. These findings reinforce its dual role as both a bioresource for material science and a nutritional ingredient with promising physiological effects.

Furthermore, the valorisation of coffee silverskin aligns with global sustainability goals by reducing agro-industrial waste through its transformation into valuable inputs for multiple industries, including energy, biopolymers, and cosmetics. Tamilselvan et al. (2024) and Pongsiriyakul et al. (2024) highlighted that the reuse of coffee residues, such as silverskin, strengthens the circular bioeconomy by transforming waste into valuable inputs for multiple industries, including energy, biopolymers, and cosmetics. Integrating silverskin into production chains reduces environmental impact and supports a transition toward cleaner technologies and responsible consumption. As a result, what was once discarded as an insignificant by-product now emerges as a versatile and sustainable raw material with high economic and environmental value.

While its biochemical richness supports nutritional and industrial applications, recent advances have also explored silverskin as a renewable precursor for advanced nanomaterials.

Beyond its biochemical and industrial applications, silverskin has recently been recognised as a precursor for nanomaterial production through carbonisation and activation processes. Studies by El-Desouky et al. (2023) and Gabisa and Ratanatamskul (2024) demonstrated its efficiency in generating photothermal and adsorptive materials for water purification. Furthermore, integrating silverskin-derived carbon with polymers has enhanced mechanical resistance and sustainability in green composites (Amena et al., 2022; Bairwan et al., 2025). These innovations consolidate silverskin's role as a renewable feedstock for advanced materials, bridging the gap between agricultural waste and nanotechnology.

3. Routes for the synthesis of nanomaterials

Future perspectives suggest hybrid synthesis routes, in which pyrolysis is combined with hydrothermal or microwave-assisted methods to maximise yield and tailor nanostructures. Moreover, emerging approaches utilise machine learning algorithms to predict optimal pyrolysis conditions, thereby accelerating the transition towards Nanotechnology 4.0 in biomass valorisation (Pongsiriyakul et al., 2024; Ko et al., 2025).

Recent advancements in the synthesis of nanomaterials from coffee waste emphasise the importance of combining green chemistry principles with high-efficiency techniques. Pyrolysis has been one of the most explored methods for converting coffee residues into carbon-based nanomaterials, producing biochar and nanocomposites with tunable surface properties (Primaz et al., 2018; Sieradzka et al., 2022). However, optimisation of temperature and heating rate is crucial to control the yield and morphology of the resulting nanostructures (Ko et al., 2025). Typically, pyrolysis temperatures range between 400 and 800 °C, with higher carbonization yields (up to 35–45 wt%) achieved under limited oxygen conditions and moderate heating rates (Mendoza Martinez et al., 2021; Gil-Gómez et al., 2024). The integration of catalytic agents and metal precursors during pyrolysis can further enhance photocatalytic and adsorption capacities, creating multifunctional materials applicable in water treatment, sensors, and energy storage (Changotra et al., 2024; Reyes-Pérez et al., 2023).

While pyrolysis mainly favours carbonaceous nanostructures, hydrothermal methods offer a complementary route for synthesising crystalline oxide-based or hybrid nanomaterials under aqueous conditions. Hydrothermal and solvothermal synthesis routes have also gained prominence due to their ability to produce nanomaterials with high crystallinity and uniform particle size under mild conditions. Zheng et al. (2025) and Lim and Foo (2024) demonstrated that the hydrothermal conversion of coffee waste, combined with natural polymers such as chitosan or cellulose, yields composites with excellent photothermal and antibacterial properties. Additionally, the hydrothermal technique allows the incorporation of magnetic nanoparticles, resulting in materials with strong adsorption and reusability potential for wastewater remediation (Rhoden et al., 2021; Mohammadpour et al., 2024). The development of these multifunctional nanocomposites aligns with the global trend of designing environmentally friendly materials for circular technological applications.

Emerging research is also exploring the coupling of conventional synthesis routes with advanced digital technologies to increase efficiency and precision. Machine learning algorithms are being applied to model and predict optimal synthesis parameters, such as temperature, pressure, and precursor ratios, leading to controlled morphology and improved functional performance (Pongsiriyakul et al., 2024). Such computational approaches mark the onset of Nanotechnology 4.0, in which data-driven optimisation enhances reproducibility and scalability in biomass-derived nanomaterial production. These integrated strategies not only improve material performance but also support sustainable innovation pathways for converting coffee waste into high-value nanostructures.

Recent work has expanded the synthesis routes to include catalytic and microwave-assisted pyrolysis, which improve carbon yield and surface area while reducing energy consumption (Sieradzka et al., 2022; Ko et al., 2025). In parallel, hybrid hydrothermal systems combining chitosane or magnetite precursors have produced multifunctional nanocomposites for photocatalysis and drug delivery (Zheng et al., 2025; Mohammadpour et al., 2024). Microwave-assisted pyrolysis shortens reaction times and ensures uniform heating, thereby improving nanocarbon yield and surface functionality compared to conventional heating methods (Islam et al., 2025; Sajini and Joseph, 2025).

Recent advances combine machine learning with hybrid pyrolysis-hydrothermal systems to predict optimal synthesis parameters and enable precise control over nanostructure formation. These data-driven strategies accelerate reproducibility, scalability, and energy efficiency, representing a shift toward Nanotechnology 4.0 in biomass-derived nanomaterial production (Ko et al., 2025; Pongsiriyakul et al., 2024).

Altogether, integrating green synthesis techniques with data-driven optimisation opens new frontiers for coffee waste valorisation, ensuring reproducibility, scalability, and reduced environmental footprint-hallmarks of Nanotechnology 4.0.

3.1 Pyrolysis and chemical activation

Pyrolysis and chemical activation are considered cornerstone processes for transforming coffee waste into highly porous carbon-based nanomaterials. In pyrolysis, the thermal decomposition of biomass occurs in an oxygen-limited environment, producing bio-oil, gases, and biochar. When optimised through controlled temperature and residence time, it becomes an excellent precursor for activated carbons and graphene-like materials. Recent studies have demonstrated that coffee-derived carbons can achieve surface areas exceeding 1.000 m^2/g when processed between 600 and 800 °C, reaching carbon yields of up to 40 wt% (Gil-Gómez et al., 2024; Mendoza Martinez et al., 2021). Thus, extremely useful in adsorption processes (Bruckman and Rhoden, 2024; Salles et al., 2023).

When coupled with chemical activation using KOH, $ZNCl_{2,}$ or H_3PO_4, the porosity and surface chemistry of these materials can be finely tuned, enhancing ion transport and adsorption performance (Stock et al., 2022; Parashar et al., 2025). Such tailored properties open up applications not only in wastewater remediation but also in energy storage technologies such as supercapacitors and lithium-ion batteries, as well as in gas storage systems. Notably, the scalability of pyrolysis reactors and their potential integration into biorefineries highlight the industrial relevance of this route (Dhiman et al., 2025).

Pyrolysis (Figure 2), a thermochemical process conducted under inert or oxygen-limited atmospheres, converts lignocellulosic coffee residues, such as spent grounds and husks, into carbonaceous materials through the release of volatile compounds and the formation of solid carbon (Sieradzka et al., 2022; Ko et al., 2025). Slow pyrolysis in fixed–bed reactors has been reported as an efficient route for producing activated carbons with adjustable porosity and large surface areas, suitable for pollutant removal (Mukherjee et al., 2022).

Figure 2: Coffee waste processing

Chemical activation using KOH or $ZnCl_2$, combined with pyrolysis, enhances the materials' porosity and increases their surface area, making them more suitable for use in energy storage devices and contaminant adsorption (Mukherjee et al., 2022; Gabisa and Ratanatamskul, 2024; Mohammadpour et al., 2024). The scalability of chemical activation and its potential integration with continuous pyrolysis systems highlight its relevance for industrial biorefineries (Islam et al., 2025; Gil-Gómez et al., 2024).

For instance, coffee-derived carbon dots have exhibited tunable photoluminescence, enabling applications in bioimaging, sensors, and optoelectronic devices. Likewise, nanocellulose produced via hydrothermal processing shows high crystallinity and mechanical strength, making it suitable for advanced composites and biodegradable packaging. Moreover, the use of water as a solvent and relatively low reaction temperatures aligns hydrothermal synthesis with green chemistry principles, reinforcing its contribution to sustainable development goals and the production of functional, eco-friendly materials.

Overall, pyrolysis and chemical activation are mature, scalable technologies capable of converting coffee waste into multifunctional carbon nanomaterials, thereby bridging sustainable waste management with advanced material design.

3.2 Hydrothermal synthesis

Hydrothermal synthesis stands out among green chemistry techniques for its versatility and environmental compatibility. It utilises water as a reaction medium under high temperatures and pressures to decompose organic components and reorganise them into nanostructured materials.

Typically, reactions are carried out at 150–250 °C under autogenous pressure, which promotes the breakdown of lignocellulosic matrices and the formation of crystalline nanostructures without toxic solvents (Gil-Gómez et al., 2024). This process is particularly advantageous for transforming biomass, such as coffee residues, into functional nanocellulose and carbon-based nanoparticles (Karunakaran et al., 2023). According to Lim and Foo (2024), this method not only facilitates the extraction of cellulose nanocrystals from coffee residue but also enables the incorporation of metal ions and other bioactive agents, resulting in multifunctional materials with both antibacterial and antifungal properties.

The tunability of hydrothermal conditions, including temperature, pH, and reaction time, enables the control of nanoparticle morphology and size distribution. Zheng et al. (2025) demonstrated that adjusting these parameters can generate hydrophobically graded porous structures, enhancing water evaporation and photothermal efficiency. Similarly, Chandran et al. (2025) reported that hydrogels derived from coffee biowaste exhibit remarkable potential for environmental cleanup and biomedical applications due to their high porosity and biocompatibility. Together, these findings underscore the versatility of hydrothermal synthesis in producing diverse nanomaterials with tailored functionalities for industrial and scientific use.

Moreover, integrating hydrothermal synthesis with other advanced methods, such as microwave-assisted or solvothermal techniques, has been shown to increase yield and reduce reaction time. Such hybrid approaches combine thermal efficiency with precise energy control, producing nanocomposites that exhibit enhanced mechanical strength, conductivity, and catalytic performance (Amena et al., 2022; Mohammadpour et al., 2024). These advancements suggest that the hydrothermal method can serve as a foundational platform for next-generation nanomaterial engineering, particularly when applied to waste valorisation and sustainable manufacturing.

However, large-scale application remains a challenge. Although pilot studies in Latin America and Asia have proven that incorporating coffee-derived carbons in wastewater treatment plants is technically feasible, economic and regulatory barriers persist (Freitas and Renó, 2022). Yusufoğlu et al. (2024) emphasise that global policy frameworks and nanomaterial safety guidelines must evolve in tandem with technological progress to ensure safe, ethical, and economically viable scaling. Establishing standardised evaluation protocols and governmental incentives could help bridge the gap between laboratory success and real-world implementation, making hydrothermal synthesis a central tool in the transition toward circular and low-impact industrial ecosystems. By coupling automation and data-driven optimisation, hydrothermal routes are poised to play a pivotal role in Nanotechnology 4.0, linking sustainable chemistry with intelligent manufacturing.

4. Environmental applications

4.1 Water and wastewater treatment

The environmental applications of coffee-derived nanomaterials are diverse and continue to expand. One of the most prominent applications is in wastewater treatment, where activated carbons and hybrid nanocomposites derived from coffee grounds have demonstrated remarkable adsorption capacities for heavy metals, dyes, pesticides, and pharmaceuticals. Beyond conventional adsorption, these materials can be engineered for photocatalytic degradation by embedding metal oxides, such as TiO_2, ZnO, or Fe_2O_3, which enables pollutant breakdown under visible or solar light (Reyes-Pérez et al., 2023; Changotra et al., 2024; Gabisa and Ratanatamskul, 2024).

Coffee-based nanocomposites exhibit high adsorption capacities for heavy metals, textile dyes, and pharmaceutical contaminants. The incorporation of metal oxides enhances photocatalytic degradation under sunlight, achieving over 90% removal efficiency for synthetic dyes such as methylene blue under optimised conditions (Reyes-Pérez et al., 2023; Changotra et al., 2024). Moreover, their chemical stability enables multiple reuse cycles without significant performance loss, a key feature for wastewater treatment plants (Barbieri et al., 2024).

Recent investigations have also explored the use of magnetic nanocomposites based on coffee-derived carbon combined with Fe_3O_4 nanoparticles. Such Fe_3O_4-combined systems enable magnetic separation after water treatment, simplifying recovery and reuse (Rhoden et al., 2021; Harres et al., 2023). Their strong adsorption behaviour toward antibiotics and pharmaceutical residues, such as Amitriptyline and Losartan, reveals high selectivity and regeneration potential (Nunes et al., 2024; Dos Santos et al., 2025). These advances suggest that coffee-based nanomaterials can be integrated into continuous flow systems and pilot reactors for real-world applications, thereby bridging the gap between laboratory efficiency and industrial feasibility.

Furthermore, the incorporation of coffee-derived nanocellulose into filtration membranes has emerged as a new frontier for sustainable water purification. Lim and Foo (2024) demonstrated that coffee-residue nanocellulose composites exhibit antibacterial and antifungal activity, thereby enhancing the longevity and safety of filtration devices. When combined with photocatalytic agents such as ZnO or TiO_2, these hybrid membranes can simultaneously filter and degrade organic pollutants under sunlight exposure, reducing maintenance costs and operational energy demands.

4.2 Air purification and CO_2 capture

Another promising environmental application involves the use of coffee-waste-derived carbon materials in air purification and CO_2 adsorption. Due to their high porosity and surface functionality, these carbons effectively capture greenhouse gases, reducing atmospheric emissions (Sieradzka et al., 2022; Ko et al., 2025). Moreover, their production via pyrolysis or hydrothermal carbonisation generates minimal by-products and can be integrated into biorefinery systems, thus enhancing the sustainability of the entire coffee value chain. The captured CO_2 can later be repurposed for producing fuels, chemicals, or algae biomass, closing the carbon loop within a circular economy framework.

4.3 Soil remediation and circular use

Beyond water and air treatment, coffee-derived nanomaterials also hold potential in soil remediation. Activated biochars and magnetised nanocomposites improve soil aeration, reduce heavy metal mobility, and promote microbial activity essential for ecosystem recovery. These eco-remediation strategies are particularly relevant for agricultural regions where coffee residues are abundant, allowing waste to be reintegrated into the same ecosystems from which it originated (Tamilselvan et al., 2024; Pongsiriyakul et al., 2024). Such circular approaches enhance soil health and reinforce the sustainable management of agricultural by-products.

4.4 Digital and predictive environmental nanotechnology

Combining artificial intelligence and machine learning with environmental nanotechnology has opened new pathways for optimising pollutant removal processes. Predictive models can estimate adsorption kinetics, regeneration cycles, and photocatalytic performance based on nanostructure morphology and surface chemistry (Pongsiriyakul et al., 2024; Wang et al., 2023). This fusion of digital and green technologies accelerates the adoption of low-cost, data-driven solutions for water and air purification, waste recovery, and carbon management.

In this context, coffee-waste-derived nanomaterials exemplify the convergence of circular economy, artificial intelligence, and sustainable innovation, positioning them as key enablers of the global transition toward a cleaner and brighter planet. These applications align with the United Nations Sustainable Development Goals (SDGs), specifically SDG 6 (Clean Water and Sanitation) and SDG 13 (Climate

Action), underscoring the social and ecological benefits of nanotechnology in the coffee sector.

5. Industrial applications

Beyond civil construction, the integration of coffee by-products into polymeric composites has shown excellent potential for developing sustainable materials for the plastics and packaging industries. Polymers reinforced with coffee husk or spent coffee grounds can exhibit improved tensile strength, stiffness, and thermal stability, making them viable for packaging, automotive parts, and consumer goods (Amena et al., 2022; Bairwan et al., 2025). The natural lignocellulosic structure of coffee residues acts as an effective filler, while phenolic compounds enhance UV resistance and provide antioxidant activity. Such materials reduce dependence on petroleum-based plastics and support the global shift toward biodegradable and recyclable alternatives (Bomfim et al., 2022).

The energy sector has also benefited from the valorisation of coffee residue, particularly through the production of biochar, biodiesel, and biomass pellets. Coffee waste-derived biochar serves as a high-calorific-value fuel and a carbon sequestration agent, contributing to cleaner energy generation and soil remediation (Kamil et al., 2019; Primaz et al., 2018). Additionally, biodiesel produced from spent coffee oil has demonstrated comparable efficiency to traditional diesel fuels while emitting lower levels of particulate matter and sulfur oxides. Integrating these energy applications into local coffee production chains strengthens rural economies and promotes decentralised renewable energy systems, reducing reliance on fossil fuels and enhancing energy security.

Another emerging industrial application lies in the field of functional coatings and smart materials. Graphene oxide and nanocellulose derived from coffee residues have been explored as fillers in conductive inks, flexible electronics, and antibacterial surface coatings (Harres et al., 2023; Mohammadpour et al., 2024). These innovative materials combine lightweight design with high electrical conductivity and antimicrobial activity, offering significant advantages for use in sensors, wearable devices, and medical packaging. The adaptability of coffee-derived nanostructures to advanced manufacturing techniques, such as 3D printing and electrospinning, further underscores their potential as next-generation biomaterials for industrial innovation.

5.1 Civil construction

Particle boards made from coffee husks and plant resins, such as castor oil, exhibit mechanical properties suitable for non-structural applications. They are also resistant to moisture and have antibacterial activity (Amena et al., 2022; Bairwan et al., 2025; Onsongo et al., 2025). These panels are ideal for furniture, interior cladding, and other decorative uses that do not require high mechanical strength.

The processing involves drying, grinding, and pressing the coffee husk into moulds with natural resins under controlled temperatures and pressures. Castor oil resin is an efficient binder due to its strong adhesion to plant particles and high resistance to humidity (Amena et al., 2022). Recent studies demonstrated that adding coffee husks to particle boards can partially or completely replace conventional wood, yielding comparable bending strength and water adsorption (Bairwan et al., 2025). Furthermore, the phenolic compounds and natural waxes present in the husk enhance its durability and impart antibacterial properties, thereby increasing its suitability for indoor environments.

Economically, the raw material is low-cost, and its reuse significantly reduces improper waste disposal. This approach aligns with the principles of the bioeconomy and circular economy by converting agro-industrial residues into high-value materials, thereby stimulating innovation in the eco-friendly construction sector (Onsongo et al., 2025).

5.2 Biodegradable packaging

Biodegradable packaging has become a global focus of research (Figure 3). Nanocellulose obtained from renewable sources, such as coffee pulp, represents an excellent additive for biodegradable polymers, improving their mechanical, thermal, and barrier properties (Karunakaran et al., 2023; Lim and Foo, 2024). Coffee pulp, a major by-product of the coffee industry, is rich in cellulose and can be transformed into nanocellulose via chemical, mechanical, or enzymatic methods.

When incorporated into polymeric matrices such as polylactic acid (PLA) or thermoplastic starch, nanocellulose enhances structural strength and reduces gas

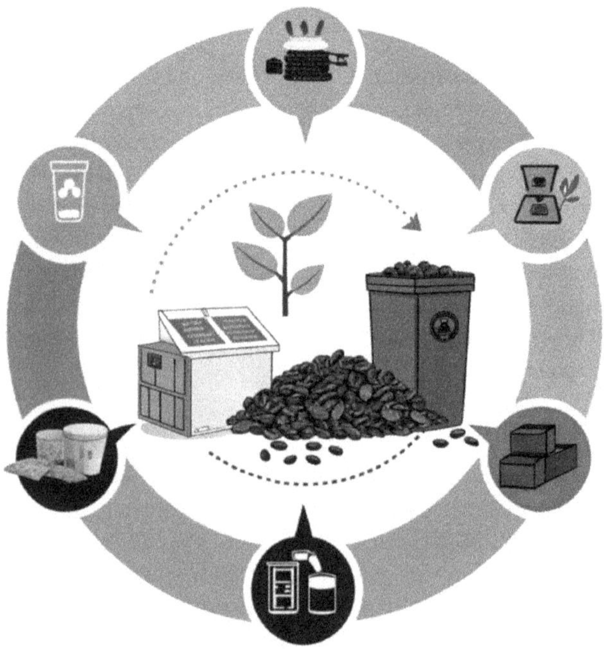

Figure 3: Sustainable reuse of waste from coffee production and consumption

permeability, key attributes for food packaging (Liu et al., 2023). It also acts as a natural barrier to moisture and oxygen, extending the shelf life of perishable goods (Oliveira et al., 2021). As modification and functionalisation techniques evolve, new applications continue to emerge, including active antimicrobial films and smart packaging that change colour in response to microbial contamination or pH variations (Da Rosa Salles et al., 2022).

Coffee residues, particularly pulp and silverskin, serve as promising feedstocks for producing nanocellulose that can be blended into biodegradable polymers such as PLA and starch. The resulting nanocomposites exhibit enhanced tensile strength, reduced gas permeability, and increased durability, making them ideal for food packaging applications. Cutting-edge developments include active packaging films containing antimicrobial or antioxidant agents derived from coffee extracts, which extend the shelf life of perishable goods.

Additionally, the biomedical sector has explored the electrochemical properties of coffee-based nanocarbons for biosensors, particularly in glucose and drug detection. This dual functionality (energy storage and biosensing) highlights the versatility of coffee-derived nanostructures (Hares et al., 2023; Barbieri et al., 2024).

5.3 Energy storage

Removing coffee waste is becoming an increasingly significant environmental issue, but it also presents opportunities to create valuable products. One of the most significant advancements in this area is the conversion of these leftovers into nanoporous carbons, which are utilised in energy storage devices such as supercapacitors (Hsieh et al., 2023; Zheng et al., 2025).

These advances are clearly aligned with the United Nations Sustainable Development Goals (SDGs), particularly SDG 6 (Clean Water and Sanitation), SDG 9 (Industry, Innovation, and Infrastructure), SDG 12 (Responsible Consumption and Production), and SDG 13 (Climate Action). Furthermore, integrating coffee waste valorisation into biorefinery models, along with other agro-industrial waste, will strengthen the circular bioeconomy, promoting a sustainable and innovation-driven future (Barreto Peixoto et al., 2023; Pyrzynska, 2025).

Residual coffee grounds contain a significant amount of carbon, and due to their natural porosity, they can be transformed into a material with a high specific surface area and a hierarchical pore distribution through processes such as controlled pyrolysis and chemical activation (Mukherjee et al., 2022; Mohammadpour et al., 2024). These characteristics are crucial for the electrochemical efficiency of supercapacitors, as they facilitate the rapid attachment and detachment of ions during charging and discharging cycles.

Recent research has shown that activated carbons made from coffee can reach specific capacities of over 300 F/g. They also exhibit remarkable stability in charge and discharge cycles, retaining more than 90% of their original capacity after 10,000 cycles (Pongsiriyakul et al., 2024; Bairwan et al., 2025).

Also, using pre-treatments with activating agents like KOH or $ZnCl_2$ lets you change the shape and surface properties of the carbon you get, which makes it work better with water- or organic-based electrolytes (Gabisa and Ratanatamskul, 2024).

This opens up new ways to utilise these materials in flexible, portable supercapacitors with high energy density.

6. Conclusion

The reuse and valorisation of waste from the coffee manufacturing chain represent an innovative and sustainable solution that bridges the principles of the circular economy and the bioeconomy. By transforming residues such as husk, grounds, and silver skin into high-value nanomaterials, it becomes possible to integrate agricultural productivity with technological innovation, creating a continuous loop of resource efficiency. This paradigm shift redefines waste not as an environmental burden but as a renewable source of functional materials, supporting both economic growth and ecological balance.

Scientific evidence demonstrates that coffee waste can serve as a promising precursor for producing nanocellulose, activated carbon, magnetic nanocomposites, and graphitic derivatives through routes such as pyrolysis, chemical activation, and hydrothermal synthesis. These materials have been successfully applied in civil construction, biodegradable packaging, wastewater treatment, drug delivery, and energy storage systems. Their multifunctional properties, including adsorption capacity, biocompatibility, and structural resilience, underscore the technological and economic potential of agro-industrial residues as next-generation materials for sustainable development.

From an environmental standpoint, the transformation of coffee residues significantly reduces organic solid waste, mitigates contamination from improper disposal, and replaces fossil-based raw materials with renewable ones. It also contributes to reducing greenhouse gas emissions and restoring soil and water systems when residues are properly redirected for reuse. By incorporating these materials into eco-efficient industrial chains, society can move closer to achieving low-carbon production cycles and resource circularity in line with global sustainability goals.

Economically, the valorisation of coffee waste stimulates the emergence of new value chains and market niches. It promotes entrepreneurship and supports the inclusion of small producers, cooperatives, and local industries, particularly in coffee-growing regions across Latin America, Africa, and Asia. This inclusive model redistributes income, fosters innovation, and strengthens the competitiveness of sustainable industries at both regional and global scales.

Socially, this process generates qualified employment opportunities and fosters multidisciplinary collaboration among academia, industry, and rural communities. It also enhances education and awareness about environmental responsibility and sustainable production, inspiring the next generation of professionals in fields such as materials science, biotechnology, and nanotechnology. Therefore, the intersection between sustainability and technology is not only an environmental necessity but also a social driver for equitable development.

Furthermore, valorising coffee waste creates synergy between traditional agricultural practices and advanced technological applications. It establishes a tangible connection among the agricultural, industrial, and research sectors—an essential term for the success of the bioeconomy. This collaborative ecosystem enables the formation of innovation clusters that transform local economies and contribute to regional sustainability transitions.

The reimagining of coffee waste as a strategic material resource also opens doors for cross-sectoral applications. Industries related to cosmetics, pharmaceuticals, energy, and environmental remediation can all benefit from the chemical versatility and renewable character of coffee-derived nanomaterials. As a result, a new industrial logic emerges, one that prioritises circularity, resilience, and sustainability without compromising technological advancement.

In summary, the valorisation of coffee residues marks a significant milestone in the development of sustainable materials and green technologies. It not only adds value to what was once discarded but also strengthens the environmental, economic, and social dimensions of sustainability. By connecting the circular economy and the bioeconomy, coffee waste becomes an emblematic model of how science and innovation can drive tangible progress toward a cleaner, more resilient, and equitable future.

7. Future perspectives

In the coming years, the development of integrated biorefineries capable of processing coffee residues into multiple nanomaterials and bioactive compounds will be a major technological focus. Combining hydrothermal, pyrolytic, and catalytic methods in modular and scalable systems could optimise energy efficiency and resource recovery. Such systems would enable the simultaneous production of bio-oil, biochar, and functional nanocellulose, reducing waste and increasing economic viability. The integration of renewable energy sources, such as solar or biomass heating, into these processes may further minimise carbon emissions, positioning coffee-based biorefineries as carbon-neutral or even carbon-negative technologies.

Digital transformation will also play a central role in the evolution of this field. Artificial intelligence and machine learning algorithms can be employed to predict optimal synthesis conditions and design customised materials with precise physicochemical properties. This data-driven approach, often referred to as Nanotechnology 4.0, will accelerate material discovery while minimising experimental costs and environmental impact. Collaborative digital platforms will further enhance knowledge sharing between research institutions and industries worldwide. In particular, predictive modelling and high-throughput simulations can help identify the most effective nanostructures before laboratory synthesis, thereby dramatically increasing research efficiency (Pongsiriyakul et al., 2024).

Another key direction involves improving the economic accessibility of nanomaterial production. Future studies should focus on the scalability of synthesis techniques, including the cost of precursor collection, purification, and transformation. By incorporating local supply chains and community-based production networks, the manufacturing of coffee-derived nanomaterials could become feasible in developing countries, where coffee is a major agricultural commodity. These decentralised production models would create employment opportunities, reduce transportation emissions, and promote regional technological autonomy (Tamilselvan et al., 2024; Yusufoğlu et al., 2024). Furthermore, establishing pilot plants in coffee-producing regions could enable continuous testing and training programs, accelerating industrial transition.

At the same time, interdisciplinary collaboration will be essential. Chemists, engineers, environmental scientists, and economists must collaborate to develop viable models for industrial applications. The future of coffee waste valorisation depends not only on scientific innovation but also on policy alignment and market readiness. Partnerships between public institutions, universities, and private enterprises can facilitate technology transfer and attract investment in green infrastructure. Moreover, the creation of open-access databases and standardised testing protocols will ensure reproducibility and accelerate commercialisation (Bairwan et al., 2025).

A further priority is the development of environmentally friendly catalysts and reagents. Current synthesis routes often rely on acids, bases, or metal salts that may generate secondary pollutants. The next generation of processes should utilise greener alternatives, such as enzymatic catalysts, deep eutectic solvents, or bio-based ionic liquids, to ensure that sustainability is maintained throughout the entire production chain (Mohammadpour et al., 2024; Zheng et al., 2025). Implementing these eco-efficient reagents would reduce toxicity, simplify post-treatment, and enhance the recyclability of the resulting nanomaterials.

From a regulatory and social perspective, establishing comprehensive frameworks for nanomaterial safety will be critical. Governments and international agencies must define exposure limits, toxicity thresholds, and environmental release criteria to protect both workers and consumers. Furthermore, certification programs for "green nanomaterials" could help industries adopt sustainable labelling practices and enhance consumer trust. Aligning these initiatives with the United Nations Sustainable Development Goals will ensure that scientific progress translates into real social and environmental benefits (Ko et al., 2025; Pongsiriyakul et al., 2024).

Finally, education and public engagement should not be overlooked. Encouraging awareness of circular bioeconomy concepts among students, farmers, and small-scale producers can promote sustainable practices and inspire a new generation of innovators. Training programs and international workshops focused on coffee residue valorisation could stimulate technological transfer and foster global collaboration. In the long term, inclusive educational strategies will be essential for building the skilled workforce required to sustain innovation and ensure the responsible use of nanotechnology worldwide.

Together, these strategies will consolidate a new era of sustainable materials derived from biomass, positioning coffee residues as one of the cornerstones of the circular bioeconomy. The convergence of advanced synthesis, digital intelligence, and global cooperation will ensure that what was once agricultural waste becomes a critical resource for addressing environmental challenges, fostering technological innovation, and supporting a greener and more equitable industrial future.

References

Ali, I., Farooq, S., Abdullah, R., Kaleem, A., Iqtedar, M., Iqbal, I., and Chen, X. (2025). Production of Phoenix dactylifera seeds coffee as an alternative source of caffeinated coffee and assessment of its bioactive and anti diabetic potential. *Journal of Food Composition and Analysis*, 140, 107273. https://doi.org/10.1016/j.jfca.2025.107273

Amena, B. T., Altenbach, H., Tibba, G. S., and Hossain, N. (2022). Investigation of mechanical properties of coffee husk–HDPE–ABS polymer composite using the injection-moulding method. *Journal of Composites Science*, 6(12), 354. https://doi.org/10.3390/jcs6120354

Bairwan, R. D., Khalil, H. P. S., Nuryawan, A., Ahmad, M. I., Kassim, M. H. M., and Ahmad, A. (2025). Enhancing biopolymer materials with coffee waste-derived reinforcements. *Polymer Engineering & Science*, 65(2), 455–477. https://doi.org/10.1002/pen.27046

Barbieri, I. A., Oliveira, M. L. S., Bruckmann, F. S., Salles, T. R., Zancanaro, L. V., Silva et al. (2024). Effective removal of hypnotic drug from the aqueous medium through adsorption on graphene oxide magnetic derivatives. *Journal of Molecular Liquids*, 393, 123657. https://doi.org/10.1016/j.molliq.2023.123657

Barreto Peixoto, J. A., Silva, J. F., Oliveira, M. B. P. P., and Alves, R. C. (2023). Sustainability issues along the coffee chain: From the field to the cup. *Comprehensive Reviews in Food Science and Food Safety*, 22(1), 287–332. https://doi.org/10.1111/1541-4337.13069

Bijla, L., Aissa, R., Laknifli, A., Bouyahya, A., Harhar, H., and Gharby, S. (2022). Spent coffee grounds: A sustainable approach toward novel perspectives of valorization. *Journal of Food Biochemistry*, 46(8), e14190. https://doi.org/10.1111/jfbc.14190

Bomfim, A., Oliveira, D., Voorwald, H., Benini, K., Dumont, M.-J., and Rodrigue, D. (2022). Valorization of spent coffee grounds as precursors for biopolymers and composite production. *Polymers*, 14(3), 437. https://doi.org/10.3390/polym14030437

Cangussu, L. B., Melo, J. C., Franca, A. S., and Oliveira, L. S. (2021). Chemical characterization of coffee husks, a by-product of *Coffea arabica* production. *Foods*, 10(12), 3125. https://doi.org/10.3390/foods10123125

Canton, H. (2021). *The Europa Directory of International Organizations 2021* (23rd ed.). Routledge. https://doi.org/10.4324/9781003179900

Chandran, A., Athulya, K. R., Rasheed, R. F., and Kumar, A. C. (2025). Biowaste to hydrogels: A sustainable solution for waste reduction with potential applications. *ES Chemistry and Sustainability*, 2, 1366. http://dx.doi.org/10.30919/escs1366

Changotra, R., Yang, J., Rajput, H., Hu, Y., and Sophia He, Q. (2024). Mechanism and performance evaluation of spent-coffee grounds-derived nanocomposite materials for highly efficient photocatalytic degradation of organic pollutant. *Journal of Industrial and Engineering Chemistry*, 133, 428–438. https://doi.org/10.1016/j.jiec.2023.12.019

Chien, H.-W., and Chen, X.-E. (2022). Spent coffee grounds as potential green photothermal materials for biofilm elimination. *Journal of Environmental Chemical Engineering*, 10(1), 107131. https://doi.org/10.1016/j.jece.2022.107131

Choe, U. (2025). Valorization of spent coffee grounds and their applications in food science. *Current Research in Food Science*, 10, 101010. https://doi.org/10.1016/j.crfs.2025.101010

Crista, D. M. A., El Mragui, A., Algarra, M., Esteves Da Silva, J. C. G., Luque, R., and Pinto Da Silva, L. (2020). Turning spent coffee grounds into sustainable precursors for the fabrication of carbon dots. *Nanomaterials*, 10(6), 1209. https://doi.org/10.3390/nano10061209

Cruz, R. M. S., Rainer, B., Wagner, I., Krauter, V., Janalíková, M., Vicente, A. A., and Vieira, J. M. (2024). Valorization of cork stoppers, coffee-grounds and walnut shells in the development and characterization of pectin-based composite films: Physical, barrier, antioxidant, genotoxic, and biodegradation properties. *Polymers*, 16(8), 1053. https://doi.org/10.3390/polym16081053

Da Rosa Salles, T., Da Silva Bruckamann, F., Viana, A. R., Krause, L. M. F., Mortari, S. R., and Rhoden, C. R. B. (2022). Magnetic nanocrystalline cellulose: Azithromycin adsorption and in vitro biological activity against melanoma cells. *Journal of Polymers and the Environment*, 30(7), 2695–2713. https://doi.org/10.1007/s10924-022-02388-3

Da Silva Bruckmann, F., and Rhoden, C. R. B. (2024). Applications of magnetic graphene oxide in water decontamination. *In:* Comprehensive Analytical Chemistry (Vol. 106, pp. 687–703). Elsevier. https://doi.org/10.1016/bs.coac.2023.10.002

De Oliveira, É. C., Da Silva Bruckmann, F., Schopf, P. F., Viana, A. R., Mortari, S. R., Sagrillo et al. (2022). In vitro and in vivo safety profile assessment of graphene oxide decorated with different concentrations of magnetite. *Journal of Nanoparticle Research*, 24(7), 150. https://doi.org/10.1007/s11051-022-05529-w

De Oliveira, M. P., Schnorr, C., Da Rosa Salles, T., Da Silva Bruckmann, F., Baumann, L., Muller et al. (2023). Efficient uptake of angiotensin-converting enzyme II inhibitor employing graphene oxide-based magnetic nanoadsorbents. *Water*, 15(2), 293. https://doi.org/10.3390/w15020293

Dhiman, S., Thakur, B., Kaur, S., Ahuja, M., Gantayat, S., Sarkar, S. et al. (2025). Closing the loop: Technological innovations in food waste valorisation for global sustainability. *Discover Sustainability*, 6(1), 258. https://doi.org/10.1007/s43621-025-01073-4

Dos Santos, A. C. P., Porciuncula, J. A., Zancanaro, L. V., Graba, B., de Oliveira, M. P., Viana, A. R., and Rhoden, C. R. B. (2025). Magnetic nanocellulose for losartan removal: Adsorption and ecotoxicity assessment. *International Journal of Biological Macromolecules*, 331, 148191. https://doi.org/10.1016/j.ijbiomac.2025.148191

El-Desouky, N., Shoueir, K., El-Mehasseb, I., and El-Kemary, M. (2023). Synthesis of silver nanoparticles using bio-valorization coffee waste extract: Photocatalytic flow-rate performance, antibacterial activity, and electrochemical investigation. *Biomass Conversion and Biorefinery*, 13(17), 15871–15885. https://doi.org/10.1007/s13399-021-02256-5

El-Naggar, M. E., Mahmoud, Y. A.-G., Abd-Elgawad, Sh. E., Zawawy, N. A. E., and Hemdan, B. A. (2024). A comprehensive review on nanoscience and nanotechnology with special emphasis on antimicrobial activities. *BioNanoScience*, 14(4), 4020–4043. https://doi.org/10.1007/s12668-024-01530-w

Franca, A. S., and Oliveira, L. S. (2022). Potential uses of spent coffee grounds in the food industry. *Foods*, 11(14), 2064. https://doi.org/10.3390/foods11142064

Freitas, L. C. F., and Renó, M. L. G. (2022). Environmental and economic evaluation of coffee residues. *Engenharia Sanitaria e Ambiental*, 27(4), 715–721. https://doi.org/10.1590/S1413-415220210159

Gabisa, E. W., and Ratanatamskul, C. (2024). Recycling of waste coffee grounds as a photothermal material modified with $ZnCl_2$ for water purification. *Scientific Reports*, 14(1), 10811. https://doi.org/10.1038/s41598-024-61768-y

Gil-Gómez, J. A., Florez-Pardo, L. M., and Leguizamón-Vargas, Y. C. (2024). Valorization of coffee by-products in the industry: A vision towards circular economy. *Discover Applied Sciences*, 6(9), 480. https://doi.org/10.1007/s42452-024-06085-9

Harres, A., Garcia, W. J. S., Salles, T. R., Bruckmann, F. S., Sulzenco, J. B., Schneider, A. D., and Rhoden, C. R. B. (2023). Magnetic properties of graphene oxide decorated with magnetite nanoparticles. *Diamond and Related Materials*, 138, 110238. https://doi.org/10.1016/j.diamond.2023.110238

Hsieh, P.-C., Chen, Y.-C., Zheng, N.-C., Mangindaan, D., and Chien, H.-W. (2023). A low-cost and environmentally-friendly chitosan/spent coffee grounds composite with high photothermal properties for interfacial water evaporation. *Journal of Industrial and Engineering Chemistry*, 126, 283–291. https://doi.org/10.1016/j.jiec.2023.06.019

Islam, K. M. O., Ahmad, N., Alboqami, F., Ahmed, U., Siddiqui, M. N., Ummer, A. C., and Jameel, A. G. A. (2025). Utilization of coffee waste for biofuel production through catalytic microwave-assisted pyrolysis approach. *Biomass Conversion and Biorefinery*, 15(7), 10585–10601. https://doi.org/10.1007/s13399-024-05854-1

Johnson, K., Liu, Y., and Lu, M. (2022). A review of recent advances in spent coffee grounds upcycle technologies and practices. *Frontiers in Chemical Engineering*, 4, 838605. https://doi.org/10.3389/fceng.2022.838605

Kamil, M., Ramadan, K. M., Awad, O. I., Ibrahim, T. K., Inayat, A., and Ma, X. (2019). Environmental impacts of biodiesel production from waste spent coffee grounds and its implementation in a compression ignition engine. *Science of the Total Environment*, 675, 13–30. https://doi.org/10.1016/j.scitotenv.2019.04.156

Karunakaran, G., Periyasamy, A. P., and Tehrani, A. (2023). Extraction of micro- and nanocrystalline cellulose and textile fibers from coffee waste. *Journal of Testing and Evaluation*, 51(5), 3042–3056. https://doi.org/10.1520/JTE20220487

Ko, H., Lee, M., Choi, J., and Oh, S. (2025). Effect of temperature and heating rate on pyrolysis characteristics of spent coffee grounds. *Korean Journal of Chemical Engineering*, 42(12), 2853–2864. https://doi.org/10.1007/s11814-024-00376-y

Koskinakis, S. E., Stergiopoulos, C., Vasileiou, C., and Krokida, M. (2025). Sustainable valorization of coffee silverskin waste: Pressurized liquid extraction of bioactive compounds. *Foods*, 14(4), 615. https://doi.org/10.3390/foods14040615

Kovalcik, A., Obruca, S., and Marova, I. (2018). Valorization of spent coffee grounds: A review. *Food and Bioproducts Processing*, 110, 104–119. https://doi.org/10.1016/j.fbp.2018.05.002

Lim, K. Y., and Foo, K. Y. (2024). Facile preparation of multifunctional cellulose nanocrystals from coffee residue via hydrothermal technique: Prolific roles on water purification, antibacterial and antifungal applications. *Journal of Water Process Engineering*, 67, 106264. https://doi.org/10.1016/j.jwpe.2024.106264

Liu, X., Sun, H., and Leng, X. (2023). Coffee silverskin cellulose-based composite film with natural pigments for food packaging: Physicochemical and sensory abilities. *Foods*, 12(15), 2839. https://doi.org/10.3390/foods12152839

Machado, M., Espírito Santo, L., Machado, S., Lobo, J. C., Costa, A. S. G., Oliveira, M. B. P. P. et al. (2023). Bioactive potential and chemical composition of coffee by-products: From pulp to silverskin. *Foods*, 12(12), 2354. https://doi.org/10.3390/foods12122354

Maraveas, C. (2024). Agriculture waste for sustainable building materials. *In:* J. Arora, A. Joshi, & R. C. Ray (Eds.). Transforming Agriculture Residues for Sustainable Development: From Waste to Wealth (pp. 321–351). Springer Nature Switzerland. https://doi.org/10.1007/978-3-031-61133-9_14

Mendoza Martinez, C. L., Saari, J., Melo, Y., Cardoso, M., De Almeida, G. M., and Vakkilainen, E. (2021). Evaluation of thermochemical routes for the valorization of solid coffee residues to produce biofuels: A Brazilian case. *Renewable and Sustainable Energy Reviews*, 137, 110585. https://doi.org/10.1016/j.rser.2020.110585

Mohammadpour, A., Kazemi, A., Baghapour, M. A., Samaei, M. R., Kannan, K., and Mousavi Khaneghah, A. (2024). Bioengineered FeZn/GA@Cu nanocomposite utilizing spent coffee ground extract and gum arabic: Enhanced nitrate removal via RSM and machine learning optimization. *International Journal of Biological Macromolecules*, 277, 134060. https://doi.org/10.1016/j.ijbiomac.2024.134060

Mukherjee, A., Okolie, J. A., Niu, C., and Dalai, A. K. (2022). Techno-economic analysis of activated carbon production from spent coffee grounds: Comparative evaluation of different production routes. *Energy Conversion and Management: X*, 14, 100218. https://doi.org/10.1016/j.ecmx.2022.100218

Ngene, G. I., and Nzihou, A. (2025). Catalytic conversion of spent coffee grounds into graphitic and nanotube biocarbon. *Materials Today Communications*, 47, 113240. https://doi.org/10.1016/j.mtcomm.2025.113240

Nolasco, A., Squillante, J., Esposito, F., Velotto, S., Romano, R., Aponte, M., and Cirillo, T. (2022). Coffee silverskin: Chemical and biological risk assessment and health profile for its potential use in functional foods. *Foods*, 11(18), 2834. https://doi.org/10.3390/foods11182834

Nunes, F. B., Sassi, R. L., Rhoden, D. S., Salles, T. R., Ramos, C. G., Santos, R. C., and Rhoden, C. R. (2024). Adsorption of amitriptyline from aqueous solutions using magnetic graphene-based material: Isotherm, kinetics, and thermodynamic study. *Colloids and Surfaces A: Physicochemical and Engineering Aspects*, 701, 134991. https://doi.org/10.1016/j.colsurfa.2024.134991

Nzekoue, F. K., Borsetta, G., Navarini, L., Abouelenein, D., Xiao, J., Sagratini, G. et al. (2022). Coffee silverskin: Characterization of B-vitamins, macronutrients, minerals and phytosterols. *Food Chemistry*, 372, 131188. https://doi.org/10.1016/j.foodchem.2021.131188

Oliveira, G., Passos, C. P., Ferreira, P., Coimbra, M. A., and Gonçalves, I. (2021). Coffee by-products and their suitability for developing active food packaging materials. *Foods*, 10(3), 683. https://doi.org/10.3390/foods10030683

Oliveira, I. G. S., Gallina, F. C., Affonso, M. R., Bimbi Junior, F. E., Souto, R. S., Lanza, M. R. V. et al. (2025). Determination of theobromine in cocoa products and tea samples: A new sensitive and selective electrochemical sensor approach using Printex L6/Au and polyaniline. *Journal of Food Composition and Analysis*, 148, 108372. https://doi.org/10.1016/j.jfca.2025.108372

Onsongo, S. K., Olukuru, J., Munyao, O. M., and Mwabonje, O. (2025). The role of agricultural ashes (rice husk ash, coffee husk ash, sugarcane bagasse ash, palm oil fuel ash) in cement production for sustainable development in Africa. *Discover Sustainability*, 6(1), 62. https://doi.org/10.1007/s43621-025-00841-6

Parashar, N., Garg, V. K., and Sharma, P. (2025). Synthesis and characterization of magnetic carbon nanocomposites for antimicrobial activity and wastewater treatment with microbial fuel cells (MFCs). *International Journal of Environmental Sciences*, 1492–1501. https://doi.org/10.64252/dn2njk54

Pongsiriyakul, K., Wongsurakul, P., Kiatkittipong, W., Premashthira, A., Kuldilok, K., Najdanovic-Visak, V. et al. (2024). Upcycling coffee waste: Key industrial activities for advancing the circular economy and overcoming commercialization challenges. *Processes*, 12(12), 2851. https://doi.org/10.3390/pr12122851

Primaz, C. T., Schena, T., Lazzari, E., Caramão, E. B., and Jacques, R. A. (2018). Influence of the temperature on the yield and composition of the bio-oil from the pyrolysis of spent coffee grounds: Characterization by comprehensive two-dimensional gas chromatography. *Fuel*, 232, 572–580. https://doi.org/10.1016/j.fuel.2018.05.097

Pyrzynska, K. (2025). Spent coffee grounds are a source of chlorogenic acid. *Molecules*, 30(3), 613. https://doi.org/10.3390/molecules30030613

Reyes-Pérez, J. A., Roa-Morales, G., De León-Condes, C. A., and Balderas-Hernández, P. (2023). Nanocomposites from spent coffee grounds and iron/zinc oxide: Green synthesis, characterization, and application in textile wastewater treatment. *Water Science & Technology*, 88(6), 1547–1563. https://doi.org/10.2166/wst.2023.285

Rhoden, C. R. B., Bruckmann, F. D. S., Salles, T. D. R., Kaufmann Junior, C. G., and Mortari, S. R. (2021). Study of the influence of magnetite on the removal of hydrochlorothiazide from aqueous solutions applying magnetic graphene oxide. *Journal of Water Process Engineering*, 43, 102262. https://doi.org/10.1016/j.jwpe.2021.102262

Sajini, T., and Joseph, J. (2025). Microwave-assisted synthesis of nanomaterials: A green chemistry perspective and sustainability assessment. *RSC Sustainability*. https://doi.org/10.1039/d5su00584a

Salles, T. D. R., Bruckmann, F., & Rhoden, C. R. B. (2023). Nanoadsorbents in the removal of organic pollutants. *In:* Adsorption through Advanced Nanoscale Materials (pp. 429–442). Elsevier. https://doi.org/10.1016/B978-0-443-18456-7.00019-5

Santos, V. P., Ribeiro, P. C. C., and Rodrigues, L. B. (2023). Sustainability assessment of coffee production in Brazil. *Environmental Science and Pollution Research*, 30(4), 11099–11118. https://doi.org/10.1007/s11356-022-22922-z

Sieradzka, M., Kirczuk, C., Kalemba-Rec, I., Mlonka-Mędrala, A., and Magdziarz, A. (2022). Pyrolysis of biomass wastes into carbon materials. *Energies*, 15(5), 1941. https://doi.org/10.3390/en15051941

Stock, S., Kostoglou, N., Selinger, J., Spirk, S., Tampaxis, C., Charalambopoulou, G. et al. (2022). Coffee waste-derived nanoporous carbons for hydrogen storage. *ACS Applied Energy Materials*, 5(9), 10915–10926. https://doi.org/10.1021/acsaem.2c01573

Tamilselvan, K., Sundarajan, S., Ramakrishna, S., Amirul, A.-A. A., and Vigneswari, S. (2024). Sustainable valorisation of coffee husk into value-added products in the context of circular bioeconomy: Exploring potential biomass-based value webs. *Food and Bioproducts Processing*, 145, 187–202. https://doi.org/10.1016/j.fbp.2024.03.008

Tsigkou, K., Demissie, B. A., Hashim, S., Ghofrani-Isfahani, P., Thomas, R., Mapinga, K. F. et al. (2025). Coffee processing waste: Unlocking opportunities for sustainable development. *Renewable and Sustainable Energy Reviews*, 210, 115263. https://doi.org/10.1016/j.rser.2024.115263

Vargas, G. O., Schnorr, C., Nunes, F. B., Da Rosa Salles, T., Tonel, M. Z., Fagan, S. B. et al. (2023). Highly furosemide uptake employing magnetic graphene oxide: DFT modeling combined with experimental approach. *Journal of Molecular Liquids*, 379, 121652. https://doi.org/10.1016/j.molliq.2023.121652

Wang, H. S.-H., and Yao, Y. (2023). Machine learning for sustainable development and applications of biomass and biomass-derived carbonaceous materials in water and agricultural systems: A review. *Resources, Conservation and Recycling*, 190, 106847. https://doi.org/10.1016/j.resconrec.2022.106847

Yusufoğlu, B., Kezer, G., Wang, Y., Ziora, Z. M., and Esatbeyoglu, T. (2024). Bio-recycling of spent coffee grounds: Recent advances and potential applications. *Current Opinion in Food Science*, 55, 101111. https://doi.org/10.1016/j.cofs.2023.101111

Zheng, N.-C., Yuan, C.-Z., Chen, Y.-C., Mangindaan, D., Lee, M., and Chien, H.-W. (2025). Environmentally friendly chitosan/spent coffee ground composites with continuous hydrophobically gradient porous structure for enhanced photothermal water evaporation. *International Journal of Biological Macromolecules*, 307, 141821. https://doi.org/10.1016/j.ijbiomac.2025.141821

CHAPTER

12

Coffee and Human Health: From Traditional Use to Pharmacological Applications

Emine Incilay Torunoglu[1*] and Arquimedes Gasparotto Junior[2]

[1] Necmettin Erbakan University, Faculty of Medicine, Department of Medical Biochemistry, Türkiye

[2] Federal University of Grande Dourados (UFGD), Faculty of Health Sciences, Laboratory of Cardiovascular Pharmacology (LaFaC), Dourados 79.804-070, MS, Brazil

This book chapter aims to provide a comprehensive evaluation of the pharmacological, biochemical, and molecular effects of coffee and to elucidate its potential contributions to health, nutrition, and disease prevention. This chapter examines the chemical composition of coffee and the biological activities of its major constituents, including caffeine, chlorogenic acids, diterpenes, trigonelline, and various phenolic compounds. Particular emphasis is placed on the roles of these bioactives in modulating oxidative stress, regulating inflammatory pathways, supporting metabolic homeostasis, and reducing cellular damage. The chapter also highlights coffee's neuroprotective potential, demonstrating its ability to influence amyloid formation, mitochondrial function, and neuroinflammation mechanisms closely associated with neurodegenerative disorders. Furthermore, the effects of coffee on cardiovascular health are discussed within the framework of a dose-dependent risk–benefit balance, with evidence indicating that moderate consumption is associated with protective outcomes. Its positive influence on metabolic disorders, obesity, and diabetes is supported through biochemical findings showing improved insulin sensitivity, reduced adipose inflammation, and attenuation of hepatic steatosis. In relation to aging biology, the chapter explores how coffee modulates key molecular mechanisms, including SIRT1 activation, autophagy, DNA repair, and mitochondrial integrity, that collectively contribute to reduced age-related oxidative damage and improved cellular resilience. Overall, this chapter demonstrates that coffee is not merely a widely consumed beverage but a complex phytochemical

*Corresponding author: torunogluemineincilay@gmail.com

DOI: 10.1201/9781003689393-12

system with significant biological activities relevant to human health. Synthesising current knowledge across experimental, clinical, and mechanistic studies provides a scientific foundation for future research and potential therapeutic applications involving coffee and its bioactive constituents.

1. Pharmacobotany and traditional use

The ethnomedicinal and pharmacobotanical applications of *Coffea* species (Rubiaceae) are extensively documented. The seeds of *Coffea arabica*, in particular, demonstrate diverse ethnomedicinal applications. For instance, beverages prepared in Indonesia are used to treat influenza (Muchtaridi et al., 2021); a hot water extract in Cuba is recognised for its aphrodisiac properties (Páez et al., 2025); in Thailand, it is employed as a cardiotonic and neurotonic agent (Bisht and Sisodia 2011); and it is utilised for the management of asthma (Yuniarta and Handayani 2024). Additionally, an infusion of *C. arabica* leaves in Haiti is reportedly used to treat anemia, edema, and asthenia (Thesnor et al., 2024). The leaf sap is also used in Africa to manage diarrhea and intestinal pain (Lamorde et al., 2010), while in Cuba, the leaves are applied to treat migraine (Matosinhos et al., 2022).

Other *Coffea* species also possess significant traditional uses. Ethnobotanical applications for *Coffea canephora* include the topical application of seeds as a massage for back pain, the use of a root extract for measles, and the oral administration of leaves for cough and jaundice (Tabuti et al., 2003). In Ethiopia, a beverage combining the pericarp of coffee berries with milk and honey is prepared to treat poisoning associated with diarrhea and nausea (Belayneh and Bussa, 2014).

Furthermore, a tea prepared from *C. robusta* leaves is known as "copi daon" in Indonesia and Ethiopia (Steger et al., 2022). A saline tea from the leaves of Liberian coffee (*C. liberica*) is reported to have a laxative effect (Belayneh and Bussa, 2014). The flowers of *Coffea benghalensis* are traditionally used in Nepal to treat menorrhagia (excessive menstrual bleeding) (Ghimire and Bastakoti, 2009). Finally, coffee charcoal is utilised in folk medicine to treat purulent wounds, pharyngitis, and stomatitis (Shahriari et al., 2020).

2 . Chemical constituents

Green *Coffea Arabica* beans contain approximately 1.67% caffeine (1,3,7-trimethylxanthine), irrespective of their geographical origin (Awwad et al., 2021). Following oral ingestion, caffeine is predominantly absorbed via the gastrointestinal tract (primarily the small intestine). It undergoes rapid systemic distribution, detectable in plasma, bile, saliva, semen, breast milk, and umbilical cord blood. Peak plasma concentrations (Cmax) are typically reached 15–120 minutes post-ingestion (Tmax), and caffeine rapidly crosses the blood-brain barrier, with detection in brain tissue possible within 5 minutes (Barcelos et al., 2020). Daily caffeine consumption has been associated with increased gray matter volume and cerebral blood flow, which may facilitate the accumulation of caffeine or its metabolites in the brain (Y. S. Lin et al., 2022). Caffeine is primarily metabolised in the liver by the cytochrome P450 1A2 (CYP1A2) enzyme system into active metabolites, including paraxanthine, theobromine, and theophylline, which are subsequently excreted in the urine (Depaula and Farah, 2019). Caffeine exerts its biological effects principally through three mechanisms: antagonism of adenosine receptors, mobilization of intracellular calcium, and inhibition of phosphodiesterases (Marcinek et al., 2024). Structurally, the purine-like configuration of caffeine and its primary metabolite, paraxanthine, enables them to act as antagonists at adenosine receptors, thereby influencing cellular energy metabolism and inflammatory pathways (Abu-Hashem et al., 2024). By inhibiting phosphodiesterase, which increases intracellular cyclic AMP (cAMP) levels, caffeine produces cardiostimulatory and antiasthmatic effects (Orbán et al., 2018). Additionally, caffeine activates ryanodine receptors (RyRs), enhancing calcium release from the sarcoplasmic reticulum in skeletal muscle and accelerating the excitation-contraction coupling process (Reggiani, 2021). Recent studies suggest that caffeine may beneficially modulate systemic metabolism, oxidative stress, inflammatory signaling, peripheral cellular signaling, and processes related to aging (Barcelos et al., 2020).

Green *Coffea Arabica* beans possess higher concentrations of chlorogenic acid (CGA) than caffeine; however, the roasting process significantly reduces CGA content (Awwad et al., 2021). Following ingestion, dietary chlorogenic acids are partially absorbed in the small intestine, where mucosal esterases hydrolyze them into quinic acid and caffeic acid, which subsequently enter systemic circulation. The fraction of CGA that remains unabsorbed transits to the colon, where it is metabolised by the resident gut microbiota (Olthof et al., 2001). The colon plays a critical role in the biotransformation of caffeic and ferulic acids into dihydroferulic acid, a metabolite that is readily absorbed. Caffeic acid can be converted into ferulic acid via catechol-O-methyltransferase (COMT). Both caffeic and ferulic acids can subsequently form various chlorogenic acid isomers through esterification with quinic acid (Lara-Guzmán et al., 2024; Shang et al., 2025). The majority of these metabolites undergo further conjugation (e.g., via transferases) and are ultimately excreted in the urine, with hippuric acid being a prominent final product (Rogozinska et al., 2023). The chemical profiles of Coffea-derived matrices that give rise to these metabolites are compiled in Table 1.

3. Antioxidant potential

The antioxidant activity of *Coffea* spp. beans have long been recognised, and numerous in vitro and in vivo studies have confirmed their high antioxidant capacity. This capacity plays a critical role in preventing the harmful effects of oxidative stress (Pintać et al., 2022), which is an important factor in the development of chronic diseases such as cardiovascular disorders (Surma et al., 2023), neurodegenerative diseases (C. Wang et al., 2023; Zidan et al., 2022), diabetes (Gonzalez-Burgos et al., 2024), and cancer (Nigra et al., 2021). Studies have shown that the strong antioxidant effect of coffee originates from its high polyphenol content (Aroufai et al., 2022). For example, coffee originating from Australia exhibited a total phenolic content (TPC) of 10.97 ± 0.74 mg GAE/g and an antioxidant capacity of 3.11 ± 0.06 mg AAE/g, whereas coffee from Colombia exhibited a TPC of 17.74 ± 0.32 mg GAE/g and an antioxidant capacity of 3.61 ± 0.06 mg AAE/g (Ali et al., 2022). Another

Table 1: Chemical compounds identified in preparations derived from *Coffea* species

Species	Analytical technique	Chemical content	References
Coffea arabica leaves L.	HPLC-ESI	Quinic acid, 3-Caffeoyl-quinic acid, Isomangiferin, Catechin/Epicatechin, Mangiferin, Kaempferol triglycoside, Caffeine, Rutin, Procyanidin B, Procyanidin C, Trigonelline	Cangeloni et al., 2022
Coffea arabica L.	HPLC-ESI-MS	Trigonelline, Gallic acid, Caffeic acid, Chlorogenic acid, Dichlorogenic acid, Caffeoyl-O-hexoside, p-Coumaroylquinic acid, p-Coumaroyl glycoside, Epicatechin, Rutin, Epicatechin, Syringic acid	Hutachok et al., 2021
Coffea arabica L.	HPLC	3-O-Caffeoylquinic acid, 5-O-Caffeoylquinic acid, 4-O-Caffeoylquinic acid, (Epi)catechin hexoside, (Epi)catechin, p-Coumaroylquinic acid, Mangiferin, 4-O-Feruloylquinic acid, 5-O-Feruloylquinic acid, Catechin dimer, 3,4-di-O-Caffeoylquinic acid, 3,5-di-O-Caffeoylquinic acid, Rutin, 4,5-di-O-Caffeoylquinic acid	Esquivel et al., 2020
Green *Coffea arabica* L.	HPLC	Apigenin, Luteolin, Quercetin, Chlorogenic acid, p-Coumaric acid, trans-Ferulic acid, Rosmarinic acid, Caffeic acid, Syringic acid, Caffeine	Doğaner et al., 2025
Coffea arabica L.	HPLC-ESI-MS/MS	Procyanidin B2, Procyanidin trimer A-type, Procyanidin C1, Procyanidin tetramer, Procyanidin pentamer	Wong-Paz et al., 2021
Coffee silverskin	HPLC-MS/MS	Caffeine, Quinine, 3,5-diCQA, 3-CQA, 5-CQA, Caffeic acid, Ferulic acid, Gallic acid, Loganic acid, p-Coumaric acid, Shikimic acid, Syringic acid, trans-Cinnamic acid, Vanillic acid, Catechin, Cyanidin 3-glucoside, Delphinidin 3,5-diglucoside, Epicatechin, Hyperoside, Kaempferol, Naringin, Quercetin dihydrate, Quercitrin, Resveratrol, Rutin Hydrate, Isogentisin, Swertiamarin, Sweroside, Gentiopicroside, Amarogentin	Nzekoue et al., 2020
Coffea canephora	LC-MS/MS	Trigonelline, 4-O-Caffeoylquinic acid, Caffeine, Erythrocentaurin, Glucopyranosylatractyligenin	Suryanti et al., 2023
Liberica sp. Coffee	HPLC	D-Glucoside, Quinic acid, (2S,3S)-2,3-Dihydro-2,3-dihydroxybenzoic acid, Penicillic acid, L-Galactose, Lucuminic acid, Deoxymiroestrol, Emmotin A, Nonoxynol-9, Digitoxigenin, L-Tryptophan, L-Isoleucine, L-Lysine	Ismail et al., 2024

(*Contd.*)

(*Contd.*)

Species	Analytical technique	Chemical content	References
Coffea benghalensis	HPLC-ESI-MS/MS.	Isocitric acid, Caffeoyl hexoside, Quinic acid derivative, Citric acid, 3-O-Caffeoylquinic acid, Procyanidin B-type dimer, 4-O-Caffeoylquinic acid, 5-O-Caffeoylquinic acid derivative, Catechin/Epicatechin, Procyanidin B-type trimer, Procyanidin B-type dimer, 5-O-Caffeoylquinic acid, 4-O-Caffeoylquinic acid, Procyanidin B-type trimer, Catechin/Epicatechin, 5-O-Coumaroylquinic acid, Mangiferin, 4-O-Feruloylquinic acid, 5-O-Caffeoylquinic acid methyl ester, 5-O-Feruloylquinic acid, 3-O-Feruloylquinic acid, 3,4-Di-O-Caffeoylquinic acid, 3,5-Di-O-Caffeoylquinic acid, 4,5-Di-O-Caffeoylquinic acid, Ferulic acid	Patay et al., 2018
Coffea racemose	HPLC	Gallic acid, Chlorogenic acid, Gentisic acid, Protocatechuic acid, p-Hydroxybenzoic acid, Vanillic acid	Pangestuti et al., 2021

study reported that green coffee beans contained a TPC of 29.3 ± 1.40 mg GAE/g and an antioxidant activity (IC50) of 146.8 ± 12.0, whereas roasted beans showed a TPC of 35.4 ± 1.49 mg GAE/g and an IC50 of 87.0 ± 5.73 (Acidri et al., 2020). However, other studies report that total phytochemicals are found at their highest concentrations in green beans, with roasting causing a decrease (Acidri et al., 2020). In coffee beans, the content of polyphenolic compounds generally decreases as the degree of roasting (from light to dark) increases. This decrease is linked to the degradation of chlorogenic, malic, and citric acids, affecting the total antioxidant activity. Consequently, lightly roasted coffee often shows higher antioxidant activity compared to dark-roasted and green (unroasted) coffee (Król et al., 2020).

Green *Coffea* sp. beans primarily contain chlorogenic acid, while roasted beans contain higher levels of caffeic acid (Król et al., 2020). Coffee beans also contain numerous bioactive compounds, including phenolics (e.g., chlorogenic acid), alkaloids (e.g., caffeine, theophylline, theobromine, trigonelline), diterpenes (e.g., cafestol, kahweol), and tocopherols (Baskaran and Radhakrishnan, 2024). Spent coffee grounds have been reported to contain protocatechuic acid, gallic acid, 5-chlorogenic acid, rutin, caffeic acid, and other polyphenols (Serna-Jiménez et al., 2023). Similarly, Torres-Valenzuela et al. (2020) reported that coffee grounds contained protocatechuic acid, gallic acid, 5-chlorogenic acid, rutin, 5-O-feruloylquinic acid, 3-O-coumaroylquinic acid, p-coumaric acid, caffeic acid, dicaffeoylquinic acids, 4-chlorogenic acid, 3-chlorogenic acid, and 3-O-feruloylquinic acid.

Coffea sp. extracts are known for their high polyphenol content, particularly chlorogenic acid, protocatechuic acid, and flavonoids, which exhibit important biological functions, such as antioxidant activity. Protocatechuic acid (3,4-dihydroxybenzoic acid) acts as a phenolic antioxidant in vitro and in vivo by enhancing enzymatic activities that reduce oxidative damage (Gao et al., 2022; Jiang et al., 2023). Gallic acid (3,4,5-trihydroxybenzoic acid), which naturally occurs in some plants, foods, and processed beverages, exhibits strong antioxidant capacity; it can induce apoptosis and protect against oxidative stress-induced damage in various tissues (Dhiman and Mukherjee, 2022; Ebaid et al., 2023). Chlorogenic acid (3-caffeoylquinic acid), sometimes referred to as coffee tannin, is one of the most common natural products in coffee (Feng et al., 2023), exhibiting antioxidant, anti-inflammatory, anti-obesity, and antitumor activities (Shi et al., 2022; Y. Xu et al., 2022). Rutin (quercetin-3-rutinoside), also known as vitamin P, is recognised for its strong antioxidant capacity (Hassane Hamadou et al., 2022; Tang et al., 2019). Caffeic acid (3,4-dihydroxycinnamic acid) has demonstrated in vitro radical scavenging activity (Chavarria et al., 2022) and exhibits immunomodulatory (Kilani-Jaziri et al., 2017), cardioprotective (Agunloye et al., 2019), anti-inflammatory (Zaitone et al., 2019), anticancer, and antimicrobial effects (Mude et al., 2022).

Oxidative stress is defined as an imbalance between the production of reactive oxygen species (ROS) and the capacity of the antioxidant defense system to neutralise them (Sharifi-Rad et al., 2020). This imbalance is a key factor in the pathophysiology of chronic diseases, including cardiovascular diseases (D. Lin et al., 2019), neurodegenerative disorders (Rekatsina et al., 2020), and cancer (Iqbal et al., 2024). Antioxidant nutrients can be obtained from foods rich in natural exogenous antioxidants, of which polyphenols are a major group (Lonati et al., 2022). Polyphenols are biologically active compounds capable of effectively neutralising free radicals (Lv et al., 2021). As coffee contains numerous phenolic compounds with antioxidant potential, it can serve as a significant source of antioxidants in the average diet and provide protective effects against ROS-mediated damage (Shaposhnikov et al., 2018).

The mechanisms underlying *Coffea* sp. effects involve interactions with phase I and phase II biotransformation enzymes, cellular signaling pathways, and antioxidant defense systems. Chlorogenic acid, for instance, exhibits strong free radical scavenging activity and increases the expression of phase II cytoprotective compounds, such as heme oxygenase-1 (HO-1), NAD(P)H: quinone oxidoreductase 1 (NQO1), glutathione, thioredoxin reductase 1, and thioredoxin 1. This neuroprotective effect is dependent on the activation of the transcription factor Nrf2 (nuclear factor erythroid 2–related factor 2) (Yao et al., 2019). Coffee is known to be a potent free radical scavenger that activates the Nrf2/antioxidant response element (ARE) signaling pathway, thereby protecting against oxidative damage (Cavin et al., 2008). Furthermore, the coffee components kahweol and cafestol have been shown to induce phase II enzyme expression even in Nrf2-/- (knockout) mouse embryonic fibroblasts, suggesting both Nrf2-dependent and -independent pathways (Higgins et al., 2008). Coffee has also been reported to induce phase I enzymes (cytochrome P450, e.g., CYP1A1 and 1A2) (Sulem et al., 2011) and other phase II enzymes (e.g., UDP-glucuronosyl transferase and glutathione S-transferase) (Huber et al., 2008; Kalthoff et al., 2010). In animal models, coffee consumption has been shown to prevent age-related oxidative stress and alveolar bone loss in rat periodontal tissues, an effect associated with Nrf2 upregulation and decreased gingival 8-hydroxydeoxyguanosine (8-OHdG) levels (Kobayashi et al., 2020).

Investigating the mechanisms underlying *Coffea* sp. protective effects, studies have shown that coffee consumption can enhance the endogenous antioxidant defense system. For example, it has been reported to increase the levels of glutathione (GSH) and the activity of antioxidant enzymes such as glutathione peroxidase (GSH-Px), glutathione S-transferase (GST), catalase (CAT), and superoxide dismutase (SOD). Concurrently, coffee consumption reduces levels of ROS and malondialdehyde (MDA), a marker of lipid peroxidation (D. Li et al., 2023).

4. Anti-inflammatory effects

The anti-inflammatory properties of *Coffea* sp. have been associated with elevated adiponectin concentrations (Osama et al. 2021). Adiponectin is an adipose tissue-derived hormone that exerts broad anti-inflammatory activity (Yamamoto et al. 2005); conversely, reduced adiponectin levels are recognised as an important risk factor for coronary artery disease (Ai et al. 2011). Coffee consumption has also been shown to lower the risk of metabolic syndrome, a condition that can aggravate inflammatory processes (Micek et al. 2018; Shang et al. 2016). As oxidative stress directly contributes to the development of inflammation (Aimo et al. 2020), it is notable that coffee intake may also enhance the expression of sirtuin-1 (SIRT1), a protein with potent anti-inflammatory effects (Gonçalinho et al. 2022).

During liver injury, levels of nitric oxide (NO), inducible nitric oxide synthase (iNOS), tumor necrosis factor-alpha (TNF-α), interleukin-6 (IL-6), interleukin-1 beta (IL-1β), and interleukin-12 (IL-12) are significantly elevated. Coffee administration has been shown to reduce these inflammatory mediators and cytokines, demonstrating marked anti-inflammatory potential. Specifically, caffeine, a key bioactive compound in coffee, suppresses NO and iNOS expression and functions as an effective inhibitor of lipopolysaccharide (LPS)-induced inflammatory responses (Hwang et al., 2016). In related research, however, coffee consumption was found not to exert a significant influence on periodontitis (Rhee et al. 2022).

Coffee consumption has been associated with reduced levels of inflammatory biomarkers, including C-reactive protein (Nieber, 2017). Moreover, evidence suggests that regular coffee intake exerts a protective influence on liver health by lowering the risk of conditions such as NAFLD, cirrhosis, and hepatocellular carcinoma (Grosso et al., 2017).

Overall, coffee consumption is linked to a lower risk of disorders in which inflammation plays a central role. Its anti-inflammatory activity is multifactorial, resulting from mechanisms that include enhanced antioxidant ~~defense~~ systems and inhibition of pro-inflammatory cytokine expression.

5. Effects on cardiovascular system

The relationship between coffee consumption and cardiovascular health is multifaceted; moderate intake has been linked to reduced risks of stroke and coronary artery disease, whereas excessive consumption may cause transient elevations in blood pressure, particularly in individuals sensitive to caffeine (Poole et al., 2017). Furthermore, due to its chlorogenic acid content, coffee has been shown to enhance insulin sensitivity and consequently lower the risk of developing type 2 diabetes (Grosso et al., 2017).

Moderate coffee consumption has been associated with a reduced risk of both stroke and coronary artery disease (Poole et al., 2017). Coffee intake has also been linked to a lower risk of several liver disorders, including non-alcoholic fatty liver disease (NAFLD), cirrhosis, and hepatocellular carcinoma (Grosso et al., 2017). In a study by Poole et al. (2017), consuming 2–3 cups of coffee per day was reported to reduce the risk of cirrhosis by approximately 40%.

Caffeic acid acts as a primary antioxidant by preventing free radical formation and as a secondary antioxidant by forming complexes with metals, thus inhibiting peroxide decomposition (Kinra et al., 2019). Moreover, coffee consumption increases hydroxycinnamic acid levels in plasma, thereby strengthening the antioxidant defense system (Agudelo-Ochoa et al., 2016).

In addition to caffeic acid, other compounds in coffee, such as chlorogenic acid, trigonelline, melanoidins, and ferulic acid, contribute to lowering blood pressure by reducing ACE activity, protecting vessels against oxidative stress, and increasing nitric oxide levels (Echeverri et al., 2010; Godos et al., 2014; Lim et al., 2020; Sharma and Sharma, 2020). Caffeine exerts antagonistic effects on adenosine receptors A1R, A2AR, and A2BR, modulating total peripheral resistance, urine output, and heart rate; these effects can either raise or lower blood pressure. However, regular consumption of 2–3 cups of coffee per day induces tolerance to caffeine, preventing its hypertensive effect in habitual coffee drinkers (Surma and Oparil, 2021). Coffee's antioxidant and anti-inflammatory properties particularly reduce CVD and inflammatory disease risks in postmenopausal women (Andersen et al., 2006).

Meta-analyses and long-term prospective studies indicate that coffee consumption significantly lowers the risk of CVD and mortality. In a meta-analysis of 26 studies, daily consumption of 3–4 cups of coffee reduced CVD incidence and related deaths (RR = 0.90; 95% CI: 0.84–0.96) (Di Maso et al., 2021). Grosso et al., in a meta-analysis of 31 studies, found that up to 4 cups of coffee per day reduced all-cause mortality by 14% and CVD-related deaths by 15%, with stronger effects in non-smokers. Similarly, Tverdal et al. followed 508,747 participants for 20 years and reported that coffee intake decreased all-cause mortality by 21% in men and 16% in women; notably, CVD-related mortality decreased by 28% in women, whereas in men, only filtered coffee consumption reduced CVD mortality by 12% (Tverdal et al., 2020). Coffee consumption also showed protective effects in individuals with a history of cardiovascular disease; in individuals without myocardial infarction, 1–6 cups of coffee reduced mortality risk by 14%, while in those with prior myocardial infarction, the risk decreased by 31%. However, no significant effect was observed in individuals with stroke history (Grosso et al., 2016).

A study in Addis Ababa involving 70 healthy, non-diabetic men and women consuming Ethiopian Arabica coffee found significant increases in serum free fatty acids (FFA) and high-density lipoprotein (HDL) levels. Coffee consumption significantly reduced triacylglycerols (TAG) but had no significant effect on total cholesterol (TC) or low-density lipoprotein (LDL), with similar results observed in both sexes (Gebeyehu et al., 2020). Another study reported that in Asian populations, coffee consumption was associated with reduced mortality risk as well as lower CVD- and cancer-related deaths (Shin et al., 2022).

Nevertheless, caffeinated coffee consumption (≥4 cups/day) has been shown to increase all-cause mortality risk in patients with cardiovascular disease (CVD), while decaffeinated coffee consumption reduced all-cause mortality risk. Additionally, an increase in levels of the caffeine metabolite 1-methylxanthine has been associated with all-cause mortality risk, suggesting that high coffee consumption may increase all-cause and CVD mortality in patients with cardiovascular disease (Zheng et al., 2022). Likewise, coffee consumption has been associated with increased CVD mortality risk in individuals with stage 2–3 hypertension; however, no such association was observed in individuals without hypertension or with stage 1 hypertension. The reported CVD mortality risk (compared to non-coffee drinkers) was 0.98 for <1 cup/day, 0.74 for 1 cup/day, and 2.05 for ≥2 cups/day. Therefore, high coffee consumption has been reported to be correlated with increased CVD mortality risk in individuals with severe hypertension (Teramoto et al., 2023).

Caffeine inhibits cAMP phosphodiesterase, increasing intracellular cAMP levels and accelerating cardiac function through positive inotropic effects (Orbán et al., 2018). Although caffeine shows a direct heart-rate–increasing effect, heart rate has been observed to slightly decrease after coffee consumption; this effect may trigger baroreflex activation and vagus nerve–mediated bradycardia as a compensatory mechanism (Butler et al., 2023).

Another important component in coffee, chlorogenic acid, may play significant roles in cardiovascular disease and related disorders. Chlorogenic acid neutralises free radicals by inhibiting NAD(P)H oxidase activity and reducing superoxide production. Additionally, it promotes NO production and inhibits plasma angiotensin-converting enzyme (Tajik et al., 2017).

Hyperlipidemia triggers inflammation through high lipid accumulation in macrophages, leading to foam cell formation, a critical step in atherosclerosis pathogenesis (Rachmawati et al., 2021). This process is associated with a significant increase in proinflammatory cytokine TNF-α levels. Literature has shown that natural compounds with anti-inflammatory effects can suppress inflammatory responses, limiting lipid accumulation and foam cell formation (Fatimatuzzahro et al., 2022). In this context, coffee, containing bioactive compounds such as polyphenols, chlorogenic acid, and diterpenes, stands out as a nutrient with anti-inflammatory and antioxidant potential. Therefore, coffee consumption may modulate lipid accumulation and inflammatory response in macrophages, limiting TNF-α elevation and reducing foam cell formation. Bioactive components of coffee may play a potential protective role in regulating hyperlipidemia-associated inflammatory processes (Rachmawati et al., 2021). For example, in a hyperlipidemic rat model, robusta coffee reduced TNF-α expression and caused less lipid accumulation in the subendothelial layer. Significant TNF-α positive staining was observed in the hyperlipidemia group, whereas robusta coffee downregulated TNF-α expression in the carotid artery of hyperlipidemic rats (Fatimatuzzahro et al., 2022).

After coffee consumption, total cholesterol, LDL cholesterol (LDL-C), and malondialdehyde (MDA), measured as thiobarbituric acid reactive substances (TBARS), significantly decreased, exerting a positive effect on cardiovascular risk (Yukawa et al., 2004). In a periodontitis-induced atherosclerotic rat model, coffee consumption showed rare occurrences of coronary atherosclerotic lesions, including atheroma, stenosis, and vascular obstruction. Coffee maintained coronary artery structure, with symmetrical intima-media thickness (IMT) and adequate lumen diameter observed in coronary arteries. Intimal collagen was intact, dense, and thick, while MMP–2 levels were significantly lower in the coffee group. Thus, coffee preserved intimal collagen, supporting the coronary artery wall and providing a protective role against coronary artery disease (CAD) (Susilawati et al., 2020).

Arterial calcification serves as a subclinical atherosclerosis marker and an important predictor of cardiovascular diseases (Criqui et al., 2014). Recent studies have shown an inverse relationship between coffee consumption and coronary artery calcification (CAC). Significant inverse relationships were observed between coffee consumption and CAC ≥100, with odds ratios (ORs) compared to non-drinkers being 0.85 for ≤1 cup/day, 0.7 for 1–3 cups/day, and 0.33 for >3 cups/day. Additionally, after stratification by smoking status, non-smokers consuming >3 cups/day showed a lower CAC OR (0.37) (Miranda et al., 2018). Compared to non-coffee drinkers, CAC scores for <1, 1–<3, 3–<5, and ≥5 cups/day were 0.77, 0.66, 0.59, and 0.81, respectively (Y. Choi et al., 2015). However, a study examining abdominal aortic calcification (AAC) reported that high coffee consumption (≥390 g/day), especially caffeinated coffee, was associated with increased AAC risk in adults with or without hypertension, diabetes, and CVD (Fan et al., 2023).

The main phenolic components of coffee, caffeic acid and ferulic acid, significantly increased HDL-mediated cholesterol efflux from THP-1 macrophages. These compounds significantly upregulated ATP-binding cassette transporter G1 (ABCG1) and scavenger receptor class B type I (SR-BI) gene and protein levels, whereas no effect was observed on ABCA1. In a study with eight healthy volunteers, blood samples were collected 30 minutes after coffee or water consumption; macrophages differentiated with autologous serum obtained after coffee intake showed increased HDL-mediated cholesterol efflux along with elevated ABCG1 and SR-BI expression compared to baseline. These findings suggest that coffee consumption strengthens HDL-mediated cholesterol transport by modulating ABCG1 and SR-BI expression via plasma phenolic acids, demonstrating potential anti-atherogenic properties (Uto-Kondo et al., 2010).

In Nω-Nitro-L-arginine methyl ester (L-NAME)–induced hypertensive rats, ACE and arginase activities are reportedly increased compared to a control group. However, pretreatment with caffeine and caffeic acid, individually and in combination, reduced ACE and arginase activities and MDA content while increasing NOx levels. The combination of caffeine and caffeic acid supports antihypertensive properties by lowering systolic blood pressure in L-NAME–induced rats (Oboh et al., 2021).

Numerous studies and meta-analyses have investigated the effects of coffee consumption on blood pressure (BP) and hypertension risk, showing that regular consumption of 2–3 cups per day does not alter arterial hypertension risk in most individuals, especially women and non-smokers. In contrast, irregular or occasional coffee consumption has been associated with increased BP and elevated hypertension risk (Surma and Oparil, 2021). A meta-analysis by Mesas et al. reported that a single intake of 200–300 mg caffeine increased systolic and diastolic BP by 8.14 mmHg and 5.75 mmHg, respectively, while long-term regular coffee consumption did not increase BP or cardiovascular disease risk (Mesas et al., 2011). Similarly, Palatini et al. reported no significant difference in 24-hour BP measurements among hypertensive individuals consuming 1–3 or >3 cups of coffee per day. Their study showed that 1–3 cups of coffee per day did not statistically increase cardiovascular event risk, whereas consumption of ≥4 cups significantly increased risk (Palatini et al., 2016).

Additionally, a study in elderly individuals with arterial hypertension reported that consuming 1–>3 cups of coffee per day could increase the risk of uncontrolled BP, whereas 2 cups had no significant effect, and coffee consumption did not affect BP control in women. Among non-smokers, coffee consumption was not significantly associated with the risk of poor BP control or a non-dipper profile. In the study by Minari et al. (2025), a total of 40 individuals (20 women and 20 men) with type 2 diabetes (T2D) and hypertension were followed for 12 months; during this period, both women and men showed a significant increase in fasting blood glucose (FBG) and HbA1c levels, as well as significant increases in body weight and body mass index (BMI). Sleep duration was significantly reduced, leading to increased nighttime awakenings. In women, even coffee sweetened with sugar showed potential beneficial effects on LDL-C levels, indicating that both possible health benefits and adverse effects in postmenopausal women should be considered.

In a study on hypercholesterolemic female rats, Liberica coffee pulp yogurt was administered for 8 weeks. In the high-dose group (25.7 mg/kg), total cholesterol, triglycerides, and LDL levels were significantly lower. Body and organ weights were lower than those of the control group; no significant changes were observed in liver and kidney function. Histology of the liver and visceral adipose tissue showed positive effects, demonstrating the beneficial effects of high-dose *Liberica* coffee pulp yogurt on hypercholesterolemia (Abidin et al., 2025).

Various studies have investigated the effects of caffeine on the cardiovascular system; however, the results have not yet provided a clear consensus. Some studies suggest that caffeine consumption may increase cardiovascular risks, whereas others report the effect as either beneficial or neutral. This indicates that cardiovascular responses depend on multiple factors, including caffeine dose, timing and frequency of consumption, absorption rate, and hepatic metabolism, and may vary among individuals (Echeverri et al., 2010).

6. Anticarcinogenic and chemoprotective effects

Epidemiological and experimental studies on the effects of coffee consumption on human health indicate that coffee generally exhibits neutral or beneficial effects (Higdon and Frei, 2006; Bøhn et al., 2014; Cornelis, 2015; Grosso et al., 2016). These effects are, at least in part, attributed to its chemical composition, including caffeine, phenolic acids, melanoidins, N-methylpyridinium, chlorogenic acid, metabolites such as quinic and caffeic acids, and diterpenes like kahweol and cafestol (Parras et al., 2007; do Carmo Carvalho et al., 2021; Bøhn et al., 2014).

Cafestol, a diterpene, exhibits anti-carcinogenic, anti-inflammatory, and pro-apoptotic properties (Lima et al., 2017). Cafestol and kahweol have been shown to increase glutathione transferase (GST) activity and several GST classes in the liver and intestine of rats and mice, as well as elevate glutathione (GSH) levels. These compounds induce conjugating and antioxidant enzymes in hepatic, intestinal, and renal tissues (Lam et al., 1987; Huber et al., 2002a,b) and provide protective effects against hydrogen peroxide-induced oxidative stress and DNA damage (Lee and Jeong, 2007).

Consumption of decaffeinated coffee has been reported to reduce colorectal cancer risk, whereas caffeinated coffee has been shown to lower the risk of rectal tumors (Um et al., 2020). Both caffeinated and decaffeinated coffee improved survival in patients with metastatic colorectal cancer. Moreover, all coffee types have been linked to a reduced risk of liver diseases and carcinoma, and daily coffee intake reduced tumor size in estrogen receptor-positive breast tumors (Mackintosh et al., 2020; Tanaka et al., 2019; Tran et al., 2019). These findings suggest that caffeine may be a critical determinant in some of coffee's biological effects (Safe et al., 2023).

The coffee components kahweol and cafestol also demonstrate anticarcinogenic properties. Kahweol suppresses tumor growth by activating the NF-κB pathway, and the combination of these two compounds affords protection against heterocyclic amine-induced malignancies (D.W. Choi et al., 2015; Oh et al., 2018). Chlorogenic acid exhibits antitumor effects by reducing cell viability and suppressing ROS production (Yamagata et al., 2018), and it inhibits adhesion molecule production in

TNF-α-stimulated endothelial cells (Chang et al., 2010). Cafestol demonstrates anti-angiogenic effects by inhibiting proliferation, migration, and tube formation in human umbilical vein endothelial cells (HUVECs) (Moeenfard et al., 2016; S. Wang et al., 2012). Concurrently, ferulic acid targets FGFR1, activates the PI3K/Akt signaling pathway, arrests the cell cycle, and induces cell death (Muchtaridi et al., 2021; Seow et al., 2021). Given these diverse effects, coffee exhibits significant anticarcinogenic potential, supported by its low cost, wide availability, and biocompatibility (Safe et al., 2023).

In specific studies, kahweol acetate and cafestol significantly inhibited the proliferation of PC-3, LNCaP, and DU145 prostate cancer cells at 24 and 48 hours, whereas caffeine only showed significant inhibition at 48 hours. In an in vivo xenograft rat model, the combined treatment with kahweol acetate and cafestol inhibited tumor growth more significantly than treatment with either agent alone (Iwamoto et al., 2019). Another study demonstrated that coffee suppressed NF-κB activity in both cell cultures and xenograft models, thereby regulating the transcription of genes related to prostate cancer-associated inflammation (Kolberg et al., 2016).

In a separate study, green coffee bean extracts (100–1000 μg/mL) decreased cell viability in human carcinoma cell lines from the colon (Caco-2), lung (A549), esophagus (OE-33), and bladder (T24), albeit while only partially inhibiting cell proliferation (Amigo-Benavent et al., 2017).

Another study investigated the protective effects of Yemeni green coffee powder against Ochratoxin A (OTA)-induced toxicity. Combined administration of coffee and OTA dose-dependently reduced oxidative stress in liver, brain, and kidney tissues. Specifically, rats treated with OTA alone exhibited increased malondialdehyde (MDA) levels and elevated catalase and glutathione reductase activities, alongside decreased superoxide dismutase (SOD) activity and reduced glutathione levels. In contrast, groups co-administered low and high doses of coffee showed increased antioxidant enzyme activities and reduced histopathological tissue damage (Nogaim et al., 2020).

Valadão Vicente et al. (2011) reported that a single dose of coffee significantly increased SOD, CAT, and GPx enzyme activities. These activities peaked (19.1%, 22.1%, and 25.1% increases, respectively) 1-hour post-administration ($p < 0.05$). The enzyme responses were dose-dependent ($p < 0.05$), and activities returned to basal levels after more than 4 hours, indicating a long-lasting effect. The total antioxidant capacity of liver tissue also increased by 6.5% at one hour, though this change was not statistically significant, potentially due to low coffee exposure. These findings suggest that coffee enhances cellular defense mechanisms against oxidative stress (Valadão Vicente et al., 2011).

A study by Kalthoff et al. (2020) investigated coffee's protective effects during irinotecan treatment. The findings showed that caffeic acid (CA) and caffeic acid phenethyl ester (CAPE) in coffee, particularly when administered together, mediated these protective effects by stimulating UGT1A gene activation. Both coffee and a CA + CAPE treatment increased UGT1A gene expression and enzyme activity in irinotecan-treated models. This accelerated SN-38 (the active metabolite of irinotecan) glucuronidation, leading to significant reductions in leukopenia, intestinal oxidative stress, and inflammation. The results identify CA and CAPE as the likely compounds responsible for coffee's regulatory effects on the UGT1A gene family, suggesting they are potent agents for mitigating irinotecan toxicity (Kalthoff et al., 2020).

In human neuroblastoma SH-SY5Y cells, pretreatment with kahweol protected against exposure to methylglyoxal (MG), a reactive compound known to impair mitochondrial function. Kahweol preserved the mitochondrial membrane potential, prevented the MG-induced decrease in ATP production, and reduced the accumulation of reactive oxygen and nitrogen species (RONS). These cytoprotective effects were regulated via the PI3K/Akt and p38 MAPK/Nrf2 signaling pathways. Thus, kahweol exerts a significant cytoprotective role by shielding mitochondria from oxidative stress and energy loss (de Oliveira et al., 2020).

The degree of roasting also appears to be a critical parameter. In a study by S. Choi et al. (2018), mice were orally administered coffee extracts for 10 days before a lipopolysaccharide (LPS) injection. Liver glutathione levels increased in mice given French roast coffee, whereas levels decreased in the LPS-only control group. LPS administration alone elevated tumor necrosis factor-alpha (TNF-α) mRNA expression; however, this increase was significantly attenuated in groups pre-treated with light and medium roast coffees. Coffee extracts suppressed interleukin-6 (IL-6) mRNA expression regardless of roasting degree, although this effect weakened with darker roasts, suggesting light and medium roasts possess higher biological protective potential (S. Choi et al., 2018).

Autoimmune diseases, characterised by a loss of self-tolerance and uncontrolled immune attacks on the body's own tissues, arise from a combination of genetic predisposition and environmental factors (Han et al., 2025; Júnior, 2020). Dietary habits, in particular, can alter gut microbiota, thereby affecting both innate and adaptive immune functions (Jiao et al., 2020).

In autoimmune thyroid diseases (AITDs), for example, diet is a key factor (Sharif et al., 2017). While no studies have directly linked human coffee consumption to the onset of AITDs, rat models show that caffeine administration alters TSH and thyroid hormone levels (Spindel et al., 1980). Research has, however, focused on coffee's impact on Hashimoto's thyroiditis management. A study by Benvenga et al. (2008) examined levothyroxine's intestinal absorption with coffee. Compared to water, coffee reduced the average and maximum increase in serum thyroxine by 36% and 30%, respectively, and delayed the time to reach peak serum concentration by 38 minutes. This demonstrates that coffee consumption can significantly impair levothyroxine absorption.

Regarding Systemic Lupus Erythematosus (SLE), recent studies suggest caffeine may influence cytokine profiles. Orefice et al. (2020) found that SLE patients with high daily caffeine intake exhibited significant reductions in serum IFN-γ, IFN-α, IL-17, and IL-6. This indicates that caffeine may modulate disease activity by suppressing these key inflammatory cytokines and potentially alleviating the SLE disease phenotype (Orefice et al., 2020).

Mechanistically, caffeine's immunomodulatory effects are linked to its role as an adenosine A2A receptor antagonist and a phosphodiesterase (PDE) inhibitor, both of which increase intracellular cyclic adenosine monophosphate (cAMP) levels. High intracellular cAMP activates protein kinase A (PKA), which suppresses proinflammatory cytokine release and inhibits broad immune responses, including T/B cell proliferation and NK cell cytotoxicity. Caffeine also modulates the expression of toll-like receptors (TLRs) and major histocompatibility complex class I (MHC I) molecules (Al Reef and Ghanem, 2018).

However, the immunomodulatory data are complex. Some studies show coffee or caffeine suppresses TNF-α and IL-6 (Capek et al., 2014; C.-W. Liu et al., 2018; Jung et al., 2017), while others report it increases cytokine expression (Green et al., 2012). Dose appears important: moderate caffeine intake was found to increase NK cell activity, but excessive doses (18 mg/kg/day) reversed this effect (Bortoluzzi et al., 2016). Overall, daily consumption of 3–4 cups of coffee is generally reported to have beneficial effects (Lire Wachamo, 2017).

Caffeine has been shown to suppress inflammatory genes (e.g., STAT1, TNF, IFNG) even at low doses, corresponding to the blood concentration after a single cup. It inhibits the STAT1 signaling pathway by reducing phosphorylated STAT1 (p-STAT1) levels, thereby controlling cytokine production (Iris et al., 2018).

Finally, coffee consumption has significant regulatory effects on immunosenescence. Both long-term and short-term coffee consumers exhibit significantly lower levels of the senescence-associated secretory phenotype (SASP) compared to non-coffee drinkers. Coffee appears to modulate systemic proinflammatory responses by suppressing JAK/STAT and MAPK signaling. Furthermore, coffee metabolites—including 1-methylxanthine, 3-methylxanthine, paraxanthine, and ceramide—were shown to reduce senescent T cells in vitro (Ruan et al., 2024).

7. Neuroprotective effects

A growing body of evidence supports the role of coffee in cognitive enhancement and neuroprotection. Caffeine has been shown to improve alertness, memory performance, and reaction time, while long-term coffee consumption may lower the risk of neurodegenerative diseases such as Alzheimer's and Parkinson's through mechanisms involving modulation of neurotransmitter activity and reduction of oxidative stress (Pérez-Hernández et al., 2022). Regular coffee intake has also been associated with improvements in memory, concentration, and mood (Cornelis, 2019a).

7.1 Effects on Alzheimer's disease

The pathology of Alzheimer's disease (AD) is characterised by the accumulation of extracellular amyloid-beta (Aβ1–42) plaques and intracellular neurofibrillary tangles in the hippocampus (Schönheit et al., 2004). Over time, these accumulations lead to severe neurodegeneration, cognitive impairment, and synaptic dysfunction (Narayan et al., 2014; Tu et al., 2014; Yau et al., 2015). Moreover, tau protein hyperphosphorylation, neuronal inflammation, oxidative stress, and cellular apoptosis also contribute to AD pathogenesis (Kamat et al., 2016). The pathophysiology of AD also affects peripheral nervous system components, such as astrocytes, microglial cells, and infiltrating cells, thereby promoting neurodegeneration (Zang et al.,

2022). A meta-analysis by Cornelis (2019a) reported that moderate coffee consumption—defined as 3–5 cups per day—was linked to a 25% reduction in Alzheimer's disease risk.

Caffeine, the main active compound in coffee, is known for its psychostimulant effect that enhances cognitive performance (Ferré, 2010). Caffeine consumption has been associated with a reduced risk of age-related neurodegenerative disorders and neurocognitive impairments linked to AD. For instance, in Mediterranean populations, elderly individuals who consume coffee and other dietary caffeine sources have been found to have better cognitive functions (Paz-Graniel et al., 2021). Furthermore, *Coffea arabica* bean extract has been reported to exhibit neuroprotective effects in a rat model of AD (Mohamed et al., 2020).

The contents of chlorogenic acid and trigonelline are higher in medium-roasted Arabica coffee compared to dark-roasted coffee. This can be explained by the thermolabile nature of these phytochemicals, which degrade during excessive roasting (Sales et al., 2020). Chlorogenic acid has been reported to exert a neuroprotective effect against toxicity (Mikami and Yamazawa, 2015). Furthermore, green coffee bean extract (GCBE) has been shown to enhance brain energy metabolism in mice (Ho et al., 2012). In another study, GCBE reduced the severity of fructose-induced AD in rats. Prophylactic administration of GCBE reduced oxidative damage by increasing cortical reduced glutathione (GSH) and superoxide dismutase (SOD) activity while decreasing malondialdehyde (MDA) levels (Mohamed et al., 2021).

A study by Zidan et al. (2022) determined that Arabica coffee consumption alleviated cognitive impairment induced by $AlCl_3$ administration in rats. It also inhibited the increase in serum $A\beta 1–42$ and TGF-β levels, thereby suppressing AD development. In the streptozotocin (STZ)-induced AD model, significant oxidative stress and inflammatory responses were observed, characterised by significantly increased MDA levels (an indicator of lipid peroxidation) and tumor necrosis factor-α (TNF-α) concentrations. Coffee administration significantly suppressed these increases, effectively preventing the progression of oxidative damage and inflammation. Furthermore, while total antioxidant capacity (TAC) was markedly reduced in the AD model, coffee administration prevented this decrease, helping to maintain endogenous antioxidant defense mechanisms. These findings support that coffee's neuroprotective effect involves mechanisms related to the attenuation of oxidative stress and the modulation of hippocampal inflammatory responses (Sedaghat et al., 2019).

7.2 Effects on Parkinson's disease

Parkinson's disease (PD) is a progressive neurodegenerative brain disorder characterised by symptoms such as bradykinesia, muscle rigidity, and resting tremor. In more advanced stages, postural instability is often observed. The main cause of these motor disorders is the selective loss of dopamine-producing neurons in the pars compacta region of the substantia nigra, leading to decreased dopamine levels in the striatum (Ross and Petrovitch, 2001). Consistent caffeine consumption has additionally been reported to reduce the likelihood of developing Parkinson's disease, largely due to its regulatory effects on dopaminergic signaling (Grosso et al., 2017).

Eicosanoyl-5-hydroxytryptamide (EHT), a compound found in coffee, has been reported to protect nigral dopaminergic neurons in the MPTP-induced PD model, limit dopamine and tyrosine hydroxylase loss in the striatum, and reduce oxidative stress and JNK activation. Furthermore, EHT has been shown to exhibit anti-inflammatory effects in microglia and astrocytes and to correct PP2A demethylation and cytotoxicity induced by MPP+ in SH-SY5Y cells. These findings support the therapeutic potential of EHT in the pathological processes associated with PD (Lee et al., 2013).

In a PD rat model, male rats were divided into five groups: a normal control, a Parkinson's control (rotenone-induced), and test groups treated with either green coffee bean extract (GCBE) or quercetin. Compared to the normal control, the Parkinson control group exhibited iron deficiency anemia, increased expression of interferon-gamma (IFN-γ), tumor necrosis factor-alpha (TNF-α), and divalent metal transporter 1 (DMT1), and elevated brain malondialdehyde (MDA) levels. Conversely, CD4 cell counts and brain reduced glutathione (GSH) levels were decreased, and adverse histopathological alterations were observed. The various treatments produced variable improvements and neuroprotective effects on inflammation, oxidative stress, the immune system, iron status, DMT1, and dopamine receptor D1 (DRD1) expression. The findings indicate that iron deficiency anemia may be associated with cytokine elevation in PD, and that decreased DRD1 expression along with increased DMT1 expression may trigger neurodegeneration (Al-Okbi et al., 2022).

In the rotenone-induced PD rat model, an oxidative and nitrosative stress condition was observed, characterised by increased lipid peroxidation and nitric oxide levels, along with decreased reduced glutathione (GSH) levels and glutathione-S-transferase (GST) and superoxide dismutase (SOD) activities in the midbrain and striatum. Rotenone also caused a reduction in acetylcholinesterase (AChE) and Na+/K+-ATPase activities in these regions, and an increase in TNF-α levels. Caffeine administration (for both protection and treatment) reportedly improved the oxidative stress and the changes in AChE and Na+/K+-ATPase activities caused by rotenone. Caffeine was shown to provide significant neuroprotective and therapeutic effects against the neurochemical, histopathological, and behavioral alterations in this rotenone-induced PD model (Khadrawy et al., 2017).

7.3 Effects on Huntington's disease

Huntington's disease (HD) is an autosomal dominant neurodegenerative disorder characterised by the accumulation of mutant huntingtin protein (mHTT), which results from a CAG repeat expansion in the huntingtin (HTT) gene (Jansen et al., 2017; Vater et al., 2025). Misfolding of mHTT leads to mitochondrial dysfunction, axonal and vesicular transport abnormalities, oxidative stress, and neuroinflammation, triggering neuronal death and clinical manifestations such as chorea, cognitive impairment, and psychiatric symptoms (Hong et al., 2021; Puigdellívol et al., 2016).

Geraniol administration in an animal model of HD resulted in significant improvements in motor and cognitive functions, a reduction in oxidative stress markers, and the restoration of antioxidant enzyme activities. Moreover, regulatory effects on neurotransmitter levels were identified. Histological analyses showed the preservation of neuronal structures in the striatum. These findings support that the neuroprotective effect of geraniol in HD occurs through the modulation of IDO1 activity, reduction of oxidative stress, and maintenance of neurotransmitter balance (Malakar et al., 2025).

7.4 Effects on amyotrophic lateral sclerosis disease

Amyotrophic lateral sclerosis (ALS) is a fatal neurodegenerative disease with a prevalence ranging from 1.6 to 11.8 per 100,000 people worldwide (Wolfson et al., 2023). One of the distinguishing features of the disease is oxidative stress. Increased reactive oxygen species (ROS) have been detected in samples from ALS patients, causing damage to proteins, lipid membranes, and DNA (Birger et al., 2019; Lopez-Gonzalez et al., 2016; Said Ahmed et al., 2000). This cellular damage has also been documented in various ALS models (Mitsumoto et al., 2020; Yang et al., 2022), demonstrating that oxidative stress plays a central role in disease progression.

The biochemical effects of caffeine treatment were reported in Wobbler mice, an animal model for ALS. Caffeine administration significantly reduced levels of ROS and restored nicotinamide adenine dinucleotide (NAD+) levels to those of healthy controls. No significant change was observed in the expression of NMNAT2 protein, which is involved in NAD+ metabolism. The study revealed that caffeine alleviates oxidative stress through alternative biochemical pathways, possibly occurring via improvements in mitochondrial function and enhancement of endogenous antioxidant defense mechanisms (Epplen et al., 2025).

8. Effects on metabolic disorders

8.1 Effects on diabetes mellitus

Diabetes induces significant disruptions in oxidative stress, inflammation, and apoptosis mechanisms by causing metabolic imbalances in multiple target tissues (Jia et al., 2019; Sadi et al., 2018). In high-fat diet and streptozotocin (HFD/STZ)-induced diabetic rat models, persistent hyperglycemia has been reported to cause significant increases in kidney weight and functional parameters (Liu et al., 2025). Severe impairments in the antioxidant defense system have been observed in diabetic rats; significant reductions in SOD, CAT, GPx, and GR activities were reported, while LPO, PCO, and NO levels increased significantly (Fernandes et al., 2016; Jia et al., 2019). Additionally, the decrease in GSH concentrations indicates that the oxidative balance is disrupted in favour of diabetes (Ahmed et al., 2019; De Brito Amaral et al., 2018).

The application of green coffee and roasted coffee extracts has been reported in various studies to restore the prooxidant/antioxidant imbalance, significantly increase antioxidant enzyme activities, and reduce oxidative damage (Priftis et al., 2018). In the HFD/STZ-induced diabetic rat model, increases in kidney weight and functional parameters, along with pronounced renal damage characterised by oxidative stress, inflammation, and apoptosis, have been reported. Green coffee water extract and metformin treatments were reported to significantly improve these impairments. Green coffee application was reported to improve kidney functions by reducing creatinine, uric acid, and aldosterone levels; it decreased oxidative stress markers (LPO, PCO, NO) while increasing antioxidant enzyme activities (SOD, CAT, GPx, GR) and their related gene expressions. Moreover, it reduced inflammation by suppressing IL-1-beta, TNF-alpha, and Nos2 levels and exerted anti-apoptotic effects by decreasing Bax and Casp3 expressions and increasing Bcl2 levels (AlAmri et al., 2020).

Bread and coffee dietary patterns showed an inverse relationship with HbA1c levels in adolescent girls, suggesting potential beneficial effects of coffee consumption on glycemic control. The literature reports that coffee consumption, due to its phenolic compounds, exhibits antioxidant and anti-inflammatory effects and improves glucose homeostasis by enhancing insulin sensitivity (Barboza et al., 2024). In another study, after 28 days of coffee treatment, rats induced with diabetes by a high-fat diet and streptozotocin showed significant decreases in fasting blood glucose levels and notable increases in serum insulin levels. Insulin resistance decreased significantly in treatment groups, and beta cell function (HOMA-B) improved significantly, particularly in coffee-treated groups. These findings demonstrate that coffee supports glycemic control, insulin production, and beta cell function (Othman et al., 2025). Another study reported that diabetic rats treated with black coffee for 45 days showed significant reductions in fasting and postprandial blood glucose and serum HbA1c levels, with marked improvement in insulin-glucose homeostasis (Butt et al., 2021).

8.2 Effects on obesity

The effects of *Coffea* sp. consumption on metabolic and molecular markers in high-fat diet-induced obesity models have been documented in various studies. Preventive coffee consumption reduced visceral white adipose tissue (WAT) hypertrophy in epididymal and retroperitoneal regions, promoting a healthier fat distribution profile (Gustafson et al., 2015; Maldonado-Ruiz et al., 2017; Vishvanath and Gupta, 2019). This effect has been attributed to increased PPARγ gene expression related to adipogenesis, thereby promoting the recruitment of new adipocytes and WAT remodeling (Gross et al., 2017; Hou et al., 2012; Santana-Oliveira et al., 2022; Unamuno et al., 2018).

Coffee also promoted the preservation of the brown adipose tissue (BAT) phenotype and enhanced browning processes by upregulating the expression of thermogenic genes such as the β3-adrenergic receptor, Ucp-1, Prdm16, and Opa-1 (Bargut et al., 2017; Chouchani et al., 2019; Santana-Oliveira et al., 2022; Velickovic et al., 2019). Coffee consumption reduced glucose intolerance, WAT hypertrophy, inflammation, and insulin resistance (Kusumah and Gonzalez de Mejia, 2022; Maki et al., 2017). This mechanism has been associated with the activation of pathways involving IRS-1 tyrosine phosphorylation, the p85/IRS-1 complex, and pAkt/PKB. Furthermore, coffee alleviated inflammation in adipocytes via the downregulation of inflammatory genes (Kusumah and Gonzalez de Mejia, 2022).

Hepatoprotective effects have also been reported; steatosis, alterations in enzyme profiles, and inflammatory gene expression induced by a high-fat diet were partially reversed by coffee consumption (Vitaglione et al., 2019). In a three-month high-fat diet-induced obesity model using C57BL/6 mice, the effects of coffee consumption on metabolic parameters and adipose tissue remodeling were examined. Preventive and therapeutic coffee administration was reported to improve glucose metabolism, reduce adipose tissue inflammation, and significantly decrease macrophage infiltration and IL-6 levels. Coffee consumption also alleviated liver steatosis and inflammation and upregulated the expression of genes associated with adaptive thermogenesis and mitochondrial biogenesis (PPARγ, Prdm16, Pcg1α, β3-adrenergic receptor, Ucp-1, and Opa-1) in the HF-CP group. Collectively, these findings suggest that coffee consumption demonstrates preventive and restorative effects on obesity-related metabolic disorders and inflammatory responses (Martins et al., 2023).

Another study reported that the administration of *Spirulina platensis* and green *Coffea* sp. bean aqueous extracts in high-fat diet-induced obese rats reduced body and liver weight, serum enzyme levels, and lipid levels while increasing insulin levels. The extracts also upregulated PGC-1α and PPARα expression while downregulating FAS expression (Hussein et al., 2020). In another study, *C. arabica* extract reduced body and organ weights in high-fat diet-induced obese rats, improved the serum lipid profile and glucose and insulin levels, increased levels of the anti-inflammatory adipokine adiponectin, and reduced adipocyte hypertrophy. RBP4 expression in adipose tissue was reduced, while GLUT4 expression was increased, indicating that *C. arabica* extract ameliorates obesity-related insulin resistance and dyslipidemia and may represent a potential natural supplement for obesity treatment (Seliem et al., 2022).

8.3 Effects on hypothyroidism

Thyroid hormones play a critical role in systemic metabolism and neurological development, and a bidirectional relationship exists between thyroid dysfunction and metabolic disorders (Walczak and Sieminska, 2021). The NHANES III study revealed that thyroid dysfunction is more prevalent in individuals with diabetes (Hollowell et al., 2002). Furthermore, metabolic disorders can exacerbate the progression of thyroid diseases, while thyroid diseases can, in turn, promote metabolic imbalances (Biondi et al., 2019).

Nutrition is another critical factor affecting thyroid function. Iodine deficiency is associated with hypothyroidism and goiter, whereas excessive iodine intake may also induce thyroid dysfunction (Leung and Braverman, 2014). Selenium supplementation has been shown to reduce autoimmune responses, alleviating the severity of Hashimoto's thyroiditis (Hu et al., 2021). On the other hand, studies investigating the relationship between caffeine intake and thyroid function are limited.

Pietzner et al. (2019) demonstrated a positive correlation between serum caffeine metabolites and serum levels of T4, T3, and the metabolite 3,5-T2. Most intervention studies have been conducted in animal models; Ahmed (2019) reported that medium and high-dose caffeine administration increased FT3 levels, whereas Clozel et al. (1993) found that in neonatal rats, caffeine stimulated both T4 and TSH and suppressed the pituitary TSH response. Ahmed et al. (2019) showed that maternal caffeine intake during pregnancy could induce maternal hyperthyroidism and fetal hypothyroidism. Caffeine may affect thyroid activity by modulating pituitary hormone secretion, a mechanism suggested to also involve regulation of the HPA axis. However, the effects of acute and chronic caffeine administration on human thyroid function remain to be fully elucidated, and the underlying mechanisms remain unclear.

9. Effects of coffee consumption on molecular mechanisms of aging

Coffee consumption has been associated with beneficial effects on both longevity and healthspan. These effects are primarily mediated through the modulation of key biological mechanisms related to aging. Among these mechanisms are the reduction of oxidative stress and chronic inflammation, improvement of mitochondrial function, enhancement of DNA repair mechanisms, stimulation of autophagy, modulation of epigenetic regulation, and regulation of cellular metabolic pathways. Each of these mechanisms plays a critical role in slowing down the aging process and reducing the incidence of age-related diseases (Fekete et al., 2024).

Studies using invertebrate models have provided valuable insights into the potential anti-aging effects of coffee and its bioactive components. Notably, *Caenorhabditis elegans* (Zheng et al., 2017; Siswanto et al., 2022; del Valle Carranza et al., 2020) and *Drosophila melanogaster* (Holvoet et al., 2023; Sotibrán et al., 2011) models have demonstrated that coffee components can extend lifespan and improve stress response. Caffeine, for instance, has been shown to extend lifespan in *C. elegans* by modulating cellular stress pathways and metabolic processes (Wang, Xiang, and Liu, 2022; Sutphin et al., 2012; Min, Youn, and Shim, 2021). However, studies in fruit flies fed with caffeine-containing diets have shown no significant lifespan extension (Nikitin et al., 2008), indicating that the effects of coffee may vary depending on the organism and experimental context.

Coffee polyphenols, particularly chlorogenic acid, exhibit notable antioxidant and anti-inflammatory properties that contribute to lifespan extension in invertebrates. Chlorogenic acid has been shown to increase stress resistance and longevity in *C. elegans* (Zheng et al., 2017; Siswanto et al., 2022; del Valle Carranza et al., 2020) and *D. melanogaster* (Sotibrán et al., 2011; Holvoet et al., 2023). Additionally, coffee flavonoids have been reported to extend lifespan in *Saccharomyces cerevisiae* by reducing oxidative stress (Czachor et al., 2020). These findings suggest that coffee components may exert similar protective effects across diverse biological systems.

Oxidative stress is a major contributor to cellular aging and the development of age-related cardiometabolic diseases (Ungvari et al., 2019; Tarantini et al., 2020). Antioxidants in coffee, particularly chlorogenic acid, can mitigate oxidative damage, thereby slowing cellular aging and modulating the pathogenesis of age-associated disorders [128, 220–226]. Nrf2 is a transcription factor crucial for cellular defense against oxidative stress and plays a central role in regulating the aging process [212, 216, 227–232]. Nrf2 controls the expression of antioxidant proteins that protect against oxidative damage caused by injury and inflammation, processes closely linked to aging. Aging reduces the efficiency of Nrf2-mediated protective responses, leading to increased oxidative damage and cellular senescence (Murai and Matsuda, 2023; Kobayashi et al., 2020; Wang et al., 2021). Importantly, chlorogenic acid, abundant in coffee, has been shown to activate the Nrf2 pathway, enhancing transcriptional activity of genes driven by the antioxidant response element (ARE) (Hada et al., 2020). This activation promotes the production of endogenous antioxidant enzymes, such as heme oxygenase-1 (HO-1), NAD(P)H quinone oxidoreductase 1 (NQO1), and glutathione S-transferase (GST), which detoxify reactive oxidants and maintain cellular redox balance, ultimately contributing to slowed aging and reduced disease risk.

Caffeine and its methylxanthine metabolites also exert anti-inflammatory effects (Barcelos et al., 2020). Chronic inflammation is a hallmark of aging and is closely linked to the progression of many age-related diseases (Ungvari et al., 2019, 2020; Fekete et al., 2024; Rebelo-Marques et al., 2018). By reducing inflammatory processes, coffee helps maintain cellular health and supports lifespan extension.

Coffee consumption has also been shown to enhance DNA repair mechanisms and preserve genomic integrity (Unno et al., 2024; Hori et al., 2014). Specifically, caffeine facilitates the repair of DNA damage, which accumulates with age and contributes to cellular senescence and age-related disease onset (Nikitina et al., 2015; Schumacher et al., 2021). Telomeres, protective structures at chromosome ends, shorten with each cell division, and telomere length is considered an indicator of cellular aging (Rizvi et al., 2014). Telomere shortening is associated with age-related diseases, including T2D and cardiovascular disorders (Zhao et al., 2013; Schneider et al., 2022). Several studies suggest that higher coffee intake may correlate with longer telomeres, indicating a potential protective effect against accelerated aging (Tucker, 2017).

Coffee also influences sirtuin-1 (SIRT1), a protein involved in cellular regulation, aging, and cardiometabolic health (Goncalinho et al., 2022; Lee et al., 2019). Polyphenols such as chlorogenic acid have been shown to enhance SIRT1 expression and activity (Ramos-Lopez, 2025). SIRT1 activation improves mitochondrial function, reduces oxidative stress, maintains cellular homeostasis, and delays the onset of age-related diseases, collectively contributing to longevity (Cohen et al., 2004; Zhou et al., 2022). Increased SIRT1 activity also lowers blood lipid levels, glucose levels, and inflammation, reducing the risk of cardiovascular diseases and T2D (Khayatan et al., 2022).

In summary, coffee and its bioactive components exert multiple cellular and molecular effects that collectively contribute to slowed aging and improved healthspan. These effects include antioxidant and anti-inflammatory activities, enhanced DNA repair, activation of SIRT1, stimulation of autophagy, and favourable modulation of metabolic pathways. By mitigating age-associated cellular damage and supporting cardiometabolic health, coffee may promote both longevity and a healthy lifespan. The findings from invertebrate models and biochemical studies provide strong mechanistic support for these protective effects, with potential implications for human health.

10. Coffee exosomes and fibrosis

Exosomes are extremely small vesicles, approximately 40 to 100 nm in diameter, that play critical roles in intercellular communication. The formation of these nanoscale structures occurs within intracellular endosomal compartments and is completed when multivesicular bodies (MVBs) fuse with the plasma membrane, releasing their contents into the extracellular space (Stoorvogel et al., 2002; Raposo and Stoorvogel, 2013). This process represents a complex biological mechanism regulated by both the physiological state of the cell and external environmental stimuli. Exosomes function as "membrane-bound cargo packages" within biological systems, as they are capable of transporting proteins, lipids, and nucleic acids to target cells (Yáñez-Mó et al., 2015). Through these properties, exosomes form a regulatory communication network in both prokaryotic and eukaryotic organisms, influencing cellular functions, metabolic processes, and signaling pathways.

An expanding body of epidemiological and experimental research demonstrates that dietary patterns and targeted nutritional interventions may play an important role in reducing the incidence of several malignancies, including hepatocellular carcinoma (HCC) (Shu et al., 2024). This growing evidence has shifted attention toward commonly consumed dietary components, among which coffee prepared as traditionally brewed or espresso-type beverages from roasted and ground seeds of *Coffea* species stands out as one of the most widely consumed plant-based products worldwide. Moreover, coffee represents a significant agricultural commodity, shaping the economic stability and trade dynamics of many producing and exporting countries (International Coffee Organization, 2018). Beyond its global cultural and economic relevance, coffee has been the focus of sustained scientific interest due to its rich phytochemical profile and potential biological activities relevant to chronic liver disease.

Epidemiological studies across diverse populations consistently report a robust inverse correlation between habitual coffee consumption and the risk of hepatic fibrogenesis and hepatocarcinogenesis (Romualdo et al., 2021; Liu et al., 2015). These findings indicate that coffee intake may confer protection across multiple stages of liver disease progression, from early fibrotic changes to advanced cirrhosis and ultimately HCC. Importantly, several large-scale prospective analyses demonstrate that consuming more than one cup of coffee per day is associated with approximately a 40% reduction in the risk of fibrosis, cirrhosis, and HCC, even after adjustment for demographic variables, lifestyle factors, alcohol use, viral hepatitis status, and other established risk modifiers (Liu et al., 2015). These observations suggest that the protective association is not simply a reflection of healthier behavioural patterns among coffee drinkers but may instead reflect specific biological activities of coffee constituents.

In contrast, investigations examining the influence of decaffeinated coffee frequently report either no significant reduction in risk or a substantially weaker protective effect than that observed with caffeinated beverages (Godos et al., 2017). This notable discrepancy has generated considerable scientific interest, raising the question of whether caffeine itself is the principal contributor to hepatoprotection or whether caffeine functions synergistically with other bioactive compounds naturally present in coffee. Coffee is chemically complex, containing not only caffeine but also chlorogenic acids, diterpenes, trigonelline, melanoidins, and numerous minor metabolites with demonstrated antioxidant, anti-inflammatory, and antifibrotic activities. Thus, the possibility that the protective effects arise from a combination of these compounds, rather than from a single bioactive molecule, warrants careful mechanistic investigation.

Although epidemiological data strongly support a beneficial relationship between coffee consumption and reduced liver disease outcomes, the underlying biochemical and molecular pathways remain only partially understood. Emerging hypotheses propose that coffee constituents may modulate oxidative stress markers, inhibit pro-inflammatory cytokine signaling, suppress hepatic stellate cell activation, enhance mitochondrial function, and influence xenobiotic and phase II detoxification pathways. Studies examining the chemical variability of coffee, such as differences in caffeine, trigonelline, and nicotinic acid concentrations across cultivars and preparation methods, further underscore the complexity of attributing specific biological effects to individual compounds (Caprioli et al., 2014). Consequently, future research integrating molecular pharmacology, metabolomics, controlled interventional trials, and experimental liver models is essential to delineate the precise mechanisms responsible for coffee's hepatoprotective effects and to determine how these findings may translate into preventive or therapeutic strategies for chronic liver disease.

The critical role of exosomes in intercellular communication can even extend across species. For example, in an experimental model, exosomes derived from mouse mast cells were transferred to human mast cells, leading to the expression of mouse proteins in the recipient cells (Valadi et al., 2007). This finding demonstrates that exosomes can mediate the transfer of biological information not only within a species but also between different species, suggesting that exosomal communication represents a highly conserved and evolutionarily ancient mechanism.

The presence of exosome-like structures in plant systems was identified well before their discovery in mammalian cells. Evidence of exosomal secretion in plant cells was reported nearly a decade before exosomes isolated from mammalian cells were recognised in the scientific literature (Halperin and Jensen, 1967). However, the isolation and detailed characterisation of edible plant-derived exosome-like nanoparticles (EPDENs) gained significant momentum starting in 2014. During this period, researchers successfully isolated EPDENs from nutritionally valuable plants such as ginger, carrot, grapefruit, and grape (Mu et al., 2014). The findings indicated that EPDENs play important regulatory roles in maintaining intestinal homeostasis by interacting with the gut epithelium, modulating immune responses, influencing inflammatory processes, and facilitating microbiota-host communication.

Chronic liver diseases—whether caused by viral infections, metabolic disturbances, or exposure to toxic substances—commonly result in progressive liver fibrosis. Liver fibrosis is characterised by excessive production and accumulation of extracellular matrix components, among which type I collagen constitutes the main structural element of fibrotic tissue (Brenner et al., 2012). In a healthy liver, only minimal amounts of type I collagen are present, and myofibroblasts are virtually absent. However, upon liver injury, myofibroblasts emerge early and play a pivotal role in initiating and sustaining fibrogenesis.

The most prominent biomarker of myofibroblasts is alpha-smooth muscle actin (α-SMA), a major actin isoform typically found in smooth muscle cells. Overexpression of α-SMA is directly associated with the progression of liver fibrosis, as this protein enhances the contractility, motility, and extracellular matrix-producing capabilities of myofibroblasts (Kawasaki et al., 2008). Therefore, α-SMA is widely used as an important indicator of fibrotic activity in both experimental and clinical assessments.

The relationship between coffee consumption and liver health has also attracted considerable scientific attention in recent years. Studies investigating the association between coffee intake and alterations in liver enzyme levels, fibrosis progression, and hepatocellular carcinoma (HCC) have reported that coffee consumption exhibits protective effects against all three outcomes (Feld et al., 2015). Specifically, increased coffee intake has been associated with reduced liver injury, diminished fibrotic processes, and a lower risk of HCC development, with a clear dose–response relationship.

Based on this information, the present study primarily focused on the isolation and characterisation of coffee-derived exosome-like nanoparticles (coffee EPDENs). The hypothesis underlying this work suggests that the beneficial effects of coffee on liver health may not solely arise from well-known chemical constituents such as caffeine and chlorogenic acid but may also involve these nanoscale biological structures. It was proposed that EPDENs isolated from roasted coffee might interact with mammalian cells and modulate specific cellular processes (Kantarcıoğlu et al., 2023).

Accordingly, the study evaluated the effects of coffee EPDENs on the proliferation of human HCC cells, as reduced proliferation may indicate potential anticancer properties. In addition, the influence of coffee-derived EPDENs on the expression of α-SMA transcripts was examined using a human hepatic stellate cell line. Because stellate cells are key regulators of liver fibrogenesis, changes in α-SMA expression provide valuable insights into fibrotic mechanisms.

This study aims to contribute to the growing body of scientific evidence suggesting that the protective effects associated with coffee consumption may be explained not only at the chemical level but also through biological nanoparticle-mediated mechanisms. The ability of plant-derived EPDENs to exert functional effects on mammalian cells offers a new perspective on the potential therapeutic roles of dietary exosome-like nanoparticles. Within this context, the study opens new research avenues not only in nutritional science but also in pharmacological biotechnology, nanoparticle-based therapeutic strategies, and the pathogenesis of liver diseases.

In this study, we present the first comprehensive and comparative investigation examining the anti-cancer effects of extracellular vesicles (EVs) isolated from unroasted and roasted *Coffea arabica* beans on SK-MEL-28 melanoma cells. Our findings demonstrate that EVs derived from both coffee types exert pronounced biological activities against melanoma cells. Specifically, unroasted and roasted coffee EVs selectively reduced SK-MEL-28 cell viability, triggered programmed cell death by inducing apoptosis, markedly suppressed the growth of three-dimensional tumor spheroids, and significantly impaired cellular migration. These multifaceted effects highlight the capacity of coffee-derived EVs to interfere with several key aspects of melanoma progression (Korkmaz et al., 2025).

Mechanistic analyses revealed that the two EV types modulate distinct oncogenic signaling pathways. Unroasted coffee EVs downregulated SERPINA1 expression and inhibited the PI3K/AKT signaling axis, a pathway critically involved in promoting cell proliferation and survival. In contrast, roasted coffee EVs reduced BRAF phosphorylation, thereby attenuating MAPK pathway activation and limiting melanoma cell growth and proliferation through a different regulatory mechanism. These complementary molecular actions indicate that coffee-derived EVs can target melanoma through multiple biochemical networks.

Overall, our results underscore that coffee-derived extracellular vesicles exhibit potent anti-melanoma properties while possessing natural origin, high biocompatibility, and suitability for oral consumption. Their ability to efficiently interact with cellular targets and modulate oncogenic pathways positions coffee EVs as promising, naturally occurring nano-carriers with significant therapeutic potential against skin cancer. This study provides a foundational framework for future research and potential translational applications of edible EV-based nanotherapeutics in melanoma treatment.

11. Conclusion

Coffee consumption exerts a multifaceted influence on human health through a broad spectrum of biochemical, molecular, physiological, and pharmacological mechanisms. Its rich composition, including caffeine, chlorogenic acids, diterpenes, trigonelline, melanoidins, and numerous phenolic compounds, modulates critical biological pathways associated with oxidative stress, inflammation, metabolic regulation, neuroprotection, and cellular aging. These effects collectively demonstrate that coffee, when consumed in moderate amounts, may serve as a functional dietary component with promising health-promoting properties. The accumulated findings from preclinical models, clinical research, metabolic investigations, and epidemiological studies consistently indicate that coffee's influence extends far beyond its traditional role as a stimulant beverage, encompassing protective effects against chronic metabolic, neurodegenerative, hepatic, inflammatory, and age-associated disorders.

Coffee's antioxidant properties, largely mediated by chlorogenic acid, caffeic acid, and other polyphenols, constitute one of its most significant biological actions. These compounds directly scavenge free radicals, activate endogenous antioxidant enzymes, and regulate phase I and II biotransformation pathways. By enhancing key components of the antioxidant defense system, such as SOD, catalase, glutathione peroxidase, NQO1, HO-1, and GST, coffee reduces lipid peroxidation, protein oxidation, and DNA damage. These protective responses, supported by Nrf2 activation, mitigate oxidative stress-driven pathologies, including cardiovascular diseases, neurodegeneration, diabetes, obesity, and liver injury. Experimental models further demonstrate that coffee prevents age-related oxidative deterioration in multiple tissues, indicating that its antioxidant actions have systemic relevance.

Coffee also exhibits strong anti-inflammatory effects by modulating cytokine expression, inhibiting inflammatory mediator release, and altering immune signaling pathways. The suppression of pro-inflammatory cytokines such as TNF-α,

IL-6, IL-1β, and IFN-γ—coupled with increases in anti-inflammatory mediators such as adiponectin – underscores coffee's ability to reduce chronic low-grade inflammation, a hallmark of metabolic and degenerative diseases. The capacity of caffeine to antagonise adenosine receptors, inhibit phosphodiesterases, and elevate intracellular cAMP further contributes to its immunomodulatory profile. These actions regulate innate and adaptive immune responses, inhibit exaggerated cytokine release, and modulate TLR and MHC expression, highlighting coffee's relevance in chronic inflammatory and autoimmune conditions. Notably, findings regarding immunosenescence emphasise that coffee may attenuate age-related inflammatory burden by influencing JAK/STAT, MAPK, and SASP pathways.

Cardiovascular effects of coffee consumption demonstrate both protective and risk-modifying influences that vary with dose, metabolic status, caffeine sensitivity, and drinking habits. Moderate consumption is repeatedly associated with reduced incidence of coronary artery disease, stroke, and cardiovascular mortality. Improvements in lipid profiles, attenuation of oxidative vascular damage, enhancement of endothelial nitric oxide availability, and inhibition of the renin angiotensin system contribute to coffee's cardioprotective potential. Chlorogenic acids, trigonelline, and hydroxycinnamic acids further modulate vascular homeostasis by influencing ACE activity, reducing arterial stiffness, and promoting HDL-mediated cholesterol efflux. Nevertheless, individuals with severe hypertension or heightened sensitivity to caffeine may exhibit adverse hemodynamic responses, emphasising the significance of personalised consumption patterns. Overall, the preponderance of evidence supports beneficial cardiovascular outcomes associated with regular, moderate coffee intake.

Anticarcinogenic and chemoprotective effects emerge as another substantial dimension of coffee's biological activity. Coffee diterpenes kahweol and cafestol demonstrate pro-apoptotic, anti-inflammatory, anti-angiogenic, and detoxifying properties by inducing glutathione-dependent pathways, modulating NF-κB activity, and inhibiting tumor cell proliferation. Phenolic compounds, particularly chlorogenic and caffeic acids, suppress ROS-mediated DNA damage, attenuate inflammatory signaling that contributes to tumor microenvironment progression, and enhance detoxification enzymes involved in xenobiotic metabolism. Evidence from clinical studies indicates that both caffeinated and decaffeinated coffee reduce the risk of colorectal, liver, breast, and endometrial cancers while improving survival outcomes in cancer patients. Coffee's protective effects during chemotherapeutic treatments, particularly via modulation of UGT1A expression and enhancement of drug detoxification, underscore its therapeutic relevance in oncology.

Neuroprotective benefits of coffee span multiple neurodegenerative disorders, including Alzheimer's disease, Parkinson's disease, Huntington's disease, and amyotrophic lateral sclerosis. Coffee constituents reduce amyloidogenic processes, suppress tau hyperphosphorylation, enhance mitochondrial function, and attenuate neuroinflammation. Caffeine's antagonism of adenosine receptors modulates dopaminergic neurotransmission and reduces neuronal vulnerability, while chlorogenic acid, trigonelline, and eicosanoyl-5-hydroxytryptamide contribute additional protective mechanisms. Experimental models demonstrate that coffee enhances cognitive performance, preserves neuronal integrity, and ameliorates behavioural impairments associated with neurodegeneration. These effects are further strengthened by reductions in oxidative stress, restoration of antioxidant capacity, and improvements in mitochondrial membrane potential.

Coffee also exerts a significant influence on metabolic health. Studies consistently demonstrate improvements in insulin sensitivity, glycemic control, lipid regulation, and adipose tissue remodeling. Coffee reduces visceral adiposity, promotes adipocyte browning, enhances thermogenic gene expression, and decreases chronic adipose inflammation. Its bioactive compounds regulate insulin signaling pathways, attenuate diet-induced metabolic disturbances, and reduce hepatic steatosis. In diabetes models, coffee normalises glucose homeostasis, increases endogenous antioxidant defenses, and suppresses inflammatory and apoptotic processes responsible for organ damage. These findings support the view that coffee contributes to metabolic resilience across multiple organ systems.

The molecular effects of coffee on aging represent a key emerging area of research. Coffee consumption modulates pathways implicated in longevity, including SIRT1 activation, Nrf2 signaling, autophagy induction, mitochondrial biogenesis, telomere maintenance, and DNA repair. Through these mechanisms, coffee attenuates age-associated oxidative damage, reduces systemic inflammation, and enhances cellular repair processes. Studies across invertebrate and mammalian models highlight that coffee bioactives extend lifespan, improve stress resistance, and preserve metabolic homeostasis. These findings suggest that coffee may contribute to healthy aging and the reduction of age-related disease risk via coordinated modulation of multiple cellular pathways.

Finally, the discovery that coffee-derived exosome-like nanoparticles participate in intercellular communication introduces a novel dimension to their biological activity. These vesicles may influence immune regulation, metabolic signaling, and fibrotic processes, indicating that coffee's health effects extend into previously unrecognised biochemical territories.

Collectively, the wide-ranging evidence presented demonstrates that coffee is not merely a dietary stimulant but a complex phytochemical system with diverse and meaningful impacts on human health. Its antioxidant, anti-inflammatory, metabolic, cardioprotective, neuroprotective, chemopreventive, and anti-aging effects position coffee as a potentially valuable component of preventive and therapeutic strategies. However, individual differences in caffeine metabolism, underlying health conditions, consumption patterns, and preparation methods must be considered to optimise health outcomes. Continued research integrating molecular, clinical, and translational perspectives will further clarify the full extent of coffee's biological significance.

References

Abidin, A. Z., Bahari, H., Othman, F., Norkhairani, N. E., and Balan, S. S. (2025). The effectiveness of consuming Liberica coffee pulp yoghurt on hypercholesterolemia induced male Sprague Dawley rats. *Natural and Life Sciences Communications*, 24(2). https://doi.org/10.12982/NLSC.2025.031

Abu-Hashem, A. A., Hakami, O., El-Shazly, M., El-Nashar, H. A. S., and Yousif, M. N. M. (2024). Caffeine and purine derivatives: A comprehensive review on the chemistry, biosynthetic pathways, synthesis-related reactions, biomedical prospectives and clinical applications. *Chemistry & Biodiversity*, 21(7), e202400050. https://doi.org/10.1002/CBDV.202400050

Acidri, R., Sawai, Y., Sugimoto, Y., Handa, T., Sasagawa, D., Masunaga, T. et al. (2020). Phytochemical profile and antioxidant capacity of coffee plant organs compared to green and roasted coffee beans. *Antioxidants*, 9(2), 93. https://doi.org/10.3390/ANTIOX9020093

Agudelo-Ochoa, G. M., Pulgarín-Zapata, I. C., Velásquez-Rodriguez, C. M., Duque-Ramírez, M., Naranjo-Cano, M., Quintero-Ortiz, M. M. et al. (2016). Coffee consumption increases the antioxidant capacity of plasma and has no effect on the lipid profile or vascular function in healthy adults in a randomized controlled trial. *The Journal of Nutrition*, 146(3), 524–531. https://doi.org/10.3945/JN.115.224774

Agunloye, O. M., Oboh, G., Ademiluyi, A. O., Ademosun, A. O., Akindahunsi, A. A., Oyagbemi, A. A. et al. (2019). Cardio-protective and antioxidant properties of caffeic acid and chlorogenic acid: Mechanistic role of angiotensin converting enzyme, cholinesterase and arginase activities in cyclosporine induced hypertensive rats. *Biomedicine & Pharmacotherapy*, 109, 450–458. https://doi.org/10.1016/J.BIOPHA.2018.10.044

Ahmed, O. M., Ali, T. M., Abdel Gaid, M. A., and Elberry, A. A. (2019). Effects of enalapril and paricalcitol treatment on diabetic nephropathy and renal expressions of TNF-α, p53, caspase-3 and Bcl-2 in STZ-induced diabetic rats. *PLOS ONE*, 14(9), e0214349. https://doi.org/10.1371/JOURNAL.PONE.0214349

Ahmed, R. G. (2019). Gestational caffeine exposure acts as a fetal thyroid-cytokine disruptor by activating caspase-3/BAX/Bcl-2/Cox2/NF-κB at ED 20. *Toxicology Research*, 8(2), 196–205.

Ai, M., Otokozawa, S., Asztalos, B. F., White, C. C., Cupples, L. A. et al. 2011. Adiponectin: An independent risk factor for coronary heart disease in men in the Framingham offspring study. *Atherosclerosis*, 217(2), 543–548 https://doi.org/10.1016/j.atherosclerosis.2011.05.035.

Aimo, A., Castiglione, V., Borrelli, C., Saccaro, L. F., Franzini, M. et al. 2020. Oxidative stress and inflammation in the evolution of heart failure: From pathophysiology to therapeutic strategies. *Eur J Prev Cardiol.*, 27(5), 494–510. https://doi.org/10.1177/2047487319870344

Al Reef, T., and Ghanem, E. (2018). Caffeine: Well-known as psychotropic substance, but little as immunomodulator. *Immunobiology*, 223(12), 818–825. https://doi.org/10.1016/J.IMBIO.2018.08.011

AlAmri, O. D., Albeltagy, R. S. M. A., Akabawy, A., Mahgoub, S., Abdel-Mohsen, D. M., Abdel Moneim, A. E. et al. (2020). Investigation of antioxidant and anti-inflammatory activities as well as the renal protective potential of green coffee extract in high fat-diet/streptozotocin-induced diabetes in male albino rats. *Journal of Functional Foods*, 71, 103996. https://doi.org/10.1016/J.JFF.2020.103996

Al-Okbi, S. Y., Mabrok, H. B., Al-Siedy, E. S. K., Mohamed, R. S., and Ramadan, A. A. (2021). Iron status, immune system, and expression of brain divalent metal transporter 1 and dopamine receptors D1 interrelationship in Parkinson's disease and the role of grape seed and green coffee bean extracts and quercetin in mitigating the disease in rats. *Journal of Herbmed Pharmacology*, 11(1), 63–74.

Amigo-Benavent, M., Wang, S., Mateos, R., Sarriá, B., and Bravo, L. (2017). Antiproliferative and cytotoxic effects of green coffee and yerba mate extracts, their main hydroxycinnamic acids, methylxanthine and metabolites in different human cell lines. *Food and Chemical Toxicology*, 106, 125–138. https://doi.org/10.1016/J.FCT.2017.05.019

Andersen, L. F., Jacobs, D. R., Carlsen, M. H., and Blomhoff, R. (2006). Consumption of coffee is associated with reduced risk of death attributed to inflammatory and cardiovascular diseases in the Iowa Women's Health Study. *The American Journal of Clinical Nutrition*, 83(5), 1039–1046. https://doi.org/10.1093/AJCN/83.5.1039

Aroufai, İ. A., Sabuncu, M., Dülger Altiner, D., and Sahan, Y. (2022). Antioxidant properties and bioaccessibility of coffee beans and their coffee silverskin grown in different countries. *Journal of Food Measurement and Characterization*, 16(3), 1873–1888. https://doi.org/10.1007/S11694-021-01271-4/FIGURES/4

Awwad, S., Issa, R., Alnsour, L., Albals, D., and Al-Momani, I. (2021). Quantification of caffeine and chlorogenic acid in green and roasted coffee samples using HPLC-DAD and evaluation of the effect of degree of roasting on their levels. *Molecule*s, 26(24). https://doi.org/10.3390/MOLECULES26247502

Barboza, B. P., Bricarello, L. P., Alves, M. de A., Tureck, C., Retondario, A., Longo, G. Z. et al. (2024). Dietary patterns and biochemical markers related to diabetes mellitus: An association analysis based on data from the study of cardiovascular risk in adolescents (ERICA). *Nutrition*, 118, 112283. https://doi.org/10.1016/J.NUT.2023.112283

Barcelos, R. P., Lima, F. D., Carvalho, N. R., Bresciani, G., and Royes, L. F. (2020). Caffeine effects on systemic metabolism, oxidative-inflammatory pathways, and exercise performance. *Nutrition Research*, 80, 1–17. https://doi.org/10.1016/j.nutres.2020.05.005

Bargut, T. C. L., Souza-Mello, V., Aguila, M. B., and Mandarim-De-Lacerda, C. A. (2017). Browning of white adipose tissue: Lessons from experimental models. *Hormone Molecular Biology and Clinical Investigation*, 31(1). https://doi.org/10.1515/HMBCI-2016-0051/MACHINEREADABLECITATION/RIS

Baskaran, K., and Radhakrishnan, M. (2024). High-pressure processing-assisted in-situ extraction of caffeine, chlorogenic acid, phenolic content, and antioxidant properties of green coffee bean. *Journal of Food Process Engineering*, 47(2), e14551. https://doi.org/10.1111/JFPE.14551

Belayneh, A., and Bussa, N. F. (2014). Ethnomedicinal plants used to treat human ailments in the prehistoric place of Harla and Dengego valleys, eastern Ethiopia. *Journal of Ethnobiology and Ethnomedicine*, 10(1), 18. https://doi.org/10.1186/1746-4269-10-18

Benvenga, S., Bartolone, L., Pappalardo, M. A., Russo, A., Lapa, D., Giorgianni, G. et al. (2008). Altered intestinal absorption of L-thyroxine caused by coffee. *Thyroid*, 18(3), 293–301. https://doi.org/10.1089/THY.2007.0222

Biondi, B., Kahaly, G. J., and Robertson, R. P. (2019). Thyroid dysfunction and diabetes mellitus: Two closely associated disorders. *Endocrine Reviews*, 40(3), 789–824. https://doi.org/10.1210/ER.2018-00163

Birger, A., Ben-Dor, I., Ottolenghi, M., Turetsky, T., Gil, Y., Sweetat, S. et al. (2019). Human iPSC-derived astrocytes from ALS patients with mutated C9ORF72 show increased oxidative stress and neurotoxicity. *EBioMedic*ine, 50, 274–289. https://doi.org/10.1016/j.ebiom.2019.11.026+

Bisht, S., and Sisodia, S. S. (2011). Protective role of coffee beans in diabetes mellitus model of rats. *Journal of Pharmacy Research*, 4(10), 3717–3720.

Bøhn, S. K., Blomhoff, R., and Paur, I. (2014). Coffee and cancer risk, epidemiological evidence, and molecular mechanisms. *Molecular Nutrition & Food Research*, 58(5), 915–930. https://doi.org/10.1002/mnfr.201300526

Bortoluzzi, A., Vincenzi, F., Govoni, M., Padovan, M., Ravani, A., Borea, P. A. et al. (2016). A2A adenosine receptor upregulation correlates with disease activity in patients with systemic lupus erythematosus. *Arthritis Research and Therapy*, 18(1), 1–11. https://doi.org/10.1186/S13075-016-1089-8/FIGURES/4

Brenner, D. A., Kisseleva, T., Scholten, D. et al. (2012). Origin of myofibroblasts in liver fibrosis. *Fibrogenesis & Tissue Repair*, 5(Suppl 1), S17. https://doi.org/10.1186/1755-1536-5-S1-S17

Butler, J. M., Frampton, C. M., Moore, G., Barclay, M. L., and Jardine, D. L. (2023). Immediate effect of caffeine on sympathetic nerve activity: Why coffee is safe? A single-centre crossover study. *Clinical Autonomic Research*, 33(6), 623. https://doi.org/10.1007/S10286-023-00967-5

Butt, J., Naeem, U., Lodhi, S., Khokhar, A., Azeem, A., and Tayyab, M. (2021). Bitter coffee for sweet diabetes: A randomized controlled trial for treatment of type 2 diabetes with black coffee in balb c albino mice. *Pakistan Journal of Physiology*, 17(1), 15–18. https://doi.org/10.69656/PJP.V17I1.1295

Cangeloni, L., Bonechi, C., Leone, G., Consumi, M., Andreassi, M., Magnani, A. et al. (2022). Characterization of extracts of coffee leaves (*Coffea arabica* L.) by spectroscopic and chromatographic/spectrometric techniques. *Foods*, 11(16), 2495. https://doi.org/10.3390/foods11162495

Capek, P., Paulovičová, E., Matulová, M., Mislovičová, D., Navarini, L., and Suggi-Liverani, F. (2014). Coffea arabica instant coffee—Chemical view and immunomodulating properties. *Carbohydrate Polymers*, 103(1), 418–426. https://doi.org/10.1016/J.CARBPOL.2013.12.068

Caprioli, G., Cortese, M., Maggi, F., Minnetti, C., Odello, L., Sagratini, G., and Vittori, S. (2014). Quantification of caffeine, trigonelline and nicotinic acid in espresso coffee: The influence of espresso machines and coffee cultivars. *International Journal of Food Sciences and Nutrition*, 65(4), 465–469.

Cavin, C., Marin-Kuan, M., Langouët, S., Bezençon, C., Guignard, G., Verguet, C. et al. (2008). Induction of Nrf2-mediated cellular defenses and alteration of phase I activities as mechanisms of chemoprotective effects of coffee in the liver. *Food and Chemical Toxicology*, 46(4), 1239–1248. https://doi.org/10.1016/J.FCT.2007.09.099

Chang, W. C., Chen, C. H., Lee, M. F., Chang, T., and Yu, Y. M. (2010). Chlorogenic acid attenuates adhesion molecules upregulation in IL-1beta-treated endothelial cells. *European Journal of Nutrition*, 49(5), 267–275. https://doi.org/10.1007/S00394-009-0083-1

Chavarria, D., Benfeito, S., Soares, P., Lima, C., Garrido, J., Serrão, P. et al. (2022). Boosting caffeic acid performance as antioxidant and monoamine oxidase B/catechol-O-methyltransferase inhibitor. *European Journal of Medicinal Chemistry*, 243, 114740. https://doi.org/10.1016/J.EJMECH.2022.114740

Choi, D. W., Lim, M. S., Lee, J. W., Chun, W., Lee, S. H., Nam, Y. H. et al. (2015). The cytotoxicity of kahweol in HT-29 human colorectal cancer cells is mediated by apoptosis and suppression of heat shock protein 70 expression. *Biomolecules & Therapeutics*, 23(2), 128. https://doi.org/10.4062/BIOMOLTHER.2014.133

Choi, S., Jung, S., and Ko, K. S. (2018). Effects of coffee extracts with different roasting degrees on antioxidant and anti-inflammatory systems in mice. *Nutrients*, 10(3), 363. https://doi.org/10.3390/NU10030363

Choi, Y., Chang, Y., Ryu, S., Cho, J., Rampal, S., Zhang, Y. et al. (2015). Coffee consumption and coronary artery calcium in young and middle-aged asymptomatic adults. *Heart*, 101(9), 686–691. https://doi.org/10.1136/HEARTJNL-2014-306663

Chouchani, E. T., Kazak, L., and Spiegelman, B. M. (2019). New advances in adaptive thermogenesis: UCP1 and beyond. *Cell Metabolism*, 29(1), 27–37. https://doi.org/10.1016/j.cmet.2018.11.002

Clozel, M., Branchaud, C. L., Tannenbaum, G. S., Dussault, J. H., and Aranda, J. V. (1983). Effect of caffeine on thyroid and pituitary function in newborn rats(40). *Pediatric Research*, 17(7), 592–595. https://doi.org/10.1203/00006450-198307000-00015;KWRD

Cornelis, M. C. (2015). Toward systems epidemiology of coffee and health. *Current Opinion in Lipidology*, 26(1), 20–29. 10.1097/MOL.0000000000000143

Criqui, M. H., Denenberg, J. O., McClelland, R. L., Allison, M. A., Ix, J. H., Guerci, A. et al. (2014). Abdominal aortic calcium, coronary artery calcium, and cardiovascular morbidity and mortality in the multi-ethnic study of atherosclerosis. *Arteriosclerosis, Thrombosis, and Vascular Biology*, 34(7), 1574–1579. https://doi.org/10.1161/ATVBAHA.114.303268/-/DC1

Czachor, J., Miłek, M., Galiniak, S., Stępień, K., Dżugan, M., and Mołoń, M. (2020). Coffee extends yeast chronological lifespan through antioxidant properties. *International Journal of Molecular Sciences*, 21(24), 9510. https://doi.org/10.3390/ijms21249510

De Brito Amaral, L. S., Souza, C. S., Volpini, R. A., Shimizu, M. H. M., De Bragança, A. C., Canale, D. et al. (2018). Previous exercise training reduces markers of renal oxidative stress and inflammation in streptozotocin-induced diabetic female rats. *Journal of Diabetes Research*, 2018(1), 6170352. https://doi.org/10.1155/2018/6170352

de Oliveira, M. R., de Souza, I. C. C., and Fürstenau, C. R. (2020). Mitochondrial protection promoted by the coffee diterpene kahweol in methylglyoxal-treated human neuroblastoma SH-SY5Y cells. *Neurotoxicity Research*, 37(1), 100–110. https://doi.org/10.1007/S12640-019-00107-W/FIGURES/9

del Valle Carranza, A., Saragusti, A., Chiabrando, G. A., Carrari, F., and Asis, R. (2020). Effects of chlorogenic acid on thermal stress tolerance in *C. elegans* via HIF-1, HSF-1 and autophagy. *Phytomedicine*, 66, 153132.

Depaula, J., and Farah, A. (2019). Caffeine consumption through coffee: Content in the beverage, metabolism, health benefits and risks. *Beverages*, 5(2), 37. https://doi.org/10.3390/BEVERAGES5020037

Dhiman, S., and Mukherjee, G. (2022). Biovalorization of fruit waste to gallic acid synthesis through the application of Bacillus haynesii SSRY4 MN031245 tannase. *Materials Today: Proceedings*. https://doi.org/10.1016/J.MATPR.2022.11.398

Di Maso, M., Boffetta, P., Negri, E., La Vecchia, C., and Bravi, F. (2021). Caffeinated coffee consumption and health outcomes in the US population: A dose–response meta-analysis and estimation of disease cases and deaths avoided. *Advances in Nutrition*, 12(4), 1160–1176. https://doi.org/10.1093/ADVANCES/NMAA177

do Carmo Carvalho, D., Brigagão, M. R. P. L., dos Santos, M. H., de Paula, F. B. A., Giusti-Paiva, A., & Azevedo, L. (2011). Organic and conventional *Coffea arabica* L.: A comparative study of the chemical composition and physiological, biochemical and toxicological effects in Wistar rats. *Plant Foods for Human Nutrition*, 66(2), 114–121. https://doi.org/10.1007/s11130-011-0221-9

Doğaner, E. N., Dilek, N. M., and Aytar, E. C. (2025). Green coffee powder as a clean-label functional ingredient in meat systems: Antioxidant performance, molecular docking mechanisms, and technological properties. *Food Biophysics*, 20(4), 162. https://doi.org/10.1007/s11483-025-10035-2

Domènech, M., Serra-Mir, M., Roth, I., Freitas-Simoes, T., Valls-Pedret, C., Cofán, M. et al. (2019). Effect of a walnut diet on office and 24-hour ambulatory blood pressure in elderly individuals. *Hypertension*, 73(5), 1049–1057. https://doi.org/10.1161/HYPERTENSIONAHA.118.12766

Du, W.Y., Chang, C., Zhang, Y., Liu, Y. Y., Sun, K. et al. 2013. High-dose chlorogenic acid induces inflammation reactions and oxidative stress injury in rats without implication of mast cell degranulation. *J Ethnopharmacol.*, 147(1), 74–83. https://doi.org/10.1016/j.jep.2013.01.042

Ebaid, H., Bashandy, S. A. E., Morsy, F. A., Al-Tamimi, J., Hassan, I., and Alhazza, I. M. (2023). Protective effect of gallic acid against thioacetamide-induced metabolic dysfunction of lipids in hepatic and renal toxicity. *Journal of King Saud University – Science*, 35(3), 102531. https://doi.org/10.1016/J.JKSUS.2022.102531

Echeverri, D., Montes, F. R., Cabrera, M., Galán, A., and Prieto, A. (2010). Caffeine's vascular mechanisms of action. *International Journal of Vascular Medicine*, 2010(1), 834060. https://doi.org/10.1155/2010/834060

Epplen, A. S. C., Rothöft, M., Stahlke, S., Theiss, C., & Matschke, V. (2025). Caffeine mitigates ROS accumulation and attenuates motor neuron degeneration in the wobbler mouse model of amyotrophic lateral sclerosis. *Cell Communication and Signaling*, 23(1), 1–17. https://doi.org/10.1186/S12964-025-02415-5/FIGURES/5.

Esquivel, P., Viñas, M., Steingass, C. B., Gruschwitz, M., Guevara, E., Carle, R. et al. (2020). Coffee (*Coffea arabica* L.) by-products as a source of carotenoids and phenolic compounds—Evaluation of varieties with different peel color. *Frontiers in Sustainable Food Systems*, 4, 590597. https://doi.org/10.3389/fsufs.2020.590597

Etaware, P. M., Jonathan, R. D., Egara, O. W., and Ekun, V. S. (2025). Pharmacological, nutritional, medicinal and industrial uses of coffee: Benefits and health risks. *J. Surgical Case Reports and Images*, 8(5), 2690-1897.

Fan, H., Xiong, Y., Huang, Y., Li, W., Xu, C., Feng, X. et al. (2023). Coffee consumption and abdominal aortic calcification among adults with and without hypertension, diabetes, and cardiovascular diseases. *Nutrition, Metabolism and Cardiovascular Diseases*, 33(10), 1960–1968. https://doi.org/10.1016/J.NUMECD.2023.06.013

Fatimatuzzahro, N., Prasetya, R. C., and Anggara, K. D. N. (2022). Robusta coffee (Coffeacanephora) down regulation TNF-α expression in carotid artery endothelial cell of hyperlipidemia rat model. *Trends in Sciences*, 19(4), 2199. https://doi.org/10.48048/tis.2022.2199

Fekete, M., Major, D., Feher, A., Fazekas-Pongor, V., and Lehoczki, A. (2024). Geroscience and pathology: A new frontier in understanding age-related diseases. *Pathology and Oncology Research*, 30, 1611623. https://doi.org/10.3389/pore.2024.1611623

Feld, J. J., Lavoie, É. G., Fausther, M., and Dranoff, J. A. (2015). I drink for my liver, Doc: Emerging evidence that coffee prevents cirrhosis. *F1000Research*, 4, 95. https://doi.org/10.12688/f1000research.6399.1

Feng, S., Zhang, Y., Fu, S., Li, Z., Zhang, J., Xu, Y. et al. (2023). Application of chlorogenic acid as a substitute for antibiotics in multidrug-resistant Escherichia coli-induced mastitis. *International Immunopharmacology*, 114, 109536. https://doi.org/10.1016/J.INTIMP.2022.109536

Fernandes, S. M., Cordeiro, P. M., Watanabe, M., da Fonseca, C. D., and Vattimo, M. de F. F. (2016). The role of oxidative stress in streptozotocin-induced diabetic nephropathy in rats. *Archives of Endocrinology and Metabolism*, 60(5), 443–449. https://doi.org/10.1590/2359-3997000000188

Ferré, S. (2010). Role of the central ascending neurotransmitter systems in the psychostimulant effects of caffeine. *Journal of Alzheimer's Disease*, 20(SUPPL. 1), 35–49. https://doi.org/10.3233/JAD-2010-1400

Gao, Y., Tian, R., Liu, H., Xue, H., Zhang, R., Han, S. et al. (2022). Research progress on intervention effect and mechanism of protocatechuic acid on nonalcoholic fatty liver disease. *Critical Reviews in Food Science and Nutrition*, 62(32), 9053–9075. https://doi.org/10.1080/10408398.2021.1939265

Gebeyehu, G. M., Feleke, D. G., Molla, M. D., and Admasu, T. D. (2020). Effect of habitual consumption of Ethiopian Arabica coffee on the risk of cardiovascular diseases among non-diabetic healthy adults. *Heliyon*, 6(9), e04886. https://doi.org/10.1016/j.heliyon.2020.e04886

Ghimire, K., and Bastakoti, R. R. (2009). Ethnomedicinal knowledge and healthcare practices among the Tharus of Nawalparasi district in central Nepal. *Forest Ecology and Management*, 257(10), 2066–2072. https://doi.org/10.1016/j.foreco.2009.01.039

Godos, J., Pluchinotta, F. R., Marventano, S., Buscemi, S., Volti, G. L., Galvano, F. et al. (2014). Coffee components and cardiovascular risk: Beneficial and detrimental effects. *International Journal of Food Sciences and Nutrition*, 65(8), 925–936. https://doi.org/10.3109/09637486.2014.940287

Gonçalinho, G. H. F., Nascimento, J. R. D. O., Mioto, B. M., Amato, R. V., Moretti, M. A., Strunz, C. M. C. et al. (2022). Effects of coffee on Sirtuin-1, Homocysteine, and cholesterol of healthy adults: Does the coffee powder matter? *Journal of Clinical Medicine*, 11(11), 2985. https://doi.org/10.3390/jcm11112985

Gonzalez-Burgos, E., Es-Safi, N. E., Chapple, B., Woodfin, S., and Moore, W. (2024). The perfect cup? Coffee-derived polyphenols and their roles in mitigating factors affecting type 2 diabetes pathogenesis. *Molecules*, 751, 29(4), 751. https://doi.org/10.3390/MOLECULES29040751

Green, T. D., Park, J., Yin, Q., Fang, S., Crews, A. L., Jones, S. L. et al. (2012). Directed migration of mouse macrophages in vitro involves myristoylated alanine-rich C-kinase substrate (MARCKS) protein. *Journal of Leukocyte Biology*, 92(3), 633–639. https://doi.org/10.1189/JLB.1211604

Gross, B., Pawlak, M., Lefebvre, P., and Staels, B. (2017). PPARs in obesity-induced T2DM, dyslipidaemia and NAFLD. *Nature Reviews Endocrinology*, 13(1), 36–49. https://doi.org/10.1038/NRENDO.2016.135;SUBJMETA

Grosso, G., Godos, J., Galvano, F., and Giovannucci, E. L. (2017). Coffee, caffeine, and health outcomes: An umbrella review. *Annual Review of Nutrition*, 37, 131–156.

Grosso, G., Micek, A., Godos, J., Sciacca, S., Pajak, A., Martínez-González, M. A. et al. (2016). Coffee consumption and risk of all-cause, cardiovascular, and cancer mortality in smokers and non-smokers: A dose-response meta-analysis. *European Journal of Epidemiology*, 31(12), 1191–1205. https://doi.org/10.1007/S10654-016-0202-2/FIGURES/3

Gustafson, B., Hedjazifar, S., Gogg, S., Hammarstedt, A., and Smith, U. (2015). Insulin resistance and impaired adipogenesis. *Trends in Endocrinology and Metabolism*, 26(4), 193–200. https://doi.org/10.1016/j.tem.2015.01.006

Hada, Y., Uchida, H. A., Otaka, N., Onishi, Y., Okamoto, S., Nishiwaki, M. et al. (2020). The protective effect of chlorogenic acid on vascular senescence via the Nrf2/HO-1 pathway. *International Journal of Molecular Sciences*, 21(12), 4527. https://doi.org/10.3390/ijms21124527

Halperin, W., and Jensen, W. A. (1967). Ultrastructural changes during growth and embryogenesis in carrot cell cultures. *Journal of Ultrastructure Research*, 18(3), 428–443. https://doi.org/10.1016/S0022-5320(67)80128-X

Han, L., Wu, T., Zhang, Q., Qi, A., and Zhou, X. (2025). Immune tolerance regulation is critical to immune homeostasis. *Journal of Immunology Research*, 2025(1), 5006201. https://doi.org/10.1155/JIMR/5006201

Hassane Hamadou, A., Zhang, J., Chao, C., and Xu, B. (2022). Stability of rutin using pectin-chitosan dual coating nanoliposomes. *LWT*, 170, 114084. https://doi.org/10.1016/J.LWT.2022.114084

Higdon, J. V., and Frei, B. (2006). Coffee and health: A review of recent human research. *Critical Reviews in Food Science and Nutrition*, 46(2), 101–123.

Higgins, L. G., Cavin, C., Itoh, K., Yamamoto, M., and Hayes, J. D. (2008). Induction of cancer chemopreventive enzymes by coffee is mediated by transcription factor Nrf2. Evidence that the coffee-specific diterpenes cafestol and kahweol confer protection against acrolein. *Toxicology and Applied Pharmacology*, 226(3), 328–337. https://doi.org/10.1016/J.TAAP.2007.09.018

Ho, L., Varghese, M., Wang, J., Zhao, W., Chen, F., Knable, L. A. et al. (2012). Dietary supplementation with decaffeinated green coffee improves diet-induced insulin resistance and brain energy metabolism in mice. *Nutritional Neuroscience*, 15(1), 37–45. https://doi.org/10.1179/1476830511Y.0000000027

Hollowell, J. G., Staehling, N. W., Dana Flanders, W., Harry Hannon, W., Gunter, E. W., Spencer, C. A. et al. (2002). Serum TSH, T4, and thyroid antibodies in the United States population (1988 to 1994): National Health and Nutrition Examination Survey (NHANES III). *The Journal of Clinical Endocrinology & Metabolism*, 87(2), 489–499. https://doi.org/10.1210/JCEM.87.2.8182

Holvoet, H., Long, D. M., Yang, L., Choi, J., Marney, L., Poeck, B. et al. (2023). Chlorogenic acids, acting via calcineurin, are the main compounds in Centella asiatica extracts that mediate resilience to chronic stress in Drosophila melanogaster. *Nutrients*, 15(18), 4016. https://doi.org/10.3390/nu15184016

Hong, E. P., MacDonald, M. E., Wheeler, V. C., Jones, L., Holmans, P., Orth, M. et al. (2021). Huntington's disease pathogenesis: Two sequential components. *Journal of Huntington's Disease*, 10(1), 35–51. https://doi.org/10.3233/JHD-200427

Hori, A., Kasai, H., Kawai, K., Nanri, A., Sato, M., Ohta, M. et al. (2014). Coffee intake is associated with lower levels of oxidative DNA damage and decreasing body iron storage in healthy women. *Nutrition and Cancer*, 66(6), 964–969. https://doi.org/10.1080/01635581.2014.932398

Hou, Y., Xue, P., Bai, Y., Liu, D., Woods, C. G., Yarborough, K. et al. (2012). Nuclear factor erythroid-derived factor 2-related factor 2 regulates transcription of CCAAT/enhancer-binding protein β during adipogenesis. *Free Radical Biology and Medicine*, 52(2), 462–472. https://doi.org/10.1016/J.FREERADBIOMED.2011.10.453

Hu, Y., Feng, W., Chen, H., Shi, H., Jiang, L., Zheng, X. et al. (2021). Effect of selenium on thyroid autoimmunity and regulatory T cells in patients with Hashimoto's thyroiditis: A prospective randomized-controlled trial. *Clinical and Translational Science*, 14(4), 1390–1402. https://doi.org/10.1111/CTS.12993

Huber, W. W., Prustomersky, S., Delbanco, E., Uhl, M., Scharf, G., Turesky, R. J. et al. (2002a). Enhancement of the chemoprotective enzymes glucuronosyl transferase and glutathione transferase in specific organs of the rat by the coffee components kahweol and cafestol. *Archives of Toxicology*, 76(4), 209–217.

Huber, W. W., Scharf, G., Rossmanith, W., Prustomersky, S., Grasl-Kraupp, B., Peter, B. et al. (2002b). The coffee components kahweol and cafestol induce γ-glutamylcysteine synthetase, the rate limiting enzyme of chemoprotective glutathione synthesis, in several organs of the rat. *Archives of Toxicology*, 75(11), 685–694. https://doi.org/10.1007/s00204-001-0295-5

Huber, W. W., Rossmanith, W., Grusch, M., Haslinger, E., Prustomersky, S., Peter-Vörösmarty, B. et al. (2008). Effects of coffee and its chemopreventive components kahweol and cafestol on cytochrome P450 and sulfotransferase in rat liver. *Food and Chemical Toxicology*, 46(4), 1230–1238. https://doi.org/10.1016/J.FCT.2007.09.094

Hussein, M. M. A., Samy, M., Arisha, A. H., Saadeldin, I. M., and Alshammari, G. M. (2020). Anti-obesity effects of individual or combination treatment with Spirulina platensis and green coffee bean aqueous extracts in high-fat diet-induced obese rats. *All Life*, 13(1), 328–338. https://doi.org/10.1080/26895293.2020.1781698

Hutachok, N., Koonyosying, P., Pankasemsuk, T., Angkasith, P., Chumpun, C., Fucharoen, S. et al. (2021). Chemical analysis, toxicity study, and free-radical scavenging and iron-binding assays involving coffee (*Coffea arabica*) extracts. *Molecules*, 26(14), 4169.

Hwang, J. H., Kim, K. J., Ryu, S. J., and Lee, B. Y. (2016). Caffeine prevents LPS-induced inflammatory responses in RAW264.7 cells and zebrafish. *Chemico-Biological Interactions*, 248, 1–7. https://doi.org/10.1016/J.CBI.2016.01.020

Iqbal, M. J., Kabeer, A., Abbas, Z., Siddiqui, H. A., Calina, D., Sharifi-Rad, J. et al. (2024). Interplay of oxidative stress, cellular communication and signaling pathways in cancer. *Cell Communication and Signaling*, 22(1), 1–16. https://doi.org/10.1186/S12964-023-01398-5/TABLES/3

Iris, M., Tsou, P. S., and Sawalha, A. H. (2018). Caffeine inhibits STAT1 signaling and downregulates inflammatory pathways involved in autoimmunity. *Clinical Immunology*, 192, 68–77. https://doi.org/10.1016/J.CLIM.2018.04.008

Ismail, N. S., Zaidan, U. H., Shamsi, S., Abd Gani, S. S., and Nillian, E. (2024). Proximate composition and phytochemical analysis of Malaysian Liberica sp. coffee bean and its pulp. *Pertanika Journal of Tropical Agricultural Science*, 47(2).

Iwamoto, H., Izumi, K., Natsagdorj, A., Naito, R., Makino, T., Kadomoto, S. et al. (2019). Coffee diterpenes kahweol acetate and cafestol synergistically inhibit the proliferation and migration of prostate cancer cells. *The Prostate*, 79(5), 468–479. https://doi.org/10.1002/PROS.23753

Jansen, A. H. P., van Hal, M., op den Kelder, I. C., Meier, R. T., de Ruiter, A. A., Schut, M. H. et al. (2017). Frequency of nuclear mutant huntingtin inclusion formation in neurons and glia is cell-type-specific. *Glia*, 65(1), 50–61. https://doi.org/10.1002/GLIA.23050

Jia, Q., Yang, R., Liu, X. F., Ma, S. F., and Wang, L. (2019). Genistein attenuates renal fibrosis in streptozotocin-induced diabetic rats. *Molecular Medicine Reports*, 19(1), 423–431. https://doi.org/10.3892/MMR.2018.9635/DOWNLOAD

Jiang, S. Q., Chen, Z. L., Zhang, S., Ye, J. L., & Wang, Y. B. (2023). Protective effects of protocatechuic acid on growth performance, intestinal barrier and antioxidant capacity in broilers challenged with lipopolysaccharide. *Animal*, 17(1), 100693. https://doi.org/10.1016/J.ANIMAL.2022.100693

Jiao, Y., Wu, L., Huntington, N. D., and Zhang, X. (2020). Crosstalk between gut microbiota and innate immunity and its implication in autoimmune diseases. *Frontiers in Immunology*, 11, 503395. https://doi.org/10.3389/FIMMU.2020.00282/FULL

Jung, S., Kim, M. H., Park, J. H., Jeong, Y., and Ko, K. S. (2017). Cellular antioxidant and anti-inflammatory effects of coffee extracts with different roasting levels. *Journal of Medicinal Food*, 20(6), 626–635. https://doi.org/10.1089/JMF.2017.3935

Júnior, D. S. T. (2020). Environmental and individual factors associated with protection and predisposition to autoimmune diseases. *International Journal of Health Sciences*, 14(6), 13. https://pmc.ncbi.nlm.nih.gov/articles/PMC7644454/

Kalthoff, S., Ehmer, U., Freiberg, N., Manns, M. P., and Strassburg, C. P. (2010). Coffee induces expression of glucuronosyltransferases by the aryl hydrocarbon receptor and Nrf2 in liver and stomach. *Gastroenterology*, 139(5), 1699–1710.e2. https://doi.org/10.1053/J.GASTRO.2010.06.048

Kalthoff, S., Paulusch, S., Rupp, A., Holdenrieder, S., Hartmann, G., and Strassburg, C. P. (2020). The coffee ingredients caffeic acid and caffeic acid phenylethyl ester protect against irinotecan-induced leukopenia and oxidative stress response. *British Journal of Pharmacology*, 177(18), 4193–4208. https://doi.org/10.1111/BPH.15162

Kamat, P. K., Kalani, A., Rai, S., Swarnkar, S., Tota, S., Nath, C. et al. (2016). Mechanism of oxidative stress and synapse dysfunction in the pathogenesis of Alzheimer's disease: Understanding the therapeutics strategies. *Molecular Neurobiology*, 53(1), 648–661. https://doi.org/10.1007/S12035-014-9053-6/TABLES/1

Kantarcıoğlu, M., Yıldırım, G., Oktar, P. A., Yanbakan, S., Özer, Z. B., Sarıca, D. Y. et al. (2023). Coffee-derived exosome-like nanoparticles: Are they the secret heroes? *The Turkish Journal of Gastroenterology*, 34(2), 161. 10.5152/tjg.2022.21895

Kawasaki, Y., Imaizumi, T., Matsuura, H. et al. (2008). Renal expression of alpha-smooth muscle actin and c-Met in children with Henoch–Schönlein purpura nephritis. *Pediatric Nephrology*, 23(6), 913–919. https://doi.org/10.1007/s00467-008-0749-6

Khadrawy, Y. A., Salem, A. M., El-Shamy, K. A., Ahmed, E. K., Fadl, N. N., and Hosny, E. N. (2017). Neuroprotective and therapeutic effect of caffeine on the rat model of Parkinson's disease induced by rotenone. *Journal of Dietary Supplements*, 14(5), 553–572. https://doi.org/10.1080/19390211.2016.1275916

Khayatan, D., Razavi, S. M., Arab, Z. N., Khanahmadi, M., Momtaz, S., Butler, A. E., ... Sahebkar, A. (2022). Regulatory effects of statins on SIRT1 and other sirtuins in cardiovascular diseases. *Life*, 12(5), 760.

Kilani-Jaziri, S., Mokdad-Bzeouich, I., Krifa, M., Nasr, N., Ghedira, K., and Chekir-Ghedira, L. (2017). Immunomodulatory and cellular anti-oxidant activities of caffeic, ferulic, and p-coumaric phenolic acids: A structure–activity relationship study. *Drug and Chemical Toxicology*, 40(4), 416–424. https://doi.org/10.1080/01480545.2016.1252919

Kobayashi, T., Maruyama, T., Yoneda, T., Miyai, H., Azuma, T., Tomofuji, T. et al. (2020). Effects of coffee intake on oxidative stress during aging-related alterations in periodontal tissue. *In vivo*, 34(2), 615–622. https://doi.org/10.21873/invivo.11815

Kolberg, M., Pedersen, S., Mitake, M., Holm, K. L., Bøhn, S. K., Blomhoff, H. K. et al. (2016). Coffee inhibits nuclear factor-kappa B in prostate cancer cells and xenografts. *The Journal of Nutritional Biochemistry*, 27, 153–163. https://doi.org/10.1016/J.JNUTBIO.2015.08.028

Korkmaz, E. D., and Temizci, B. (2025). Coffee bean-derived extracellular vesicles exhibit anti-cancer effects on neuroblastoma cells. *Journal of High School Science*, 9(3), 306–322.

Król, K., Gantner, M., Tatarak, A., and Hallmann, E. (2020). The content of polyphenols in coffee beans as roasting, origin and storage effect. *European Food Research and Technology*, 246(1), 33–39. https://doi.org/10.1007/S00217-019-03388-9/FIGURES/1

Kusumah, J., and Gonzalez de Mejia, E. (2022). Coffee constituents with antiadipogenic and antidiabetic potentials: A narrative review. *Food and Chemical Toxicology*, 161, 112821. https://doi.org/10.1016/J.FCT.2022.112821

Lam, L. K., Sparnins, V. L., & Wattenberg, L. W. (1987). Effects of derivatives of kahweol and cafestol on the activity of glutathione S-transferase in mice. *Journal of Medicinal Chemistry*, 30(8), 1399–1403.

Lamorde, M., Tabuti, J.R.S., Obua, C., Kukunda-Byobona, C., Lanyero, H. et al. (2010). Medicinal plants used by traditional medicine practitioners for the treatment of HIV/AIDS and related conditions in Uganda. *J Ethnopharmacol*, 130(1), 43–53. https://doi.org/10.1016/j.jep.2010.04.004

Lara-Guzmán, O. J., Arango-González, Á., Rivera, D. A., Munoz-Durango, K., and Sierra, J. A. (2024). The colonic polyphenol catabolite dihydroferulic acid (DHFA) regulates macrophages activated by oxidized LDL, 7-ketocholesterol, and LPS switching from pro- to anti-inflammatory mediators. *Food & Function*, 15(20), 10399–10413. https://doi.org/10.1039/D4FO02114B

Lee, K. J., and Jeong, H. G. (2007). Protective effects of kahweol and cafestol against hydrogen peroxide-induced oxidative stress and DNA damage. *Toxicology Letters*, 173(2), 80–87. https://doi.org/10.1016/j.toxlet.2007.06.008

Lee, K. W., Im, J. Y., Woo, J. M., Grosso, H., Kim, Y. S., Cristovao, A. C. et al. (2013). Neuroprotective and anti-inflammatory properties of a coffee component in the MPTP model of Parkinson's disease. *Neurotherapeutics*, 10(1), 143–153. https://doi.org/10.1007/s13311-012-0165-2

Leung, A. M., and Braverman, L. E. (2014). Consequences of excess iodine. *Nature Reviews Endocrinology*, 10(3), 136–142. https://doi.org/10.1038/nrendo.2013.251

Li, D., Li, Z., Dong, L., Zhang, Y., Lu, Y., Wang, J. et al. (2023). Coffee prevents IQ-induced liver damage by regulating oxidative stress, inflammation, endoplasmic reticulum stress, autophagy, apoptosis, and the MAPK/NF-κB signaling pathway in zebrafish. *Food Research International*, 169, 112946. https://doi.org/10.1016/j.foodres.2023.112946

Li, Y., Nishihara, E., and Kakudo, K. (2011). Hashimoto's thyroiditis: Old concepts and new insights. *Current Opinion in Rheumatology*, 23(1), 102–107. https://doi.org/10.1097/BOR.0b013e328341378c

Lim, D., Chang, J., Ahn, J., and Kim, J. (2020). Conflicting effects of coffee consumption on cardiovascular diseases: Does coffee consumption aggravate pre-existing risk factors? *Processes*, 8(4), 438. https://doi.org/10.3390/pr8040438

Lima, C. S., Spindola, D. G., Bechara, A., Garcia, D. M., Palmeira-dos-Santos, C., Peixoto-da-Silva, J. et al. (2017). Cafestol, a diterpene molecule found in coffee, induces leukemia cell death. *Biomedicine & Pharmacotherapy*, 92, 1045–1054. https://doi.org/10.1016/j.biopha.2017.05.109

Lin, D., Wang, L., Yan, S., Zhang, Q., Zhang, J. H., and Shao, A. (2019). The role of oxidative stress in common risk factors and mechanisms of cardio-cerebrovascular ischemia and depression. *Oxidative Medicine and Cellular Longevity*, 2019, 2491927. https://doi.org/10.1155/2019/2491927

Lin, Y. S., Weibel, J., Landolt, H. P., Santini, F., Garbazza, C., Kistler, J. et al. (2022). Time to recover from daily caffeine intake. *Frontiers in Nutrition*, 8, 787225. https://doi.org/10.3389/fnut.2021.787225

Liu, C.-W., Tsai, H.-C., Huang, C.-C., Tsai, C.-Y., Su, Y.-B., Lin, M.-W. et al. (2018). Effects and mechanisms of caffeine to improve immunological and metabolic abnormalities in diet-induced obese rats. *American Journal of Physiology-Endocrinology and Metabolism*, 314(5), E433–E447. https://doi.org/10.1152/ajpendo.00094.2017

Liu, X., Zhai, X., Wang, X., Zhu, X., Wang, Z., Jiang, Z., ... Chen, Z. (2025). Nuclear factor erythroid 2 related factor 2 activator ddo-1039 ameliorates podocyte injury in diabetic kidney disease via suppressing oxidative stress, inflammation, and ferroptosis. *Antioxidants & Redox Signaling*, 42(13-15), 787–806.

Lonati, E., Carrozzini, T., Bruni, I., Mena, P., Botto, L., Cazzaniga, E. et al. (2022). Coffee-derived phenolic compounds activate Nrf2 antioxidant pathway in I/R injury in vitro model: A nutritional approach preventing age-related damages. *Molecules*, 27(3), 1049. https://doi.org/10.3390/molecules27031049

Lopez-Gonzalez, R., Lu, Y., Gendron, T. F., Karydas, A., Tran, H., Yang, D. et al. (2016). Poly(GR) in C9ORF72-related ALS/FTD compromises mitochondrial function and increases oxidative stress and DNA damage in iPSC-derived motor neurons. *Neuron*, 92(2), 383–391. https://doi.org/10.1016/j.neuron.2016.09.015

Lv, Q. Z., Long, J. T., Gong, Z. F., Nong, K. Y., Liang, X. M., Qin, T. et al. (2021). Current state of knowledge on the antioxidant effects and mechanisms of action of polyphenolic compounds. *Natural Product Communications*, 16(7), 1934578X211027745. https://doi.org/10.1177/1934578X211027745

Mackintosh, C., Yuan, C., Ou, F. S., Zhang, S., Niedzwiecki, D., Chang, I. W. et al. (2020). Association of coffee intake with survival in patients with advanced or metastatic colorectal cancer. *JAMA Oncology*, 6(11), 1713–1721. https://doi.org/10.1001/jamaoncol.2020.3938

Maki, C., Funakoshi-Tago, M., Aoyagi, R., Ueda, F., Kimura, M., Kobata, K. et al. (2017). Coffee extract inhibits adipogenesis in 3T3-L1 preadipocyes by interrupting insulin signaling through the downregulation of IRS1. *PLOS ONE*, 12(3), e0173264. https://doi.org/10.1371/JOURNAL.PONE.0173264

Malakar, V. K., Roy, D., Malakar, C. C., Khetmalis, Y. M., Mali, P. C., and Poddar, N. K. (2025). Targeting IDO1 in Huntington's disease: Network pharmacology and preclinical evidence from Coffea arabica. *Neurochemical Research*, 50(4), 1–23. https://doi.org/10.1007/S11064-025-04509-5/FIGURES/18

Maldonado-Ruiz, R., Montalvo-Martínez, L., Fuentes-Mera, L., and Camacho, A. (2017). Microglia activation due to obesity programs metabolic failure leading to type two diabetes. *Nutrition and Diabetes*, 7(3), e254–e254. https://doi.org/10.1038/NUTD.2017.10;TECHMETA

Marcinek, K., Luzak, B., and Rozalski, M. (2024). The effects of caffeine on blood platelets and the cardiovascular system through adenosine receptors. *International Journal of Molecular Sciences*, 25(16), 8905. https://doi.org/10.3390/IJMS25168905

Martins, B. C., Soares, A. C., Martins, F. F., Resende, A. de C., Inada, K. O. P., Souza-Mello, V. et al. (2023). Coffee consumption prevents obesity-related comorbidities and attenuates brown adipose tissue whitening in high-fat diet-fed mice. *The Journal of Nutritional Biochemistry*, 117, 109336. https://doi.org/10.1016/J.JNUTBIO.2023.109336

Matosinhos, R.C., Bezerra, J.P., Barros, C.H., Fernandes Pereira Ferreira, Bernardes, A.C., Coelho, G.B. et al. (2022). Coffea arabica extracts and their chemical constituents in a murine model of gouty arthritis: How they modulate pain and inflammation. *J Ethnopharmacol*, 284, 114778.

Mesas, A. E., Leon-Muñoz, L. M., Rodriguez-Artalejo, F., and Lopez-Garcia, E. (2011). The effect of coffee on blood pressure and cardiovascular disease in hypertensive individuals: A systematic review and meta-analysis. The *American Journal of Clinical Nutrition*, 94(4), 1113–1126. https://doi.org/10.3945/AJCN.111.016667

Micek, A., Grosso, G., Polak, M., Kozakiewicz, K., Tykarski, A. et al. 2018. Association between tea and coffee consumption and prevalence of metabolic syndrome in Poland – Results from the WOBASZ II study (2013–2014). *Int J Food Sci Nutr*., 69(3), 358–368. https://doi.org/10.1080/09637486.2017.1362690

Mikami, Y., and Yamazawa, T. (2015). Chlorogenic acid, a polyphenol in coffee, protects neurons against glutamate neurotoxicity. *Life Sciences*, 139, 69–74. https://doi.org/10.1016/J.LFS.2015.08.005

Min, H., Youn, E., and Shim, Y. H. (2021). Long-term caffeine intake exerts protective effects on intestinal aging by regulating vitellogenesis and mitochondrial function in an aged Caenorhabditis elegans model. *Nutrients*, 13(8), 2517. https://doi.org/10.3390/nu13082517

Minari, T. P., Vilela-Martin, J. F., Yugar-Toledo, J. C., and Pisani, L. P. (2025). Correlations between coffee intake, glycemic control, cardiovascular risk, and sleep in type 2 diabetes and hypertension: A 12-month observational study. *Biomedicines*, 13(8), 1875. https://doi.org/10.3390/BIOMEDICINES13081875/S1

Miranda, A. M., Steluti, J., Goulart, A. C., Benseñor, I. M., Lotufo, P. A., and Marchioni, D. M. (2018). Coffee consumption and coronary artery calcium score: Cross-sectional results of ELSA-Brasil (Brazilian longitudinal study of adult health). *Journal of the American Heart Association*, 7(7). https://doi.org/10.1161/JAHA.117.007155

Mitsumoto, H., Garofalo, D. C., Santella, R. M., Sorenson, E. J., Oskarsson, B., Fernandes, J. et al. (2020). Plasma creatinine and oxidative stress biomarkers in amyotrophic lateral sclerosis. *Amyotrophic Lateral Sclerosis and Frontotemporal Degeneration*, 21(3–4), 263–272. https://doi.org/10.1080/21678421.2020.1746810

Moeenfard, M., Cortez, A., Machado, V., Costa, R., Luís, C., Coelho, P. et al. (2016). Anti-angiogenic properties of cafestol and kahweol palmitate diterpene esters. *Journal of Cellular Biochemistry*, 117(12), 2748–2756. https://doi.org/10.1002/JCB.25573

Mohamed, H. E., Asker, M. E., Shaheen, M. A., Eissa, R. G., & Younis, N. N. (2021). Alleviation of fructose-induced Alzheimer's disease in rats by pioglitazone and decaffeinated green coffee bean extract. *Journal of Food Biochemistry*, 45(5), e13715. https://doi.org/10.1111/jfbc.13715

Mohamed, H. E., Asker, M. E., Younis, N. N., Shaheen, M. A., and Eissa, R. G. (2020). Modulation of brain insulin signaling in Alzheimer's disease: New insight on the protective role of green coffee bean extract. *Nutritional Neuroscience*, 23(1), 27–36. https://doi.org/10.1080/1028415X.2018.1468535

Mu, J., Zhuang, X., Wang, Q. et al. (2014). Interspecies communication between plant and mouse gut host cells through edible plant-derived exosome-like nanoparticles. *Molecular Nutrition & Food Research*, 58(7), 1561–1573. https://doi.org/10.1002/mnfr.201300729

Muchtaridi, M., Lestari, D., Khairul Ikram, N. K., Gazzali, A. M., Hariono, M., and Wahab, H. A. (2021). Decaffeination and neuraminidase inhibitory activity of arabica green coffee (*Coffea arabica*) beans: Chlorogenic acid as a potential bioactive compound. *Molecules*, 26(11), 3402. https://doi.org/10.3390/molecules26113402

Mude, H., Maroju, P. A., Balapure, A., Ganesan, R., and Ray Dutta, J. (2022). Water-soluble caffeic acid–dopamine acid–base complex exhibits enhanced bactericidal, antioxidant, and anticancer properties. *Food Chemistry*, 374, 131830. https://doi.org/10.1016/j.foodchem.2021.131830

Murai, T., and Matsuda, S. (2023). The chemopreventive effects of chlorogenic acids, phenolic compounds in coffee, against inflammation, cancer, and neurological diseases. *Molecules*, 28(5), 2381. https://doi.org/10.3390/molecules28052381

Narayan, P., Holmström, K. M., Kim, D. H., Whitcomb, D. J., Wilson, M. R., St. George-Hyslop, P. et al. (2014). Rare individual amyloid-β oligomers act on astrocytes to initiate neuronal damage. *Biochemistry*, 53(15), 2442–2453. https://doi.org/10.1021/bi401606f

Nieber, K. (2017). The impact of coffee on health. *Planta Medica*, 83(16), 1256–1263. https://doi.org/10.1055/s-0043-115007

Nigra, A. D., de Almeida Bauer Guimarães, D., Prucca, C. G., Freitas-Silva, O., Teodoro, A. J., and Gil, G. A. (2021). Antitumor effects of freeze-dried robusta coffee (*Coffea canephora*) extracts on breast cancer cell lines. *Oxidative Medicine and Cellular Longevity*, 2021, 5572630. https://doi.org/10.1155/2021/5572630

Nikitin, A. G., Navitskas, S., and Gordon, L. A. N. (2008). Effect of varying doses of caffeine on life span of *Drosophila melanogaster*. *The Journals of Gerontology: Series A, Biological Sciences and Medical Sciences*, 63(2), 149–150. https://doi.org/10.1093/gerona/63.2.149

Nikitina, D., Chen, Z., Vallis, K., Poll, A., Ainsworth, P., Narod, S. A., et al. (2015). Relationship between caffeine and levels of DNA repair and oxidative stress in women with and without a BRCA1 mutation. *Journal of Nutrigenetics and Nutrigenomics*, 8(4–6), 174–184. https://doi.org/10.1159/000439110

Nogaim, Q. A., Bugata, L. S. P., Reddy, U. A., and Kumari, I. (2020). Protective effect of Yemeni green coffee powder against the oxidative stress induced by Ochratoxin A. *Toxicology Reports*, 7, 142–148.

Nzekoue, F. K., Angeloni, S., Navarini, L., Angeloni, C., Freschi, M., Hrelia, S. et al. (2020). Coffee silverskin extracts: Quantification of 30 bioactive compounds by a new HPLC-MS/MS method and evaluation of their antioxidant and antibacterial activities. *Food Research International*, 133, 109128. https://doi.org/10.1016/j.foodres.2020.109128

Oboh, G., Ojueromi, O. O., Ademosun, A. O., Omayone, T. P., Oyagbemi, A. A., Ajibade, T. O. et al. (2021). Effects of caffeine and caffeic acid on selected biochemical parameters in L-NAME-induced hypertensive rats. *Journal of Food Biochemistry*, 45(3), e13384. https://doi.org/10.1111/jfbc.13384

Oh, S. H., Hwang, Y. P., Choi, J. H., Jin, S. W., Lee, G. H., Han, E. H. et al. (2018). Kahweol inhibits proliferation and induces apoptosis by suppressing fatty acid synthase in HER2-overexpressing cancer cells. *Food and Chemical Toxicology*, 121, 326–335. https://doi.org/10.1016/j.fct.2018.09.008

Olthof, M. R., Hollman, P. C. H., and Katan, M. B. (2001). Chlorogenic acid and caffeic acid are absorbed in humans. *The Journal of Nutrition*, 131(1), 66–71. https://doi.org/10.1093/jn/131.1.66

Orbán, C., Vásárhelyi, Z., Bajnok, A., Sava, F., and Toldi, G. (2018). Effects of caffeine and phosphodiesterase inhibitors on activation of neonatal T lymphocytes. *Immunobiology*, 223(11), 627–633. https://doi.org/10.1016/j.imbio.2018.07.008

Orefice, V., Ceccarelli, F., Barbati, C., Lucchetti, R., Olivieri, G., Cipriano, E. et al. (2020). Caffeine intake influences disease activity and clinical phenotype in systemic lupus erythematosus patients. *Lupus*, 29(11), 1377–1384. https://doi.org/10.1177/0961203320941920

Osama, H., Abdelrahman, M. A., Madney, Y. M., Harb, H. S., Saeed, H., and Abdelrahim, M. E. A. (2021). Coffee and type 2 diabetes risk: Is the association mediated by adiponectin, leptin, C-reactive protein, or interleukin-6? A systematic review and meta-analysis. *International Journal of Clinical Practice*, 75(6), e13983. https://doi.org/10.1111/ijcp.13983

Othman, N. H., Ismawi, H. R., Zainudin, M. M., & Fuaat, A. A. (2025). Antidiabetic effects of coffee enriched with maca and marine collagen peptide (Blackbelt®) in a type 2 diabetes mellitus rat model. *IIUM Medical Journal Malaysia*, 24(1). https://doi.org/10.31436/imjm.v24i01.2622

Páez, E., Galvis-Pérez, Y., Pereañez, J. A., Preciado, L. M., and Henao-Castañeda, I. C. (2025). *Coffea arabica* extracts and metabolites with potential inhibitory activity of the major enzymes in *Bothrops asper* venom. *Pharmaceuticals*, 18(8), 1151. https://doi.org/10.3390/ph18081151

Palatini, P., Fania, C., Mos, L., Garavelli, G., Mazzer, A., Cozzio, S. et al. (2016). Coffee consumption and risk of cardiovascular events in hypertensive patients: Results from the HARVEST study. *International Journal of Cardiology*, 212, 131–137. https://doi.org/10.1016/j.ijcard.2016.03.006

Pangestuti, R., Haq, M., Rahmadi, P., and Chun, B. S. (2021). Nutritional value and biofunctionalities of two edible green seaweeds (*Ulva lactuca* and *Caulerpa racemosa*) from Indonesia by subcritical water hydrolysis. *Marine Drugs*, 19(10), 578. https://doi.org/10.3390/md19100578

Parras, P., Martínez-Tomé, M., Jiménez, A. M., and Murcia, M. A. (2007). Antioxidant capacity of coffees of several origins brewed following three different procedures. *Food Chemistry*, 102(3), 582–592. https://doi.org/10.1016/j.foodchem.2006.05.037

Patay, É. B., Alberti, Á., Csernák, O., Stranczinger, S., and Papp, N. (2018). Comparative phytochemical analysis of *Coffea benghalensis*, *Coffea arabica*, and *Coffea liberica*. *Asian Pacific Journal of Tropical Medicine*, 11(8), 480–485. https://doi.org/10.4103/1995-7645.240084

Paz-Graniel, I., Babio, N., Becerra-Tomás, N., Toledo, E., Camacho-Barcia, L., Corella, D. et al. (2021). Association between coffee consumption and total dietary caffeine intake with cognitive functioning in an elderly Mediterranean population. *European Journal of Nutrition*, 60(5), 2381–2396. https://doi.org/10.1007/s00394-020-02415-w

Pietzner, M., Köhrle, J., Lehmphul, I., Budde, K., Kastenmüller, G., Brabant, G., ... Homuth, G. (2019). A thyroid hormone-independent molecular fingerprint of 3,5-Diiodothyronine suggests a strong relationship with coffee metabolism in humans. *Thyroid*, 29(12), 1743–1754.

Pintać, D., Bekvalac, K., Mimica-Dukić, N., Rašeta, M., Anđelić, N., Lesjak, M. et al. (2022). Comparison study between popular brands of coffee, tea and red wine regarding polyphenols content and antioxidant activity. *Food Chemistry Advances*, 1, 100030. https://doi.org/10.1016/j.focha.2022.100030

Poole, R., Kennedy, O. J., Roderick, P., Fallowfield, J. A., and Hayes, P. C. (2017). Coffee consumption and health: Umbrella review of meta-analyses of multiple health outcomes. *BMJ*, 359, j5024. https://doi.org/10.1136/bmj.j5024

Priftis, A., Panagiotou, E. M., Lakis, K., Plika, C., Halabalaki, M., Ntasi, G. et al. (2018). Roasted and green coffee extracts show antioxidant and cytotoxic activity in a cell-specific manner. *Food and Chemical Toxicology*, 114, 119–127. https://doi.org/10.1016/j.fct.2018.02.029

Puigdellívol, M., Saavedra, A., and Pérez-Navarro, E. (2016). Cognitive dysfunction in Huntington's disease: Mechanisms and therapeutic strategies beyond BDNF. *Brain Pathology*, 26(6), 752–771.

Rachmawati, E., Sargowo, D., Rohman, M. S., and Kalsum, U. (2022). In silico modelling, regulation of cell viability and anti atherosclerotic effect in macrophage by decaffeinated coffee and green tea extract. *Pharmacognosy Journal*, 14(1), 46–55.

Ramos-Lopez, O. (2025). Personalizing dietary polyphenols for health maintenance and disease management: A nutrigenetic approach. *Current Nutrition Reports*, 14(1), 29.

Raposo, G., and Stoorvogel, W. (2013). Extracellular vesicles: Exosomes, microvesicles, and friends. *Journal of Cell Biology*, 200(4), 373–383. https://doi.org/10.1083/jcb.201211138

Rebelo-Marques, A., De Sousa Lages, A., Andrade, R., Ribeiro, C. F., Mota-Pinto, A., Carrilho, F. et al. (2018). Aging hallmarks: The benefits of physical exercise. *Frontiers in Endocrinology*, 9, 258. https://doi.org/10.3389/fendo.2018.00258

Reggiani, C. (2021). Caffeine as a tool to investigate sarcoplasmic reticulum and intracellular calcium dynamics in human skeletal muscles. *Journal of Muscle Research and Cell Motility*, 42(2), 281–289. https://doi.org/10.1007/s10974-020-09574-7

Rekatsina, M., Paladini, A., Piroli, A., Zis, P., Pergolizzi, J. V., and Varrassi, G. (2020). Pathophysiology and therapeutic perspectives of oxidative stress and neurodegenerative diseases: A narrative review. *Advances in Therapy*, 37(1), 113–139. https://doi.org/10.1007/s12325-019-01148-5

Rhee, Y., Choi, Y., Park, J., Park, H. R., Kim, K., and Kim, Y. H. (2022). Association between coffee consumption and periodontal diseases: A systematic review and meta-analysis. *BMC Oral Health*, 22(1), 272. https://doi.org/10.1186/s12903-022-02310-2

Rizvi, S., Raza, S. T., and Mahdi, F. (2014). Telomere length variations in aging and age-related diseases. *Current Aging Science*, 7(3), 161–167. https://doi.org/10.2174/1874609808666150122153151

Rogozinska, M., Lisiecki, K., Czarnocki, Z., & Biesaga, M. (2023). Antioxidant activity of sulfate metabolites of chlorogenic acid. *Applied Sciences*, 13(4), 2192. https://doi.org/10.3390/app13042192

Romualdo, G. R., de Souza, I. P., de Souza, L. V., Prata, G. B., de Campos Fraga-Silva, T. F., Sartori, A., ... Barbisan, L. F. (2021). Beneficial effects of anthocyanin-rich peels of Myrtaceae fruits on chemically-induced liver fibrosis and carcinogenesis in mice. *Food Research International*, 139, 109964.

Ross, G. W., and Petrovitch, H. (2001). Current evidence for neuroprotective effects of nicotine and caffeine against Parkinson's disease. *Drugs & Aging*, 18(11), 797–806. https://doi.org/10.2165/00002512-200118110-00001

Ruan, P., Yang, M., Lv, X., Shen, K., Chen, Y., Li, H. et al. (2024). Metabolic shifts during coffee consumption refresh the immune response: Insight from comprehensive multiomics analysis. *MedComm*, 5(7), e617. https://doi.org/10.1002/mco2.617

Sadi, G., Şahin, G., and Bostanci, A. (2019). Modulation of renal insulin signaling pathway and antioxidant enzymes with streptozotocin-induced diabetes: Effects of resveratrol. *Medicina*, 55(1), 3. https://doi.org/10.3390/medicina55010003

Safe, S., Kothari, J., Hailemariam, A., Upadhyay, S., Davidson, L. A., and Chapkin, R. S. (2023). Health benefits of coffee consumption for cancer and other diseases and mechanisms of action. *International Journal of Molecular Sciences*, 24(3), 2706. https://doi.org/10.3390/ijms24032706

Said Ahmed, M., Hung, W. Y., Zu, J. S., Hockberger, P., and Siddique, T. (2000). Increased reactive oxygen species in familial amyotrophic lateral sclerosis with mutations in SOD1. *Journal of the Neurological Sciences*, 176(2), 88–94. https://doi.org/10.1016/s0022-510x(00)00317-8

Sales, A. L., Depaula, J., Mellinger Silva, C., Cruz, A., Lemos Miguel, M. A., and Farah, A. (2020). Effects of regular and decaffeinated roasted coffee (*Coffea arabica* and *Coffea canephora*) extracts and bioactive compounds on in vitro probiotic bacterial growth. *Food & Function*, 11(2), 1410–1424. https://doi.org/10.1039/c9fo02589h

Santana-Oliveira, D. A., Fernandes-Da-Silva, A., Miranda, C. S., Martins, F. F., Mandarim-De-Lacerda, C. A., & Souza-Mello, V. (2022). A PPAR-alpha agonist and DPP-4 inhibitor mitigate adipocyte dysfunction in obese mice. *Journal of Molecular Endocrinology*, 68(4), 225–241. https://doi.org/10.1530/jme-21-0084

Schneider, C. V., Schneider, K. M., Teumer, A., Rudolph, K. L., Hartmann, D., Rader, D. J. et al. (2022). Association of telomere length with risk of disease and mortality. *JAMA Internal Medicine*, 182(3), 291–300. https://doi.org/10.1001/jamainternmed.2021.7804

Schönheit, B., Zarski, R., and Ohm, T. G. (2004). Spatial and temporal relationships between plaques and tangles in Alzheimer pathology. *Neurobiology of Aging*, 25(6), 697–711. https://doi.org/10.1016/j.neurobiolaging.2003.09.009

Schumacher, B., Pothof, J., Vijg, J., and Hoeijmakers, J. H. (2021). The central role of DNA damage in the ageing process. *Nature*, 592(7856), 695–703. https://doi.org/10.1038/s41586-021-03307-7

Sedaghat, G., Mirshekar, M. A., Amirpour, M., Montazerifar, F., Miri, S., and Shourestani, S. (2019). Sub-chronic administration of brewed coffee on rat behavior, cognition, and oxidative stress in an Alzheimer's disease model. *Clinical Nutrition Experimental*, 28, 62–73. https://doi.org/10.1016/j.yclnex.2019.10.005

Seliem, E. M., Azab, M. E., Ismail, R. S., Nafeaa, A. A., Alotaibi, B. S., and Negm, W. A. (2022). Green coffee bean extract normalizes obesity-induced alterations of metabolic parameters in rats by upregulating adiponectin and GLUT4 levels and reducing RBP-4 and HOMA-IR. *Life*, 12(5), 693. https://doi.org/10.3390/life12050693

Seow, L. J., Shamlan, S., and Seow, E. K. (2021). Influence of roasting degrees on the antioxidant and anti-angiogenic effects of *Coffea liberica*. *Journal of Food Measurement and Characterization*, 15(5), 4030–4036. https://doi.org/10.1007/s11694-021-00987-7

Serna-Jiménez, J. A., Torres-Valenzuela, L. S., Sanín Villarreal, A., Roldan, C., Martín, M. A., Siles, J. A. et al. (2023). Advanced extraction of caffeine and polyphenols from coffee pulp: Comparison of conventional and ultrasound-assisted methods. *LWT*, 177, 114571. https://doi.org/10.1016/j.lwt.2023.114571

Shahriari, R., Tamri, P., Harchegani, A. L., and Nourian, A. (2020). Green coffee bean hydroalcoholic extract accelerates wound healing in full-thickness wounds in rabbits. *Traditional Medicine Research*, 5(6), 433–441. https://doi.org/10.12032/tmr20200603191

Shang, F., Li, X., and Jiang, X. (2016). Coffee consumption and risk of the metabolic syndrome: A meta-analysis. *Diabetes & Metabolism*, 42(2), 80–87. https://doi.org/10.1016/j.diabet.2015.09.001

Shang, H. Q., Yang, Q. B., Qiang, S., Zheng, R., Zhang, C. Q., Hu, C. Y. et al. (2025). Engineering caffeic acid O-methyltransferase for efficient de novo ferulic acid synthesis. *Engineering in Life Sciences*, 25(4), e70018. https://doi.org/10.1002/elsc.70018

Shaposhnikov, S., Hatzold, T., El Yamani, N., Stavro, P. M., Lorenzo, Y., Dusinska, M. et al. (2018). Coffee and oxidative stress: A human intervention study. *European Journal of Nutrition*, 57(2), 533–544. https://doi.org/10.1007/s00394-016-1336-4

Sharif, K., Watad, A., Bragazzi, N. L., Adawi, M., Amital, H., and Shoenfeld, Y. (2017). Coffee and autoimmunity: More than a mere hot beverage! *Autoimmunity Reviews*, 16(7), 712–721. https://doi.org/10.1016/j.autrev.2017.05.007

Sharifi-Rad, M., Anil Kumar, N. V., Zucca, P., Varoni, E. M., Dini, L., Panzarini, E. et al. (2020). Lifestyle, oxidative stress, and antioxidants: Back and forth in the pathophysiology of chronic diseases. *Frontiers in Physiology*, 11, 552535. https://doi.org/10.3389/fphys.2020.00694

Sharma, H., and Sharma, H. (2020). A detailed chemistry of coffee and its analysis. In: *Coffee – Production and Research*. IntechOpen. https://doi.org/10.5772/intechopen.91725

Shi, J., Cui, Y. F., Zhou, G., Li, N., Sun, X., Wang, X. et al. (2022). Covalent interaction of soy protein isolate and chlorogenic acid: Effect on protein structure and functional properties. *LWT*, 170, 114081. https://doi.org/10.1016/j.lwt.2022.114081

Shin, S., Lee, J. E., Loftfield, E., Shu, X. O., Abe, S. K., Rahman, M. S. et al. (2022). Coffee and tea consumption and mortality from all causes, cardiovascular disease and cancer: A pooled analysis of prospective studies from the Asia Cohort Consortium. *International Journal of Epidemiology*, 51(2), 626–640. https://doi.org/10.1093/IJE/DYAB161

Shu, W., Liu, L., Jiang, J., and Yao, Q. (2024). Dietary patterns and hepatocellular carcinoma risk: A systematic review and meta-analysis of cohort and case–control studies. *Nutrition & Metabolism*, 21(1), 47.

Siswanto, F. M., Sakuma, R., Oguro, A., and Imaoka, S. (2022). Chlorogenic acid activates Nrf2/SKN-1 and prolongs the lifespan of Caenorhabditis elegans via the Akt-FOXO3/DAF16a-DDB1 pathway and activation of DAF16f. *The Journals of Gerontology: Series A*, 77(8), 1503–1516. https://doi.org/10.1093/gerona/glac062

Sotibrán, A. N. C., Ordaz-Téllez, M. G., and Rodríguez-Arnaiz, R. (2011). Flavonoids and oxidative stress in *Drosophila melanogaster*. *Mutation Research/Genetic Toxicology and Environmental Mutagenesis*, 726(1), 60–65. https://doi.org/10.1016/j.mrgentox.2011.08.005

Spindel, E., Arnold, M., Cusak, B., and Wurtman, R. J. (1980). Effects of caffeine on anterior pituitary and thyroid function in the rat. *Journal of Pharmacology and Experimental Therapeutics*, 214(1), 58–62. https://doi.org/10.1016/s0022-3565(25)32130-0

Steger, M. C., Rigling, M., Blumenthal, P., Segatz, V., Quintanilla-Belucci, A., Beisel, J. M. et al. (2022). Coffee leaf tea from El Salvador: On-site production considering influences of processing on chemical composition. *Foods*, 11(17), 2553. https://doi.org/10.3390/foods11172553

Stoorvogel, W., Kleijmeer, M. J., Geuze, H. J., and Raposo, G. (2002). The biogenesis and functions of exosomes. *Traffic*, 3(5), 321–330. https://doi.org/10.1034/j.1600-0854.2002.30502.x

Sulem, P., Gudbjartsson, D. F., Geller, F., Prokopenko, I., Feenstra, B., Aben, K. K. H. et al. (2011). Sequence variants at CYP1A1–CYP1A2 and AHR associate with coffee consumption. *Human Molecular Genetics*, 20(10), 2071–2077. https://doi.org/10.1093/HMG/DDR086

Surma, S., & Oparil, S. (2021). Coffee and arterial hypertension. *Current Hypertension Reports*, 23(7), 1–11. https://doi.org/10.1007/S11906-021-01156-3/TABLES/3

Surma, S., Sahebkar, A., and Banach, M. (2023). Coffee or tea: Anti-inflammatory properties in the context of atherosclerotic cardiovascular disease prevention. *Pharmacological Research*, 187, 106596. https://doi.org/10.1016/J.PHRS.2022.106596

Suryanti, E., Retnowati, D., Prastya, M. E., Ariani, N., Yati, I., Permatasari, V. et al. (2023). Chemical composition, antioxidant, antibacterial, antibiofilm, and cytotoxic activities of robusta coffee extract (Coffea canephora). *HAYATI Journal of Biosciences*, 30(4), 632–642. https://doi.org/10.4308/hjb.30.4.632-642

Susilawati, I. D. A., Suryono, S., Purwanto, P., Burlakovs, J., and Yaro, A. (2020). Coffee protects cardiovascular health by maintaining the structure of coronary arterial wall intimal collagen. *Coffee Science*, 15(1), 1–6. https://doi.org/10.25186/.V15I.1637

Sutphin, G. L., Bishop, E., Yanos, M. E., Moller, R. M., and Kaeberlein, M. (2012). Caffeine extends life span, improves healthspan, and delays age-associated pathology in Caenorhabditis elegans. *Longevity & Healthspan*, 1(1), 9. https://doi.org/10.1186/2046-2395-1-9

Tabuti, J. R., Lye, K. A., and Dhillion, S. S. (2003). Traditional herbal drugs of Bulamogi, Uganda: Plants, use and administration. *Journal of Ethnopharmacology*, 88(1), 19–44. https://doi.org/10.1016/S0378-8741(03)00161-2

Tajik, N., Tajik, M., Mack, I., and Enck, P. (2017). The potential effects of chlorogenic acid, the main phenolic components in coffee, on health: A comprehensive review of the literature. *European Journal of Nutrition*, 56(7), 2215–2244. https://doi.org/10.1007/S00394-017-1379-1

Tanaka, K., Tamakoshi, A., Sugawara, Y., Mizoue, T., Inoue, M., Sawada, N. et al. (2019). Coffee, green tea and liver cancer risk: An evaluation based on a systematic review of epidemiologic evidence among the Japanese population. *Japanese Journal of Clinical Oncology*, 49(10), 972–984. https://doi.org/10.1093/JJCO/HYZ097

Tang, R., Luo, J., Wang, W., Liu, D., Wang, G., and Guo, X. (2019). Rutin's natural source Flos Sophorae as potential antioxidant and improver of fungal community in Chinese sausages. *LWT*, 101, 435–443. https://doi.org/10.1016/J.LWT.2018.11.064

Tarantini, S., Valcarcel-Ares, N. M., Yabluchanskiy, A., Fulop, G. A., Hertelendy, P., Gautam, T. et al. (2018). Treatment with the mitochondrial-targeted antioxidant peptide SS-31 rescues neurovascular coupling responses and cerebrovascular endothelial function and improves cognition in aged mice. *Aging Cell*, 17(2), e12731. https://doi.org/10.1111/acel.12731

Teramoto, M., Yamagishi, K., Muraki, I., Tamakoshi, A., and Iso, H. (2023). Coffee and green tea consumption and cardiovascular disease mortality among people with and without hypertension. *Journal of the American Heart Association*, 12(2), 26477. https://doi.org/10.1161/JAHA.122.026477

Thesnor, V., Cheremond, Y., Sylvestre, M., Meffre, P., Cebrián-Torrejón, G., and Benfodda, Z. (2024). Survey on the traditional use of medicinal herbs in Haiti: A study on knowledge, practices, and efficacy prevention. *Plants*, 13(17), 2383. https://doi.org/10.3390/plants13172383

Torres-Valenzuela, L. S., Ballesteros-Gómez, A., and Rubio, S. (2020). Supramolecular solvent extraction of bioactives from coffee cherry pulp. *Journal of Food Engineering*, 278, 109933. https://doi.org/10.1016/J.JFOODENG.2020.109933

Tran, K. T., Coleman, H. G., McMenamin, Ú. C., and Cardwell, C. R. (2019). Coffee consumption by type and risk of digestive cancer: A large prospective cohort study. *British Journal of Cancer*, 120(11), 1059–1066. https://doi.org/10.1038/S41416-019-0465-Y;SUBJMETA

Tu, S., Okamoto, S. Ichi, Lipton, S. A., and Xu, H. (2014). Oligomeric Aβ-induced synaptic dysfunction in Alzheimer's disease. *Molecular Neurodegeneration*, 9(1), 48. https://doi.org/10.1186/1750-1326-9-48/FIGURES/1

Tucker, L. A. (2017). Caffeine consumption and telomere length in men and women of the National Health and Nutrition Examination Survey (NHANES). *Nutrition & Metabolism*, 14(1), 10. https://doi.org/10.1186/s12986-017-0162-x

Tverdal, A., Selmer, R., Cohen, J. M., and Thelle, D. S. (2020). Coffee consumption and mortality from cardiovascular diseases and total mortality: Does the brewing method matter? *European Journal of Preventive Cardiology*, 27(18), 1986–1993. https://doi.org/10.1177/2047487320914443

Um, C. Y., McCullough, M. L., Guinter, M. A., Campbell, P. T., Jacobs, E. J., and Gapstur, S. M. (2020). Coffee consumption and risk of colorectal cancer in the Cancer Prevention

Study-II Nutrition Cohort. *Cancer Epidemiology*, 67. https://doi.org/10.1016/j.canep.2020.101730

Unamuno, X., Gómez-Ambrosi, J., Rodríguez, A., Becerril, S., Frühbeck, G., and Catalán, V. (2018). Adipokine dysregulation and adipose tissue inflammation in human obesity. *European Journal of Clinical Investigation*, 48(9), e12997. https://doi.org/10.1111/ECI.12997

Ungvari, Z., Tarantini, S., Nyúl-Tóth, Á., Kiss, T., Yabluchanskiy, A., Csipo, T. et al. (2019). Nrf2 dysfunction and impaired cellular resilience to oxidative stressors in the aged vasculature: From increased cellular senescence to the pathogenesis of age-related vascular diseases. *Geroscience*, 41(6), 727–738. https://doi.org/10.1007/s11357-019-00107-w

Unno, K., Taguchi, K., Hase, T., Meguro, S., and Nakamura, Y. (2024). DNA mutagenicity of hydroxyhydroquinone in roasted coffee products and its suppression by chlorogenic acid, a coffee polyphenol, in oxidative-damage-sensitive SAMP8 mice. *International Journal of Molecular Sciences*, 25(2), 720. https://doi.org/10.3390/ijms25020720

Uto-Kondo, H., Ayaori, M., Ogura, M., Nakaya, K., Ito, M., Suzuki, A. et al. (2010). Coffee consumption enhances high-density lipoprotein-mediated cholesterol efflux in macrophages. *Circulation Research*, 106(4), 779–787. https://doi.org/10.1161/CIRCRESAHA.109.206615

Valadão Vicente, S. J., Ishimoto, E. Y., Cruz, R. J., Seabra Pereira, C. D., and Torres, E. A. F. D. S. (2011). Increase of the activity of phase II antioxidant enzymes in rats after a single dose of coffee. *Journal of Agricultural and Food Chemistry*, 59(20), 10887–10892. https://doi.org/10.1021/JF202390X

Valadi, H., Ekström, K., Bossios, A., Sjöstrand, M., Lee, J. J., and Lötvall, J. O. (2007). Exosome-mediated transfer of mRNAs and microRNAs is a novel mechanism of genetic exchange between cells. *Nature Cell Biology*, 9(6), 654–659. https://doi.org/10.1038/ncb1596

Vater, M., Rost, N., Eckstein, G., Sauer, S., Tontsch, A., Erhardt, A. et al. (2025). Huntingtin CAG repeat size variations below the Huntington's disease threshold: Associations with depression, anxiety and basal ganglia structure. *European Journal of Human Genetics*, 33(5), 624–632. https://doi.org/10.1038/S41431-024-01737-1;TECHMETA

Velickovic, K., Wayne, D., Leija, H. A. L., Bloor, I., Morris, D. E., Law, J. et al. (2019). Caffeine exposure induces browning features in adipose tissue in vitro and in vivo. *Scientific Reports*, 9(1), 1–11. https://doi.org/10.1038/S41598-019-45540-1;TECHMETA

Vishvanath, L., and Gupta, R. K. (2019). Contribution of adipogenesis to healthy adipose tissue expansion in obesity. *The Journal of Clinical Investigation*, 129(10), 4022–4031. https://doi.org/10.1172/JCI129191

Vitaglione, P., Mazzone, G., Lembo, V., D'Argenio, G., Rossi, A., Guido, M. et al. (2019). Coffee prevents fatty liver disease induced by a high-fat diet by modulating pathways of the gut–liver axis. *Journal of Nutritional Science*, 8, e15. https://doi.org/10.1017/JNS.2019.10

Wachamo, H. L. (2017). Review on health benefit and risk of coffee consumption. *Medicinal & Aromatic Plants*, 6(4), 1–12. https://doi.org/10.4172/2167-0412.1000301

Walczak, K., and Sieminska, L. (2021). Obesity and thyroid axis. *International Journal of Environmental Research and Public Health*, 18(18). https://doi.org/10.3390/IJERPH18189434

Wang, D., Hou, J., Wan, J., Yang, Y., Liu, S., Li, X. et al. (2021). Dietary chlorogenic acid ameliorates oxidative stress and improves endothelial function in diabetic mice via Nrf2 activation. *Journal of International Medical Research*, 49(1), 0300060520985363. https://doi.org/10.1177/0300060520985363

Wang, C., Zhou, C., Guo, T., Jiaerken, Y., Yang, S., Huang, P. et al. (2023). Association of coffee consumption and striatal volume in patients with Parkinson's disease and healthy controls. *CNS Neuroscience and Therapeutics*, 29(10), 2800–2810. https://doi.org/10.1111/CNS.14216;WGROUP:STRING:PUBLICATION

Wang, S., Yoon, Y. C., Sung, M. J., Hur, H. J., and Park, J. H. (2012). Antiangiogenic properties of cafestol, a coffee diterpene, in human umbilical vein endothelial cells. *Biochemical and Biophysical Research Communications*, 421(3), 567–571. https://doi.org/10.1016/J.BBRC.2012.04.046

Wang, Y., Xiang, Y. F., and Liu, A. L. (2022). Comparative and combined effects of epigallocatechin-3-gallate and caffeine in reducing lipid accumulation in Caenorhabditis elegans. *Plant Foods for Human Nutrition*, 77(2), 279–285. https://doi.org/10.1007/s11130-022-00978-z

Wolfson, C., Gauvin, D. E., Ishola, F., and Oskoui, M. (2023). Global prevalence and incidence of amyotrophic lateral sclerosis. *Neurology*, 101(6), e613–e623. https://doi.org/10.1212/WNL.0000000000207474

Wong-Paz, J. E., Guyot, S., Aguilar-Zárate, P., Muñiz-Márquez, D. B., Contreras-Esquivel, J. C., and Aguilar, C. N. (2021). Structural characterization of native and oxidized procyanidins (condensed tannins) from coffee pulp (Coffea arabica) using phloroglucinolysis and thioglycolysis-HPLC-ESI-MS. *Food Chemistry*, 340, 127830.

World Health Organization (WHO) updates on cardiovascular disease. (2016). Retrieved from http://www.who.int/cardiovascular_diseases/about_cvd/en/

Xu, X., Ma, R., Zhang, X., Guo, H., Keerman, M., Wang, X. et al. (2024). Association between renal function trajectories and risk of cardiovascular disease: A prospective cohort study. *Annals of Medicine*, 56(1), 2427907. https://doi.org/10.1080/07853890.2024.2427907

Xu, Y., Zhang, J., Pan, T., Ren, F., Luo, H., and Zhang, H. (2022). Synthesis, characterization and effect of alkyl chain unsaturation on the antioxidant activities of chlorogenic acid derivatives. *LWT*, 162, 113325. https://doi.org/10.1016/J.LWT.2022.113325

Yamagata, K., Izawa, Y., Onodera, D., and Tagami, M. (2018). Chlorogenic acid regulates apoptosis and stem cell marker-related gene expression in A549 human lung cancer cells. *Molecular and Cellular Biochemistry*, 441(1–2), 9–19. https://doi.org/10.1007/S11010-017-3171-1/FIGURES/7

Yamamoto, K., Kiyohara, T., Murayama, Y., Kihara, S., Okamoto, Y., Funahashi, T. et al. (2005). Production of adiponectin, an anti-inflammatory protein, in mesenteric adipose tissue in Crohn's disease. *Gut*, 54(6), 789–796. https://doi.org/10.1136/gut.2004.046516

Yáñez-Mó, M., Siljander, P. R., Andreu, Z. et al. (2015). Biological properties of extracellular vesicles and their physiological functions. *Journal of Extracellular Vesicles*, 4, 27066. https://doi.org/10.3402/jev.v4.27066

Yang, Y., Liang, S., Geng, H., Xiong, M., Li, M., Su, Q. et al. (2022). Proteomics revealed the crosstalk between copper stress and cuproptosis, and explored the feasibility of curcumin as anticancer copper ionophore. *Free Radical Biology and Medicine*, 193(Pt 2), 638–647. https://doi.org/10.1016/j.freeradbiomed.2022.11.023

Yao, J., Peng, S., Xu, J., and Fang, J. (2019). Reversing ROS-mediated neurotoxicity by chlorogenic acid involves its direct antioxidant activity and activation of Nrf2-ARE signaling pathway. *BioFactors*, 45(4), 616–626. https://doi.org/10.1002/BIOF.1507;WGROUP:STRING:PUBLICATION

Yau, W. Y. W., Tudorascu, D. L., McDade, E. M., Ikonomovic, S., James, J. A., Minhas, D. et al. (2015). Longitudinal assessment of neuroimaging and clinical markers in autosomal dominant Alzheimer's disease: A prospective cohort study. *The Lancet Neurology*, 14(8), 804–813. https://doi.org/10.1016/S1474-4422(15)00135-0

Yukawa, G. S., Mune, M., Otani, H., Tone, Y., Liang, X. M., Iwahashi, H. et al. (2004). Effects of coffee consumption on oxidative susceptibility of low-density lipoproteins and serum lipid levels in humans. *Biochemistry* (Moscow), 69(1), 70–74. https://doi.org/10.1023/B:BIRY.0000016354.05438.0F/METRICS

Yuniarta, T. A., and Handayani, R. (2024). An initial investigation of the potential of Robusta coffee, Arabica coffee, and caffeine in asthma treatment through the evaluation of 5-lipoxygenase inhibition activity. *Borneo Journal of Pharmacy*, 7(1), 80–88. https://doi.org/10.33084/bjop.v7i1.44

Zaitone, S. A., Ahmed, E., Elsherbiny, N. M., Mehanna, E. T., El-Kherbetawy, M. K., ElSayed, M. H. et al. (2019). Caffeic acid improves locomotor activity and lessens inflammatory burden in a mouse model of rotenone-induced nigral neurodegeneration: Relevance to Parkinson's disease therapy. *Pharmacological Reports*, 71(1), 32–41. https://doi.org/10.1016/J.PHAREP.2018.08.004/METRICS

Zang, X., Chen, S., Zhu, J. Y., Ma, J., and Zhai, Y. (2022). The emerging role of central and peripheral immune systems in neurodegenerative diseases. *Frontiers in Aging Neuroscience*, 14, 872134. https://doi.org/10.3389/FNAGI.2022.872134/FULL

Zhao, J., Miao, K., Wang, H., Ding, H., and Wang, D. W. (2013). Association between telomere length and type 2 diabetes mellitus: A meta-analysis. *PLoS ONE*, 8(11), e79993. https://doi.org/10.1371/journal.pone.0079993

Zheng, H., Lin, F., Xin, N., Yang, L., and Zhu, P. (2022). Association of coffee, tea, and caffeine consumption with all-cause risk and specific mortality for cardiovascular disease patients. *Frontiers in Nutrition*, 9, 842856. https://doi.org/10.3389/FNUT.2022.842856/BIBTEX

Zheng, S. Q., Huang, X. B., Xing, T. K., Ding, A. J., Wu, G. S., and Luo, H. R. (2017). Chlorogenic acid extends the lifespan of Caenorhabditis elegans via insulin/IGF-1 signaling pathway. *Journals of Gerontology Series A: Biomedical Sciences and Medical Sciences*, 72(4), 464–472. https://doi.org/10.1093/gerona/glw105

Zidan, N. S., Omran, A. M. E., Rezk, S. M., Atteia, H. H., and Sakran, M. I. (2022). Anti-Alzheimer's disease potential of Arabian coffee versus Date palm seed extract in male rats. *Journal of Food Biochemistry*, 46(1). https://doi.org/10.1111/JFBC.14017

Index

About the Editor

Erdi Can Aytar is an academic researcher at Uşak University, Türkiye. He received his bachelor's degree in biology from Adnan Menderes University, his master's degree in biology from Gazi University, and completed his Ph.D. in Biology at Ondokuz Mayıs University. His research focuses on biochemistry, plant secondary metabolites, natural product chemistry, nanotechnology, carbon quantum dots, and in silico molecular modelling. His work primarily investigates the chemical characterization of bioactive natural compounds, evaluation of their biological activities, and elucidation of molecular interaction mechanisms through computational and molecular modelling approaches. He has authored more than 100 peer-reviewed international scientific publications and has contributed to over 10 international book chapters. His research encompasses antioxidant, antimicrobial, anticancer, neuroprotective, and metabolic disease-related bioactivities, as well as the biomedical applications of nanomaterials. In addition, he actively serves as a reviewer for more than 100 international scientific journals and is a member of the editorial boards of various international journals. He engages in interdisciplinary research integrating biotechnology, pharmaceutical biochemistry, and computational biology, and participates in national and international collaborative scientific projects.